AF576052

Forest Management and Biodiversity Conservation

Forest Management and Biodiversity Conservation

Editors

Lucian Dinca
Miglena Zhiyanski

Basel • Beijing • Wuhan • Barcelona • Belgrade • Novi Sad • Cluj • Manchester

Editors

Lucian Dinca Forest Research and Management Institute Brasov Station, Romania	Miglena Zhiyanski Forest Research Institute Sofia, Bulgaria

Editorial Office
MDPI
St. Alban-Anlage 66
4052 Basel, Switzerland

This is a reprint of articles from the Special Issue published online in the open access journal *Diversity* (ISSN 1424-2818) (available at: https://www.mdpi.com/journal/diversity/special_issues/for_manag_biodivers).

For citation purposes, cite each article independently as indicated on the article page online and as indicated below:

Lastname, A.A.; Lastname, B.B. Article Title. *Journal Name* **Year**, *Volume Number*, Page Range.

ISBN 978-3-0365-9414-9 (Hbk)
ISBN 978-3-0365-9415-6 (PDF)
doi.org/10.3390/books978-3-0365-9415-6

Cover image courtesy of Lazar Gabriel

Contents

Editorial

Forest Management and Biodiversity Conservation: Introduction to the Special Issue

Lucian Dinca [1],* and Miglena Zhiyanski [2]

1 Forest Research and Management Institute, 13 Closca Street, 500040 Brasov, Romania
2 Forest Research Institute—Bulgaria Academy of Science, 132 Kl. Ohridski Blvd., 1756 Sofia, Bulgaria; zhiyanski@abv.bg
* Correspondence: brasov@icas.ro

Forest ecosystems contribute to human wellbeing and the economy through the complex ecosystem services they provide [1–6]. The sustainable and regular supply of ecosystem services by forests requires excellent knowledge of the functional and biological diversity in these complex ecosystems [7–10]. This is related to the management of forests, which evolves over time in order to face contemporary challenges [11–15]. One of the major challenges in forest management is the sustainability of the resource itself, while the challenge for the conservation of biological diversity is to secure a minimum set of strategically located primary forests in representative areas with high diversity and endemism [16–21].

This Special Issue contains a total of 18 articles from many different countries, including Brazil, Bulgaria, Ecuador, Germany, Malaysia, Peru, Poland, Romania, Slovakia, and Slovenia situated on three continents (Figure 1).

Figure 1. Countries where the research articles were conducted.

These research articles are highly varied and can be classified into eight sub-domains that are representative of the chosen domain: natural and human disturbances, genetics, site conditions, tropical forest, peri-urban forest, forest soils, forest reserves, and mountain ecosystems (Figure 2).

Citation: Dinca, L.; Zhiyanski, M. Forest Management and Biodiversity Conservation: Introduction to the Special Issue. *Diversity* **2023**, *15*, 1078. https://doi.org/10.3390/d15101078

Received: 15 September 2023
Revised: 7 October 2023
Accepted: 10 October 2023
Published: 11 October 2023

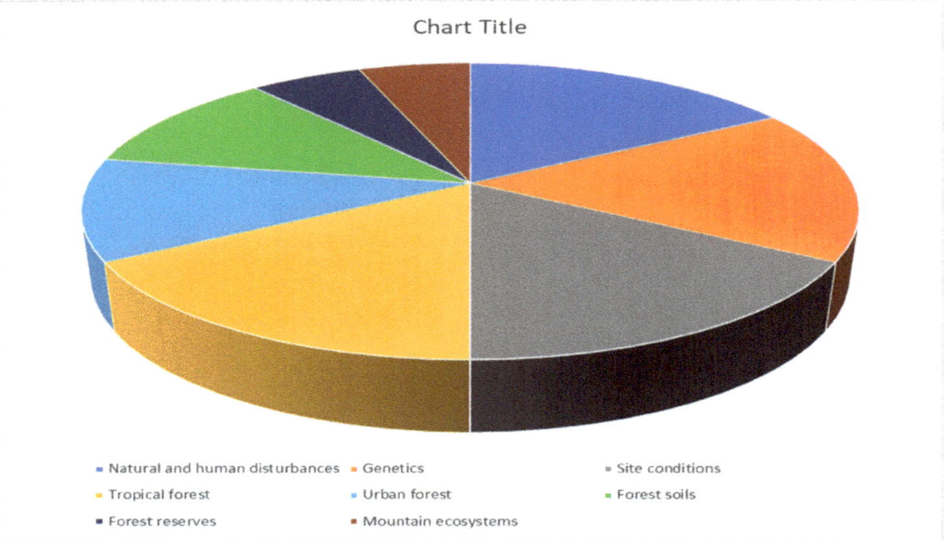

Figure 2. Research domains.

Ecosystems and individuals or local communities are important aspects in the recent research conducted to assess such relations and their effects. In this Special Issue, there are four papers that focus on this research topic.

Peri-urban forests [22–25], like urban forests [20,26–30], play an important role for urban agglomerations. This theme is highlighted in this Special Issue via the analysis of tree species, the position and shapes of trees, and the biometric characteristics of trees (diameter at breast height). The results indicate that the stand will be highly stable for future generations, meaning that the human–nature interactions, such as recreational and outdoor activities, are secure.

In addition to its relation to human activities, the main objective of ecotourism, i.e., ecological tourism [31–33], is to promote sustainable development by visiting natural environments to minimize the negative impact of traditional tourist activities and support the conservation efforts of those regions [34–37]. The second paper in this sub-domain discusses the how local communities can use such knowledge related to plant diversity and cultural–archaeological offerings. The authors of this study conducted a survey which detected 384 species of vascular plants with 220 genera and 69 families, the main proportion of which comprised *Asteraceae*, *Poaceae*, and *Fabaceae*. Through applying methods centered around biological, ecosystem, and cultural values, this study revealed the potential mitigation measures that local authorities should implement.

A paper in this sub-domain [38] investigated an urbanization gradient of a tropical city where 96 woody plant species belonging to 71 genera and 42 families were present in the research area. This study reports remarkable results that emphasize that when urbanization spread from wildland areas to suburban areas, a 67.6% reduction in the native species took place, whereas the non-native species remained stable.

The last paper in this sub-domain also focuses on human factors through describing the perceptions of local inhabitants on the used land management systems in the rainforest of Ecuador [39]. In this study, natural forests were the most positively rated type of forest, while the managed ones were the least positively rated, revealing an important trend among the participants, and human intervention was not the foremost landscape-related factor affecting this perception.

Also, research papers concerning management, non-management practices, tree growth, or disturbances that may occur have made an important contribution to the understanding of such practices [40–43]. Five papers in this Special Issues address these topics through collecting data from various forests, from the Amazon area to European forests. The first of these papers deals with the effects of management abandonment on vegetation dynamics in a non-native Douglas fir forest in Germany. It is worth noting that such analyses are of great importance to understand the impact of some non-native species on biodiversity. This paper's survey showed consistent development after management abandonment while also showing that the species became less diverse and more shade-tolerant.

In addition to human intervention in the context of forests, natural phenomena can also have important impacts, as noted by the authors of [44] in their analysis of species turnover. For this paper, five plots and one control plot were analyzed in the southern part of Slovenia. The total number of species recorded in the gaps was 184, with the highest number (106) being recorded for the largest forest gap and 58 species being recorded for the control one. Based on their findings, the authors determined that a forest gap represents a significant habitat patch, especially for plant species which were not present there before.

Economical aspects are also import in a forestry system [45–47]. The third paper in this sub-domain proposes a silvicultural management system that has the potential to recover and improve the productivity of an intensively logged tropical forest due to the fact that the applied management techniques are intended for natural forests. The targeted area spanned 535.6 ha (selected trees with dbh $\geq$ 25), and two treatments were designed. Applying the harvesting criteria resulted in a positive cost–benefit ratio, which was superior to the control treatment in all scenarios. This can favor the maintenance of biodiversity, promote the expansion of populations of low density species, and improve the quality of forests.

The fourth paper in this sub-domain documents the impact that forest management can have on the germination and growth [48] of seeds and seedlings from different geographic provenances rich in *Robinia pseudoacacia*. In total, eight Romanian provenances were selected for the study. The researchers applied water-soaked seeds and heat/cold treatment on one side and sulfuric acid on the other side. The results highlighted that one provenance (Satu Mare) had the lowest germination with both treatments, while the highest germination occurred in the Bihor provenance (68.2%). Such aspects are of great importance to forest managers, as they can help inform their decisions to ensure they apply the correct method when preparing new seedlings.

The last paper in this sub-domain focuses on examining black locust (*Robinia pseudoacacia* L.) from a silvicultural point of view; for this study, the authors conducted a literature review [49] with special emphasis on Romania wherein aspects such as species propagation, stand management, and vulnerability issues were covered and addressed to highlight the knowledge accumulated by Romanian foresters and researchers. Aspects such as ecological adaptability, CO_2 sequestration, and biomass yield are highlighted as positive, while short lifespan, invasiveness, and even dieback in drought were listed as negative aspects. However, aspects such as genetics, invasive potential, and adaptation to climate change require more research.

Also, another paper in this Special Issue analyzes harvesting operations and their outcomes from a practical point of view [50]. Damage and tolerability thresholds for the remaining trees was established for specific stands from southwest Romania. It has been well documented that damages to remnant trees will cause health deterioration and rot development. Equations were created to determine the tolerance threshold, and record values of 0.09 for thinning, cuttings, and final cuttings from shelterwood were obtained; a final value of 0.10 was obtained for the first intervention cuttings, as well as for the preparatory and seed cuttings.

Trees can be affected by biotic factors that can harm the health and productivity of certain species. In order to fight the effects of such aspects, certain genotypes can be selected [51–53] over time to improve the presence of the species. In another study, *Ophiostoma novo-ulmi* in Romania after 1990 was extensively studied for three years through conducting tests on *Ulmus minor*, *Ulmus glabra*, and *Ulmus laevis* in 38 provenances. The authors of this study artificially inoculated a local strain of *O. novo-ulmi* to observe the outcomes. New observations materialized after almost 30 years, and a new hybrid form was identified between *O. novo-ulmi* ssp. *americana* x *O. novo-ulmi* ssp. *novo-ulmi*. Of the three elm species, European white elm showed a constant tolerance to the disease, while Wych elm was extremely sensitive to it.

Three papers in this Special Issue analyze soil aspects. The first paper focused on microbial abundance in post-bauxite mining land [54]. Mining activities leave a footprint on the environment for decades; even though restorative measures can be taken, they require a great amount of energy consumption. Soils are a reservoir for bacteria, and in this study, the bacterial potential was calculated by using the bacterial soil quality index (BSQI), while the Shannon diversity index and the Jaccard distance was used to show the level of bacterial diversity for the two studied plots. The results of this study are promising; the chemical and microbiological parameters determined in the adjacent area indicated similar soil conditions to the site that had been ecologically reconstructed 15 years earlier.

Even though no anthropic activities were observed, importantly, the authors of [55] researched the changes made to the soil microbiota by planting different tree species. The analyses took place in central Slovakia. The researchers created special areas wherein pasture land was afforested to observe the possible changes in soil properties that took place after decades to gain insights into the relation between soil and trees. After applying multivariate physico-chemical analyses to the soil, there was an overlap in terms of soil between Douglas fir and spruce areas but a clear separation of beech from sycamore. It is notable that microbial activity and diversity were highest under Douglas fir, followed by sycamore, with the beech and spruce having the lowest values.

Relief and soils were the focus of the last article from this sub-section [56], in which environmental and stand conditions were analyzed for silver fir (*Abies alba* Mill.). The database used in this study consisted of 77,251 stands covering an area of 211,954 ha. Data were computed by using MATLAB scripts (The MathWorks Inc. (2022)); eight factors—altitude, field aspect, field slope, soil type, participation percentage, road distance, structure, and consistency—were processed. It is well known that forest owners and managers wish to maximize the potential of afforested areas. This study revealed that the highest silver fir productivity is found at altitudes of up to 1200 m, on mid and upper slopes, on NW field aspects, and on eutric cambisols and dystric cambisols, with a 10–20% participation in stand composition among relatively even aged, fully consistent stands.

The environmental conditions in the tropical rain forest are studied in one paper in this Special Issue. In this specific paper, a non-metric multidimensional analyses was conducted to determine the correlation between plot altitude and stand characteristics. Support was provided by the Biological Reserve of San Francisco, and a 13 ha area was monitored. By using statistical methods like CCA (canonical correspondence analysis) and "Four Corners" analysis, the hypothesis that altitude and some stand characteristics are the key factors for the formation of the two studied forest types were validated.

Elsewhere in this Special Issue, climate change aspects were analyzed in the Rila Mountain, Bulgaria, with the help of modern technologies such as remote sensing vegetation indices from a period spanning 42 years, Copernicus High-Resolution Layer products, and climate change reanalysis data from a period spanning 40 years. A series of trends in ecosystem extent and functioning were found, and new candidate indicators that are suitable for the remote monitoring of climate change effects were defined. Climate change is an important topic as its effects affect both communities and nature.

The final two papers in this Special Issue deal with genetics, analyzing in depth information that normally cannot be perceived by the human eye.

Conservation can ensure the timely propagation of tree species, and the authors of the penultimate paper in this Special Issue observed genetic processes in the largest national forest park in Poland. The focus of this study was the core mother pine stand from the protected area, with its progeny generation occurring on the basis of its chloroplast DNA (cpDNA). The degree of variations observed declined by generation, with the results showing that the significant genetic diversity of the studied stands was found to be reduced over the course of generations.

The authors of the final paper in this Special Issue utilized noninvasive genetic monitoring, sampling, and collection techniques to achieve desirable results. They reviewed 148 genetic research papers that had specific content pertinent to the geographic region. In North America, hair samples were collected in favor of feces, while in Europe, both types of samples are recommended, though there is more focus on feces. Also, methods like Isohelix can be applied on a national level, while in the field, trained dogs for feces detection could be used, along with specialized personnel during the autumn and winter periods. Due to the field's difficulties in obtaining these samples, the large-scale noninvasive genetic monitoring of large bear populations represents a challenge; however, if this challenge can be alleviated valuable insights into biodiversity monitoring and climate change could be gained.

The papers contained in this Special Issue address many research topics, from the bacterial world to trees, animals, communities, and climate change. Even though we may initially think that some research topics are not linked to each other, as is the case in nature, everything is connected. For example, by understanding genetics, we can apply valid conservation methods and promote and preserve genetic heritage; by improving the awareness of local authorities, sustainable management can be applied; by understanding diseases, we can mitigate their effects; by understanding what is in the Earth's soils, we can draw correlations regarding above-ground vegetation. The main goal of all of these research endeavors is to promote the understanding and wellbeing of the ecosystem.

Conflicts of Interest: The authors declare no conflict of interest.

References

1. Bull, J.W.; Jobstvogt, N.; Böhnke-Henrichs, A.; Mascarenhas, A.; Sitas, N.; Baulcomb, C.; Koss, R. Strengths, Weaknesses, Opportunities and Threats: A SWOT analysis of the ecosystem services framework. *Ecosyst. Serv.* **2016**, *17*, 99–111. [CrossRef]
2. Perrings, C.; Naeem, S.; Ahrestani, F.; Bunker, D.E.; Burkill, P.; Canziani, G.; Elmqvist, T.; Ferrati, R.; Fuhrman, J.; Jaksic, F.; et al. Ecosystem services for 2020. *Science* **2010**, *330*, 323–324. [CrossRef] [PubMed]
3. Wang, L.; Zheng, H.; Wen, Z.; Liu, L.; Robinson, B.E.; Li, R.; Kong, L. Ecosystem service synergies/trade-offs informing the supply-demand match of ecosystem services: Framework and application. *Ecosyst. Serv.* **2019**, *37*, 100939. [CrossRef]
4. Raihan, A. A review on the integrative approach for economic valuation of forest ecosystem services. *J. Environ. Sci. Econ.* **2023**, *2*, 1–18. [CrossRef]
5. Taye, F.A.; Folkersen, M.V.; Fleming, C.M.; Buckwell, A.; Mackey, B.; Diwakar, K.C.; Saint Ange, C. The economic values of global forest ecosystem services: A meta-analysis. *Ecol. Econ.* **2021**, *189*, 107145. [CrossRef]
6. Snäll, T.; Triviño, M.; Mair, L.; Bengtsson, J.; Moen, J. High rates of short-term dynamics of forest ecosystem services. *Nat. Sustain.* **2021**, *4*, 951–957. [CrossRef]
7. Niemelä, J.; Saarela, S.R.; Söderman, T.; Kopperoinen, L.; Yli-Pelkonen, V.; Väre, S.; Kotze, D.J. Using the ecosystem services approach for better planning and conservation of urban green spaces: A Finland case study. *Biodivers. Conserv.* **2010**, *19*, 3225–3243. [CrossRef]
8. Brockerhoff, E.G.; Barbaro, L.; Castagneyrol, B.; Forrester, D.I.; Gardiner, B.; González-Olabarria, J.R.; Jactel, H. Forest biodiversity, ecosystem functioning and the provision of ecosystem services. *Biodivers. Conserv.* **2017**, *26*, 3005–3035. [CrossRef]
9. Bengtsson, J.; Nilsson, S.G.; Franc, A.; Menozzi, P. Biodiversity, disturbances, ecosystem function and management of European forests. *For. Ecol. Manag.* **2000**, *132*, 39–50. [CrossRef]
10. Angelstam, P.; Elbakidze, M.; Lawrence, A.; Manton, M.; Melecis, V.; Perera, A.H. Barriers and bridges for landscape stewardship and knowledge production to sustain functional green infrastructures. *Ecosyst. Serv. For. Landsc. Broadscale Consid.* **2018**, 127–167. [CrossRef]
11. Frank, S.; Fürst, C.; Pietzsch, F. Cross-sectoral resource management: How forest management alternatives affect the provision of biomass and other ecosystem services. *Forests* **2015**, *6*, 533–560. [CrossRef]
12. Knoke, T.; Stimm, B.; Ammer, C.; Moog, M. Mixed forests reconsidered: A forest economics contribution on an ecological concept. *For. Ecol. Manag.* **2005**, *213*, 102–116. [CrossRef]
13. Nunes LJ, R.; Meireles CI, R.; Pinto Gomes, C.J.; de Almeida Ribeiro, N.M.C. Socioeconomic aspects of the forests in Portugal: Recent evolution and perspectives of sustainability of the resource. *Forests* **2019**, *10*, 361. [CrossRef]
14. Erdoğan, S.; Çakar, N.D.; Ulucak, R.; Danish, A.; Kassouri, Y. The role of natural resources abundance and dependence in achieving environmental sustainability: Evidence from resource-based economies. *Sustain. Dev.* **2021**, *29*, 143–154. [CrossRef]

15. Gea-Izquierdo, G.; Pastur, G.M.; Cellini, J.M.; Lencinas, M.V. Forty years of silvicultural management in southern Nothofagus pumilio primary forests. *For. Ecol. Manag.* **2004**, *201*, 335–347. [CrossRef]
16. Bawa, K.S.; Rai, N.D.; Sodhi, N.S. Rights, governance, and conservation of biological diversity. *Conserv. Biol.* **2011**, *25*, 639–641. [CrossRef]
17. Gering, J.C.; Crist, T.O.; Veech, J.A. Additive partitioning of species diversity across multiple spatial scales: Implications for regional conservation of biodiversity. *Conserv. Biol.* **2003**, *17*, 488–499.
18. Yoccoz, N.G.; Nichols, J.D.; Boulinier, T. Monitoring of biological diversity in space and time. *Trends Ecol. Evol.* **2001**, *16*, 446–453. [CrossRef]
19. Varma, V.K.; Ferguson, I.; Wild, I. Decision support system for the sustainable forest management. *For. Ecol. Manag.* **2000**, *128*, 49–55. [CrossRef]
20. Duncker, P.S.; Barreiro, S.M.; Hengeveld, G.M.; Lind, T.; Mason, W.L.; Ambrozy, S.; Spiecker, H. Classification of forest management approaches: A new conceptual framework and its applicability to European forestry. *Ecol. Soc.* **2012**, *17*, 51. Available online: https://www.jstor.org/stable/26269224 (accessed on 14 September 2023). [CrossRef]
21. Bauhus, J.; Puettmann, K.J.; Kühne, C. Close-to-nature forest management in Europe: Does it support complexity and adaptability of forest ecosystems? In *Managing Forests as Complex Adaptive Systems*; Routledge: Milton Park, UK, 2013; pp. 187–213.
22. Tu, G.; Abildtrup, J.; Garcia, S. Preferences for urban green spaces and peri-urban forests: An analysis of stated residential choices. *Landsc. Urban Plan.* **2016**, *148*, 120–131. [CrossRef]
23. Beckmann-Wübbelt, A.; Fricke, A.; Sebesvari, Z.; Yakouchenkova, I.A.; Fröhlich, K.; Saha, S. High public appreciation for the cultural ecosystem services of urban and peri-urban forests during the COVID-19 pandemic. *Sustain. Cities Soc.* **2021**, *74*, 103240. [CrossRef]
24. Cueva, J.; Yakouchenkova, I.A.; Fröhlich, K.; Dermann, A.F.; Dermann, F.; Köhler, M.; Saha, S. Synergies and trade-offs in ecosystem services from urban and peri-urban forests and their implication to sustainable city design and planning. *Sustain. Cities Soc.* **2022**, *82*, 103903. [CrossRef]
25. Lawrence, A.; De Vreese, R.; Johnston, M.; Van Den Bosch, C.C.K.; Sanesi, G. Urban forest governance: Towards a framework for comparing approaches. *Urban For. Urban Green.* **2013**, *12*, 464–473. [CrossRef]
26. Morgenroth, J.; Östberg, J. Measuring and monitoring urban trees and urban forests. In *Routledge Handbook of Urban Forestry*; Routledge: Milton Park, UK, 2017; pp. 33–48.
27. Nesbitt, L.; Hotte, N.; Barron, S.; Cowan, J.; Sheppard, S.R. The social and economic value of cultural ecosystem services provided by urban forests in North America: A review and suggestions for future research. *Urban For. Urban Green.* **2017**, *25*, 103–111. [CrossRef]
28. Yan, P.; Yang, J. Species diversity of urban forests in China. *Urban For. Urban Green.* **2017**, *28*, 160–166. [CrossRef]
29. Paquette, A.; Sousa-Silva, R.; Maure, F.; Cameron, E.; Belluau, M.; Messier, C. Praise for diversity: A functional approach to reduce risks in urban forests. *Urban For. Urban Green.* **2021**, *62*, 127157. [CrossRef]
30. Endreny, T.; Santagata, R.; Perna, A.; De Stefano, C.; Rallo, R.F.; Ulgiati, S. Implementing and managing urban forests: A much needed conservation strategy to increase ecosystem services and urban wellbeing. *Ecol. Model.* **2017**, *360*, 328–335. [CrossRef]
31. Brandt, J.S.; Buckley, R.C. A global systematic review of empirical evidence of ecotourism impacts on forests in biodiversity hotspots. *Curr. Opin. Environ. Sustain.* **2018**, *32*, 112–118. [CrossRef]
32. Khaledi Koure, F.; Hajjarian, M.; Hossein Zadeh, O.; Alijanpour, A.; Mosadeghi, R. Ecotourism development strategies and the importance of local community engagement. *Environ. Dev. Sustain.* **2023**, *25*, 6849–6877. [CrossRef]
33. Xiang, C.; Yin, L. Study on the rural ecotourism resource evaluation system. *Environ. Technol. Innov.* **2020**, *20*, 101131. [CrossRef]
34. Powell, R.B.; Ham, S.H. Can ecotourism interpretation really lead to pro-conservation knowledge, attitudes and behaviour? Evidence from the Galapagos Islands. *J. Sustain. Tour.* **2008**, *16*, 467–489. [CrossRef]
35. Oleśniewicz, P.; Pytel, S.; Markiewicz-Patkowska, J.; Szromek, A.R.; Jandová, S. A model of the sustainable management of the natural environment in national parks—A case study of national parks in Poland. *Sustainability* **2020**, *12*, 2704. [CrossRef]
36. Pittelkow, C.M.; Liang, X.; Linquist, B.A.; Van Groenigen, K.J.; Lee, J.; Lundy, M.E.; Van Kessel, C. Productivity limits and potentials of the principles of conservation agriculture. *Nature* **2015**, *517*, 365–368. [CrossRef] [PubMed]
37. Bell, S.; Tyrväinen, L.; Sievänen, T.; Pröbstl, U.; Simpson, M. Outdoor recreation and nature tourism: A European perspective. *Living Rev. Landsc. Res.* **2007**, *1*, 1–46. [CrossRef]
38. Alue, B.; Salleh Hudin, N.; Mohamed, F.; Mat Said, Z.; Ismail, K. Plant Diversity along an Urbanization Gradient of a Tropical City. *Diversity* **2022**, *14*, 1024. [CrossRef]
39. Gavilanes Montoya, A.; Castillo Vizuete, D.; Borz, S. Perceptions of Local Inhabitants towards Land Management Systems Used in the Rainforest Area of Ecuador: An Evaluation Based on Visual Rating of the Main Land Use Types. *Diversity* **2021**, *13*, 592. [CrossRef]
40. Jandl, R.; Spathelf, P.; Bolte, A.; Prescott, C.E. Forest adaptation to climate change—Is non-management an option? *Ann. For. Sci.* **2019**, *76*, 48. [CrossRef]
41. Kovac, M.; Hladnik, D.; Kutnar, L. Biodiversity in (the Natura 2000) forest habitats is not static: Its conservation calls for an active management approach. *J. Nat. Conserv.* **2018**, *43*, 250–260. [CrossRef]
42. Suárez-Muñoz, M.; Bonet-García, F.J.; Navarro-Cerrillo, R.; Herrero, J.; Mina, M. Forest management scenarios drive future dynamics of Mediterranean planted pine forests under climate change. *Landsc. Ecol.* **2023**, *38*, 2069–2084. [CrossRef]

43. Sutherland, L.A.; Huttunen, S. Linking practices of multifunctional forestry to policy objectives: Case studies in Finland and the UK. *For. Policy Econ.* **2018**, *86*, 35–44. [CrossRef]
44. Ravnjak, B.; Bavcon, J.; Čarni, A. Plant Species Turnover on Forest Gaps after Natural Disturbances in the Dinaric Fir Beech Forests (*Omphalodo-Fagetum sylvaticae*). *Diversity* **2022**, *14*, 209. [CrossRef]
45. Cambero, C.; Sowlati, T. Assessment and optimization of forest biomass supply chains from economic, social and environmental perspectives–A review of literature. *Renew. Sustain. Energy Rev.* **2014**, *36*, 62–73. [CrossRef]
46. Karvonen, J.; Halder, P.; Kangas, J.; Leskinen, P. Indicators and tools for assessing sustainability impacts of the forest bioeconomy. *For. Ecosyst.* **2017**, *4*, 2. [CrossRef]
47. Farrell, E.P.; Führer, E.; Ryan, D.; Andersson, F.; Hüttl, R.; Piussi, P. European forest ecosystems: Building the future on the legacy of the past. *For. Ecol. Manag.* **2000**, *132*, 5–20. [CrossRef]
48. Roman, A.; Truta, A.; Viman, O.; Morar, I.; Spalevic, V.; Dan, C.; Sestras, R.; Holonec, L.; Sestras, A. Seed Germination and Seedling Growth of Robinia pseudoacacia Depending on the Origin of Different Geographic Provenances. *Diversity* **2022**, *14*, 34. [CrossRef]
49. Ciuvăț, A.; Abrudan, I.; Ciuvăț, C.; Marcu, C.; Lorenț, A.; Dincă, L.; Szilard, B. Black Locust (*Robinia pseudoacacia* L.) in Romanian Forestry. *Diversity* **2022**, *14*, 780. [CrossRef]
50. Cântar, I.; Ciontu, C.; Dincă, L.; Borlea, G.; Crişan, V. Damage and Tolerability Thresholds for Remaining Trees after Timber Harvesting: A Case Study from Southwest Romania. *Diversity* **2022**, *14*, 193. [CrossRef]
51. Li, Y.; Suontama, M.; Burdon, R.D.; Dungey, H.S. Genotype by environment interactions in forest tree breeding: Review of methodology and perspectives on research and application. *Tree Genet. Genomes* **2017**, *13*, 60. [CrossRef]
52. Lindsey, H.A.; Gallie, J.; Taylor, S.; Kerr, B. Evolutionary rescue from extinction is contingent on a lower rate of environmental change. *Nature* **2013**, *494*, 463–467. [CrossRef]
53. Khadivi-Khub, A.; Ebrahimi, A.; Mohammadi, A.; Kari, A. Characterization and selection of walnut (*Juglans regia* L.) genotypes from seedling origin trees. *Tree Genet. Genomes* **2015**, *11*, 54. [CrossRef]
54. Oneț, A.; Brejea, R.; Dincă, L.; Enescu, R.; Oneț, C.; Besliu, E. Evaluation of Biological Characteristics of Soil as Indicator for Sustainable Rehabilitation of a Post-Bauxite-Mining Land. *Diversity* **2022**, *14*, 1087. [CrossRef]
55. Finkel, O.M.; Castrillo, G.; Paredes, S.H.; González, I.S.; Dangl, J.L. Understanding and exploiting plant beneficial microbes. *Curr. Opin. Plant Biol.* **2017**, *38*, 155–163. [CrossRef] [PubMed]
56. Dinca, L.; Marin, M.; Radu, V.; Murariu, G.; Drasovean, R.; Cretu, R.; Georgescu, L.; Timiș-Gânsac, V. Which Are the Best Site and Stand Conditions for Silver Fir (*Abies alba* Mill.) Located in the Carpathian Mountains? *Diversity* **2022**, *14*, 547. [CrossRef]

Review

Noninvasive Genetics Knowledge from the Brown Bear Populations to Assist Biodiversity Conservation

Iulia Baciu [1,†], Ancuta Fedorca [1,2,†] and Georgeta Ionescu [1,2,*]

1 Wildlife Department, National Institute for Research and Development in Forestry Marin Dracea, 077190 Voluntari, Romania; iuliaa.baciu@yahoo.com (I.B.); ancutacotovelea@yahoo.com (A.F.)
2 Silviculture Department, Faculty of Silviculture and Forest Engineering, Transilvania University of Brasov, 500036 Brasov, Romania
* Correspondence: titi@icaswildlife.ro; Tel.: +40-744377574
† These authors contributed equally to this work.

Abstract: Genetic monitoring has proven helpful in estimating species presence and abundance, and detecting trends in genetic diversity, to be incorporated in providing data and recommendations to management authorities for action and policy development. We reviewed 148 genetics research papers conducted on the bear species worldwide retrieved from Web of Science, SCOPUS, and Google Scholar. This review aims to reveal sampling methodology and data collection instructions, and to unveil innovative noninvasively genetic monitoring techniques that may be integrated into the genetic monitoring of a large bear population. In North American studies, hair samples were collected more often than faeces, whereas in Europe, both faeces and hair samples surveys are recommended, usually focusing on faeces. The use of the Isohelix sample collection method, previously tested locally and, if suitable, applied at the national level, could generate numerous advantages by reducing shortcomings. Additionally, dogs trained for faeces sampling could be used in parallel with hunting managers, foresters, and volunteers for sample collection organised during autumn and winter. It was stated that this is the best period in terms of cost-efficiency and high quality of the gathered samples. We conclude that large-scale noninvasive genetic monitoring of a large bear population represents a challenge; nevertheless, it provides valuable insights for biodiversity monitoring and actions to respond to climate change.

Keywords: noninvasive genetic sampling; brown bear; management plan

Citation: Baciu, I.; Fedorca, A.; Ionescu, G. Noninvasive Genetics Knowledge from the Brown Bear Populations to Assist Biodiversity Conservation. *Diversity* **2022**, *14*, 121. https://doi.org/10.3390/d14020121

Academic Editors: Lucian Dinca, Miglena Zhiyanski and Luc Legal

Received: 30 December 2021
Accepted: 6 February 2022
Published: 8 February 2022

1. Introduction

Some anthropogenic activities harm the environment [1,2]; however, reducing these disturbances in human-dominated landscapes is a challenge for humanity [2]. The negative impact on the environment can be easily observed in the species' biodiversity, which is also negatively impacted [3]. Species loss due to human impact is documented in various studies and books [3–6]. Biodiversity is also correlated to climatic changes [7,8]. The challenge of predicting the complex action of climate evolution makes biodiversity conservation even more difficult [8]. However various models have been developed to predict changes in climate [8,9]; therefore, diminishing potential climate change impacts on species loss is possible [10–12].

Wildlife populations are disturbed in various ways by humanised landscapes, through poaching [13], habitats' alteration under different climatic [7] and vegetation type conditions [14], loss of habitats [3,15], and gene flow limitation, thus reducing landscape connectivity [16–18]. The consequences following these threats affect the survival of wildlife populations [19–21]. At the same time, there is a clear need for considering climate change adaptation and long-term sustainability that is well anchored in very effective policies [22,23]. Wildlife conservation is closely related to population management [24], and appropriate management of wildlife populations implies constant species monitoring [25] and conservation measures specific to targeted species, or, even more specifically, sex-biased conservation measures, if needed [26]. However, genetic monitoring has played a significant role in the conservation and management of species, and the understanding of their ecology [27,28].

Most of the worldwide genetic studies related to bear species are mainly targeted at the regional level. Previous studies, such as in North America, Scandinavia, and the Life DINALP BEAR project (across Croatia, Slovenia, Austria, and Italy), delivered valuable information from comprehensive genetic studies, including interconnected transboundary bears' populations. Therefore, the conservation actions should consider the particularities of species conservation statuses at each country level, with a high focus on the shared transboundary populations [29]. Moreover, the periodic monitoring of threatened large carnivores' populations reveals changes in population conservation status, documented in the European Habitats Directive Annexes 92/43/EEC [30,31]. Genetic methods used for population size estimation and population monitoring are essential for the effective long-term management of wildlife populations [28]. Consequently, each country should allocate resources for establishing permanent genetic monitoring programs [32]. Accurate monitoring efforts of the large carnivores usually require high costs [33–37]. These are also difficult to conduct for wide-ranging species [38], especially because the individuals have greater ecological importance than economic importance [39].

Ursids are of great importance worldwide; however, in addition to their charisma [40], the bear has an ecological and economic value in its habitats [39,41,42]. The brown bear (*Ursus arctos*) has the widest distribution worldwide; it ranges from North America to Eurasia [41]. The species is known for its opportunistic behaviour regarding diet [43]. However, the intraspecific competition for food can be high [44], even if their diet is omnivorous [43,45]. Some of the populations, depending on the living area, have a great preference for plants and a low interest in a carnivorous diet [45]. Interesting facts have been concluded following this behaviour, such as the bears becoming more aggressive towards humans due to higher intraspecific predation [44]. For some studied bears, the feeding locations during the hyperphagia period impacted the selection of their den locations during winter [46].

The brown bear is listed for protection and conservation by several international acts and regulations (Bern Convention, Washington Convention on International Trade in Endangered Species of Wild Fauna and Flora) [41]. Large carnivores, including brown bears, have high priority in conservation across the European continent [47]; the species was declared to have a community interest that needs high protection (it was included in Annex IV of the Habitats Directive), so that the bear population conservation required a declaration of SAC within the Natura 2000 network (Annex II of the Habitats Directive) [48]. Therefore, conservation efforts were considered when bear hunting was banned for some states [49], better-quality habitats were modelled [50], and reintroduction actions of brown bears in specific ecosystems were accomplished [51].

Twenty-two European countries share ten brown bear populations with a permanent species presence, mostly native [47]. The Carpathian population is shared between the Czech Republic, Slovakia, Eastern Serbia, Romania, Ukraine, and Poland [52]. The roaming of individuals between these countries is facilitated by favourable and untouched habitats, the lowest fragmentation rate in Europe, and primarily by the high rate of human acceptance [52].

In Romania, this apex predator population registers approximately 6000 individuals [53], and human–bear conflict across the country has continuously increased in recent years (according to the A2 action in the frame of the LIFE FOR BEAR). The National Action Plan (NAP) (http://www.forbear.icaswildlife.ro/wp-content/uploads/2018/05/plan.pdf last accessed on the 21 December 2021) was approved in 2018 and establishes the main direction for species management and preservation of the favourable conservation status as defined by the last report to the European Commission under Article 17 of the Habitat Directive (https://ec.europa.eu/environment/nature/knowledge/rep_habitats/index_en.htm last accessed on the 19 December 2021). Proper implementation of the NAP requires rigorous monitoring of the population in Romania, as part of the intensive population monitoring objective. However, by analysing the relative strengths and weaknesses of different monitoring field methods, noninvasive genetics could contribute in choosing the optimal strategy and the more efficient allocation of resources for monitoring the Romanian brown bear population [27]. In this way, the human errors when collecting genetic samples (faeces and hair samples) will be reduced to the minimum. At the same time, proper genetic methods will deliver useful biological information, such as that regarding genetic diversity, demography, population bottlenecks, inbreeding, gene flow, or isolated populations [27,53–55]. Moreover, a well-established noninvasive genetic monitoring programme will improve data and information about the bear population's evolutionary history, connectivity, and genetic health [56].

Consequently, this research review aims to provide: (1) relevant insights to develop and improve the methodology for organising the sampling in the Romanian brown bear population, which may lead to a successful research development, with promising results enhancing the institutional (National Institute for Research and Development in Forestry Marin Dracea) and national capacity (Ministry of Environment, Water and Forests); (2) to build upon clear instructions for rigorous training concerning the methodology and instructions for data collection; and (3) to unveil innovative noninvasive better-adapted genetic monitoring techniques. These objectives will contribute to establishing the noninvasive genetic monitoring method that best fits the Romanian brown bear population. Moreover, our approach will impact on future management actions taken for the bear population, and thus increase the degree of conservation of the brown bear population in Romania.

2. Materials and Methods

This research was conducted following the guidelines provided by the Preferred Reporting Items for Systematic Reviews and Meta-Analyses (PRSIMA) (Figure 1). Two of the most popular scientific databases—Web of Science and SCOPUS [57]—were used for identifying relevant scientific literature. Moreover, the web-based database Google Scholar was used, achieving good results in combination with the Web of Science database [58]. All three authors searched within the mentioned databases using the keyword searches *"noninvasive"* AND *"genetic"* AND *"bear"/"brown bear"*. Sometimes, the country was specified as a keyword to receive the necessary results from the searched databases. The authors worked independently and shared their results periodically. The ending date for the search was December 2021.

The first assessment was performed to determine whether to include the records in this review. Therefore, before screening the full text, the studies considered appropriate were selected by (1) title, (2) abstract, and (3) keywords. As the first exclusion, documents from the results that were not on bear species or within the objective of this review were excluded. The grey literature was not an exclusion criterion, even if peer-reviewed literature was preferred; hence, reports following national bear projects, guidelines for noninvasive genetic methods, and national management action plans for bear populations were considered important for this literature review (Table 1).

Table 1. The categories of the reviewed literature related to noninvasive genetics on bear populations.

Category	Peer-Reviewed	Guidelines	Reports	Action Plans
N	140	3	3	2
Total		148		

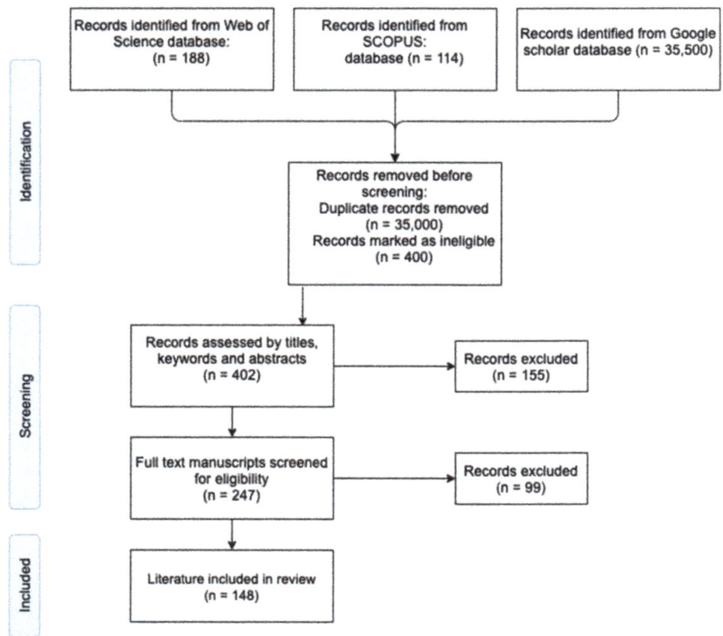

Figure 1. PRISMA literature search flow diagram. The number of works (n) screened, included and excluded in our review are identified at each step of the process.

In the second step, we applied the inclusion/exclusion criteria. The inclusion criteria were (a) studies focused on noninvasive genetic sampling methods, which include sampling detailing, samples' collection and storage, and quality assessment of the sample; (b) efficacy of noninvasive genetic sampling (cost and resources effectiveness); (c) research performed on bear populations across the world; (d) studies published in English (from Europe, North America, and Asia); (e) studies that had significant results/recommendations for developing a good workflow for future research. The exclusion criteria were: (A) studies in fields other than wildlife research; (B) studies that only contained recommendations regarding the laboratory work; (C) studies that used the same datasets and had no new information about the noninvasive genetic sampling; (D) studies that only analysed samples collected invasively.

Although documents from the first results that were not on bear species or within the objective of this review were excluded, 6 of the included studies addressed other species and obtained promising results that we assumed we could adapt to our species of interest. Moreover, a total of 9 studies included in this review used the same samples' database (partially or completely) from another research.

This review encompasses significant information about sampling and sample collection methods used by authors in their studies to monitor bear populations with noninvasive genetics. Additionally of interest were the studies that assessed the quality of the collected samples depending on the bears' diet, the location of the collected samples, the weather conditions when the samples were collected, human error, and other factors [59]. From all the studies identified as appropriate for the review, data were extracted as follows: (1) sampling scheme, including location and bear population type/size, (2) data collection and storage, and (3) significant results and recommendations. Thereby, most of the topic-specific studies should be covered by the chosen web-based databases.

3. Results

Using the three databases, 35,802 papers (peer-reviewed and grey literature) were retrieved, from which 247 were considered relevant for full text screening according to the first exclusion step (Figure 1). Table 1 shows the quantity of the scientific literature according to its classification, based on the inclusion/exclusion criteria after completing the second step, namely, the screening. Finally, the present review consisted of 148 relevant studies, of which the majority are peer-reviewed articles (n = 137), with the remainder being grey literature (three guidelines, three reviews, three reports, and two national action plans).

The final database on the noninvasive genetic sampling (nNGS) of different bear populations worldwide is represented in Figure 2. It includes Europe (Cantabrian, Pyrenees, Apennine, Dinaric—Pindos, East Balkan, Carpathian, and Scandinavian brown bear populations), North America (black bear and grizzly brown bear populations), and Asia (brown bear and Asiatic black bear populations, Malayan sun bear, and Gobi bear).

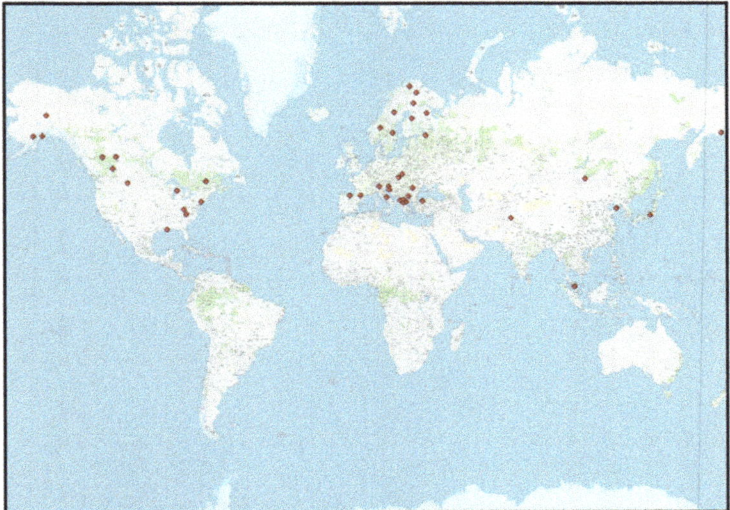

Figure 2. Locations referred to in the retrieved literature.

3.1. Sampling Scheme Including Location and Bear Population Size

The genetic sampling scheme is one of the preliminary stages when conducting genetic research and monitoring [60]. Pilot studies are recommended for establishing the proper sampling strategy (including sampling scheme, timespan between investigations, trap spacing, and subsampling, if necessary) [61]. Table 2 reveals essential information of the evaluated studies worldwide related to the nNGS of the bear population. It includes the sampling scheme used by the authors, the type and number of samples that were analysed, the bear population, and the location of the study.

Table 2. Sampling schemes retrieved from the reviewed literature.

Location	Population	Sampling Scheme *1	N Samples *2	Temporal Extent	Study
Europe					
Northern Europe (E-W)	Scandinavian brown bear	M, SS	3365 F, H, T	2001, 2002, 2004, 2006	[62]
Northern Europe (Se, Norw, Fi, Karelia)	Scandinavian brown bear	Dataset obtained from [63]		2005–2017	[64]
Northern Europe	Scandinavian brown bear	Dataset obtained from [65]			[66]
Northern Europe	Scandinavian brown bear	Dataset obtained from regional monitoring programs		2006–2013	[65]
Northern Europe	Scandinavian brown bear	Dataset obtained from regional monitoring programs		2006–2012	[67]
Sweden	Scandinavian brown bear	Dataset obtained from [55]		2001–2002	[68]
Sweden	Scandinavian brown bear	OS	1904 F	2001–2002	[63]
Sweden	Scandinavian brown bear	OS	5185 F		[69]
Slovenia (south)	Dinaric brown bear	Dataset obtained from a pilot study		2004–2007	[59]
Slovenia	Dinaric brown bear	SS	1053 F	2007–2008	[70]
	M, SS, OS		4687	2015	[71]
Slovenia	Dinaric brown bear	Dataset obtained from regional and national studies		2007	[72]
Slovenia	Dinaric brown bear	CM, OS, SS		2007	[73]
Carpathian brown bear		M, OS	339 T, B, F, H, bones	2004–2009	[74]
Slovakia	Carpathian brown bear		76 H, F	2005–2006	[75]
Slovakia	Carpathian brown bear		140 F, H	2007–2008, 2010	[76]
Romania	Carpathian brown bear	HT, OS, SS	1426 F, H	2017–2018	[77]
Poland	Carpathian brown bear	HT, SS	858 H	2010	[78]
Bulgaria	Eastern Balkan brown bear	HT, CM, OS, M	355 F, H	2004–2008, 2009–2012	[79]
Greece	Eastern Balkan brown bear	HT, TS, CM, M, SS	382 H, F, B	2006–2010	[80]
Greece	Eastern Balkan brown bear	HT, SS	860 H	2007–2010	[81]
Greece (Kastoria)	Eastern Balkan brown bear	HT, OS	232 H, F, B	2011	[82]
GR, FYROM, ALB	Eastern Balkan brown bear	HT, SS	191 H		[83]
FYR Macedonia	Eastern Balkan brown bear	HT, OC, SS	106 H	2008–2009	[84]

Table 2. *Cont.*

Location	Population	Sampling Scheme *1	N Samples *2	Temporal Extent	Study
Albania	Eastern Balkan brown bear	HT, M, TS, SS	12 H	2008–2009	[85]
	HT, OC, OS, SS		643 H	2011	[86]
Italy	Apennine Brown Bear	TS, M, OS	80 H, F, T	1991–2002	[87]
	HT, TS, OS		1164 F, H	2003–2004	[33]
Italy (Alps)	Alp Brown Bear	HT, TS, CM, M, OS, SS	2781 F, H	2002, 2003–2008	[51]
Spain		TS, SS	96 F	1990–1992	[88]
Spain		CM, M, OS, SS	133 F, H, B, T	2004–2006	[89]
Spain	Cantabrian Brown Bear	TS, SS	151 F, H	2017	[31]
France	Pyrenean Brown Bear	TS, OS, SS	153 F	2014–2019	[90]
North America					
Alberta, Canada	Grizzly	HT, TS, OS, SS	183 F, 958 H	2016	[91]
	TS, SS		880 F	1999, 2001	[92]
Alberta, Canada	Grizzly	HT, SS	3363 H	2004	[93]
BNP, Canada	Grizzly and American black bear	HT, CM	6236 H, T	2006–2008	[36]
BC, Canada	American black and brown bears	HT, SS	447 H	1995	[94]
Quebec, Canada	American black bear	HT, SS	411 H	2005	[95]
Alaska	Brown bear	HT, SS	2245 H	2014 - 2017	[96]
Alaska	Grizzly bear	HT, SS	466 H	2002–2003	[97]
	HT, SS		345 H	2003–2005	[98]
Alaska	Brown bear	TS, OS, SS	428 F, saliva	2014	[99]
Montana, USA	Grizzly bear	HT, SS	33741 H	2004	[100]
Northern New York, USA	American black bear	HT, SS		2006	[101]
Louisiana, USA	Louisiana black bear	HT, SS	922 H	1999	[102]

Table 2. *Cont.*

Location	Population	Sampling Scheme *1	N Samples *2	Temporal Extent	Study
Louisiana, USA	Louisiana black bear	OS	448 H	1999	[103]
NLP, Michigan, USA	American black bear	HT, SS	1564 H, T	2003	[104]
Kentucky–Virginia, USA	American black bear	HT, SS	1503 H	2012–2013	[105]
New York, USA	American black bear	HT, SS	1985 H	2012	[106]
North Carolina, USA	American black bear	HT, SS	468 H	2001–2002	[61]
Asia					
Pakistan	Brown bear	TS, SS	136F	2004	[107]
	HT, OS		272 H	2008	[37]
Mongolia	Gobi bear	HT	200 H	1996–1998	[108]
GKM, Turkey	Brown bear	CM, M, OS	154 H, T	2008–2014	[109]
Malaysia	Malayan sun bear	HT	69 H	2017, 2019	[110]
Japan	Asiatic black bear	OC	99 corn-bite samples	2004	[111]

*1 HT = hair trapping; TS = transect sampling; CM = capture for management; M = mortalities (including hunting and corpses); OS = opportunistic sampling; OC = occasional collection; SS = systematic sampling. *2 F = faeces sample; H = hair sample; T = tissue sample; U = urine sample; S = saliva sample; B = blood sample.

The samples used for the genetic analyses were mainly faeces (scat samples) and hair samples collected with different hair traps (further information is presented in Section 3.2). Sometimes, to improve the data collection, samples from accidentally killed bears were also included [72]. Moreover, in the regions where bears were hunted, most studies included tissue samples, typically in Northern Europe [38,62,67,112] and North America [104]. Tissue samples collected from GPS-collared bears and from legally killed individuals following a derogation have also been used in Bulgaria [79].

The sampling methods were correlated to the size of the bear population, the objectives of the study, and the expected results. Both opportunistic and systematic samplings were used in the literature listed in Table 2 (in 57 peer-reviewed research studies).

The studies covering large areas with large bear populations generally adopted the systematic sampling scheme in combination with other sample-collection methods to enlarge the capture probability, as was the case of the Northern Europe population [62,69], the Dinaric—Pindos population, including the transboundary population [71], the Carpathian population [113], and the North America bear populations [36,91,104].

Specific regions where bears usually live were opportunistically surveyed [62,66,77], but not as a single sampling scheme, mainly for the small bear populations. In this situation, the samples can be dispersed across the range, and sample collection may be challenging [86,88,114,115] The research from Italy on a sized population using various monitoring methods has proven that not a single scheme has managed to identify all the individuals. The sampling methods combined several techniques, namely, hair trapping, opportunistic collection of faeces and hair samples, and the transect method. In addition, this study (Table 2) suggested that opportunistic samples are usually helpful when gathering evidence following bear damage [33]. In addition, samples gathered through opportunistic methods usually had lower genotyping success than the samples from hair traps [51].

Sampling a large bear population is challenging; the number of the collected samples has a significant impact on the quality of the study [68]. The large population implies many samples need to be collected and many resources are required. Data collection is critical, and a highly intensive noninvasive genetic sampling can provide good results for large bear populations, as was previously successfully conducted in LIFE DINALP BEAR, in which the target of 3000 samples was exceeded by over 56%, and samples were collected with "good temporal and spatial coverage" [71]. In another study (sampling in November–December 2017), 128 bear faeces samples were collected from the eastern part of the small Cantabrian bear population using 25 km^2 plots. As a result of the sampling of 624.5 km of transects and 151 collected samples (faeces and hair), a minimum of 33 individual bears were identified [31].

Generally, known areas with permanent bear presence have been sampled using equal grids [31,36,86,91,100] or by having the entire range surveyed [65,71,88,104]. The grids can be geographically biased, e.g., Bulgaria, where they were unequally scattered in the targeted mountain regions with a bear presence [79]. Sometimes the sporadic presence of bears was also targeted [72]. Furthermore, areas where bears were known not to exist were included in the sampling scheme in Sweden [69].

When sampling, it is essential to consider the bear behaviour, particularly when hair collection is the goal of the sampling scheme, and the method includes natural rubs [78,81,83,85]. Usually, when sampling for hair collection, the area was divided into grids so that at least one hair trap was placed in an established grid [33,62,78,81,86,116]. The fixed points were specific for hair-collection methods, while hair samples were also opportunistically collected from areas with bear frequency, such as certain power poles used for rubbing [83–85,117], rub trees [78], or buckthorn patches during the berries ripening season [86].

In North America, hair samples were collected more often than faeces, whereas in Europe, the surveys generally included both faeces and hair samples, especially when monitoring rare and elusive species [118]. Some authors even recommended using both types of samples for accurate results following the monitoring of a bear population [113].

Choosing the best sampling scheme is essential; therefore, combining more types of sampling may lead to identifying more individuals or improving the capture mark recapture (CMR) results and resolution.

3.2. Samples' Collection and Storage

Samples' collection and storage significantly influence the genetic analyses' results, in addition to the sampling scheme. A suitable protocol requires a low human error; thereby, it may contribute to accurate DNA extraction and consequently to high qualitative study results [93].

The samples' collection can be conducted by people with different backgrounds as long as they follow a well-established set of instructions. Therefore, pre-season training for the personnel is needed, along with regular updates on results [33,59,71]. In Sweden, for instance, one monitoring technique collected samples opportunistically by moose hunters together with volunteers and personnel from the Scandinavian Brown Bear Research Project [69]. Data collection using volunteers in the framework of bear monitoring programmes has been successfully implemented in other studies [55,69,71,73]. A clear advantage is to monitor the sampling effort in real time by plotting samples on maps; in this way, the "blank areas" are avoided, and forces can be concentrated in particular areas [71].

The number of collected samples plays a significant role, and it is recommended to establish the number of samples to be collected (faeces and/or hair samples) from the beginning. However, 2.5–3 times more scat samples should be collected relative to the number of assumed bear individuals from the study area because it is supposed that 20–30% of the samples will not be genotyped [68].

Regarding the samples' labelling, most of the studies included the geographic location and additional data about the sample registered from the field, while several studies used georeferenced samples [31], including location and date [69,81]. Additionally, the name of the team/person who collected the sample can be recorded [69] to monitor the operators' work. Thereby, remarks can be made, if necessary, which will address the specific issue the operator is facing; thus, human errors may decrease, resulting in a more qualitative study with reduced costs.

As mentioned previously, the sample type (faeces or hair samples) determines the protocol for a noninvasive sample collection from the field. The following subsections present recommendations and relevant remarks for collecting bear faeces and hair samples retrieved from the reviewed articles.

3.2.1. Faeces Collection

First, the age of each faeces sample must be approximated according to the specific smell, visual appearance, and presence of mucous and insect larvae to decide if the sample should be gathered. According to other similar studies, subjective age estimation must be considered for the success of DNA genotyping [59,97,107,113]. Usually, it is suggested not to collect scat samples older than five days because the chances of being properly genotyped are small in a sample of this age [59]. Moreover, the sampling season increases the effectiveness of future noninvasive genetic studies on European bear populations if this is planned correctly [59]. In addition, laboratory costs may also be lower if only higher-quality DNA samples are genotyped, which usually means faeces collected during the autumn period [33].

Approximately 1 cm^3 of scat sample [90,119] was collected with a wood stick [107] and stored in a collection tube/bag [69] in most of the reviewed studies. According to the instructions, the scat sample should be collected from the outer layer (no ground contact) and not from the scat top because the DNA from there could have been washed by rain [59].

In many cases, the sample was collected in solutions with different percentages of ethanol [89,120], from 95% [33,68,90,107] to 96% [31,59,79]. Similarly, it is recommended to use a scat-detection dog, as the mean number of bear scats collected per year may increase significantly; for instance, in the study from the French Pyrenees it was four times higher, and the costs were much lower, because of the validation of the bear scat. The dog also managed to detect the bear cub scats, which are the most difficult to collect by humans because they can be easily confused with faeces from other species. Moreover, a better understanding of the bear's diet is possible using this method. Human-only teams collected 337 scat samples, and the trained dog indicated 239 of them; however, using trained dogs is also cost-effective [90,121].

Another innovative method for faeces collection was revealed in elephant monitoring from Africa, whereby a swab to collect the sample was used [122]. This swab was initially used for collecting saliva samples [123]. The swab was used to scrub the entire surface of the dung pile without touching the ground and the sample was stored in a lysis buffer (Isohelix), in a 2 mL Eppendorf safe-lock tube. This nNGS collection method was efficiently used for elephant monitoring in Gabon, Central Africa, and may be feasible for European brown bear population monitoring. The swab method provides several advantages: the storage tube is smaller and safer than the classic 50 mL tube, the operator error decreases in this case, and the samples may be stored at ambient temperature for 1–4 weeks. A limitation of this practice is that the scat sample is preferably fresh in the field [124] (not older than 3 days) or at least in good condition. Moreover, it should be taken into account that this method does not fit every bear population, regardless of its size, because finding fresh faeces samples in a low-density bear population is difficult [125].

Similarly, in another study from Oman and the United Arab Emirates, a combination of the Isohelix DNA Isolation Kit (provided by Cell Projects Ltd., Harrietsham, Kent, UK) and the QIAamp DNA Stool Mini Kit (provided by Qiagen Ltd., Germantown, MD, USA) was used for faeces collection and DNA extraction from the endangered Arabian tahr [126]. During the research on phylogenetic evidence for the ancient Himalayan wolf, the wolf scat samples were swabbed and stored in Isohelix solution [124]. The swabbing technique provided a higher quality of DNA concentration following the extraction [122].

Following the evaluated studies, some recommendations for faeces collection can be made. The sampling period has a significant impact on the results [68]; thus, it should be chosen accordingly. Therefore, it is recommended to sample bears during autumn and winter because these seasons overlap with the hyperphagia behaviour and it does not interfere with the cubs' period (April–June) [31,127]. Hence, in this period, the number of samples may increase significantly.

In addition to the sampling month, the bear's diet can negatively influence genetic scat analyses. For instance, scats with beech nuts have a high genotyping success rate, but scat age estimates may bias this because faeces with beech nut content may look older than they are [59]. Several studies suggest that plants in the bear diet will affect DNA extraction and inhibit PCR reactions [63,128]. However, other studies obtained acceptable results from faeces samples composed predominantly of plants [107,125].

In addition, rainfall and sunlight exposure were considered other factors of DNA degradation [59,129,130]. However, if the scats are exposed to sunlight and rainfall, they may look older and be excluded by the sampler, as documented in recent research [59].

3.2.2. Hair Samples Collection

Hair sample collection methods from the reviewed research studies are presented in this section in terms of the sample characteristics, followed by the description of the collection and storage techniques. The types of hair traps that may be suitable for monitoring the Carpathian brown bear are mentioned, along with the recommended sampling periods for their installation, and the guidance regarding improving the DNA quality.

The hair traps can be baited, or passive (unbaited); further information regarding the baits and/or lures that can be used as attractants is presented in this chapter [131].

First, it is also necessary to mention the types of hair traps that are successfully used for noninvasive samples collection from ursids. Therefore, Table 3 presents the studies in which several types of hair traps have been used, with their specific characteristics, if they placed bait/lure, and where the research was conducted.

Table 3. Types of bear hair traps.

Hair Trap	Specifications	Bait and/or Lure	Location	Study
Hair corral	At least a single strand of barbed wire stretched around 4 or more trees at 50–55 cm above ground	yes	Italy, Poland, Malaysia, Turkey, California, Michigan, Montana, Alberta, BC, Quebec (Canada)	[33,35,36,78,86,94,109, 110,132–134]
Adhesive rub stations	Tree trunk or wooden blocks wrapped with duct tape	yes	Malaysia	[110]
Power poles	Covered with barbed wire	yes, not on purpose	Greece, Albania, FYR Macedonia, Turkey, Montana	[81–83,85,109,117,135]
Natural rubs (bear rub trees)	equipped hair snagging devices (e.g., barbed wire)	no	Italy, Greece, Bulgaria, Romania, Poland, Alberta, Montana, BC, California, Alaska, Mongolia, Japan, Russian Far East,	[37,77,78,108,132,133, 135,136]
Path traps	Barbed wire installed across known bear travel routes or at feeding routes	no	Italy, Poland, Alaska, Yellowstone Lake,	[37,77–79,108,133]
Modified hair snares	Barbed wire constructed in such way, that allows the bear to escape but keeps hair samples while doing it, and it disables after the process	no	Southeast Alaska	[97]

There are different types of hair traps, and it has been proved that this has a significant influence on trapping success [78]. The relevant studies are described in Table 3, which illustrates the type of bear hair traps and the operating procedures from different study locations. Some studies have used natural rubs, taking advantage of the bears' natural behaviour to rub on wooden and/or power poles [81,83–85,117]. The use of power poles as hair traps is not feasible everywhere, e.g., in Albania; although some of are made of wood, others are concrete, and bears do not rub on them [85]. In addition, in Turkey, most of the poles were exposed to sunlight, so that their genotyping success was very low in comparison to the rub trees [109].

Hair corrals were the most common hair traps, consisting of a single or even double-strand (designed for cubs [33]) stretched around four or more trees at the height of 50–53 cm [36,37,94] above the ground, enclosing a pile of branches and woody debris in the centre, frequently with attractants [35,36,78,94,131]. The height of the strand usually depends on the field personnel who installs it.

These were followed by natural rubs. In Eurasia, it seems that hair traps based on the rubbing behaviour (the natural rubs) are more effective than corrals [37], in comparison to North America [78]. Habituation may be a severe issue; therefore, traps' movement is recommended ($\geq$1 km) within each sampling grid [37,86] every session or at the middle of the sampling session [33]. It is underlined that previously detected bears are likely to be captured again, lured by the baited station. Moving traps is recommended in other similar studies [93,101].

Other studies have used path traps. Barbed wire was installed with different methods in areas where bears were frequently seen (ungulated feeding sites, locations where salmon aggregates during spawning, known den locations, feeding points). Usually, the locations of the hair traps were chosen according to the experts' opinion [36,93] The subjective sampling increases the capture probability [137]. For instance, in a study in Banff National Park, these traps were placed 1 km apart in the vicinity of seasonal food sources or known wildlife paths, far from heavy anthropic areas and tourists' trails [36]. Being far from anthropic areas, the chances of the trap being vandalised were minimal [104]. A study from Poland concluded that the natural rubs had the highest efficiency, followed by smola tree traps. The path traps were considered to be the most ineffective in this situation [78] and were recommended to only be used in known important feeding sites, where bears aggregate [138]. Conversely, these have been successfully used across known spawning streams in Lake Aleknagik (Alaska). The traps have been placed across salmon spawning streams where bears tend to agglomerate. In addition, camera traps were installed nearby to evaluate the bears' behaviour regarding the passive path traps [116]. Some of the researchers have used camera traps to assess the bears' behaviour when reaching the hair trap to make future remarks and recommendations [78,116], an action that is helpful for future studies. Unbaited path traps were also installed in Kenai Fjords National Park, in south-central Alaska, in targeted locations near salmon spawning streams or dense berry patches. This forces the operator to deploy the traps during specific periods (late July–early August, to coincide with the berry productivity and salmon runs) [98]. The passive traps are preferred, especially in CMR studies, because they may increase the sample size without affecting the species' behaviour [98,118].

Bear's attitude regarding the barbed wire was studied, and it was concluded that female bears accompanied by cubs are shyer when contacting the barbed wire compared to the males [116,139]. Additionally, single individuals were more intimidated by the wire than in groups, and cubs were the most curious. However, the percentage of bears that contacted the wire was still high (80.9%), of which 28 were females and 90 males. In another study from Alaska, the bears were less reluctant to contact the wire at night (44.9%), followed by daylight (23.7%), dusk (17.1%), and, finally, dawn (14.3%) [116].

a. Hair clumps: qualitative collection and storage.

Only hair clumps that match the bear hair should be chosen following the sampling. The operator's experience is crucial because he must identify the species hair from the barbed wire. A hair clump is considered a sample worth collecting when it contains a minimum of five underfur hairs on a set of barbs [37,81]. The number of hair follicles from a sample is usually positively associated with the amplification success [63,69,109,110,117,134]. The bunch of hair from one wire is usually considered an individual [78] and is kept separate from others [89]. If the operator succeeds in identifying hair clumps from different individuals, it would be a significant advantage in terms of the costs and quality of the study. In addition, the contamination of the sample with other DNA (from other individuals or humans) should be avoided as much as possible [97].

Many studies collected the hair samples with sterilised forceps or latex gloves [33]. These are generally maintained in the dark in paper envelopes [90,97] and, in some situations, with silica gel packs [37,77,86,109,135,140,141]. In this way, they can be preserved at room temperature [31,79,109]. Prior to the laboratory analyses, some studies recommend freezing the hair samples [142]. In Macedonia, the envelopes were placed in Ziplock bags with silica gel [84], whereas in Slovakia and Italy, the hair material was stored either in paper envelopes or in 70% [108] or 95% ethanol [54]. In the studies on the Pasvik population and in Banff National Park, the hair material was stored dry in paper envelopes [36,62]; some were labelled with barcodes [36,100].

The trap checking interval is crucial for the quality of the collected sample. The hair traps were visited at least once per month [84], at 14-day intervals [36,37,78], or even as often as 7 days [61,104,110], to prevent contamination between individuals' DNA. It is recommended to have the trap checked once a week, or at least every 14 days [97]. After the samples' collection, the barbed wire must be flamed [36,86,100,102,105]. In addition, the functionality of the trap should also be checked, so that the trap's effectiveness is high [78].

Concerning the samples' analysis prioritisation, the hair samples must be examined as soon as possible due to quick DNA degradation [143]. It should be noted that the hair samples are sensitive to sunlight and moisture. When wet samples were collected because of the traps' location (path traps placed near streams; Table 3), they were dried near a heat source [97]. This fact was also observed in another study when hair samples from grizzly and black bears that were not in direct sunlight or exposed to moisture produced high quality genotypes [135,140].

The bear activity season did not substantially impact the hair-trapping success in the situation from Poland [78]. However, the laboratory costs may be lower if the hair sampling period is favourable, and only samples with high-quality DNA are processed, which usually means hair collected during spring [33].

b. Attractants.

The use of attractants is vital to create effective unnatural hair traps. When surveying wildlife, bait and lure have different meanings. There are two types of attractants: natural and unnatural. The first are objects from the habitat which are used by the bear species, such as the rub trees or snags. The unnatural attractants can be divided into rewarding attractants, such as bait (any type of food preferred by ursids), and non-rewarding attractants, such as a scent lure (attractive smells for bears) [131]. There are cases when the bears are involuntarily attracted to some places that are man-made, such as the wooden power poles (Table 3) from Greece and FYR Macedonia that were treated with creosote. This was enough to trigger the bears' rubbing behaviour [81–84,117]. Equivalent to creosote effects, the smola trees from Poland have been used due to their effectiveness as hair traps [78].

The sampling plots for the natural traps depend on the most visited sites where the bears rubbed in the past [81]. The rubbing behaviour is easily observed in the field through the presence of bites and claw signs [84]. Buckthorn was also used as natural rubs [86], even if further studies indicate that bears tend to prefer coniferous trees for rubbing, mainly fir and spruce [78,136,144] with large diameters [78,136,145,146].

For food bait, a mixture of peanut butter and oats or bacon was used [134]. For ursids, raw chicken, fish (also canned), meat (also rotten), carcasses, honey, fruit jam, maple syrup, livestock blood, fruits and vegetables, and pastries [105] are recommended to be used as baits [131].

As scent lures, cattle blood or/and rotten fish may be used. In Poland, ~300 mL of inedible liquid of cattle blood and rotten fish juice (3:1) was used as a lure [91]. In Italy, aged cattle blood (~5–6 L) and decomposed fish oil (2:1) were used as a lure placed on the pile of rotten wood from the centre of a barbed wire encasement (hair corral trap type from Table 3). The pile was covered with leaves, moss, and other forest debris [86,94,131]. Scents of milk, eggs, canned fish, and food scraps (one month old) were used in another study as non-rewarding liquid scent lures [37]. Scent lures and food bait were also used in Michigan's northern Lower Peninsula study. Cherry syrup and scat from black bear individuals at a zoo were used as attractants [134]. Another study from North America used bacon and anise extract to lure bears to the installed hair snares [104].

It is important to replace the bait and refresh lures after the hair sample collection, and especially after precipitation [37].

c. Effectiveness of hair sampling.

It has been revealed that hair sampling had a higher success (86%) in identifying individual grizzly bears; the hair samples' DNA identified nearly two times more unique bears than the DNA from the scat samples. In addition, gender identification also had higher success from hair sampling than scat sampling [91]. Compared to other European studies (from Italy and Sweden), the scat sampling success was significantly lower in the present situation [51,91]. However, when studying large bear populations, competition can also affect the efficiency of hair traps, e.g., the study conducted in Banff National Park indicates that detection varies between the two studied bear populations (grizzly versus black bear), whereas differences between genders' behaviour due to the same variation in the hair trap type have also been observed [36].

Nevertheless, in the case of the small Apennine bear population, future recommendations include the hair snagging method as a primary sampling method, together with at least two different secondary types of hair traps [86]. Hair trapping and opportunistic sampling may provide vital information for the other small bear population from the Italian Alps, following a two-year pilot study where different methods have been tested [33]. These two methods are the most feasible and cost-effective for monitoring this bear population; moreover, it was observed that the hair trapping cost per bear sample may decrease if sampling occurs from the end of May to mid-August, when capture probability is maximised [33,93]. In North America, the standard monitoring method for the grizzly bear population is hair snag sampling, but faeces sampling is promising [91].

On the contrary, the pilot study conducted in the Southern Carpathians on the large Romanian bear population recommends that the sampling is focused on faeces collection in future studies, and that hair samples be considered an alternative [113]. The limitation of the hair traps in this situation is that they can be male-biased, especially during the mating season, and this drawback is also revealed in other research studies and guidelines [37,78,83,100,116,131,147,148].

Regardless of these factors, the capture probability is a limitation when using hair traps. Almost every reviewed study recommends improving the capture probability [34,61,98] or also including other sample-collection methods [33,61,77].

4. Noninvasive Genetics in Bear Conservation and Management

Following noninvasive genetic monitoring of the bear populations, valuable findings can improve management and conservation actions. It was proven that further research on the Gobi bear and implementation of conservation management actions are necessary, due to a noninvasive study conducted to determine its status [108].

Genetic analysis always facilitates an appropriately developed management plan [89,104]. It has been proven that a bear population's genetic structure at a large spatial and temporal scale may be investigated using noninvasively collected genetic data, and even the future shape of the population can be predicted based on the results following the genetic study [62,66,149–151]; however, if needed, the restocking demand can be assessed [87].

Particular attention has been drawn to the increasing bear population from Sweden and its consequences, particularly regarding the coexistence of humans and bears. The bear population has been monitored, including using noninvasive genetic methods, and potential upcoming management issues were identified. A solution was found by the specialists, namely, zoning the targeted areas' carrying capacity [152].

Following the Slovakian genetic study, the recommendations were that the continuous brown bear habitat must be kept to ensure the genetic diversity of West Carpathian bears [108]. However, genetic diversity is not the only condition for long-term population survival. This is also influenced by other factors, such as environment and life history factors [67].

The nNGS proves the need to study transboundary bear populations to provide interesting outcomes [64,71,79,84,153,154]. Three individuals' DNA from FYR Macedonia was also detected in Greece, and this demonstrates the presence of a single interconnected population [84]. The research on the Bulgarian bear population revealed that the apparent mitochondrial lineage separation shows a pattern of male-triggered gene flow. Sample material from two male bears was found following the noninvasive genetic method in both Bulgarian regions; hence, this is considered evidence that narrow corridors exist [79,155].

Female philopatry is demonstrated in this study, as observed in other studies [66,74,79,156]. Dispersal behaviour is mainly exhibited, regardless of sex, due to the increased population density [95].

However, it is suggested that the increased population size and individuals' dispersal do not necessarily imply an increase in the gene flow [65]. The results of an older study indicated some limitations of gene flow between the population from the eastern and western part of Northern Europe [62], even if in recent decades the Scandinavian brown bear population has recovered substantially [65].

Additionally, the results from noninvasive studies may be used as forensic samples. In Norway, DNA extracted in a research study matched with a bear from an illegal hunting case, and the outcome of the study was used as evidence in the trial [112].

Most of the genetic studies conducted on brown bear populations from Europe estimated the number of individuals, their density, and sex ratio to reveal the species' conservation status and distribution [69,81]. Some authors made assumptions and demonstrated the current need for connectivity between bear populations starting from these indicators [56,64,65,71,76,79,80,84,87,89,107] or the sex-biased philopatry [53,65,74,79,95,156–158].

A total of 10 European studies conducted in Northern Europe were retrieved following the results of their noninvasive genetic sampling. Sweden and Norway have developed national bear monitoring plans in which both noninvasive (scats and hairs) and invasive samples are collected permanently (tissue and/or blood from legally harvested bears) [159].

Usually, CMR estimations are difficult to obtain because of the large sample sizes needed, especially if there are extensive areas that must be covered [69,160]. The opportunistic monitoring and the national bear monitoring program provide the management authorities of Northern Europe with a significant amount of data regarding the Scandinavian bear population [159]. The genetic monitoring programs could include the Y chromosome, which is essential for male survival [161], e.g., a study from Northern Europe included the Y-SNP (UAY318.2C839) [67], and this has also been found among the East European bear population [85]. Other studies included additional samples to increase the recapture probability (e.g., tissues from legally harvested bears [104]) and build a diverse dataset. It is acknowledged that bear populations are dynamic, and the "ideal" population size will change depending on the landscape alteration and human dimension [73]. For this reason, a permanent, long-term genetic monitoring program is valuable for every state that hosts a bear population [81,86].

Sustainable genetic monitoring programmes that involve local volunteers are effective. Some of them were found to be rewarding [42], whereas others were not but were still successfully accomplished [55,69,71,73,91,117]. It is suggested that the monitoring effort may increase by having volunteers and standardised data-collection techniques [69,71]. Therefore, the developed database contains essential information that could be, and already is, used in other research studies [59,64–68,72,112], e.g., two different genetic studies concluded that, in addition to the small population effects and habitat quality, mortality thresholds are critical when recovering a large carnivore population [133].

The LIFE DINALP BEAR project provided important outcomes following an exhaustive effort replicated with the help of volunteers. It concluded that the size of the brown bear population from Slovenia increased 41.3% in the last eight years. A total of 1962 samples (including transboundary individuals) were successfully genotyped, resulting in 599 unique bears (545–655 individuals). A total of 552 participants were involved in this intensive sampling with a high recapture rate of 69.5% [71]. This study was challenging in terms of resources, but provided valuable results to be transferred into the species' management and conservation and into the noninvasive genetic knowledge.

For nuisance bear identification, saliva samples were collected by swabbing the surface of a corn-bite sample from a damaged corn field. However, from 99 corn-bite samples, only 30% contained sufficient DNA, and a minimum number of 21 individuals responsible for the damage were detected. Consequently, this study from Japan demonstrated that if agricultural damage samples are collected as quickly as possible, individual bears can be identified [111]. Additionally, saliva samples from partially consumed salmon carcasses were proven to be more cost-effective and have higher quality than the faeces samples in another study [99].

In addition to the genetic approaches, conventional monitoring techniques are also recommended to be used. When data are available from telemetry, direct observations, harvest records, and presence signs, they should be combined with the genetic data to achieve accurate maps of species distribution and relative local population densities, which can be further analysed [34,36,65,152]. Another study concluded that noninvasive genetic sampling and the CMR modelling approach are promising tools when monitoring large carnivore species at a regional scale [31]. Additionally, in another study it is mentioned that, currently, traditional monitoring of bear populations is frequently supplemented by the genetic identification of the species based on samples collected noninvasively [51,162]. Moreover, noninvasively genetic collected data were successfully complemented by the traditional bear population monitoring in the Cantabrian Mountains, namely, the direct observation method through counting females with cubs of the year [163].

Very often, it is suggested to use both noninvasive DNA sampling and the photo-trapping method. If combined with traditional monitoring techniques, the genetic approach will improve the quality of the population study. Consequently, the noninvasive genetic approach does not exclusively assess the individual attributes (age, body condition, reproduction status, distinct signs) [38]. Moreover, it was concluded that the camera trap was more likely to detect grizzly bears than the hair trap in the sampling sites where they both were deployed. The same situation occurred for sun bears, when the camera detected more visits than the hair trap [110].

Another limitation in noninvasive population evaluation is individual heterogeneity (IH), which must be carefully considered. This problem can be solved by developing better laboratory and field protocols. It is even recommended to conduct a pilot study to assess genotyping error rates [164] and detect IH bias sources in the study area. Biological and ecological knowledge and information should be included to validate a model [165]. The use of two capture methods (e.g., hair traps and tissue from hunting individuals) contributes to minimising the individual capture heterogeneity [104]. Avoiding DNA contamination is also vital. Maintaining rigorous conduct is very important while manipulating the sample, beginning with collecting, transporting, storage, and laboratory analysis [70,97,104,106]. For this reason, guidance from similar studies should be considered. In addition to the quality, the low DNA quantity can negatively influence accurate genetic typing; hence, the laboratory personnel should be further instructed to be careful about the amount of DNA extracted from field samples [75].

Noninvasive genetic monitoring requires some resources in the field and the laboratory. From all the reviewed methods, opportunistic sampling showed lower costs and was less challenging for a small bear population. During this research, the opportunistic approach was considered to be the most affordable, at approximately 600 euros per individual [33], which is higher than the Romanian 2021 minimum wage of about 500 euros. Due to budgets being constrained in some situations, it is recommended that only high-quality samples are genotyped [61].

The interpretation of the results after a noninvasive genetic study is essential. The bear population estimation following the study conducted in Bulgaria was not considered to be reliable for the decision-making process when discussing management actions. It was deemed that cubs and individuals from neighbouring countries were included in the three-year sampling period (e.g., migrants from Greece or other mountainous regions). Establishing a wide area may achieve more accurate population size estimation and an improved dataset, including sampling of genetic material, sampling across state boundaries [71,84,153], and a standardised capture–recapture design [79]. The importance of sampling neighbouring countries was also proven in a recent study that underlined the need to closely monitor the Fennoscandian bear, including a regular sampling of bear populations from Finland, Sweden, and Norway. The researchers recommended the holistic approach with a regular sampling of noninvasive genetic material to assess the bear populations [64].

5. Conclusions

A comprehensive noninvasive genetic study can be developed following this review, which evaluated former research studies related to the nNGS on bear populations across the world. The three databases (WoS, SCOPUS and Google Scholar) provided the needed literature; thus, the 148 screened papers allowed the collation of important information related to noninvasive sampling schemes, faeces and hair sample collection, and recommendations from the worldwide studies on bear populations. Depending on the size of the bear population, the location of the study, and the available resources, various suggestions were extracted to enhance the knowledge about noninvasive genetic sampling of the bear population.

The insights from the peer-reviewed and grey literature cited in this manuscript highlight that the minimum number of individuals in the Romanian brown bear population could be estimated through a rigorous genetic study, with considerable efforts and coordination during sampling, and will contribute to the proper implementation of the NAP. Following this review, the development and improvement of the methodology and instructions for genetic sampling are achievable; innovative data collection is vital for large bear population monitoring. The minimum number of individuals and species' genetic diversity and evolutionary potential could be effectively assessed through a large-scale noninvasive genetic monitoring study that embraces large bear habitats. However, this goal represents a significant challenge.

A number of main ideas to be applied at the national scale can be derived from this review:

- The Isohelix method to collect the samples needs to be tested in a local study and, if suitable, applied nationally, by considering the numerous advantages of this method and the high number of people involved in sample gathering, in addition to the shortcomings associated with the storage of such large quantities of samples.
- Trained dogs for faeces gathering should be used across the brown bear distribution, in parallel with hunting managers, foresters, and volunteers.
- Both faeces and hair samples should be collected using the systematic and opportunistic schemes, with a large focus on faeces.
- Samples following damage should be gathered without allowing much time to pass after the damaging event (these can be used for further forensic analysis).
- Sampling should be organised during autumn and winter because these seasons overlap with the hyperphagia behaviour, and it does not interfere with the cub's period.

Therefore, we advocate the noninvasive genetics approach for the establishment of a permanent monitoring program with multi-year coverage. This is recommended as the primary mechanism for better understanding of functional connectivity, gathering species' presence and abundance, and detecting trends in genetic diversity. These strategies provide rich information to be incorporated in recommendations to management authorities for action and policy development, which are crucial for the species survival.

Author Contributions: All authors contributed to the study conception and design. I.B. and A.F. performed material preparation. I.B., A.F. and G.I., data collection and analysis. I.B. wrote the first draft of the manuscript, and A.F. and G.I. commented on previous versions of the manuscript. I.B. and A.F. contributed equally. All authors have read and agreed to the published version of the manuscript.

Funding: This research was funded by POIM Ministry of Investment and European Projects within the project Implementarea planului național de acțiune pentru conservarea populației de urs brun din România—Cod SMIS 2014+: 136899, Nucleu Programme (PN19070601) funded by the Romanian National Authority for Scientific Research and Innovation and "Creșterea capacității și performanței instituționale a INCDS "Marin Drăcea" în activitatea de CDI—CresPerfInst" (Contract nr. 34PFE. /30 December 2021) finanţat de Ministerul Cercetării, Inovării şi Digitalizării prin Programul 1—Dezvoltarea sistemului naţional de cercetare—dezvoltare, Subprogram 1.2—Performanţă instituţională—Proiecte de finanţare a excelenţei în CDI.

Institutional Review Board Statement: Not aplicable.

Informed Consent Statement: Not aplicable.

Data Availability Statement: Data available on request.

Conflicts of Interest: The authors declare no conflict of interest. The funders had no role in the design of the study; in the collection, analyses, or interpretation of data; in the writing of the manuscript, or in the decision to publish the results.

References

1. Pimm, S.L.; Lawton, J.H. Planning for Biodiversity. *Science* **1998**, *279*, 2068–2069. [CrossRef]
2. Zeng, H.; Sui, D.Z.; Wu, B. Human Disturbances on Landscapes in Protected Areas: A Case Study of the Wolong Nature Reserve. *Ecol. Res.* **2005**, *20*, 487–496. [CrossRef]
3. Gaston, K.J.; Spicer, J.I. *Biodiversity: An Introduction*, 2nd ed.; Blackwell Publishing Ltd.: Malden, Massachusetts; Oxford, UK; Victoria, Australia, 2004; ISBN 1-4051-1857-1.
4. Munguía, M.; Trejo, I.; González-Salazar, C.; Pérez-Maqueo, O. Human Impact Gradient on Mammalian Biodiversity. *Glob. Ecol. Biogeogr.* **2016**, *6*, 79–92. [CrossRef]
5. Díaz, S.; Fargione, J.; Chapin, F.S.; Tilman, D. Biodiversity Loss Threatens Human Well-Being. *PLoS Biol.* **2006**, *4*, 277. [CrossRef]
6. Weitzman, M. Diversity Functions. In *Biodiversity Loss: Economic and Ecological Issues*; Perrings, C., Maeler, K.G., Folke, C., Holling, C.S., Jansson, B.O., Eds.; Cambridge University Press: Cambridge, UK, 1995; pp. 21–43.
7. Quratulann, S.; Muhammad Ehsan, M.; Rabia, E.; Sana, A. Review on Climate Change and Its Effect on Wildlife and Ecosystem. *J. Environ. Biol.* **2021**, 008–014. [CrossRef]
8. Pimm, S.L. Biodiversity: Climate Change or Habitat Loss—Which Will Kill More Species? *Curr. Biol.* **2008**, *18*, R117–R119. [CrossRef]
9. Dinca, L.; Badea, O.; Guiman, G.; Braga, C.; Crisan, V.; Greavu, V.; Murariu, G.; Georgescu, L. Monitoring of Soil Moisture in Long-Term Ecological Research (LTER) Sites of Romanian Carpathians. *Ann. For. Res.* **2018**, *61*, 171–188. [CrossRef]
10. Pearson, R.G.; Dawson, T.P. Predicting the Impacts of Climate Change on the Distribution of Species: Are Bioclimate Envelope Models Useful? *Glob. Ecol. Biogeogr.* **2003**, *12*, 361–371. [CrossRef]
11. Araújo, M.B.; Rahbek, C. How Does Climate Change Affect Biodiversity? *Science* **2006**, *313*, 1396–1397. [CrossRef]
12. Austin, M.P.; van Niel, K.P. Improving Species Distribution Models for Climate Change Studies: Variable Selection and Scale. *J. Biogeogr.* **2011**, *38*, 1–8. [CrossRef]
13. Khalidah, K.N.; Wahdaniyah, S.; Kamarudin, N.; Lechner, A.M.; Azhar, B. Spared from Poaching and Natural Predation, Wild Boars Are Likely to Play the Role of Dominant Forest Species in Peninsular Malaysia. *For. Ecol. Manag.* **2021**, *496*, 119458. [CrossRef]
14. Fischer, J.; Lindenmayer, D.B. Landscape Modification and Habitat Fragmentation: A Synthesis. *Glob. Ecol. Biogeogr.* **2007**, *16*, 265–280. [CrossRef]
15. Scanes, C.G. Human Activity and Habitat Loss: Destruction, Fragmentation, and Degradation. *Anim. Hum. Soc.* **2018**, 451–482. [CrossRef]
16. Coulon, A.; Cosson, J.F.; Angibault, J.M.; Cargnelutti, B.; Galan, M.; Morellet, N.; Petit, E.; Aulagnier, S.; Hewison, A.J.M. Landscape Connectivity Influences Gene Flow in a Roe Deer Population Inhabiting a Fragmented Landscape: An Individual-Based Approach. *Mol. Ecol.* **2004**, *13*, 2841–2850. [CrossRef]
17. Baguette, M.; Blanchet, S.; Legrand, D.; Stevens, V.M.; Turlure, C. Individual Dispersal, Landscape Connectivity and Ecological Networks. *Biol. Rev.* **2013**, *88*, 310–326. [CrossRef]
18. Anderson, C.D.; Epperson, B.K.; Fortin, M.J.; Holderegger, R.; James, P.M.A.; Rosenberg, M.S.; Scribner, K.T.; Spear, S. Considering Spatial and Temporal Scale in Landscape-Genetic Studies of Gene Flow. *Mol. Ecol.* **2010**, *19*, 3565–3575. [CrossRef]
19. Frid, A.; Dill, L. Human-Caused Disturbance Stimuli as a Form of Predation Risk. *Ecol. Soc.* **2002**, *6*. [CrossRef]
20. Guanghun, J.; Jianzhang, M.; Minghai, Z. Spatial Distribution of Ungulate Responses to Habitat Factors in Wandashan Forest Region, Northeastern China. *J. Wildl. Manag.* **2006**, *70*, 1470–1476. [CrossRef]
21. Paudel, P.K.; Kindlmann, P. Human Disturbance Is a Major Determinant of Wildlife Distribution in Himalayan Midhill Landscapes of Nepal. *Anim. Conserv.* **2012**, *15*, 283–293. [CrossRef]
22. Tudose, N.C.; Cremades, R.; Broekman, A.; Sanchez-Plaza, A.; Mitter, H.; Marin, M. Mainstreaming the Nexus Approach in Climate Services Will Enable Coherent Local and Regional Climate Policies. *Adv. Clim. Chang. Res.* **2021**, *12*, 752–755. [CrossRef]
23. Marin, M.; Clinciu, I.; Tudose, N.C.; Ungurean, C.; Adorjani, A.; Mihalache, A.L.; Davidescu, A.A.; Davidescu, Ş.O.; Dinca, L.; Cacovean, H. Assessing the Vulnerability of Water Resources in the Context of Climate Changes in a Small Forested Watershed Using SWAT: A Review. *Environ. Res.* **2020**, *184*, 109330. [CrossRef]

24. Algotsson, E. Wildlife Conservation through People-Centred Approaches to Natural Resource Management Programmes and the Control of Wildlife Exploitation. *Local Environ.* **2006**, *11*, 79–93. [CrossRef]
25. Hodgson, J.C.; Mott, R.; Baylis, S.M.; Pham, T.T.; Wotherspoon, S.; Kilpatrick, A.D.; Raja Segaran, R.; Reid, I.; Terauds, A.; Koh, L.P. Drones Count Wildlife More Accurately and Precisely than Humans. *Methods Ecol. Evol.* **2018**, *9*, 1160–1167. [CrossRef]
26. García-Sánchez, M.P.; González-Ávila, S.; Solana-Gutiérrez, J.; Popa, M.; Jurj, R.; Ionescu, G.; Ionescu, O.; Fedorca, M.; Fedorca, A. Sex-Specific Connectivity Modelling for Brown Bear Conservation in the Carpathian Mountains. *Landsc. Ecol.* **2021**. [CrossRef]
27. Swenson, J.E.; Taberlet, P.; Bellemain, E. Genetics and Conservation of European Brown Bears Ursus Arctos. *Mamm. Rev.* **2011**, *41*, 87–98. [CrossRef]
28. Antao, T.; Pérez-Figueroa, A.; Luikart, G. Early Detection of Population Declines: High Power of Genetic Monitoring Using Effective Population Size Estimators. *Evol. Appl.* **2011**, *4*, 144–154. [CrossRef]
29. Bartoń, K.A.; Zwijacz-Kozica, T.; Zięba, F.; Sergiel, A.; Selva, N. Bears without Borders: Long-Distance Movement in Human-Dominated Landscapes. *Glob. Ecol. Conserv.* **2019**, *17*, e00541. [CrossRef]
30. Epstein, Y.; Jos, J.; Vicente, J.; Opez-Bao, L.; Chapron, G.; Fischer, J. A Legal-Ecological Understanding of Favorable Conservation Status for Species in Europe. *Conserv. Lett.* **2016**, *9*, 81–88. [CrossRef]
31. López-Bao, J.V.; Godinho, R.; Rocha, R.G.; Palomero, G.; Blanco, J.C.; Ballesteros, F.; Jiménez, J. Consistent Bear Population DNA-Based Estimates Regardless Molecular Markers Type. *Biol. Conserv.* **2020**, *248*. [CrossRef]
32. Bischof, R.; Brøseth, H.; Gimenez, O. Wildlife in a Politically Divided World: Insularism Inflates Estimates of Brown Bear Abundance. *Conserv. Lett.* **2016**, *9*, 122–130. [CrossRef]
33. De Barba, M.; Waits, L.P.; Genovesi, P.; Randi, E.; Chirichella, R.; Cetto, E. Comparing Opportunistic and Systematic Sampling Methods for Non-Invasive Genetic Monitoring of a Small Translocated Brown Bear Population. *J. Appl. Ecol.* **2010**, *47*, 172–181. [CrossRef]
34. Boulanger, J.; Stenhouse, G.; Munro, R. Sources of Heterogeneity Bias When DNA Mark-Recapture Sampling Methods Are Applied to Grizzly Bear (Ursus Arctos) Populations. *J. Mammal.* **2004**, *85*, 618–624. [CrossRef]
35. Boulanger, J.; Kendall, K.C.; Stetz, J.B.; Roon, D.A.; Waits, L.P.; Paetkau, D. Multiple Data Sources Improve DNA-Based Mark-Recapture Population Estimates of Grizzly Bears. *Ecol. Appl.* **2008**, *18*, 577–589. [CrossRef] [PubMed]
36. Sawaya, M.A.; Stetz, J.B.; Clevenger, A.P.; Gibeau, M.L.; Kalinowski, S.T. Estimating Grizzly and Black Bear Population Abundance and Trend in Banff National Park Using Noninvasive Genetic Sampling. *PLoS ONE* **2012**, *7*. [CrossRef]
37. Latham, E.; Stetz, J.B.; Seryodkin, I.; Miquelle, D.; Gibeau, M.L. Non-Invasive Genetic Sampling of Brown Bears and Asiatic Black Bears in the Russian Far East: A Pilot Study. *Ursus* **2012**, *23*, 145–158. [CrossRef]
38. Bischof, R.; Swenson, J.E. Linking Noninvasive Genetic Sampling and Traditional Monitoring to Aid Management of a Trans-Border Carnivore Population. *Ecol. Appl.* **2012**, *22*, 361–373. [CrossRef]
39. Van Vliet, N.; Cornelis, D.; Beck, H.; Lindsey, P.A. Meat from the Wild: Extractive Uses of Wildlife and Alternatives for Sustainability. *Curr. Trends Wildl. Res.* **2016**. [CrossRef]
40. Albert, C.; Luque, G.M.; Courchamp, F. The Twenty Most Charismatic Species. *PLoS ONE* **2018**, *13*, e0199149. [CrossRef]
41. Servheen, C.; Herrero, S.; Peyton, B. *Bears. Status Survey and Conservation Action Plan*; IUCN/SSC Bear and Polar Bear Specialist Groups, IUCN: Cland, Switzerland; Cambridge, UK; ISBN 2-8317-0462-6.
42. Singh, N.J.; Danell, K.; Edenius, L.; Ericsson, G. Tackling the Motivation to Monitor: Success and Sustainability of a Participatory Monitoring Program. *Ecol. Soc.* **2014**, *19*. [CrossRef]
43. Collins, D.M. Ursidae. In *Fowler's Zoo and Wild Animal Medicine*; Elsevier: Amsterdam, The Netherlands, 2015; Volume 8, pp. 498–508.
44. Kolchin, S.A.; Volkova, E.V.; Pokrovskaya, L.V.; Zavadskaya, A.V. Consequences of a Sockeye Salmon Shortage for the Brown Bear in the Basin of Lake Kurilskoe, Southern Kamchatka. *Nat. Conserv. Res.* **2021**, *6*, 53–65. [CrossRef]
45. Ogurtsov, S.S. The Diet of the Brown Bear (Ursus Arctos) in the Central Forest Nature Reserve (West-European Russia), Based on Scat Analysis Data. *Biology* **2018**, *45*, 1039–1054. [CrossRef]
46. Seryodkin, I.V.; Paczkowski, J.; Goodrich, J.M.; Petrunenko, Y.K. Locations of Dens with Respect to Space Use, Pre-and Post-Denning Movements of Brown Bears in the Russian Far East. *Nat. Conserv. Res.* **2021**, *6*, 97–109. [CrossRef]
47. Chapron, G.; Kaczensky, P.; Linnell, J.D.C.; von Arx, M.; Huber, D.; Andrén, H.; Vicente López-Bao, J.; Adamec, M.; Álvares, F.; Anders, O.; et al. Recovery of Large Carnivores in Europe's Modern Human-Dominated Landscapes. *Science* **2014**, *346*, 1517–1519. [CrossRef]
48. Trouwborst, A. Managing the Carnivore Comeback: International and EU Species Protection Law and the Return of Lynx, Wolf and Bear to Western Europe. *J. Environ. Law* **2010**, *22*, 347–372. [CrossRef]
49. Darimont, C.T.; Hall, H.; Eckert, L.; Mihalik, I.; Artelle, K.; Treves, A.; Paquet, P.C. Large Carnivore Hunting and the Social License to Hunt. *Conserv. Biol.* **2021**, *35*, 1111–1119. [CrossRef]
50. Ogurtsov, S.S. Brown Bear (Ursus Arctos) Ecological Niche and Habitat Suitability Modeling in the Southern Taiga Subzone Using the Method of GNESFA. *Nat. Conserv. Res.* **2020**, *5*, 86–113. [CrossRef]
51. De Barba, M.; Waits, L.P.; Garton, E.O.; Genovesi, P.; Randi, E.; Mustoni, A.; Groff, C. The Power of Genetic Monitoring for Studying Demography, Ecology and Genetics of a Reintroduced Brown Bear Population. *Mol. Ecol.* **2010**, *19*, 3938–3951. [CrossRef]
52. Management and Action Plan for the Bear Population in Romania. Available online: http://www.mmediu.ro/app/webroot/uploads/files/17Management_Action_Plan.pdf" (accessed on 29 December 2021).

53. Cotovelea, A.; Sofletea, N.; Ionescu, G.; Ionescu, O. Genetic Approaches for Romanian Conservation. *Bull. Trans. Univ. Bras. Ser. II* **2013**, *6*, 18–26.
54. Kohn, M.H.; Wayne, R.K. Facts from Feces Revisited. *Trends Ecol. Evol.* **1997**, *12*, 223–227. [CrossRef]
55. Bellemain, E.; Swenson, J.E.; Tallmon, D.; Taberlet, P. Estimating Population Size of Elusive Animals with DNA from Hunter-Collected Feces: Four Methods for Brown Bears. *Conserv. Biol.* **2005**, 150–161. [CrossRef]
56. Tumendemberel, O.; Zedrosser, A.; Proctor, M.F.; Reynolds, H.V.; Adams, J.R.; Sullivan, J.M.; Jacobs, S.J.; Khorloojav, T.; Tserenbataa, T.; Batmunkh, M.; et al. Phylogeography, Genetic Diversity, and Connectivity of Brown Bear Populations in Central Asia. *PLoS ONE* **2019**, *14*, 1–23. [CrossRef]
57. Pranckutė, R. Web of Science (Wos) and Scopus: The Titans of Bibliographic Information in Today's Academic World. *Publications* **2021**, *9*, 12. [CrossRef]
58. Bramer, W.M.; Rethlefsen, M.L.; Kleijnen, J.; Franco, O.H. Optimal Database Combinations for Literature Searches in Systematic Reviews: A Prospective Exploratory Study. *Syst. Rev.* **2017**, *6*, 1–12. [CrossRef]
59. Skrbinšek, T. Effects of Different Environmental and Sampling Variables on the Genotyping Success in Field-Collected Scat Samples: A Brown Bear Case Study. *Acta Biol. Slov.* **2020**, *63*, 89–98.
60. Schwartz, M.K.; McKelvey, K.S. Why Sampling Scheme Matters: The Effect of Sampling Scheme on Landscape Genetic Results. *Conserv. Genet.* **2008**, *10*, 441–452. [CrossRef]
61. Tredick, C.A.; Vaughan, M.R.; Stauffer, D.F.; Simek, S.L.; Eason, T. Sub-Sampling Genetic Data to Estimate Black Bear Population Size: A Case Study. *Ursus* **2007**, *18*, 179–188. [CrossRef]
62. Schregel, J.; Kopatz, A.; Hagen, S.B.; Broseth, H.; Smith, M.E.; Wikan, S.; Wartiainen, I.; Aspholm, P.E.; Aspi, J.; Swenson, J.E.; et al. Limited Gene Flow among Brown Bear Populations in Far Northern Europe? Genetic Analysis of the East-West Border Population in the Pasvik Valley. *Mol. Ecol.* **2012**, *21*, 3474–3488. [CrossRef]
63. Tallmon, D.A.; Bellemain, E.; Swendon, J.E.; Taberlet, P. Genetic Monitoring of Scandinavian Brown Bear Effective Population Size and Immigration. *J. Wildl. Manag.* **2004**, *68*, 960–965. [CrossRef]
64. Kopatz, A.; Kleven, O.; Kojola, I.; Aspi, J.; Norman, A.J.; Spong, G.; Gyllenstrand, N.; Dalén, L.; Fløystad, I.; Hagen, S.B.; et al. Restoration of Transborder Connectivity for Fennoscandian Brown Bears (Ursus Arctos). *Biol. Conserv.* **2021**, *253*, 11–25. [CrossRef]
65. Schregel, J.; Kopatz, A.; Eiken, H.G.; Swenson, J.E.; Hagen, S.B. Sex-Specific Genetic Analysis Indicates Low Correlation between Demographic and Genetic Connectivity in the Scandinavian Brown Bear (Ursus Arctos). *PLoS ONE* **2017**, *12*, e0180701. [CrossRef]
66. Schregel, J.; Remm, J.; Eiken, H.G.; Swenson, J.E.; Saarma, U.; Hagen, S.B. Multi-Level Patterns in Population Genetics: Variogram Series Detects a Hidden Isolation-by-Distance-Dominated Structure of Scandinavian Brown Bears Ursus Arctos. *Methods Ecol. Evol.* **2018**, *9*, 1324–1334. [CrossRef]
67. Schregel, J.; Eiken, H.G.; Grøndahl, F.A.; Hailer, F.; Aspi, J.; Kojola, I.; Tirronen, K.; Danilov, P.; Rykov, A.; Poroshin, E.; et al. Y Chromosome Haplotype Distribution of Brown Bears (Ursus Arctos) in Northern Europe Provides Insight into Population History and Recovery. *Mol. Ecol.* **2015**, *24*, 6041–6060. [CrossRef] [PubMed]
68. Solberg, K.H.; Bellemain, E.; Drageset, O.M.; Taberlet, P.; Swenson, J.E. An Evaluation of Field and Non-Invasive Genetic Methods to Estimate Brown Bear (Ursus Arctos) Population Size. *Biol. Conserv.* **2006**, *128*, 158–168. [CrossRef]
69. Kindberg, J.; Swenson, J.E.; Ericsson, G.; Bellemain, E.; Miquel, C.; Taberlet, P. Estimating Population Size and Trends of the Swedish Brown Bear Ursus Arctos Population. *Wildl. Biol.* **2011**, *17*, 114–123. [CrossRef]
70. Skrbinšek, T.; Jelenčič, M.; Waits, L.; Kos, I.; Trontelj, P. Highly Efficient Multiplex PCR of Noninvasive DNA Does Not Require Pre-Amplification. *Mol. Ecol. Resour.* **2010**, *10*, 495–501. [CrossRef]
71. Final Report <LIFE13 NAT/SI/000550 Life DINALP BEAR>. Available online: https://dinalpbear.eu/wp-content/uploads/LIFE-DINALP-BEAR_final-report_web.pdf (accessed on 9 June 2021).
72. Jerina, K.; Jonozovič, M.; Krofel, M.; Skrbinšek, T. Range and Local Population Densities of Brown Bear Ursus Arctos in Slovenia. *Eur. J. Wildl. Res.* **2013**, *59*, 459–467. [CrossRef]
73. Skrbinšek, T.; Luštrik, R.; Majić-Skrbinšek, A.; Potočnik, H.; Kljun, F.; Jelenčič, M.; Kos, I.; Trontelj, P. From Science to Practice: Genetic Estimate of Brown Bear Population Size in Slovenia and How It Influenced Bear Management. *Eur. J. Wildl. Res.* **2019**, *65*. [CrossRef]
74. Straka, M.; Paule, L.; Ionescu, O.; Štofík, J.; Adamec, M. Microsatellite Diversity and Structure of Carpathian Brown Bears (Ursus Arctos): Consequences of Human Caused Fragmentation. *Conserv. Genet.* **2012**, *13*, 153–164. [CrossRef]
75. Janiga, M.; Fečková, M.; Korňan, J. Preliminary Results on Genetic Tracking of the Brown Bear (Ursus Arctos) Individuals in the Malá Fatra National Park (Slovakia). *Oecologia* **2006**, *15*, 24–26.
76. Graban, J.; Kisková, J.; Pepich, P.; Rigg, R. Genetic Analysis for Geographic Isolation Comparison of Brown Bears Living in the Periphery of the Western Carpathians Mountains with Bears Living in Other Areas. *Open J. Genet.* **2013**, *3*, 174–182. [CrossRef]
77. Iosif, R.; Skrbinšek, T.; Jelenčič, M.; Boljte, B.; Konec, M.; Erich, M.; Sulică, B.; Moza, I.; Ungureanu, L.; Rohan, R.; et al. Report on Monitoring Brown Bears Using Non-Invasive DNA Sampling in the Romanian Carpathians Bear Report FOUNDATION CONSERVATION CARPATHIA. Available online: https://www.carpathia.org/wp-content/uploads/2021/10/FCC-Report-on-monitoring-brown-bear-using-non-invasive-DNA-sampling-in-the-Romanian-Carpathians.pdf (accessed on 29 December 2021).
78. Berezowska-Cnota, T.; Luque-Márquez, I.; Elguero-Claramunt, I.; Bojarska, K.; Okarma, H.; Selva, N. Effectiveness of Different Types of Hair Traps for Brown Bear Research and Monitoring. *PLoS ONE* **2017**, *12*, e0186605. [CrossRef]

79. Frosch, C.; Dutsov, A.; Zlatanova, D.; Valchev, K.; Reiners, T.E.; Steyer, K.; Pfenninger, M.; Nowak, C. Noninvasive Genetic Assessment of Brown Bear Population Structure in Bulgarian Mountain Regions. *Mamm. Biol.* **2014**, *79*, 268–276. [CrossRef]
80. Pylidis, C.; Anijalg, P.; Saarma, U.; Dawson, D.A.; Karaiskou, N.; Butlin, R.; Mertzanis, Y.; Giannakopoulos, A.; Iliopoulos, Y.; Krupa, A.; et al. Multisource Noninvasive Genetics of Brown Bears (Ursus Arctos) in Greece Reveals a Highly Structured Population and a New Matrilineal Contact Zone in Southern Europe. *Ecol. Evol.* **2021**, *11*, 6427–6443. [CrossRef]
81. Karamanlidis, A.A.; de Hernando, M.G.; Krambokoukis, L.; Gimenez, O. Evidence of a Large Carnivore Population Recovery: Counting Bears in Greece. *J. Nat. Conserv.* **2015**, *27*, 10–17. [CrossRef]
82. Tsaparis, D.; Karaiskou, N.; Mertzanis, Y.; Triantafyllidis, A. Non-Invasive Genetic Study and Population Monitoring of the Brown Bear (Ursus Arctos) (Mammalia: Ursidae) in Kastoria Region–Greece. *J. Nat. Hist.* **2015**, *49*, 393–410. [CrossRef]
83. Karamanlidis, A.A.; Drosopoulou, E.; de Hernando, M.G.; Georgiadis, L.; Krambokoukis, L.; Pllaha, S.; Zedrosser, A.; Scouras, Z. Noninvasive Genetic Studies of Brown Bears Using Power Poles. *Eur. J. Wildl. Res.* **2010**, *56*, 693–702. [CrossRef]
84. Karamanlidis, A.A.; Stojanov, A.; de Gabriel Hernando, M.; Ivanov, G.; Kocijan, I.; Melovski, D.; Skrbinšek, T.; Zedrosser, A. Distribution and Genetic Status of Brown Bears in FYR Macedonia: Implications for Conservation. *Acta Theriol.* **2014**, *59*, 119–128. [CrossRef]
85. Karamanlidis, A.A.; Pllaha, S.; Krambokoukis, L.; Shore, K.; Zedrosser, A. Preliminary Brown Bear Survey in Southeastern Albania. *Ursus* **2014**, *25*, 1–7. [CrossRef]
86. Ciucci, P.; Gervasi, V.; Boitani, L.; Boulanger, J.; Paetkau, D.; Prive, R.; Tosoni, E. Estimating Abundance of the Remnant Apennine Brown Bear Population Using Multiple Noninvasive Genetic Data Sources. *J. Mammal.* **2015**, *96*, 206–220. [CrossRef]
87. Lorenzini, R.; Posillico, M.; Lovari, S.; Petrella, A. Non-Invasive Genotyping of the Endangered Apennine Brown Bear: A Case Study Not to Let One's Hair Down. *Anim. Conserv.* **2004**, *7*, 199–209. [CrossRef]
88. Clevenger, A.P.; Purroy, F.J. Sign Surveys for Estimating Trend of a Remnant Brown Bear Ursus Arctos Population in Northern Spain. *Wildl. Biol.* **1996**, *2*, 275–281. [CrossRef]
89. Pérez, T.; Vázquez, F.; Naves, J.; Fernández, A.; Corao, A.; Albornoz, J.; Domínguez, A. Non-Invasive Genetic Study of the Endangered Cantabrian Brown Bear (Ursus Arctos). *Conserv. Genet.* **2009**, *10*, 291–301. [CrossRef]
90. Sentilles, J.; Vanpé, C.; Quenette, P.-Y. Benefits of Incorporating a Scat-Detection Dog into Wildlife Monitoring: A Case Study of Pyrenean Brown Bear. *J. Vertebr. Biol.* **2021**, *69*. [CrossRef]
91. Phoebus, I.; Boulanger, J.; Eiken, H.G.; Fløystad, I.; Graham, K.; Hagen, S.B.; Sorensen, A.; Stenhouse, G. Comparison of Grizzly Bear Hair-Snag and Scat Sampling along Roads to Inform Wildlife Population Monitoring. *Wildl. Biol.* **2020**, *2020*. [CrossRef]
92. Wasser, S.K.; Davenport, B.; Ramage, E.R.; Hunt, K.E.; Parker, M.; Clarke, C.; Stenhouse, G. Scat Detection Dogs in Wildlife Research and Management: Application to Grizzly and Black Bears in the Yellowhead Ecosystem, Alberta, Canada. *Can. J. Zool.* **2004**, *82*, 475–492. [CrossRef]
93. Boulanger, J.; Proctor, M.; Himmer, S.; Stenhouse, G.; Paetkau, D.; Cranston, J. An Empirical Test of DNA Mark-Recapture Sampling Strategies for Grizzly Bears. *Ursus* **2006**, *17*, 149–158. [CrossRef]
94. Woods, J.G.; Paetkau, D.; Lewis, D.; McLellan, B.N.; Proctor, M.; Strobeck, C. Genetic Tagging of Free-Ranging Black and Brown Bears. *Wild. Soc. Bull.* **1999**, *27*, 616–627.
95. Roy, J.; Yannic, G.; Côté, S.D.; Bernatchez, L. Negative Density-Dependent Dispersal in the American Black Bear (Ursus Americanus) Revealed by Noninvasive Sampling and Genotyping. *Ecol. Evol.* **2012**, *2*, 525–537. [CrossRef]
96. Wirsing, A.J.; Quinn, T.P.; Adams, J.R.; Waits, L.P. Optimizing Selection of Brown Bear Hair for Noninvasive Genetic Analysis. *Wild. Soc. Bull.* **2020**, *44*, 94–100. [CrossRef]
97. Beier, L.R.; Lewis, S.B.; Flynn, R.W.; Pendleton, G.; Schumacher, V.T. A Single-Catch Snare to Collect Brown Bear Hair for Genetic Mark–Recapture Studies. *Wild. Soc. Bull.* **2005**, *33*, 766–773. [CrossRef]
98. Robinson, S.J.; Waits, L.P.; Martin, I.D. Estimating Abundance of American Black Bears Using DNA-Based Capture-Mark-Recapture Models. *Ursus* **2009**, *20*, 1–11. [CrossRef]
99. Wheat, R.E.; Allen, J.M.; Miller, S.D.L.; Wilmers, C.C.; Levi, T. Environmental DNA from Residual Saliva for Efficient Noninvasive Genetic Monitoring of Brown Bears (Ursus Arctos). *PLoS ONE* **2016**, *11*, e0165259. [CrossRef] [PubMed]
100. Kendall, K.C.; Stetz, J.B.; Boulanger, J.; Macleod, A.C.; Paetkau, D.; White, G.C. Demography and Genetic Structure of a Recovering Grizzly Bear Population. *J. Wildl. Manag.* **2009**, *73*, 3–17. [CrossRef]
101. Gardner, B.; Royle, J.A.; Wegan, M.T.; Rainbolt, R.E.; Curtis, P.D. Estimating Black Bear Density Using DNA Data From Hair Snares. *J. Wildl. Manag.* **2010**, *74*, 318–325. [CrossRef]
102. Triant, D.A.; Pace, R.M.; Stine, M. Abundance, Genetic Diversity and Conservation of Louisiana Black Bears (Ursus Americanus Luteolus) as Detected through Noninvasive Sampling. *Conserv. Genet.* **2004**, *5*, 647–659. [CrossRef]
103. Boersen, M.R.; Clark, J.D.; King, T.L. Estimating Black Bear Population Density and Genetic Diversity at Tensas River, Louisiana Using Microsatellite DNA Markers. *Wildl. Soc. Bull.* **2003**, *31*, 197–207. [CrossRef]
104. Dreher, B.P.; Winterstein, S.R.; Scribner, K.T.; Lukacs, P.M.; Etter, D.R.; Rosa, G.J.M.; Lopez, V.A.; Libants, S.; Filcek, K.B. Noninvasive Estimation of Black Bear Abundance Incorporating Genotyping Errors and Harvested Bear. *J. Wildl. Manag.* **2007**, *71*, 2684–2693. [CrossRef]
105. Murphy, S.M.; Cox, J.J.; Augustine, B.C.; Hast, J.T.; Guthrie, J.M.; Wright, J.; McDermott, J.; Maehr, S.C.; Plaxico, J.H. Characterizing Recolonization by a Reintroduced Bear Population Using Genetic Spatial Capture–Recapture. *J. Wildl. Manag.* **2016**, *80*, 1390–1407. [CrossRef]

106. Sun, C.C.; Fuller, A.K.; Hare, M.P.; Hurst, J.E. Evaluating Population Expansion of Black Bears Using Spatial Capture-Recapture. *J. Wildl. Manag.* **2017**, *81*, 814–823. [CrossRef]
107. Bellemain, E.; Nawaz, M.A.; Valentini, A.; Swenson, J.E.; Taberlet, P. Genetic Tracking of the Brown Bear in Northern Pakistan and Implications for Conservation. *Biol. Conserv.* **2007**, *134*, 537–547. [CrossRef]
108. Mccarthy, T.M.; Waits, L.P.; Mijiddorj, B. Status of the Gobi Bear in Mongolia as Determined by Noninvasive Genetic Methods. *Ursus* **2009**, *20*, 30–38. [CrossRef]
109. Ambarli, H.; Mengüllüoğlu, D.; Fickel, J.; Förster, D.W. Population Genetics of the Main Population of Brown Bears in Southwest Asia. *PeerJ* **2018**, *2018*, e5660. [CrossRef]
110. Tee, T.L.; Lai, W.L.; Ju Wei, T.K.; Shern, O.Z.; van Manen, F.T.; Sharp, S.P.; Wong, S.; Chew, J.; Ratnayeke, S. An Evaluation of Noninvasive Sampling Techniques for Malayan Sun Bears. *Ursus* **2020**, *2020*, 1–12. [CrossRef]
111. Saito, M.; Yamauchi, K.; Aoi, T. Individual Identification of Asiatic Black Bears Using Extracted DNA from Damaged Crops. *Ursus* **2008**, *19*, 162–167. [CrossRef]
112. Andreassen, R.; Schregel, J.; Kopatz, A.; Tobiassen, C.; Knappskog, P.M.; Hagen, S.B.; Kleven, O.; Schneider, M.; Kojola, I.; Aspi, J.; et al. A Forensic DNA Profiling System for Northern European Brown Bears (Ursus Arctos). *Forensic Sci. Int. Genet.* **2012**, *6*, 798–809. [CrossRef]
113. Skrbinšek, T.; Jelenčič, M.; Boljte, B.; Konec, M.; Erich, M.; Iosif, R.; Moza, I.; Promberger, B. Report on Analysis of Genetic Samples Collected in 2017–2018 on Brown Bears (Ursus Arctos), Eurasian Lynx (Lynx Lynx) and Grey Wolf (Canis Lupus) in a Pilot Area in Southern Carpathians, Romania. Available online: https://www.carpathia.org/wp-content/uploads/2019/09/FCC2017.2018.FinalReport.Ver1_.1.pdf (accessed on 29 December 2021).
114. Lynam, A.J.; Rabinowitz, A.; Myint, T.; Maung, M.; Latt, K.T.; Po, S.H.T. Estimating Abundance with Sparse Data: Tigers in Northern Myanmar. *Popul. Ecol.* **2009**, *51*, 115–121. [CrossRef]
115. Breck, S. Sampling Rare or Elusive Species: Concepts, Designs, and Techniques for Estimating Population Parameters by William L. Thompson. *Wildl. Soc. Bull.* **2006**, 897–998. [CrossRef]
116. Wold, K.; Wirsing, A.J.; Quinn, T.P. Do Brown Bears Ursus Arctos Avoid Barbed Wires Deployed to Obtain Hair Samples? A Videographic Assessment. *Wildl. Biol.* **2020**, *2020*. [CrossRef]
117. Karamanlidis, A.A.; Youlatos, D.; Sgardelis, S.; Scouras, Z. Using Sign at Power Poles to Document Presence of Bears in Greece. *Ursus* **2007**, *18*, 54–61. [CrossRef]
118. Waits, L.P.; Paetkau, D. Noninvasive Genetic Sampling Tools for Wildlife Biologists: A Review of Applications and Recommendations for Accurate Data Collection. *J. Wildl. Manag.* **2005**, *69*, 1419–1433. [CrossRef]
119. De Barba, M.; Miquel, C.; Lobréaux, S.; Quenette, P.Y.; Swenson, J.E.; Taberlet, P. High-Throughput Microsatellite Genotyping in Ecology: Improved Accuracy, Efficiency, Standardization and Success with Low-Quantity and Degraded DNA. *Mol. Ecol. Resour.* **2016**, *17*, 492–507. [CrossRef]
120. Nsubuga, A.M.; Robbins, M.M.; Roeder, A.D.; Morin, P.A.; Boesch, C.; Vigilant, L. Factors Affecting the Amount of Genomic DNA Extracted from Ape Faeces and the Identification of an Improved Sample Storage Method. *Mol. Ecol.* **2004**, *13*, 2089–2094. [CrossRef]
121. Sentilles, J.; Delrieu, N.; Quenette, P.Y. Un Chien Pour La Détection de Fèces: Premiers Résultats Pour Le Suivi de l Ours Brun Dans Les Pyrénées. *Faune Sauvag.* **2016**, *312*, 22–26.
122. Stephanie, B.; Jenny, K.; Helen, S.; Nils, B.; Kathryn, J.J.; Etienne, F.; Akomo, O.; Rob Ogden, R.M. Improving Cost-Efficiency of Faecal Genotyping_ New Tools for Elephant Species. *PLoS ONE* **2019**. [CrossRef]
123. Petchey, A.; Gray, A.; Andrén, C.; Skelton, T.; Kubicki, B.; Allen, C.; Jehle, R. Characterisation of 9 Polymorphic Microsatellite Markers for the Critically Endangered Lemur Leaf Frog Agalychnis Lemur. *Conserv. Genet. Res.* **2014**, *6*, 971–973. [CrossRef]
124. Werhahn, G.; Senn, H.; Kaden, J.; Joshi, J.; Bhattarai, S.; Kusi, N.; Sillero-Zubiri, C.; Macdonald, D.W. Phylogenetic Evidence for the Ancient Himalayan Wolf: Towards a Clarification of Its Taxonomic Status Based on Genetic Sampling from Western Nepal. *R. Soc. Open Sci.* **2017**, *4*, 170186. [CrossRef]
125. Murphy, M.A.; Waits, L.P.; Kendall, K.C. The Influence of Diet on Faecal DNA Amplification and Sex Identification in Brown Bears (Ursus Arctos). *Mol. Ecol.* **2003**, *12*, 2261–2265. [CrossRef]
126. Ross, S.; Costanzi, J.M.; al Jahdhami, M.; al Rawahi, H.; Ghazali, M.; Senn, H. First Evaluation of the Population Structure, Genetic Diversity and Landscape Connectivity of the Endangered Arabian Tahr. *Mamm. Biol.* **2020**, *100*, 659–673. [CrossRef]
127. Planella, A.; Jiménez, J.; Palomero, G.; Ballesteros, F.; Blanco, J.C.; López-Bao, J.V. Integrating Critical Periods for Bear Cub Survival into Temporal Regulations of Human Activities. *Biol. Conserv.* **2019**, *236*, 489–495. [CrossRef]
128. Huber, S.; Bruns, U.; Arnold, W. Sex Determination of Red Deer Using Polymerase Chain Reaction of DNA from Feces on JSTOR. *Wildl. Soc. Bull.* **2002**, *30*, 208–212.
129. Murphy, M.A.; Kendall, K.C.; Robinson, A.; Waits, L.P. The Impact of Time and Field Conditions on Brown Bear (Ursus Arctos) Faecal DNA Amplification. *Conserv. Genet.* **2007**, *8*, 1219–1224. [CrossRef]
130. Brinkman, T.J.; Schwartz, M.K.; Person, D.K.; Pilgrim, K.L.; Hundertmark, K.J. Effects of Time and Rainfall on PCR Success Using DNA Extracted from Deer Fecal Pellets. *Conserv. Genet.* **2009**, *11*, 1547–1552. [CrossRef]
131. Kendall, K.C.; McKelvey, K.S. Hair Collection Methods. In *Noninvasive Survey Methods for North American Carnivores*; Island Press: Washington, DC, USA, 2008; pp. 135–176. ISBN 978-1-59726-119-7.

132. Fusaro, J.L.; Conner, M.M.; Conover, M.R.; Taylor, T.J.; Kenyon, M.W. Best Management Practices in Counting Urban Black Bears. *Hum. Wildl. Interact.* **2017**, *11*, 64–77. [CrossRef]
133. McLellan, M.L.; McLellan, B.N.; Sollmann, R.; Lamb, C.T.; Apps, C.D.; Wittmer, H.U. Divergent Population Trends Following the Cessation of Legal Grizzly Bear Hunting in Southwestern British Columbia, Canada. *Biol. Conserv.* **2019**, *233*, 247–254. [CrossRef]
134. Gurney, S.M.; Smith, J.B.; Etter, D.R.; Williams, D.M. American Black Bears and Hair Snares: A Behavioral Analysis. *Ursus* **2020**, *2020*, 1–9. [CrossRef]
135. Stetz, J.B.; Seitz, T.; Sawaya, M.A. Effects of Exposure on Genotyping Success Rates of Hair Samples from Brown and American Black Bears. *J. Fish. Wildl. Manag.* **2015**, *6*, 191–198. [CrossRef]
136. Sato, Y.; Kamiishi, C.; Tokaji, T.; Mori, M.; Koizumi, S.; Kobayashi, K.; Itoh, T.; Sonohara, W.; Takada, M.B.; Urata, T. Selection of Rub Trees by Brown Bears (Ursus Arctos) in Hokkaido, Japan. *Acta Theriol.* **2014**, *59*, 129–137. [CrossRef]
137. Mowat, G.; Strobeck, C. Estimating Population Size of Grizzly Bears Using Hair Capture, DNA Profiling, and Mark-Recapture Analysis. *J. Wildl. Manag.* **2000**, *64*, 183. [CrossRef]
138. Haroldson, M.A.; Gunther, K.A.; Reinhart, D.P.; Podruzny, S.R.; Cegelski, C.; Waits, L.; Wyman, T.; Smith, J. Changing Numbers of Spawning Cutthroat Trout in Tributary Streams of Yellowstone Lake and Estimates of Grizzly Bears Visiting Streams from DNA. *Ursus* **2005**, *16*, 167–180. [CrossRef]
139. Evans, M.J.; Hawley, J.E.; Rego, P.W.; Rittenhouse, T.A.G. Hourly Movement Decisions Indicate How a Large Carnivore Inhabits Developed Landscapes. *Oecologia* **2018**, *190*, 11–23. [CrossRef]
140. Paetkau, D.; Amstrup, S.C.; Born, E.W.; Calvert, W.; Derocher, A.E.; Garner, G.W.; Messier, F.; Stirling, I.; Taylor, M.K.; Wiig, O.; et al. Genetic Structure of the World's Polar Bear Populations. *Mol. Ecol.* **1995**, *9*, 84. [CrossRef]
141. Robinson, S.J.; Waits, L.P.; Martin, I.D. Evaluating Population Structure of Black Bears on the Kenai Peninsula Using Mitochondrial and Nuclear DNA Analyses. *J. Mammal.* **2007**, *88*, 1288–1299. [CrossRef]
142. Shih, C.C.; Wu, S.L.; Hwang, M.H.; Lee, L.L. Evaluation on the Effects of Ageing Factor, Sampling and Preservation Methods on Asiatic Black Bear (Ursus Thibetanus) Noninvasive DNA Amplification. *Taiwania* **2017**, *62*, 363–370. [CrossRef]
143. Skrbinšek, T. *Collecting Lynx Noninvasive Genetic Samples: Instruction Manual for Field Personnel and Volunteers*; Biotechnical Faculty, University of Ljubljana: Ljubljana, Slovenia, 2017.
144. Puchkovskiy, S.V. Selectivity of Tree Species as Activity Target of Brown Bear in Taiga. *Contemp. Probl. Ecol.* **2009**, *2*, 260–268. [CrossRef]
145. Green, G.I.; Mattson, D.J. Tree Rubbing by Yellowstone Grizzly Bears Ursus Arctos. *Wildl. Biol.* **2003**, *9*, 1–9. [CrossRef]
146. Clapham, M.; Nevin, O.T.; Ramsey, A.D.; Rosell, F. The Function of Strategic Tree Selectivity in the Chemical Signalling Ofbrown Bears. *Anim. Behav.* **2013**, *85*, 1351–1357. [CrossRef]
147. Lamb, C.T.; Walsh, D.A.; Mowat, G. Factors Influencing Detection of Grizzly Bears at Genetic Sampling Sites. *Ursus* **2016**, *27*, 31–44. [CrossRef]
148. Rogers, L.L. Effects of Food Supply and Kinship on Social Behavior, Movements, and Population Growth of Black Bears in Northeastern Minnesota. *Wildl. Monogr.* **1987**, *97*, 3–72.
149. Gabrielsen, C.G.; Kovach, A.I.; Babbitt, K.J.; McDowell, W.H. Limited Effects of Suburbanization on the Genetic Structure of an Abundant Vernal Pool-Breeding Amphibian. *Conserv. Genet.* **2013**, *14*, 1083–1097. [CrossRef]
150. Gorospe, K.D.; Karl, S.A. Genetic Relatedness Does Not Retain Spatial Pattern across Multiple Spatial Scales: Dispersal and Colonization in the Coral, Pocillopora Damicornis. *Mol. Ecol.* **2013**, *22*, 3721–3736. [CrossRef]
151. Keller, D.; Holderegger, R.; van Strien, M.J. Spatial Scale Affects Landscape Genetic Analysis of a Wetland Grasshopper. *Mol. Ecol.* **2013**, *22*, 2467–2482. [CrossRef]
152. Nellemann, C.; Støen, O.G.; Kindberg, J.; Swenson, J.E.; Vistnes, I.; Ericsson, G.; Katajisto, J.; Kaltenborn, B.P.; Martin, J.; Ordiz, A. Terrain Use by an Expanding Brown Bear Population in Relation to Age, Recreational Resorts and Human Settlements. *Biol. Conserv.* **2007**, *138*, 157–165. [CrossRef]
153. Piédallu, B.; Quenette, P.-Y.; Jordana, I.A.; Bombillon, N.; Gastineau, A.; Jato, R.; Miquel, C.; Muñoz, P.; Palazón, S.; Solà de la Torre, J.; et al. Better Together: A Transboundary Approach to Brown Bear Monitoring in the Pyrenees. *bioRxiv* **2016**, 075663. [CrossRef]
154. Matosiuk, M.; Śmietana, W.; Czajkowska, M.; Paule, L.; Štofík, J.; Krajmerová, D.; Bashta, A.T.; Jakimiuk, S.; Ratkiewicz, M. Genetic Differentiation and Asymmetric Gene Flow among Carpathian Brown Bear (Ursus Arctos) Populations—Implications for Conservation of Transboundary Populations. *Ecol. Evol.* **2019**, *9*, 1501–1511. [CrossRef]
155. Zingstra, H.; Kovachev, A.; Kitnaes, K.; Tzonev, R.; Dimova, D.; Tzvetkov, P. Guidelines for Assessing Favorable Conservation Status of Natura 2000 Species and Habitat Types in Bulgaria 2009. Available online: https://library.wur.nl/WebQuery/wurpubs/fulltext/247891 (accessed on 29 December 2021).
156. Taberlet, P.; Swenson, J.E.; Sandegren, F.; Bjarvall, A. Localization of a Contact Zone between Two Highly Divergent Mitochondrial DNA Lineages of the Brown Bear Ursus Arctos in Scandinavia. *Conserv. Biol.* **1995**, *9*, 1255–1261. [CrossRef]
157. Bidon, T.; Janke, A.; Fain, S.R.; Eiken, H.G.; Hagen, S.B.; Saarma, U.; Hallström, B.M.; Lecomte, N.; Hailer, F. Brown and Polar Bear Y Chromosomes Reveal Extensive Male-Biased Gene Flow within Brother Lineages. *Mol. Biol. Evol.* **2014**, *31*, 1353–1363. [CrossRef]
158. Støen, O.G.; Zedrosser, A.; Sæbø, S.; Swenson, J.E. Inversely Density-Dependent Natal Dispersal in Brown Bears Ursus Arctos. *Oecologia* **2006**, *148*, 356–364. [CrossRef]

159. Bischof, R.; Milleret, C.; Dupont, P.; Chipperfield, J.; Tourani, M.; Ordiz, A.; de Valpine, P.; Turek, D.; Andrew Royle, J.; Gimenez, O.; et al. Estimating and Forecasting Spatial Population Dynamics of Apex Predators Using Transnational Genetic Monitoring. *Proc. Natl. Acad. Sci. USA* **2020**, *117*, 30531–30538. [CrossRef]
160. Mills, L.S.; Citta, J.J.; Lair, K.P.; Schwartz, M.K.; Tallmon, D.A. Estimating Animal Abundance Using Noninvasive DNA Sampling: Promise and Pitfalls. *Ecol. Appl.* **2000**, *10*, 283. [CrossRef]
161. Bellott, D.W.; Hughes, J.F.; Skaletsky, H.; Brown, L.G.; Pyntikova, T.; Cho, T.-J.; Koutseva, N.; Zaghlul, S.; Graves, T.; Rock, S.; et al. Mammalian Y Chromosomes Retain Widely Expressed Dosage-Sensitive Regulators. *Nature* **2014**, *508*, 494–499. [CrossRef]
162. Hollerbach, L.; Heurich, M.; Reiners, T.E.; Nowak, C. Detection Dogs Allow for Systematic Non-Invasive Collection of DNA Samples from Eurasian Lynx. *Mamm. Biol.* **2018**, *90*, 42–46. [CrossRef]
163. Gonzalez, E.G.; Blanco, J.C.; Ballesteros, F.; Alcaraz, L.; Palomero, G.; Doadrio, I. Genetic and Demographic Recovery of an Isolated Population of Brown Bear Ursus Arctos L., 1758. *PeerJ* **2016**, *2016*, e1928. [CrossRef] [PubMed]
164. Goossens, B.; Waits, L.P.; Taberlet, P. Plucked Hair Samples as a Source of DNA: Reliability of Dinucleotide Microsatellite Genotyping. *Mol. Ecol.* **1998**, *7*, 1237–1241. [CrossRef] [PubMed]
165. Ebert, C.; Knauer, F.; Storch, I.; Hohmann, U. Individual Heterogeneity as a Pitfall in Population Estimates Based on Non-Invasive Genetic Sampling: A Review and Recommendations. *Wildl. Biol.* **2010**, *16*, 225–240. [CrossRef]

MDPI

Review

Black Locust (*Robinia pseudoacacia* L.) in Romanian Forestry

Alexandru Liviu Ciuvăț [1], Ioan Vasile Abrudan [2], Cristiana Georgeta Ciuvăț [1], Cristiana Marcu [1], Adrian Lorenăț [1], Lucian Dincă [1] and Bartha Szilard [3,*]

1 National Institute for Research and Development in Forestry "Marin Drăcea", 077190 Bucharest, Romania
2 Department of Silviculture, Faculty of Silviculture and Forest Engineering, Transilvania University of Brasov, 500123 Brașov, Romania
3 Department of Forestry and Forest Engineering, University of Oradea, 410048 Oradea, Romania
* Correspondence: barthaszilard10@yahoo.com; Tel.: +40-727307277

Abstract: This paper presents a literature review of black locust (*Robinia pseudoacacia* L.) and the knowledge accumulated by Romanian foresters and researchers, covering species propagation, stand management, and vulnerability issues. As highlighted by numerous authors, black locust manifests dual features, both as an exogenous species and one that is already naturalized. The main drivers for this species' expansion in Romania is its ecological adaptability on degraded lands, fast growth, and high biomass yields, in addition to other economic benefits. Black locust plantations and coppices also offer an important range of ecosystem services such as CO_2 sequestration, landscape reclamation, fuel wood, or maintaining traditional crafts in regions with little to no forest cover. Highlighted disadvantages include short lifespan, invasiveness when introduced on fertile sites, and dieback in drought/frost prone areas. The results of extensive research and studies are captured in technical norms, although aspects such as species genetics, invasive potential, and adaptation to climate change dynamics call for more research and optimizing in species management. As Romania rallies its efforts with those of the international community in order to address climate change and desertification, black locust stands out as a proven solution for reclaiming degraded lands when native species are not an alternative.

Keywords: black locust; silviculture; ecology; management; risks; uses; impact

Citation: Ciuvăț, A.L.; Abrudan, I.V.; Ciuvăț, C.G.; Marcu, C.; Lorent, A.; Dincă, L.; Szilard, B. Black Locust (*Robinia pseudoacacia* L.) in Romanian Forestry. *Diversity* **2022**, *14*, 780. https://doi.org/10.3390/d14100780

Academic Editor: Rüdiger Wittig

Received: 8 August 2022
Accepted: 16 September 2022
Published: 21 September 2022

Publisher's Note: MDPI stays neutral with regard to jurisdictional claims in published maps and institutional affiliations.

1. Introduction

Black locust (*Robinia pseudoacacia* L.) is a North American species that was introduced in Romania around the end of the 17th century [1], and it was first used in large-scale afforestation of degraded lands in the year 1852 [2]. Due to its remarkable adaptability, fast growth, and vigorous sprouting capacity, it has become one of the most widely spread exotic species in Romania [3]. Using black locust for afforestation, Romanian foresters successfully reclaimed large areas of abandoned agricultural land degraded by "flying sands" in southwest of the country [4], thus contributing to mitigation of the aridization phenomena [4].

Driving the extensive use of the species was its durable and versatile wood, much appreciated by rural communities, complemented with major benefits from crop fields and settlement protection against wind/sand deflation [5], as well as economically viable byproducts. In addition to the immediate improvements to local microclimate, afforestations also contributed to long-term climate change mitigation by sequestering atmospheric CO_2 [6–9] in the carbon pools (living tree biomass, soil organic matter, litter) as well as downstream wood products (e.g., furniture). Nevertheless, in the past decade, awareness has been raised towards its invasive potential in protected areas.

The authors aim at highlighting the peculiarities of this species in Romanian forestry with regard to its ecology and management, but also its economic importance and potential environmental impact.

2. Materials and Methods

This work aims for a comprehensive review of Romanian research and studies about the silviculture, management, and impact of black locust. The literature on the subject was found by consulting the archive of the National Institute for research and Development in Forestry "Marin Dracea" (approx. 150 publications), and by using the Google search engine. The largest pool of publications occurred at a national level between 1950 and 1990, regarding the silviculture and genetics of the species. The literature survey shows a poor post-1990 focus on this species, with a limited number of papers, especially regarding management, damaging factors, and risks.

3. Results

3.1. Black Locust in Romania—Areal and Ecological Behavior

Robinia pseudoacacia is a fast-growing species, reaching heights above 30 m and ages in excess of 100 years, while maturing at early ages of 5–7 years old. It regenerates both from stump sprouts and root suckers, and it can be easily propagated by cutting or grafting, although in Romanian forestry natural regeneration from root suckers and plantation of seedlings are almost exclusively used. It is reported to become invasive if inadequately planted in sites with high productivity potential [10], but also on poor sites due to its intolerance to other large tree species, so much that it forms pure stands. Nevertheless, it was also reported in association with shrubs and a few other compatible tree species whenever adequate planting schemes are applied [10,11]. Being a sun loving species, it prefers areas with long summers (mean annual temperatures above 10 °C), and annual rainfall between 400 and 600 mm. According to Ivanschi et al. [12] it grows well on sandy and sandy loam soils, with loose to lightly compaction, deep, with a medium humidity regime; conversely, it does not grow/survive on compact soils with calcium carbonates or soluble salts in top layers of soil or in the case of excess of humidity. Calcium carbonates ($CaCO_3$) in the top layers of soil (<40 cm) inhibit growth. Due to its nutritional particularities, it fixes nitrogen (N) in the soil (in symbiosis with N fixing bacteria) but it also consumes large quantities of soil minerals. It has a hard and durable wood, comparable with that of oak, which makes it very appreciated by rural populations.

Geographically, black locust stands occupy about 5% of the national forested area (250,000 ha), concentrated in the southwest, west, and in the east of the country [13].

The largest areas of compact stands and also the most valuable ones are located in the southwest of the country in the Oltenia region (Figure 1), which was also the location of its first introduction in Romania. Currently, the species is considered as a sub-spontaneous one, its range spreading from the plains to the lower mountain regions [3].

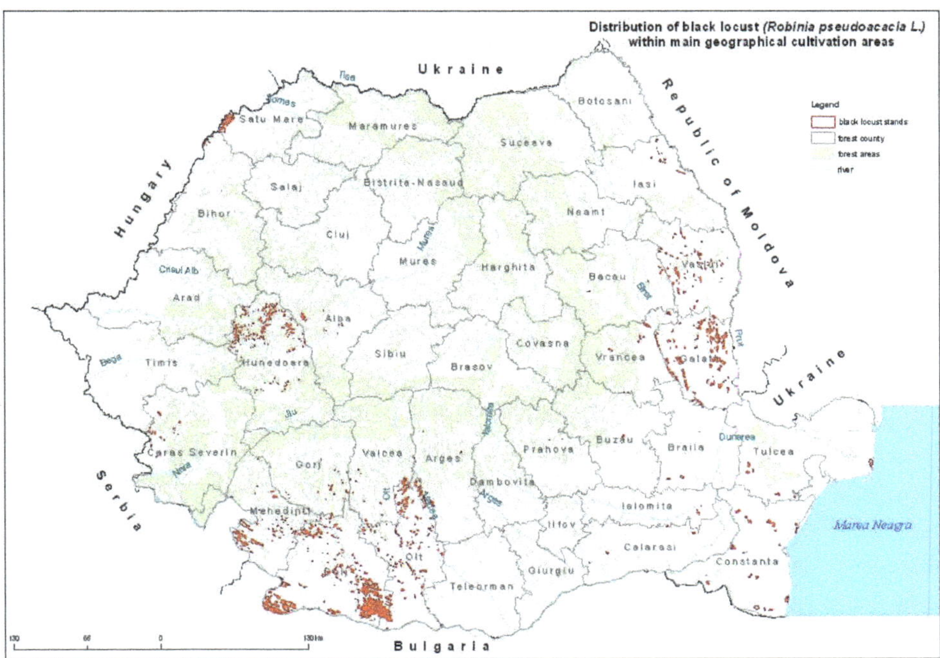

Figure 1. Distribution of black locust stands in the main culture regions of Romania [10].

3.2. *Overview of Management Approaches*

3.2.1. Forest Regeneration and Establishment of Plantations

Although black locust produces annually a very large number of seeds, because of their thick tegument, successful natural stand regeneration is missing almost completely [13]. Exceptionally isolated trees can be recognized as being from seeds in very particular sites, such as upper sides of river banks [14] or at pile burning sites where forest residues are burned [15]. Therefore, vegetative regeneration of harvested stands by mechanical stimulation of root sprouting and artificial regeneration by plantation of nursery seedlings in reforestation and afforestation of non-forest lands are preferred and largely practiced across the country [2].

3.2.2. Vegetative Regeneration

While the harvesting technique is always clear cutting, stand regeneration occurs via sprouts on the aboveground part of the stump or suckers by mechanical stimulation of roots by: (a) ploughing of the soil among the stumps at 20 cm depth; or (b) digging out the stump together with the roots around it in a radius of 50 cm combined with additional ploughing. When regeneration targets root-suckers, the sprouts must be removed through early tending operations, so as to not overwhelm the slow growing root suckers [13]. After two cycles of vegetative regeneration from sprouts, the stumps weaken and lose their sprouting capacity (due to fungal impact and soil nutrient depletion); and by the third generation, the stand's productivity drops to 40% of that of plantations [16]. That is why, after a second generation of sprouts, regeneration must be ensured from root-suckers [17]. In any case, after the third generation of vegetative regeneration, all of the stumps must be removed from the harvested area and further regeneration is ensured via the planting of seedlings. Recent studies have showed that, after clearcutting and stump removal, the density of root suckers can reach 50,000 plants per hectare [13].

3.2.3. Generative Regeneration

The recommended technique for seedling production includes harvesting the seeds directly from trees (late in the winter or early January) or separated from litter and mechanical or chemical 'forcing' before sowing next spring. Seedlings are almost without exception ready for transplantation after one season of growth (e.g., root collar diameter of 4 mm). After a second year in the nursery, they generally surpass the planting dimension standards which are associated with substantial risk of not surviving summer drought. Although, for afforestation in such areas, specific standards apply to ensure more robust plants are used—e.g., root collar diameter of 8 mm [17]. The seeds for nursery production of seedlings are often collected from seed orchards and have to meet standard requirements for purity, technical germination, and mass [18].

3.2.4. Afforestation Techniques

According to technical norms in force in Romania [19], afforestation with black locust is exclusively carried out by planting bare-root seedlings in hand-dug holes. This method has proved so successful in practice that there has been no more research on other methods, although associated costs may be prohibitive in an expensive labor market. The literature states that sowing the previously treated seeds can be used to create black locust stands [13]. Planting can be done both in spring as in autumn. Spring planting ensures adequate soil humidity (common in upper altitude regions), while planting in autumn (usually in the southern ad low-altitude regions) ensures root growth over winter and avoids spring flooding and early droughts [20].

The technical norms state that, in the case of sandy soils or degraded lands, optimum planting density for black locust is 5000 seedlings per hectare—with 2.0 m of distance between rows and 1.0 m between seedlings—and the planting pit is 40 cm wide and 40 cm deep. In the case of non-degraded lands, the planting density for black locust is 4000 seedlings per hectare—with 2.0 m of distance between rows and 1.25 m between seedlings—and the planting pit width, length, and depth is 40 cm. In order to ensure the highest survival rate, a key technical intervention after planting seedlings is trimming of their stems 1 cm above the soil before growing season starts [20]. Further on, to ensure success of plantations, regular maintenance involves gap filling in the first 2–3 years after planting, if necessary. The survival rate of plantations is checked yearly, with dead seedlings being replaced until canopy closure (2–4 years).

3.2.5. Tending Operations

Being a fast-growing species, canopy closure takes place early (1–3 years following planting), therefore tending operations (especially cleaning and thinning) need to be made at shorter periods of time (3–5 years) depending on the tree origin (seedlings or root-suckers) and density of the stand. Periodicity of tendings varies in function of stand development (size of trees) and influences its productivity (stand density influences the rate of growth) as well as its protective function, in which case higher stand densities are preferable only on degraded lands due to the slenderness coefficient [17].

Research and experiments have allowed development of technical norms to guide forest operations according to site features (Table 1). First cleaning should be applied starting at 3–5 years for normal sites and, slightly later, 4–7 years (depending on site density) in the case of plantations on degraded sites.

The second intervention should occur after 3–5 years. Adequate results will be obtained when thinning operations are performed before stands reach an average diameter of about 20 cm (e.g., around 8–10 years old stands), and then this process is repeated every 4–6 years [21]. Intensity of tending operations (% of extracted trees), and especially that of thinning, plays an important role in the stand's further development with respect to growth and wood quality. Experimenting on the effects of type and intensity of thinning, Armasescu et al. [22] concluded that black locust reacts to thinning even at ages > 20; natural self-thinning continues after thinning operations; classical thinning (low and high) does not lead to significant increase in total stand volume. The study also showed that very intensive thinning (>35% of total volume extracted) that lowers the stand density index below 0.8 actually leads to a decrease in stand production of up to 10%, in comparison to moderate thinning (25–30% of total volume extracted from all Kraft categories across the entire stand) which leads to an increase in stand productivity of up to 25%.

Table 1. Tending operations for black locust stands on degraded lands [20].

Tending Operation Characteristics Depending on Stand Density	**Normal Tending Operations (Trees x ha^{-1})**		**Delayed Tending Operations (Trees x ha^{-1})**	
	>5000 Trees x ha^{-1}			
Age of stand when tending is done (years)	4–6	6–7	10–20	10–20
Tending operation and intensity of extraction (% of volume or basal area)	Cleaning I	Cleaning I	Cleaning I	Cleaning I
	strong/very strong (16–30%)	strong (16–25%)	strong (16–25%)	moderate/strong (10–20%)
Trees per ha after the extraction Periodicity (years)	4000–5000 3–5	3000–5000 3–5	2000–3500 3–5	2500–3000 3–5
Tending operation and intensity of extraction (% of volume or basal area)	Cleaning II	Cleaning II	Cleaning II/Thinning I	Cleaning II/Thinning I
	strong (16–25%)	strong (16–25%)	strong (16–20%)	moderate (6–15%)
Trees per ha after the extraction Periodicity (years)	2500–3000 5–7	2000–2500 5–7	2000–2500 5–7	2000–2500 5–7
Tending operation and intensity of extraction (% of volume or basal area)	Cleaning III	Cleaning III	Cleaning III/Thinning II	Cleaning III/Thinning II
	moderate (6–15%)	moderate (6–15%)	moderate (6–15%)	moderate (6–15%)
Trees per ha after the extraction	1500–2000	1500–2000	1500–2000	1500–2000

3.2.6. Stand Management

Forest management planning sets either productive or protective objectives for black locust forests, following strict management indicators: land use, silvicultural regime, stand structure, rotation, and harvesting technique. When dealing with such technical requirements, the forest management plan has to follow ecological, social, and economic local needs. While some stands' fate is wood production (mainly timber and construction wood), many stands have additional strong protective functions (e.g., degraded land reclamation, protection of soils and infrastructure, etc.).

3.2.7. Silvicultural Regime

In Romanian forestry law and official technical norms [19], the recommended form of management system for black locust stands is the coppice. Two forms can be differentiated: the simple one applied extensively and the selection coppice applied locally and occasionally. The simple coppice consists of clear-cutting of the stand followed by vegetative regeneration or seedling plantations. In the coppice selection system, a part of the sprouts/root-suckers from each stump can be preserved, eliminating only the crooked ones and those that have reached the targeted diameter. It can be applied with experimental purpose in the case of plantations on severely degraded lands or with small privately owned forests. In the past, some stands were managed as coppices with reserves, meaning that a percentage of the old stands were held until the following cycle in order to obtain trees with larger diameters [23].

3.2.8. Stand Composition and Structure

Black locust generally achieves large areas of continuous pure stands. Being a typical exclusivist species, stands are generally pure and even aged, their intraspecific competition being poor [24]. Nevertheless, on degraded lands, it is planted together with other exotic species.

Black locust can tolerate other species, most commonly honey locust (*Gleditsia triacanthos* L.) on sandy soils in the south and east of the country, or with black pine (*Pinus nigra* L.) on eroded slopes in the east of Romania [24,25]. It behaves as an exclusivist species for ground vegetation, black locust stands have very poor herbaceous diversity (e.g., frequent are *Urtica* sp., *Sambucus ebulus*).

3.2.9. Harvesting and Rotation/Cycle Length

Black locust stands have the shortest rotation length in Romanian forestry. Rotation length differs as a function of the forest primary purpose (wood production or protection). If the stand primarily has a production purpose (e.g., timber or construction wood), harvesting age varies between 15 years for coppices classified under the lowest production class to 35 years for the planted stands classified as the highest production class. If classified as having protection function (e.g., steep terrains prone to soil erosion), the harvesting age can reach 40 years in stands [19]. Wood harvesting occurs through clear-cuts in plots with a maximum area of 5 ha.

3.3. Genetics, Selection, and Tree Breeding

An outstanding natural variety—*Robinia pseudoacacia* var. *oltenica*—was identified in the stands located in the south-west part of Romania (Oltenia region) by Bârlănescu, Costea, and Stoiculescu in 1966 [26]. Due to its valuable auxological and morphological characteristics [27,28], this variety was the subject of tree selection and breeding by grafting/cutting [29,30] and in vitro micropropagation. Later on, eco-physiological research by Bolea at al. [31] showed that this variety has both higher intensity of photosynthesis and tolerance to droughts, due to larger leaf area index (LAI), compared to common stands. Conservation and expansion of the Oltenica variety was continued in recent years by producing seedlings from cuttings, as maternal lineages of identified plus-trees [13].

Genetics research prior to 1990 focused on tree selection and breeding for enhancement of productivity and wood standing volume [29,32,33], as well as selection of valuable forest and beekeeping genotypes by hybridization. Trials consisting in testing different provenances in comparative cultures, identified "plus" trees and enhanced the number of clones and establishing vegetative propagation methods (e.g., grafting) for the valuable forms [34]. Currently, Romania has six qualified seed orchards with a combined area of 27 ha [35].

Recently, in vitro micropropagation—as the most modern technology used in tree genetics—achieved important results in selecting valuable forms of black locust via organogenesis and somatic embryogenesis [36,37]. Over the last two decades, black locust genetic research also focused on isolation, culture, and regeneration of protoplasts [37], and the influence of endogenic and exogenic factors on somatic embryogenesis [38,39] determined the genetic parameters of half-sib (free pollinated) and full-sib (controlled pollinated) black locust descendants [40]. Mirancea I. [36] tested black locust multiplication and rooting in vitro phases on different cultural media and hormonal balances, concluding that the optimum concentrations for which explants can multiply is 6 g/L for NaCl and 100 g/L for CaCO3. More recently, Băbeanu et al. [41] studied the isoperoxidase pattern of in vitro culture of *R. pseudoacacia* var. *oltenica* on various media, learning how isoperoxidase activity can be modulated through media. Further tests have showed the resistance of the plantlets obtained by micropropagation to severe ecological conditions of *R. pseudoacacia* var. *oltenica*, emphasizing how deuterium-depleted water-based media amplifies the caulogenesis.

3.4. *Vulnerability of Black Locust*

Systematic records of biotic and non-biotic factors and damage data in Romanian forests have been collected through the Forest National Survey and Forecast System operations since 1958. Based on such data, the main cryptogamic diseases affecting *Robinia pseudoacacia* L. were listed as follows [18]: virosis, *Fusarium* sp. and *Cuscuta* sp. attacks on seedlings, mildew (*Oidium* sp.), sooty mold (*Coniothyrium* sp., *Alternaria* sp., Cladosporium sp.), leaf spots (*Phleospora robiniae*) and tar spots on leaves (*Ectostroma* sp.), twig blight (*Pseudovalsa* sp.) and *Chorostate* sp. on young shoots, black spots (*Cucurbitaria* sp.) on old shoots, and wood-destroying fungi (*Hironela* sp, *Trametes* sp., *Phellinus* sp.).

A comprehensive inventory in the main black locust biotopes (seedling nurseries, young plantations, and mature stands), as well as complete description of biology and adequate means of control of all harmful insects of black locusts was achieved by [42]. Most infestation power was assessed for defoliating *Lymantria dispar* L. and fruit damaging *Etiella zinckenella* Tr. Later on, Trantescu et al. [43] concluded that it is not economically justified to spray insecticide except in the case of very valuable stands (e.g., seed orchards). *Semiothisa alternaria* Hb. (*Macaria alternaria* Hb.) was identified as an important pest for black locust [44] and it was estimated that the damage critical number by *Lymantria dispar* L. is 2–6 times higher for black locust stands than oak [45]. Information was brought forward on novel attacks in the south-western part of Romania of three leaf-miner moths of North American origin (*Parectopa robinella* and *Phyllonorycter robiniella* Clemens), accidentally introduced in Europe and apparently expanding their area [46,47].

Older stands show tree dieback in drought-prone areas, a fact apparently linked to higher cycles of vegetative regeneration and stands on poor sites [25,46]. Furthermore, dieback of black locust trees was reported in industrial zones, one example being near the town of Copșa Mică (in central Romania) due to soil and air pollution with sulfur and heavy metal compounds generated by industry (carbon black smoke) until the early 1990s. The main symptom was leaf necrosis, with the effects diminishing after the pollution source was removed in past decades [48–52].

Black locust bark leaves and young shoots are occasionally consumed by herbivores, such as game species, especially European brown hares and deer (roe-deer and red-deer). Domestic livestock (e.g., sheep, goats) consume leaves and young shoots on an occasional basis, with no national references regarding toxicity on animals or milk products. Locally, near villages—especially in private owned forests—black locust stands are under multiple human pressures such as illegal logging, erratic wood collection, and grazing [49].

Because plantations occur more often in low lands and drought prone areas, close to agricultural crops locations, black locust stands are subject to wildfires in which trees—even younger ones—survive while litter is burnt completely [51]. Long periods with very little/no rainfall were assumed as the main cause for dieback and decline of stands in 1980s and 1990s in the south and east of the country [53].

3.5. *Wood Products and Other Uses of Black Locust*

3.5.1. Wood Production

As a result of high percentage of coppice stands [54], the size of the black locust round wood usually does not exceed 20 cm in diameter at thin end, and quality of timber is relatively poor compared to key native species (Table 2). In most of the Romanian regions, short longevity because of the tree dieback is the main cause for rather small tree dimensions comparative to local broadleaved tree species. Therefore, majority of wood generated by black locust stands is mostly used as firewood or in rural construction (e.g., fences, sheds, poles) or less for props for gardening and vineyards [55]. The good quality wood can also be used for interior and garden furniture, parquet and floors or woodchip boards while more common uses are for fenceposts, poles, railroad crossties, stakes and fuel wood [56–58]. Traditionally, its wood is very appreciated for making barrels, handles for tools (e.g., axes, shovels, etc.).

Table 2. Black locust yield compared to main broadleaf species in Romania (for average yield production class III, pure, and even aged stands at 20 years old) [59].

Species	Stand Structure	Age (Years)	Mean Diameter (cm)	Average Height (m)	Standing Volume (m^3/ha)
R. pseudoacacia	coppice	20	13.7	15.9	185
R. pseudoacacia	plantation	20	13.8	16.1	195
Quercus robur	high stand	20	8.0	8.3	97
Populus sp.	High stand	20	18.4	14.4	226
Fagus sylvatica	high stand	20	5.2	7	56

3.5.2. Non-Wood Products

Black locust is so appreciated by foresters and farmers alike due to a range of non-wood products that are available already at an early age of stands [60]. The most valued non-wood byproduct is honey, considered of the highest quality [61], which makes it the most expensive on the market. Compared to the traditional melliferous species (linden/lime, black locust, sunflower, rapeseed), black locust is the first to bloom (in May); therefore, beekeepers start the pastoral beekeeping in the black locust forests. Black locust stands have a high melliferous potential in Romania with up to 697,000 tonnes of honey per year. In order to ensure that the full benefit is reached, an application to support the planning of the pastoral beekeeping was developed based on forest maps, as a tool for decision makers at a national level in the planning of pastoral activity of the beekeepers. Black locust flowers show medicinal use in teas or infusions for digestive and pulmonary effects, and they have a calming effect on the nervous system [62–64]. A mix of flowers and leaves can be used as a tea drink to treat stomach pains and migraine. Generally, they have to be used in small quantities due to slight toxicity [65]. The seeds and bark of black locust are not used in traditional medicine because they contain substances that are toxic to humans [66].

3.6. Landscape Improvement Contribution

In Romania, black locust was very successful in reclaiming degraded lands [67] by exercising its anti-erosion role together with phytoremediation [68], and biomass accumulation in site conditions strongly prohibitive to other species [69]. On industrial or mining dumps, black locust plantations reached volumes of up to 73 m^3/ha among 8-year-old stands [70]. It was also successfully used in the ecological remediation of historically heavily polluted lands [71]. In a Kyoto Protocol project of afforestation implemented in Romania, about 2500 ha were afforested with black locust in the south of the country to reclaim degraded and marginal agricultural lands [72]. Ten years after their afforestation, black locust plantations have reached heights of 14 m and basal diameters of 16 cm (Figure 2).

Biomass production and CO_2 sequestration of young trees was highlighted in recent years through allometric models of growth [73]. Black locust plantation biomass production on degraded lands in the south-west of Romania can reach 9.4 tons of dry mass per hectare at age 4, while coppiced stands average about 7.0 t dry mass ha^{-1} [74].

Short rotation crops (SRC) for biomass energy use based on *Robinia pseudoacacacia* represent a viable solution for degraded lands across Romania; however, a lack of state subsidies for afforestation investments by private land owners kept this branch relatively undeveloped [75]. Across Europe, recent studies have highlighted the increased use of black locust in SRC for bioenergy compared with the traditional species used such as poplar and willow. Among advantages of using Robinia are large biomass yields, high density wood, low moisture, and greater calorific output [76–79].

The potential of *Robinia pseudoacacia* plantations is significant in the efforts to mitigate the effects of climate change [74]. In the short term, the largest amount of C accumulation occurs in tree biomass, because the C is stored at least for the rotation period/life span of the forest, or even longer when manufactured into furniture, while litter and organic matter of mineral soils also act as steady growing carbon pools. Whether planted on degraded agricultural lands [80,81] or as agroforestry shelterbelts [82], black locust C accumulation capacity is surpassed only by that of native and hybrid poplars (Table 3).

(a) (b)

Figure 2. (**a**) Four-year-old black locust plantation; (**b**) Ten-year-old plantations of oak mix (left) and black locust (right).

Table 3. Carbon (tC ha^{-1}) stock in woody living biomass in plantations/shelterbelts for black locust and poplar.

Age (Years)	C stock in Plantations on Degraded Agricultural Lands [82]		C stock in Forest Shelterbelts [4]	
	Robinia pseudoacacia	*Populus* sp.	*Robinia pseudoacacia*	*Populus x euroamericana*
	(tC ha^{-1})	(tC ha^{-1})	(tC ha^{-1})	(tC ha^{-1})
5	3.52	1.66	5.47	3.71
10	16.90	10.20	17.02	14.56
15	23.79	17.12	-	29.89
20	32,73	39,77	41	43.38
25	-	-	53.93	54.51
30	-	-	60.51	63.21

3.7. Ecosystem Services

When used on degraded lands, social benefits include improving local microclimate and mitigating the negative effects of climate change [83,84]. Shortly after planting, the stands start ensuring improvement of local biodiversity by offering shelter and food sources for birds, mammals, and other species. The economic and ecological role of black locust in agroforestry is represented by the forest shelterbelts to protect crop fields in the south of Romania [85–88], showing that the presence of shelterbelts in the Oltenia region led to an increase in different crop productions (e.g., wheat) of up to 130%, as opposed to unprotected fields. The shelterbelts are also a source of wood, honey, and they offer shelter for game and bird species. As one of the main species used in degraded land restoration, black locust plays an important role at both local and regional levels [13] in the effort to adapt and mitigate climate change by diversifying local supply for wood and feedstock or revenues from improved land use. Under recent more advanced ecosystem services payments, afforestation of degraded lands allowed further economic benefits, by trading greenhouse gas emissions reductions generated by tree plantations as financial instruments provided by the Kyoto Protocol [81]. Black locust is also appreciated in landscaping or gardening, for its decorative grape-like—sweet perfumed—white flowers, and also for its robustness and adaptability [89].

3.8. Invasiveness and Control of Black Locust

In some European countries, the species is considered invasive [90]. Caused in general by inappropriate silvicultural decision-making and practices, the invasive character of black locust manifests itself as a result of the species great adaptability, highly developed vegetative regeneration (especially sprouting), soil condition alteration, and fast rate of growth compared to native species. When planted in close mixtures with slower growing species (e.g., oak) it overwhelmed them, after which it was extremely difficult and expensive to substitute it [91]. In this respect, Romanian forestry law and technical norms provide for the use of black locust only on degraded lands that prohibit the use of native species (severe site conditions). Its invasion on non-forest lands—e.g., orchards and vineyards especially—has occurred on lands abandoned after 1990. While, in other cases, it expands especially because of soil disturbance and stimulation of root-sucker growth. According to Enescu and Dănescu [92] "black locust should be regarded more as a very useful multi-purpose tree species with a high potential for forest land reclamation, rather than a dangerous invasive neophyte. Nevertheless, the presence of this species should be carefully monitored around nature reserves and fragile landscapes in nutrient-poor and dry locations."

The official status of black locust in the European Union is addressed in the EU Regulation 1143/2014 on invasive alien species [93] where black locust is included in the list of 80 invasive alien species (IAS) provided by the European Commission, but each EU country has to manage this species according to their national specificity.

4. Discussion

The knowledge regarding black locust has steadily progressed as the experience of Romanian foresters in cultivating this species increased. Nevertheless, research has brought forward the complexity and also vulnerability of this species that was considered to be the answer to more problems than it could solve.

The main strong point of black locust management in Romania is the successful use of the species to reclaim large areas of degraded lands [93–95], thus improving local landscape, biodiversity, and socioeconomic aspects (e.g., fuel and construction wood, beekeeping) for many rural communities [2,10,17,18].

Further improvement through selection/tree breeding is still needed in order to increase the quality of the plantation stands; however, unfortunately there is no ongoing program on this matter. Furthermore, forest managers have to take into account the importance and effects of periodic tending operations in black locust stands [28,29,35,37]. The demand on the market for material resulted from tending operations making the latter economically viable, thus it was implemented at the required development stages and intervals which in turn had a positive influence on stand growth and development. In these respects, applying cleaning and thinning operations at appropriate intervals could lead to increases in stand productivity of up to 25% [17,23].

The management mistakes of the past have revealed both the invasive ability of the species when planted extensively on productive sites in the detriment of native species, and also its limitations if planted in inappropriate site conditions (e.g., soils with excess humidity, high calcium carbonates) [10,92].

Being a relatively new species in the Romanian flora, black locust has few major threats; although this seems to be changing, especially due to more frequent migration of some allochthonous insects accidentally released in Europe [42–47].

The present work highlights the results of research carried out at national level on black locust management, vulnerability, and uses in the past 150 years in Romania. Although most papers cited in this review were published before 1990, this does not necessarily mean that this species presents less interest today for forestry researchers and practitioners.

Having in mind the large areas of degraded lands still existent in Romania, as well as the international efforts to support the use of renewable energy sources, black locust stands out as one of the most suitable species for the phytoremediation of these unproductive areas, thus helping to mitigate the negative effects of global climate change.

Author Contributions: Conceptualization, A.L.C.; Methodology, A.L.C.; Software, A.L. and C.M.; Validation, I.V.A. and L.D.; Formal analysis, A.L.C.; Investigation, A.L.C.; Data curation, A.L.C.; Writing—original draft preparation, A.L.C.; Writing—review and editing, C.G.C.; Visualization, C.G.C. and B.S.; Supervision, I.V.A. and L.D. All authors have read and agreed to the published version of the manuscript.

Funding: This research received no external funding.

Institutional Review Board Statement: Not applicable.

Data Availability Statement: Not applicable.

Acknowledgments: This paper was supported by the Sectoral Operational Programme Human Resources Development (SOP HRD), ID76945 financed from the European Social Fund and by the Romanian Government.

Conflicts of Interest: The authors declare no conflict of interest.

References

1. Drăcea, M. *Beiträge zur Kenntnis der Robinie in Rumänien unter besonderer Berücksichtigung ihrer Kultur auf Sandböden in Oltenien [Contributions to the Study of Black Locusts in Romania with a Special Emphasis on Culture on Sandy Soils in the Region of Oltenia];* Tipografia Lucia: Bucureşti, Romania, 1928; 112p.
2. Costea, A.; Lăzărescu, C.; Bârlănescu, E.; Ivanschi, T.; Spârchez, Z. Studii asupra tipurilor de culturi de salcâm. In *Cercetări Privind Cultura Salcâmului;* Costea, A., Ed.; Editura Agrosilvică Bucureşti: Bucureşti, Romania, 1969; pp. 63–78.
3. Şofletea, N.; Curtu, L. *Dendrologie;* Editura Universităţii Transylvania: Bucureşti, Romania, 2007; 540p.
4. Nuţă, S. Cercetări Privind Influenţa Complexă a Perdelelor Forestiere din Sudul Olteniei. Ph.D. Thesis, Universitatea de Ştiinte Agronomice şi Medicină Veterinară, Bucureşti, Romania, 2007.
5. Pienaru, A.; Iancu, P.; Căzănescu, S. Desertification and Its Effects on Environment and Agricultural Production in Romania. *Annals. Food Sci. Technol.* **2009**, *10*, 624–629.
6. Blujdea, V.; Bird, D.N.; Kapp, G.; Burian, M.; Nuta, I.S.; Ciuvat, L. *Robinia pseudoacacia* Stump Feature Based Methodology for In Situ Forest Degradation Assessment. *Mitig. Adapt. Strateg. Glob. Change* **2010**, *16*, 463–476. [CrossRef]
7. Dincă, L.; Spârchez, G.; Dincă, M.; Blujdea, V. Organic Carbon Concentrations and Stocks in Romanian Mineral Forest Soils. *Ann. For. Res.* **2012**, *55*, 229–241.
8. Dincă, L.; Dincă, M.; Vasile, D.; Spârchez, G.; Holonec, L. Calculating organic carbon stock from forest soils. *Not. Bot. Horti Agrobot. Cluj-Napoca* **2015**, *43*, 568–575. [CrossRef]
9. Ciuvăţ, A.L.; Abrudan, I.V.; Blujdea, V.; Dutcă, I.; Nuţă, S.; Edu, E. Biomass Equations and Carbon Content of Young Black Locust (*Robinia pseudoacacia* L.) Trees from Plantations and Coppices on Sandy Soils in South-Western Romanian Plain. *Not. Bot. Horti Agrobot. Cluj-Napoca* **2013**, *41*, 590–592. [CrossRef]
10. Ciuvăţ, A.L.; Abrudan, I.V.; Blujdea, V.; Enescu, M.; Marcu, C.; Dinu, C. Distribution and Particularities of Black Locust in Romania. *Rev. Silvic. Şi Cinegetică Anul XVIII* **2013**, *32*, 76–85.
11. Vlad, R.; Constandache, C.; Dincă, L.; Tudose, N.C.; Sidor, C.G.; Popovici, L.; Ispravnic, A. Influence of climatic, site and stand characteristics on some structural parameters of scots pine (Pinus sylvestris) forests situated on degraded lands from east Romania. *Range Manag. Agrofor.* **2019**, *40*, 40–48.
12. Ivanschi, T.; Costea, A.; Bîrlănescu, E.; Mărcoiu, A.; Nonuţe, I. Cercetări privind stabilirea staţiunilor apte pentru cultura salcâmului. In *Cercetări Privind Cultura Salcâmului;* Costea, A., Ed.; Editura Agrosilvică Bucureşti: Bucureşti, Romania, 1969; pp. 11–55.
13. Nicolescu, V.-N.; Buzatu-Goanta, C.; Bartlett, D.; Iacob, N. Regeneration and Early Tending of Black Locust (*Robinia pseudoacacia* L.) Stands in the North-West of Romania. *South-East Eur.* **2019**, *10*, 97–105. [CrossRef]
14. Dincă, L.; Timiș-Gânsac, V.; Breabăn, I.G. Forest stands from direct accumulation and natural lakes slopes from the Southern Carpathians. *Present Environ. Sustain. Dev.* **2020**, *14*, 211–218. [CrossRef]
15. Oneț, A.; Dincă, L.; Teușdea, A.; Crișan, V.; Bragă, C.; Enescu, R.; Oneț, C. The influence of fires on the biological activity of forest soils in Vrancea, Romania. *Environ. Eng. Manag. J.* **2019**, *18*, 2643–2654.
16. Bârlănescu, E.; Costea, A.; Armăşescu, S.; Tănăsescu, S.; Vârbănescu, M.; Ruţă, S.; Baciu, V. *Experimentări Privind Aplicarea Răriturilor în Salcâmetele din Oltenia. Cercetări Privind Cultura Salcâmului;* Editura Agrosilvică Bucureşti: Bucureşti, Romania, 1969; pp. 111–154.
17. Costea, A.; Lăzărescu, C.; Bârlănescu, E.; Ivanschi, T.; Armăşescu, S.; Trantescu Gr Latiş, L.; Pârvu, E. *Recomandări Privind Cultura Salcâmului—Robinia pseudoacacia L.;* Institutul de Cercetări Studii şi Proiectări Silvice: Bucureşti, Romania, 1969; 39p.

18. Hernea, C.; Poşta, D.S.; Dragomir, P.I.; Corneanu, M.; Sărac, I. Aspects Concerning Germination Tests of *Robinia pseudoacacia* var. *Oltenica*. *J. Hortic. For. Biotechnol.* **2010**, *14*, 138–142.
19. *Technical Norms for Forestry*; Ministry of Environment, Water and Forests: Bucharest, Romania, 2000.
20. Abrudan, I.V. *Impăduriri*; Editura Universitaţii Transilvania din Braşov: Brașov, Romania, 2000; 200p, ISBN 973-635-688-4.
21. Florescu, I.; Nicolescu, N.V. *Silvicultură, Vol. I, Studiul Pădurii*; Editura EX LIBRIS: Braşov, Romania, 1996.
22. Armăşescu, S.; Ţabrea, A.; Decei, I.; Bârlănescu, E.; Olănescu, L.; Diaconu, E.; Tabană, D.; Ichimescu, D. *Cercetări Asupra Creşterii Producţiei şi Calităţii Arboretelor de Salcâm*; Bucureşti, A., Ed.; Agrosilvica: Bucuresti, Romania, 1969; pp. 111–154.
23. Nicolescu, V.N. *Silvicultura. Vol. II Silvotehnica*; Editura Transilvania Brasov: Braşov, Romania, 1998; 194p.
24. Popa, B. Analiza lucrărilor de substituire şi refacere în arboretele din Podişul Covurlui. *Rev. Pădurilor* **2003**, *2*, 13–17.
25. Constandache, C.; Sanda, N.; Virgil, I. Împădurirea terenurilor degradate ineficiente pentru agricultura din sud-estul ţării. *Ann. For. Res.* **2006**, *49*, 187–204.
26. Bîrlănescu, E.; Aurelian, C.; Cristian, S. O nouă varietate de salcîm identificată în România—Robinia pseudacacia L. var oltenica Birl. Bost. et Stoic. *Rev. Pădurilor* **1966**, *89*, 483–486.
27. Hernea, C.; Netoiu, C.; Corneanu, M. *Auxological Research Concerning Robinia Pseudoacacia var. Oltenica*; Editura Editura Universitaria Craiova: Craiova, Romania, 2008.
28. Corneanu, M.; Corneanu, G.C.; Iliev, I.; Danci, O.; Stefanescu, I.; Popa, M. Micropropagation of *Robinia pseudoacacia* var. oltenica Selected Stress Resistant Clones on Media with Deuterium Depleted Water. *J. Hortic. For. Biotechnol.* **2010**, *14*, 141–145.
29. Lăzărescu, C. Rezultatele privind înmulţirea salcâmului prin butăşire repetată. *Rev. Pădurilor* **1968**, *83*, 57–59.
30. Costea, A.; Lăzărescu, C.; Bârlănescu, E. Contribuţii privind selecţia salcâmului şi tipurile de cultură cu salcâm în terenuri forestiere. *Ann. For. Res.* **1970**, *27*, 163–168.
31. Bolea, V.; Catrina, I.; Popa, A.; Afrenie, F.; Cioloca, N.; Nicolescu, L. Particularităţi ecologice ale salcîmului—*Robinia pseudacacia* L.—relevate prin mediul variaţiei sezoniere a fotosintezei în raport cu factorii de mediu. *Rev. Pădurilor* **1995**, *110*, 33–41.
32. Enescu, V.; Bîrlănescu, E.; Costea, A. *Selection de Certaines Populations D'elite du Robinier Faux Acacia et Possibilityes de Leur Multiplication par Voie Vegetative*; FAO—Consultation Mondiale sur la Genetique Forestiere et l'Amelioration des arbres, Stokholm; FAO FORGEN: Rome, Italy, 1963; pp. 63–64.
33. Mihai, G.; Dincă, L. In situ conservation of forest genetic resources from the Southern Carpathians. *Int. J. Conserv. Sci.* **2010**, *11*, 1051–1058.
34. Bârlanescu, E.; Diaconu, M.; Costea, A.; Cojocaru, I. Cercetări privind ameliorarea salcâmului. *Ann. For. Res.* **1977**, *34*, 41–54.
35. Nicolescu, V.; Cornelia, H.; Beatrix, B.; Zsolt, K.; Borbála, A.; Károly, R. Black locust (*Robinia pseudoacacia* L.) as a multi-purpose tree species in Hungary and Romania: A review. *J. For. Res.* **2018**, *29*, 1449–1463. [CrossRef]
36. Enescu, V.; Mirancea, I. O tehnologie nouă de micropropagare in vitro a salcîmului (*Robinia pseudoacacia* L.). *Rev. Pădurilor* **1991**, *106*, 2–3.
37. Mirancea, I. Cercetări privind selecţia in vitro a unor genitipuri de arbori forestieri rezistenţi la stres abiotic. *Ann. For. Res.* **2002**, *45*, 31–38.
38. Ioniţă, L. Regenerarea protoplaştilor din diferite tipuri de explante iniţiale de salcâm (*Robinia pseudoacacia* L.). *Rev. Pădurilor* **1998**, *113*, 2–6.
39. Ioniţă, L. Cercetări privind influenţa diferiţilor factori endogeni şi exogeni asupra embriogenezei somatice la salcîm (*Robinia pseudacacia* L.). *Rev. Pădurilor* **1999**, *114*, 22–26.
40. Man, G. *Determinarea Principalilor Parametri Genetici la Larice şi Salcâm*; Scientific Report (Unpublished); Institute for Research and Management in Forestry: Bucureşti, Romania, 2001; 137p.
41. Băbeanu, C.; Ciobanu, G.; Corneanu, M.; Marinescu, G.; Corneanu, C.G. Isoperoxidases Pattern of *Robinia pseudoacacia* var. Oltenica in vitro culture on Different Media. *Bul. USAMV-Cj. Napoca Ser. Agric.* **2006**, *62*, 25–28. [CrossRef]
42. Trântescu, G.; Ceianu Ig Badea, N.; Vîrbănescu, M.; Baicu, V. *Cercetări Privind Biologia şi Combaterea Dăunătorilor Salcâmului. Cercetări Privind Cultura Salcâmului*; Editura Agrosilvică Bucureşti: Bucureşti, Romania, 1969; pp. 257–328.
43. Trântescu, G.; Vîrbănescu, M.; Popescu, I. Contribuţii la cunoaşterea biologiei şi combaterii dăunătorilor fructificaţiei salcâmului Etiella zinchenella Tr. *Ann. For. Res.* **1971**, *28*, 220–250.
44. Dissescu, G. Semiothisa alternaria HB. (*Macaria alternaria* Hb.) (fam. Geometridae), un dăunător important al salcîmului. *Rev. Pădurilor* **1987**, *102*, 29–30.
45. Dissescu, G. Cercetări asupra dăunătorilor *Lymantria dispar* L. pe salcâm. *Ann. For. Res.* **1971**, *28*, 151–172.
46. Neţoiu, C.; Tomescu, R. Moliile miniere ale salcâmului (*Parectopa robinella* Clemens—1863 şi *Phyllonorycter robiniella* Clemens—1859, Lepidoptera, Gracillaridae). *Ann. For. Res.* **2006**, *49*, 119–132.
47. Fora, C.G.; Moatăr, M.M.; Camen, D.D. Potential biotic risks for non-native species *Robinia pseudoacacia* L. In Proceedings of the COST Action FP1403 NNEXT—International Conference Non-Native Tree Species for European Forests, Vienna, Austria, 12–14 September 2018.
48. Vechiu, E.; Dinca, L.; Breabăn, I.G. Old Forests from Dobrogea's Plateau. *Present Environ. Sustain. Dev.* **2021**, *15*, 171–178. [CrossRef]
49. Ianculescu, M.; Ionescu, M.; Lucaci, D.; Neagu, Ş.; Măcărescu, C.M. Dynamics of Pollutants Concentration in the Forest Stands from Copşa Mică Industrial Area. *Ann. For. Res.* **2009**, *52*, 207–226.
50. Crișan, V.E.; Dincă, L.; Decă, S. Analysis of chemical properties of forest soils from Bacau County. *Rev. Chim.* **2020**, *71*, 81–86. [CrossRef]

51. Cântar, I.C.; Ciontu, C.I.; Dincă, L.; Borlea, G.F.; Crișan, V.E. Damage and Tolerability Thresholds for Remaining Trees after Timber Harvesting: A Case Study from Southwest Romania. *Diversity* **2022**, *14*, 193. [CrossRef]
52. Oneaţă, M. Studiul proceselor fiziologice ale arborilor din ecosistemele forestiere afectate de poluare în zona Copşa Mică. *Rev. Pădurilor* **2010**, *125*, 17–25.
53. Barbu, I.; Popa, I. Cartarea teritoriului României în raport cu lungimea medie a perioadelor de secetă și uscăciune. *Bucov. For.* **2002**, *10*, 13–23.
54. Milescu, I.; Armăşescu, S. Unele particularităţi dendrometrice ale arboretelor de salcâm în raport cu provenienţa. *Rev. Pădurilor* **1960**, *75*, 414–417.
55. Holban, C. Culturi specializate de salcâm pe soluri aluviale. *Rev. Pădurilor* **1969**, *84*, 359–360.
56. Bularca, M. Cercetări cu privire la utilizarea lemnului de salcîm şi salcie şi a unor materiale lignocelulozice (coarde de viţă de vie şi puzderii de in) ca materie primă la fabricarea plăcilor din fibre de lemn (PFL)—procedeu umed. *Rev. Pădurilor* **1985**, *36*, 193–200.
57. Porojan, M. Contribuţii la Studiul Proprietatilor Fizico-Mecanice si Tehnologice ale Lemnului de Salcâm. Ph.D. Thesis, Universitatea Transilvania din Braşov, Facultatea de Industria Lemnului, Brașov, Romania, 2007.
58. DeGomez, T.; Wagner, M.R. Culture and Use of Black Locust. *HortTechnology* **2001**, *11*, 279–288. [CrossRef]
59. Giurgiu, V.; Decei, I.; Armasescu, S. *Biometria Arborilor si Arboretelor din Romania, Tabele Dendrometrice*; Ed. Ceres: Bucuresti, Romania, 1972; 1155p.
60. Pleșca, I.M.; Blaga, T.; Dincă, L.; Breabăn, I.G. Prioritizing the potential of non-wood forest products from Arad County by using the analytical huerarchy process. *Present Environ. Sustain. Dev.* **2019**, *13*, 225–233. [CrossRef]
61. Mădaş, M.N. Cercetări privind indicii de calitate şi autenticitate ai mierii de salcâm (*Robinia pseudoacacia* L.). Doctoral Thesis, Facultatea de Zootehnie şi Biotehnologii, Universitatea de Ştiinţe Agricole şi Medicină Veterinară, Cluj-Napoca, Romania, 2013; 34p.
62. Chirilă, M.; Chirilă, P. *Tratament Homeopatic. Îndreptar de Simptome şi Semne*; Editura Ştiinţifică şi Enciclopedică Bucureşti: Bucuresti, Romania, 1986; 253p.
63. Pârvu, C. *Universul Plantelor—Mică Enciclopedie, Ediţia a III a*; Editura Enciclopedica: Bucuresti, Romania, 2000.
64. Vasile, D.; Enescu, M.; Dincă, L. Which are the main medicinal plants that could be harvested from Eastern Romania? *Sci. Pap. Ser. Manag. Econ. Eng. Agric. Rural. Dev.* **2018**, *18*, 523–528.
65. Bojor, O. *Ghidul Plantelor Medicinale şi Aromatice de la A la Z*; Editura FIAT LUX: Bucuresti, Romania, 2003.
66. Drăgulescu, C. *Plantele Alimentare din Flora Spontană a României*; Editura Sport—Turism: Bucuresti, Romania, 1991; 191p, ISBN 973-41-0145-5.
67. Constandache, C.; Dincă, L.; Tudor, C. The bioproductive potential of fast-growing forest species on degraded lands. *Sci. Papers. Ser. E. Land Reclam. Earth Obs. Surv. Environ. Eng.* **2020**, *IX*, 87–93.
68. Neţoiu, C.; Hernea, C.; Tomescu, R. Vegetaţia forestieră din bazinul mijlociu al Jiului. In *Bazinul Mijlociu al Jiului. Implicaţii de mediu şi Sociale ale Industriei Extractive şi Energetice. Studiu Monographic*; Editura Universitaria Craiova: Craiova, Romania, 2011; pp. 191–219.
69. Mantovani, D.; Veste, M.; Freese, D. Black locust (*Robinia pseudoacacia* L.) ecophysiological and morphological adaptations to drought and their consequence on biomass production and water-use efficiency. *New Zealand J. For. Sci.* **2014**, *44*, 29. [CrossRef]
70. Cântar, I.C.; Chisăliță, I.; Dincă, L. Structure of some stands installed on tailing dumps: Case study from Moldova Noua, Romania. *Int. Multidiscip. Sci. GeoConference SGEM* **2018**, *18*, 757–764.
71. Ianculescu, M. *Studiu Privind Reconstrucţia Ecologică a Pădurilor Afectate de Poluare din Zona Copşa Mică*; Tema ICAS nr. 6.6.—A.38; Scientific report (Unpublished); Arhiva ICAS Bucureşti: Bucuresti, Romania, 1994.
72. Abrudan, I.V.; Blujdea, V.; Pahonţu, C. Împădurirea terenurilor degradate în contextul eforturilor de diminuare a schimbărilor climatice. *Rev. Pădurilor* **2002**, *117*, 1–5.
73. Stankova, T.; Gyuleva, V.; Kalmukov, K.; Dimitrova, P.; Velizarova, E.; Dimitrov, D.; Hristova, H.; Andonova, E.; Kalaydziev, I.; Velinova, K. Biometric models for the aboveground biomass of juvenile black locust trees. *Silva Balc.* **2016**, *17*, 21–30.
74. Ciuvăţ, A.L. Producţia de biomasă şi stocarea carbonului în arborete tinere de salcâm din sudul României [Biomass Production and Carbon Stocking in Young Black Locust Plantations in Southern Romania]. Ph.D. Thesis, Universitatea Transilvania din Braşov, Braşov, Romania, 2013; 145p.
75. Iordache, E.; Popa, B.; Derczeny, R. The investments opportunities in wood energy plantations in Romania. *Balt. For.* **2014**, *20*, 352–357.
76. Böhm, C.; Quinkenstein, A.; Freese, D. Yield prediction of young black locust (*Robinia pseudoacacia* L.) plantations for woody biomass produc¬tion using allometric relations. *Ann. For. Res.* **2011**, *54*, 215–227.
77. Rédei, K.; Veperdi, I.; Tomé, M.; Soares, P. Black Locust (*Robinia pseudoacacia* L.) Short-Rotation Energy Crops in Hungary: A Review. *Silva Lusit.* **2010**, *18*, 217–223.
78. Rédei, K.; Csiha, I.; Keserü, Z. Black locust (*Robinia pseudoacacia* L.) Short-Rotation Crops under Marginal Site Conditions. *Acta Silv. Lign. Hung.* **2011**, *7*, 125–132.
79. Klašnja, B.; Orlović, S.; Galić, Z. Comparison of Different Wood Species as Raw Materials for Bioenergy. *South-East Eur.* **2013**, *4*, 81–88. [CrossRef]
80. Blujdea, V.; Pilli, R.; Dutca, I.; Ciuvat, L.; Abrudan, I.V. Allometric biomass equations for young broadleaved trees in plantations in Romania. *For. Ecol. Manag.* **2012**, *264*, 172–184. [CrossRef]

81. Ciuvăţ, A.L.; Abrudan, I.V.; Blujdea, V. Biomass Production in Young Black Locust Plantations in Southern Romania. In Proceedings of the Proceedings International Conference Ecology-Interdisciplinary Science and Practice, Sofia, Bulgaria, 25–26 October 2012; pp. 70–74, ISBN 978-954-749-096-3.
82. Nuţă, S. Caracteristici structurale şi funcţionale ale perdelelor forestiere de protecţie a câmpului agricol din Sudul Olteniei. *Ann. For. Res.* **2005**, *48*, 161–169.
83. Ciuvăţ, A.L. *Monitorizarea Proiectului de Împădurire a Terenurilor Degradate, Estimarea Acumulării de Carbon, Raportarea şi Resimularea Acumulării—Referat Ştiinţific*; Scientific Report (Unpublished); Institute for Research and Management in Forestry Bucharest: Bucuresti, Romania, 2011.
84. Ducci, F.; De Rogatis, A.; Proietti, R.; Curtu, L.A.; Marchi, M.; Belletti, P. Establishing a baseline to monitor future climate change-effects on peripheral populations of Abies alba in central Apennines. *Ann. For. Res.* **2021**, *64*, 33–66. [CrossRef]
85. Kutnar, L.; Kermavnar, K.; Pintar, A.M. Climate change and disturbances will shape future temperate forests in the transitionzone between Central and SE Europe. *Ann. For. Res.* **2021**, *64*, 67–86.
86. Ianculescu, M. Perdelele forestiere de protecţie în contextual majorării suprafeţei pădurilor şi al atenuării modificărilor climatic. In *Silvologie Vol. IVA*; Giurgiu, V., Ed.; Editura Academiei Române Bucureşti: Bucharest, Romania, 2005; pp. 201–223.
87. Costăchescu, C.; Dănescu, F.; Mihăilă, E. *Perdele Forestiere de Protecţie*; Editura Silvică: Bucharest, Romania, 2010; 262p, ISBN 978-606-8020.
88. Mihăilă, E.; Costăchescu, C.; Dănescu, F. *Sisteme Agrosilvice*; Editura Silvică: Bucharest, Romania, 2010; 190p, ISBN 978-606-8020-06-8.
89. Purcelean, Ş. Specii şi varietăţii decorative de Robinia indicate pentru spaţii verzi. *Rev. Pădurilor* **1954**, *69*, 369–372.
90. Cierjacks, A.; Kowarik, I.; Joshi, J.; Hempel, S.; Ristow, M.; Von der Lippe, M.; Weber, E. Biological Flora of the British Isles: Robinia pseudoacacia. *J. Ecol.* **2013**, *101*, 1623–1640. [CrossRef]
91. Constandache, C.; Peticilă, A.; Dincă, L.; Vasile, D. The usage of Sea Buckthorn (*Hippophae rhamnoides* L.) for improving Romania's degraded lands. *AgroLife Sci. J.* **2016**, *5*, 50–58.
92. Enescu, C.M.; Danescu, A. *Black Locust (Robinia pseudoacacia L.)—An Invasive Neophyte in the Conventional Land Reclamation Flora in Romania*; Bulletion of the Transilvania University from Brasov; Series II. 6(55); Transilvania University: Brașov, Romania, 2013; pp. 23–30.
93. Regulation (EU) No 1143/2014 of the European Parliament and of the Council of 22 October 2014 on the Prevention and Management of the Introduction and Spread of Invasive Alien Species. 21p. Available online: https://www.eea.europa.eu/policy-documents/ec-2014-regulation-eu-no (accessed on 12 September 2022).
94. Nicolescu, V.; Cornelia, H. Black Locust (*Robinia pseudoacacia* L.), a Key Species for the Forest-Related Rural Economy in the West of Romania. In Proceedings of the Conference, Coppice FORESTS in Europe: A Traditional Resource with Great Potential, Limoges, France, 21 June 2017; Available online: https://www.eurocoppice.uni-freiburg.de/intern/pdf/conference-limoges-2017/limoges-pres-nicolescu (accessed on 12 September 2022).
95. Cornelia, H.; Corneanu, M.; Neţoiu, C.; Buzatu, A.; Lăcătușu, A.R.; Cojocaru, L. Black Locust stand structure on the sterile dump in the Middle Basin of Jiu River (Romania). *Reforesta* **2020**, *9*, 9–19.

Article

Silvicultural Management System Applied to Logged Forests in the Brazilian Amazon: A Case Study of Adaptation of Techniques to Increase the Yield and Diversity of Species Forestry

Agust Sales [1,2,*], Marco Antonio Siviero [2], Sabrina Benmuyal Vieira [2], Jorge Alberto Gazel Yared [2], Ademir Roberto Ruschel [3] and Márcio Lopes da Silva [1]

1 Department of Forest Engineering, Universidade Federal de Viçosa, Viçosa 36570-000, MG, Brazil; marlosil@ufv.br
2 Department of Research and Innovation, Grupo Arboris, Dom Eliseu 68633-000, PA, Brazil; marco.siviero@grupoarboris.com.br (M.A.S.); sabrina.benmuyal@grupoarboris.com.br (S.B.V.); jagyared@gmail.com (J.A.G.Y.)
3 Department of Research and Development in Forestry and Forest Management, Embrapa Amazônia Oriental, Belém 66095-100, PA, Brazil; ademir.ruschel@embrapa.br
* Correspondence: agust.sales@ufv.br

Citation: Sales, A.; Siviero, M.A.; Vieira, S.B.; Yared, J.A.G.; Ruschel, A.R.; da Silva, M.L. Silvicultural Management System Applied to Logged Forests in the Brazilian Amazon: A Case Study of Adaptation of Techniques to Increase the Yield and Diversity of Species Forestry. *Diversity* **2021**, *13*, 509. https://doi.org/10.3390/d13110509

Academic Editors: Lucian Dinca, Miglena Zhiyanski and Michael Wink

Received: 14 September 2021
Accepted: 12 October 2021
Published: 20 October 2021

Abstract: The existence of degraded forests is common in the Eastern Amazon. The maintenance of these forests standing and the recovery of their productivity play an important role in the conservation of biodiversity, storage and carbon sequestration. However, the management techniques currently employed are designed for natural forests in the first harvest cycle or lightly explored and do not apply adequately to forests that have gone through several harvest cycles. Therefore, adaptations and the establishment of new management criteria that take into account other characteristics of these types of forests are necessary to ensure their sustainability. The objective of this study was to propose a silvicultural management system that has the potential to recover and perpetuate the productivity of an intensively logged tropical forest. A forest census was carried out on 535.6 ha for trees with dbh $\geq$ 25 cm. With these data, the following two treatments were designed: (1) criteria: the BDq method was applied from B = 9.8 m^2 ha^{-1}, D = 100 cm and q = 2. The criteria for standing wood commercialization were, in this order, first, Health; second, Tree Stem; third, Tree Density and fourth, dbh $\geq$ 105 cm. (2) Control: the planning was in accordance with Brazilian regulations. For the cost–benefit and sensitivity analysis, the Net Present Value (NPV) was used and a projection of $\pm$20% was made in the commercial price of standing wood. In the criteria treatment, a higher number of trees and species destined for the commercialization of standing wood was verified in relation to the control treatment, showing a greater diversity of species. In the criteria treatment, NPV was positive and superior to the control treatment in all scenarios. The proposed silvicultural management system with an object of an explored and enriched forest, with criteria for harvesting trees with a minimum cut diameter of 25 cm, proved to be viable to generate economic returns and with conservationist potential for the continuous supply of forest products and maintaining biodiversity.

Keywords: BDq method; Eastern Amazon; degraded tropical forest; forest restoration; forest economy; logging selection criteria; sustainable forest management

1. Introduction

Procedures for forest management in the Brazilian Amazon have not been regulated based on technical parameters appropriated to all types of forests. There are also questions that normative instructions, government policies and forest management are the main barriers to make the management effective, efficient and sustainable [1–6].

It is estimated that 4.8% (4.5 million hectares) of natural forests classified as degraded, that is, forests in which successive logging cycles have occurred, are in the Legal Amazon [7]. These forests have a high richness of tree species with a diversified floristic composition, with potential for biodiversity conservation. A promising aspect is that, in recent decades, research in the area of forest restoration has been intensified for the profile of exploited or degraded forest, as an attempt to encourage the use of techniques that can keep the forest standing [8–10].

A forest restoration and management system proved to be viable to generate economic returns with continuous supply of wood products [11,12] and potential for the maintenance of the diversity of species forestry [13,14] as well contribute to mitigating the effects of climate change [15].

One of the key elements for forest restoration and natural forest management is understanding the structure of natural forests, especially the use of indicators such as floristic composition and diametric structure [3,16,17]. The diversity of species and their distribution in different classes of diameters are important factors, among many others, that provide stability to ecosystems [18]. The "J-inverted" shape, diametric distribution or negative exponential distribution for uneven-aged forests, indicates that populations of a forest naturally recompose themselves through the balance between mortality and entrance of individuals [19,20]. However, in order to maintain the forest's balanced diametric structure, it is necessary to apply management techniques to lead it to a "balanced" distribution [21,22].

The diametric distribution models that estimate growth and production provide support for forest planning, making it possible to identify the trees that make up the forest by diameter classes and predict forest production [18,23,24]. They also determine the interventions in the forest in a way to guarantee the ecological and economic sustainability of the stands [25].

The advance in management techniques in uneven-aged forests has led to the development of the BDq (B = remaining basal area, D = maximum diameter and q = De Liocourt constant) method of selection [19]. This method was also used in forests in Brazil [20,26,27] to determine the balanced harvest intensity in the diametric classes [28]. This method is conditioned to the values of the remaining basal area, the maximum intended diameter, the number of individuals per class of diameter per hectare and the choice of species to be harvested [29]. Associated with the balanced distribution model, the application of post-harvest silvicultural treatments, such as conduction of regeneration, thinning and enrichment planting, has been an alternative to recover the forest structure as well as the populations of species of interest [11,12,30–35].

The consolidation of new technical parameters is important to make forest management effectively sustainable from an environmental, social and economic point of view [36,37]. This case study deals with the use of the BDq Method and ecological and economic criteria for tree selection for planning the harvest and commercialization of standing wood in an intensively logged tropical forest.

The objective of this case study is to provide technical and economic information to promote a sustainable silvicultural management system for the commercialization of wood in an intensely logged tropical forest and make it desirable for environmental protection and conservation as a perpetual financial asset, with the maintenance of biodiversity and expansion of populations of low-density species and the quality of the forest.

2. Materials and Methods

2.1. Study Area

The study was conducted in the forest management area (535.6 ha) of the Fazenda Shet farm, located in the municipality of Dom Eliseu, State of Pará, Brazil (altitude, 320 m; 4°30′48″ S and 47°39′36″ W) (Figure 1), owned by the Arboris Group®.

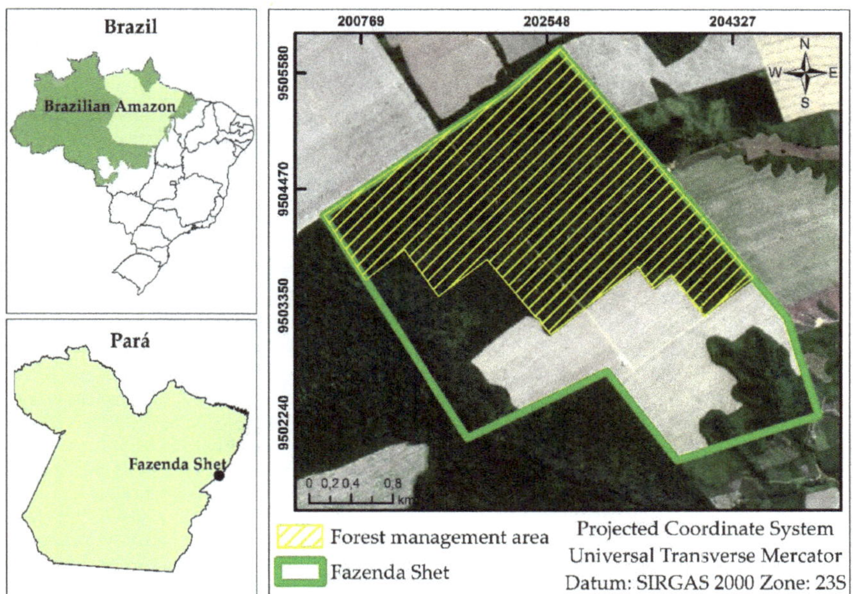

Figure 1. Forest management area of Fazenda Shet, Dom Eliseu, State of Pará, Brazil.

The region's climate is Aw (Köpen), tropical with summer rains, with an average annual rainfall of 2500 mm [38,39]. The average annual temperature is 25 °C. The municipality's vegetation type is Dense Ombrophilous Submontane Emerging Canopy Forest [40]. The predominant soils are dystrophic Yellow Latosol and dystrophic Red–Yellow Argisol [41].

The forest management area of Fazenda Shet is represented by an exploited natural forest characteristic of the region of the arch of deforestation in the Amazon [42]. This forest underwent successive forest exploitation processes that took place between the 1970s and the 1990s, and the volume of wood extracted is unknown. After the establishment of legal regulations for management, the first logging in the area, permitted by environmental agencies, occurred in 1993 and 1994. The average harvest volume was 65 m^3 ha^{-1}. The current regulation that prescribes a maximum harvest of 30 m^3 ha^{-1} for the Brazilian Amazon only began in 2006 [6]. In the clearings formed by logging, enrichment planting with direct seeding of *Schizolobium parahyba* var. *amazonicum* was performed by [11].

2.2. Forest Census

A forest census was carried out in 2008, considering the methodology presented in Technical Guidelines for Exploration of Reduced Impact in Forest Operations of Terra Firme in the Brazilian Amazon [43] for all trees with a dbh (diameter at 1.3 m high) ≥25 cm. On that occasion, the following procedures were performed: botanical identification procedures, cutting vines, fixing a numbered label at the base of the trunk, measuring the dbh, visual estimate of the commercial height, classification of the quality of the tree stem (tree stem 1—straight and cylindrical tree stem; tree stem 2—slightly tortuous and/or corrugated; tree stem 3—crooked, strongly corrugated or forked) and sanity (rot, senescence, broken, dead and/or fallen canopy).

The species were divided into three groups of market value. For groups 1, 2 and 3, 13, 41 and 52 species are included, respectively, in which the species in group 1 represent the highest economic value, followed by the species in group 2 and the species in group 3 representing the lowest economic value (Table 1 and Supplementary Materials).

Table 1. Species (popular and scientific name) classified by market value group, forest management area of Fazenda Shet (535.6 ha), Dom Eliseu, State of Pará, Brazil.

Group 1
Angelim-pedra (*Hymenolobium petraeum* Ducke); cedro (*Cedrela odorata* L.); copaíba (*Copaifera* Ducke); cumaru (*Dipteryx odorata* (Aubl.) Willd.); freijó-cinza (*Cordia goeldiana* Huber); ipê-amarelo (*Handrohanthus serratifolius* (Vahl) S. Grose); jatobá (*Hymenaea courbaril* L.); jatobá-curuba (*Hymenaea parvifolia* Huber); louro-canela (*Nectandra* sp.); maçaranduba (*Manilkara elata* (Allemão ex Miq.) Monach.); muiracatiara (*Astronium lecointei* Ducke); roxinho (*Peltogyne lecointei* Ducke) and tatajuba (*Bagassa guianensis* Aubl.).
Group 2
Amapá (*Brosimum guianense* (Aubl.) Huber); amarelão (*Apuleia leiocarpa* (Vogel) J. F. Macbr.); amescla/breu (*Trattinnickia burseraefolia* Mart. Willd.); amesclão (*Trattinnickia rhoifolia* Willd.); amesclinha (*Protium altissimum* (Aubl.)) Marchand); angico/timborana (*Pseudopiptadenia suaveolens* (Miq.) J. W. Grimes); caju (*Anacardium giganteum* W. Hancock ex Engl.); caneleiro (*Cenostigma tocantinum* Ducke); casca seca (*Licania* sp. Aubl.); catuaba (*Lacmellea aculeata* (Ducke) Monach.); cedrorana (*Vochysia maxima* Ducke); coco-pau (*Coupeia robusta* Huber); cupiúba (*Goupia glabra* Aubl.); axixá/envira-quiabo (*Sterculia pruriens* (Aubl.) K. Schum.); envira/envira-preta (*Guatteria punctata* (Aubl.) R. A. Howard); escorrega-macaco (*Albizia pedicellaris* (DC.) L. Rico); estopeiro/tauari (*Couratari* sp. Aubl.); farinha-seca (*Ampelocera edentula* Kuhlm.); faveira (*Parkia multijuga* Benth.); goiabão (*Pouteria bilocularis* (H. K. A. Winkl.) Baehni); inharé (*Helicostylis pedunculata* Benoist); jarana (*Lecythis lurida* (Miers) S.A. Mori); louro-pimenta (*Ocotea* sp.); louro-vermelho (*Sextonia rubra* (Mez) van der Werff); mandiocão/morototó (*Didymopanax morototoni* (Aubl.) Decne. & Planch.); marupá (*Simarouba amara* Aubl.); orelha-de-macaco (*Enterolobium schomburgkii* (Benth.) Benth.); paricá (*Schizolobium parahyba* var. *amazonicum* (Huber ex Ducke) Barneby); pau-santo (*Zollernia paraensis* Huber); pequiá (*Caryocar vilosum* (Aubl.) Pers.); pequiarana (*Caryocar glabrum* (Aubl.) Pers.); quina (*Geissospermum sericeum* Miers); quina-rosa (*Quiina amazonica* A.C.Sm.); sapucaia (*Lecythis pisonis* Cambess.); seringarana (*Ecclinusa guianensis* Eyma); sumaúma (*Ceiba pentandra* (L.) Gaertn.); tanibuca (*Terminalia tanibouca* Rich.); itaúba (*Mezilaurus itauba* (Meisn.) Taub. ex Mez); tauari (*Couratari* ssp. /*Eschweilera coriacea* (DC.) S. A. Mori) and uxi (*Endopleura uchi* (Huber) Cuatrec.).
Group 3
Amarelinho (*Neoraputia paraensis* (Ducke) Emmerich ex Kallunki); andirobarana (*Guarea kunthiana* A. Juss.); ata (*Annona* sp.); atraca (*Ficus* sp.); baço-de-boi (*Myrocarpus venezuelensis* Rudd); bicuíba/ucuúba-da-terra-firme (*Virola michelii* Heckel); Buranju (*Neea floribunda* Poepp. & Endl.); Cacau (*Theobroma speciosa* Willd. ex Spreng.); canafístula (*Senna multijuga* (Rich.) H. S. Irwin & Barneby); capa-bode (*Bauhinia acreana* Harms.); conduru (*Cynometra bauhiniifolia* Benth.); cravinho/goiabarana (*Myrcia paivae* O.Berg); embaúba (*Cecropia distachya* Huber./*C. sciadophylla* Mart./*C. palmata* Willd.; *Pourouma guianensis* Aubl.); freijó-branco (*Cordia alliodora* (Ruiz & Pav.) Cham.); Gabiroba (*Campomanesia grandiflora* (Aubl.) Sagot); gema-de-ovo (*Amphiodon effusus HuberPoecilanthe*); goiabinha (*Eugenia lambertiana* DC.); inajarana (*Quararibea guianensis* Aubl.); ingá (*Inga* spp.; *Inga alba* (Sw.) Willd.; jaca-braba (*Abarema campestres* (Spruce ex Benth.) Barneby & J. W. Grimes); jambo/muúba (*Bellucia grossularioides* (L.) Triana); jiboião/matamatá-preto (*Eschweilera grandiflora* (Aubl.) Sandwith); jurema (*Senna polyphylla* (Jacq.) H. S. Irwin & Barneby); juruparana (*Gustavia augusta* L.); limãozinho (*Zanthoxylum rhoifolia* Lam /*Z. ekmanii* (Urb.) Alain); mangaba/abiu-mangabarana (*Micropholis guyanensis* (A. DC.) *Pierre*); mangue (*Buchenavia capitata* (Vahl) Eichler); maria-preta (*Ziziphus cinnamomum* Triana & Planch.); matamata/matamata-jibóia (*Eschweilera ovata* (Cambess.) Mart. ex Miers); mirindiba (*Glycydendron amazonicum* Ducke); moreira (*Maclura tinctoria* (L.) D. Don ex Steud.); mutamba (*Guazuma umifolia* Lam.); pele de sapo (*Pausandra trianae* (Müll.Arg.) Baill.); pitomba (*Talisia* sp.); seringueira (*Hevea brasiliensis* (Willd. ex A Juss) Mull. Arg.); tamburil (*Enterolobium maximum* Ducke); taxi/taxi-branco (*Tachigali vulgaris* L. G. Silva & H. C. Lima /*Tachigali glauca* Tul.) and tuturubá/abiurana (*Pouteria guianensis Aubl.* /*Pouteria venosa* subsp. amazonica T. D. Penn).

Based on the forest census data of forest management area (535.6 ha) of Fazenda Shet, the following two scenarios in harvest planning were designed for the same area of the case study: (1) criteria treatment: trees with a dbh $\geq$ 25 cm, quantifying 46,010 trees and 106 species; and (2) control treatment: trees with a dbh $\geq$ 40 cm, accounting for 13,638 trees and 91 species.

2.3. Criteria Treatment

BDq method: to plan the wood harvest with the aim of maintaining a balanced structure of the forest, we performed the BDq method of selection [19], which was first used in Brazil in the Atlantic forest [26]. The diametric structure was characterized by the distribution of the number of trees, the basal area and the volume, per hectare and by diameter class. To perform this analysis, trees with a dbh $\geq$ 25 cm were grouped according to classes of diameter, previously fixing the class width of 10 cm.

The remaining basal area (B = 9.8 m^2 ha^{-1}) was obtained through the forest census; a maximum desirable diameter was established (D = 100 cm, amplitude of $\pm$5 cm) because generally trees with a dbh > 105 cm have already fulfilled their reproductive role [12] and reached the culmination (asymptote) growth, presenting an increase in volume relatively lower than that trees of predecessor diameter class [44]. Trees with diameters above 100 cm may eventually remain in the area, if they are rare species or if they have very dense crowns, since they can promote large clearings in their fall. The quotient (q = 2) of Liocourt was adjusted in order to minimize the diameter classes with deficits and to allocate to commercialization a larger number of standing trees of a specific diameter class in relation to the successor diameter class.

From the BDq parameters, coefficients β_0 and β_1 were calculated using the following equations:

$$\beta_1 = \frac{Ln\ (q)}{X_i - X_{i+1}}\ \beta_0 = Ln.\frac{40000.B}{\pi.\sum_{i=1}^{n} dbh_i^2\ .e^{\beta_1.dbh_i}} \tag{1}$$

The parameters were included in the function for the calculation of the remaining frequencies, thus the classes with surplus or deficit in number of trees were defined using the following equation:

$$Y = e^{\beta_0+\beta_1.X_i} + e_i \tag{2}$$

where Y = the estimated number of trees per diameter class per hectare; X = the diameter class center; e = the base of the Neperian logarithm; e_i = random error and β_0 and β_1 = function parameters.

The number of trees destined for commercialization of standing wood (harvest) within each dbh class was obtained by subtracting the number of trees in the "forest" by the number of "remaining" trees [20]. In this way, balancing curves of the forest were generated with the number of trees, per hectare, observed (forest), estimated by the BDq method (remaining) and destined for commercialization of standing wood (harvest) by dbh class center.

Criteria for commercialization of standing wood: considering the number of trees by diameter class center, with surplus or destined for commercialization of standing wood (harvest) using the BDq method, a selection was made, considering the silvicultural conditions and the tree density by species, in the following order:

- 1st, Health: trees identified with rot, senescence, broken crown, signs of disease or that were dead were selected for sale of standing wood.

Traditionally, trees with compromised health are not selected for harvest. However, the permanence of these trees causes negative influences on the quality of the future forest, as they are more susceptible to pests and diseases, facilitating their proliferation. Furthermore, trees in these conditions use growth resources (space, light, water and nutrients) that could be made available to healthy and productive trees. In general, trees with compromised health have a reduced life cycle. Branches of different sizes and leaves remain in the forest, which allows for the survival of other living organisms that have a habitat or feed on this type of material.

- 2nd, Tree Stem: trees with quality of tree stem 2 and 3 were selected for commercialization of standing wood, aiming at maintaining trees with quality tree stem 1;
- 3rd, Density: trees of the 15 species with the higher tree density values located in the management area were selected to be commercialized as standing wood. This

procedure aims to conserve low density species and to maintain biodiversity. The density was obtained using the following equation [45]:

$$Da = \frac{n_i}{A} \tag{3}$$

where Da = Absolute density; ni = the number of inventoried trees of the i-th species and A = total area sampled, in hectares.

- 4th, dbh > 105 cm: trees with a dbh > 105 cm were selected to be commercialized as standing wood. This criterion aimed at reducing the number of senescent trees and increasing the population of trees with smaller diameters. The current harvest of trees with larger diameters aims to make the future industrial plant compatible, which will be adequate for a larger number of trees with smaller diameters.
- 5th, abundance: prioritized the selection of the most abundant species. These species have characteristics of pioneers, fast growing, with greater ease of propagation and were less pressured in previous harvests.

2.4. Control Treatment

Harvest planning in control treatment was carried out in the same forest management area (535.6 ha) of Fazenda Shet that was used for the criteria treatment. The difference is that the control treatment was conducted according to the following Brazilian regulations [2,6,46]:

- Minimum interval of 12 years from the last logging;
- Maximum harvest (m^3 ha^{-1}) determined by multiplying the constant "0.86 m^3 ha^{-1}", which is the average annual increase in volume of the portfolio of commercial species, by the number of years that have elapsed since the last logging [6];
- Forest census for all trees with a dbh $\geq$ 40 cm, from which trees with dbh $\geq$ 50 and $\leq$200 cm were destined for the commercialization of standing wood;
- Minimum maintenance of 3 trees per species, per 100 hectares (or 0.03 trees ha^{-1}) with a dbh $\geq$ 50 cm;
- Exclusion of prohibited species or of those with harvesting restriction;
- Selection of trees with tree stem form quality 1 and 2 for commercialization of standing timber.

2.5. Cost–Benefit and Sensitivity Analysis

The Net Present Value (NPV) was the indicator used to estimate the financial viability of applying the BDq method and criteria for commercialization of standing wood in forests explored in the Amazon (i.e., the criteria treatment), and the control treatment as well. The NPV is a tool to calculate the profitability of projects through the analysis of discounted cash flow [47].

$$\mathrm{NPV} = \sum_{t=0}^{n} \frac{B_t}{(1+r)^t} - \frac{C_t}{(1+r)^t} \tag{4}$$

where t is the year when revenue or cost occurs, B_t is the income from the commercialization of standing wood in the year t, C_t is the total cost in the year t, r is the discount rate per year and n is the time demanded for incomes. Only an NPV greater than zero indicates a return on investment [48]. The profitability of standing timber sale was calculated based on the 13 years of forest growth, after the last logging at Fazenda Shet. This period was defined as a function of monitoring the productivity of the site, carried out in permanent sample plots considering the growth of the species planted in the enrichment of clearings, of species with a short life cycle that regenerated and the abundance of other species in the area, making an assortment of 30 m^3 ha^{-1} harvest, which is the maximum production allowed for harvesting according to Brazilian regulations [2,6,46].

The cash flow was based on field spreadsheets provided by the Arboris Group. Costs totaled USD 29.70 ha^{-1} for criteria treatment and USD 23.31 ha^{-1} for control, resulting from the sum of the annual cost, forest census and administration.

The incomes were calculated from the volumes of each tree destined for the commercialization of standing wood from the criteria and control treatments. The calculated volumes were multiplied by the price determined by the business group for the inventoried species, according to the market value group: 1 (USD 183.76 m^{-3}), 2 (USD 74.25 m^{-3}) and 3 (USD 62.25 m^{-3}).

Profitability per hectare (NPV ha^{-1}) was calculated based on a yearly nominal interest of 7%. This is the current interest rate on the capital loan adopted by the *Banco do Brasil* Commercial Forest Planting Program (PROPFLORA) and by other banks accredited by the National Bank for Economic and Social Development (BNDES) for forest investment and production.

Different price scenarios for the wood commercialization were built through a sensitivity analysis to project the profitability of the commercialization of standing trees. In addition to the wood commercialization values established for the groups of the species 1, 2 and 3, two other scenarios were built varying from −20% to +20% on the wood commercialization price.

3. Results

3.1. Criteria and Control Treatment

BDq method: the dbh class centers with surpluses and deficits in the number of trees ha^{-1} in relation to the adjusted balanced curve (remaining curve) are shown in Figure 2. By means of balancing, the commercialization of trees in the surplus classes (harvest curve) until they reach the desired remaining number and the maintenance of the trees in the deficit classes was proposed. In the dbh class centers of 30, 40 and 50 cm, there were surpluses of 10.414, 6.985 and 0.732 trees ha^{-1}, respectively, totaling 18.131 trees ha^{-1}, which were destined for the commercialization of standing wood. For the other dbh class centers, deficits ranging from 0.134 to 1.231 ha^{-1} were found.

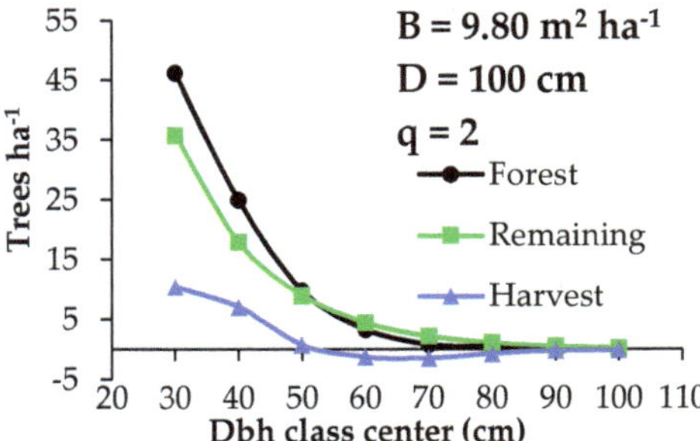

Figure 2. Balancing curves of the forest with the number of trees, per hectare, observed (forest), estimated using the BDq method (remaining) and destined for the commercialization of standing wood (harvest) by dbh class center, management area of the Fazenda Shet, Dom Eliseu, State of Pará, Brazil.

In the criteria treatment's forest census, 85.907 trees ha^{-1} with a dbh $\geq$ 25 cm were inventoried, corresponding to 106 species (535.6 ha). Of the inventoried population, 29.78% are trees of the species *S. parahyba* var. *amazonicum* (12.880 trees ha^{-1}) and *C. obtusa* (12.705 trees ha^{-1}). When applying the BDq method and the criteria for the commercialization of standing wood, 18.332 trees ha^{-1} were selected, referring to 77 species (535.6 ha). Some trees were selected by more than one criterion, totaling 1.752 trees ha^{-1} for compromised health, 10.951 trees ha^{-1} by tree stem (quality two and three), 16.173 trees ha^{-1} by tree density and 0.224 trees ha^{-1} per dbh > 105 cm.

Concerning the population destined for the commercialization of standing wood in the criteria treatment, *S. parahyba* var. *amazonicum* and *C. obtusa* contributed with 6.614 trees ha^{-1}, the largest number of trees being selected for harvest by the density criterion. In addition to these species, *Inga* sp., *J. spinosa*, *T. burseraefolia*, *P. densiflora*, *P. guianensis*, *C. tocantinum*, *P. effusa*, *S. parviflora*, *S. pruriens*, *Z. rhoifolium*, *G. sericeum*, *C. aliodora* and *Nectandra* sp. had trees selected for standing wood commercialization by the density criterion. The sum of the trees of these species corresponded to 16.173 trees ha^{-1}, corresponding to the 15 species with the highest tree density.

The trees selected by the health criterion represented 1.752 trees ha^{-1}, with 44 verified species (535.6 ha). The species with the highest frequency of trees with compromised health destined for the sale of standing wood were *C. obstusa*, *Inga* sp., *T. burseraefolia* and *P. guianensis*, adding up to 1.205 trees ha^{-1}. Regarding the tree stem form criterion (quality two and three), 10.951 trees ha^{-1} of 74 species (535.6 ha) were destined to be commercialized as standing timber. The species selected by the tree stem criterion with the highest frequency of trees were *J. spinosa*, *Inga* sp., *T. burseraefolia* and *C. obstusa*, totaling 5.804 trees ha^{-1}.

Considering the dbh criterion >105 cm, 0.224 trees ha^{-1} of 26 species (535.6 ha) were destined for the commercialization of standing wood. The species most frequently found in trees with a dbh >105 cm were *C. tocantinum*, *P. suaveolens*, *G. sericeum* and *T. burseraefolia*, totaling 0.127 trees ha^{-1}.

In the control treatment, 25.463 trees ha^{-1} were inventoried, which presented a dbh $\geq$40 cm and belonged to 91 species. Of the inventoried population, 27.79% were trees of the species *S. parahyba* var. *amazonicum* (3.462 trees ha^{-1}) and *N. paraensis* (3.615 trees ha^{-1}). Following Brazilian regulations, 2.112 trees ha^{-1} were selected for the commercialization of standing wood, which refer to 41 species (535.6 ha).

Trees and species destined for the sale of standing wood by the dbh class center and market value group for the criteria and control treatments are shown in Figure 3. In the criteria treatment, a higher quantity of trees ha^{-1} and species destined for the commercialization of wood was verified in relation to the control treatment, with the largest quantity present in the dbh class centers of 30, 40 and 50 cm (Figure 3a).

3.2. Cost–Benefit and Sensitivity Analysis

The treatments (criteria and control) were profitable for the commercialization of standing wood under the nominal interest rate of 7% per year, since its NPVs were greater than zero in all scenarios (Figure 4a). The differences, however, were in favor of criteria treatment in all scenarios. The scenarios with a variation of ±20% for the commercialization price of standing wood showed that the criteria treatment is more profitable than conducting this forest profile based on the technical guidelines contained in the Brazilian regulations (i.e., the control treatment).

The scenarios indicated an NPV variation from 156.27 to USD 263.84 ha^{-1} in the criteria treatment and 124.80 to USD 198.74 ha^{-1} in the control treatment, under the nominal interest rate of 7% per year. In the criteria treatment, total production (m^3 ha^{-1}) was found to be 44% higher than the control treatment, with higher production observed in market value groups two and three (Figure 4b).

For the control treatment, the highest production was found in the market value group one, indicating the preference that is generally destined for the species with the greatest economic value. However, the highest trade demand for the use of a reduced number of species, even though they have greater commercial value, does not guarantee greater profitability as shown in Figure 4b, besides the fact that loss in species diversity may occur due to high selective logging pressure on certain and few species.

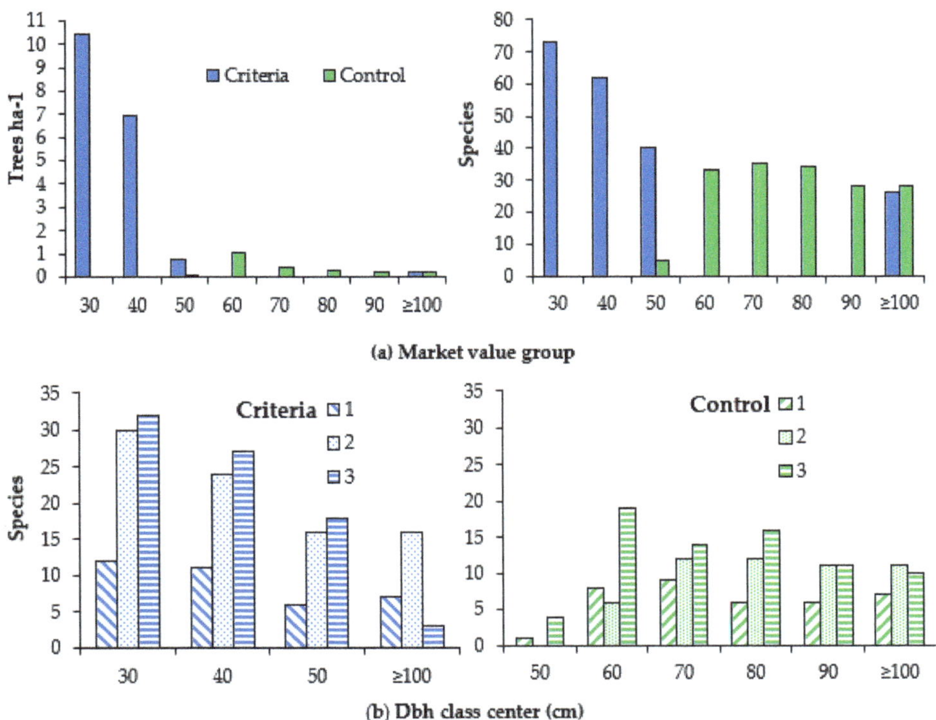

Figure 3. Trees and species destined for the sale of standing wood by the market value group (**a**) and dbh class center (**b**) for the criteria and control treatments, forest management area of Fazenda Shet, Dom Eliseu, State of Pará, Brazil.

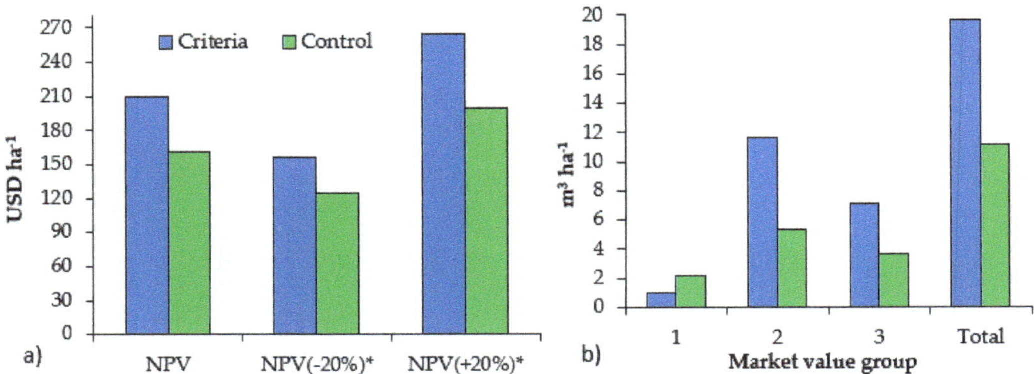

Figure 4. Financial indicators (NPV ha^{-1}) (**a**) and production (m^3 ha^{-1}) by the market value group of species destined for the commercialization of standing wood (**b**) for the criteria and control treatments, forest management area of Fazenda Shet, Dom Eliseu, State of Pará, Brazil. * NPV result of the projection of ±20% in the commercialization price of standing wood.

4. Discussion

4.1. Criteria and Control Treatment

In exploited natural forests, similar to the one present in this study, it is common to have an imbalance in relation to the abundance or lack of trees in certain centers of diameter classes, reflecting the result in general of human action [19,27,49].

However, it is observed that the studied forest maintains its structure in the inverted-J form, although it may change slightly after harvesting and decrease the abundance of species. Additionally, it presents a sufficient tree stock to be recovered, as shown in other studies [11,44]. Considering the minimum dbh of 25 cm and the regulation of the remaining stock of trees, the application of the BDq method in this forest profile demonstrates a trend in improving the future stock of wood with greater commercial attractiveness, which should promote greater industry sustainability in the short, medium and long terms [12].

Our findings demonstrate that there is greater diversity and possibilities in the choice of species, as an advantageous result of the application of the BDq method, just as of the criteria for the commercialization of standing wood in these dbh class centers, an alternative that would not be possible when following the Brazilian regulations for this forest profile. This fact is confirmed by the number of species per market value group verified according to the criteria treatment (Figure 3b), which shows a higher number of species selected for the commercialization of standing wood, distributed in dbh class centers, in relation to the control treatment.

The application of the BDq method and the criteria for the commercialization of standing wood could be seen as a silvicultural treatment of thinning, in which the selected trees (all trees with a dbh ≥25 cm) from the enrichment planting and from natural regeneration are commercialized. For instance, at Fazenda Shet, there was a high trees stock of various species not included in the first line of commercialization (those with a dbh ≥25 cm) that could serve as an assortment for the harvest and generate revenue in successive intermediate periods of the harvest cycle. Therefore, with forest growth monitoring, it is possible to identify the peculiarities of each species for proper management. Fast-growing species could be commercialized in shorter periods of time, while slower-growing species would be part of a longer-term crop assortment. Therefore, silvicultural systems that take into account the management by species or group of species with similar characteristics are an option potentially applicable to the reality of small, medium and large areas of forest management.

The authorization for logging in Amazon natural forests is subject to regulations for harvesting trees with a dbh ≥50 cm. With this regulation, trees of larger diameters and high commercial value are prioritized, establishing greater pressure under a reduced number of species after consecutive harvests. This fact compromises the richness of the forest and the perpetuation of timber productivity [6,13,50,51].

The commercialization of trees with a dbh ≥25 cm will make possible the inclusion of new species in the market that do not reach a dbh ≥50 cm due to biological characteristics, as they would be excluded according to Brazilian regulations. These species are represented by a high density of trees in the forest in this studied region, as is the example of *C. obtusa* [52]. On the other hand, species from enrichment sowing in clearings, such as the species *S. parahyba* var. *amazonicum* in this study, also present a high tree density [11].

In addition to this, it is preferable to allocate species with high tree abundance for commercialization, with the objective of favoring the maintenance of biodiversity and the regeneration of low abundance species that, in some cases, have been widely pressured in previous logging.

Enrichment sowing in a clearing with *S. parahyba* var. *amazonicum* is a consolidated practice in the areas of the studied company. This practice has made it possible to accelerate the recovery of the wood stock in a shorter time than indicated by Brazilian regulations [6,11,34].

The application of the BDq method and the definition of criteria for the commercialization of standing wood imply rational management aimed at maintaining the richness of species in the forest, based on the conservation of species and the use of trees with compromised health and senescent, promoting the species of greater and lowest value in the forest. Still, the application of these criteria for harvesting promotes continuous forest production, as this is an essential factor for this type of forest to become a desirable asset for conservation.

It is worth mentioning that, in general, the mortality rate in forest ecosystems is influenced by the health of the trees [53]. The mortality rate of trees for health reasons has been wrongly treated with little environmental and economic relevance. Although the importance of compromised health trees for nutrient cycling has been questioned, the presence of these trees in the forest promotes the function of hosts and precursors of pests and diseases, in addition to the poor reproductive quality [54].

The use of silvicultural practices aiming at profitability in the short, medium and long term tends to make the management of tropical forests truly sustainable in order to maintain wealth and productivity. This study demonstrates the application of technical and economic tools with the potential to increase the health and productivity of these forest profiles, contributing significantly to breaking paradigms and encouraging effective sustainable management practices of tropical forests exploited in recent decades.

Trees destined for the commercialization of standing wood by the health criterion represent a relatively low percentage. However, harvesting trees with compromised health is an efficient and profitable method to eliminate host trees for pests and diseases, and to make more growth resources available (space, light, water and nutrients) for the remaining healthy and productive trees [12]. Harvesting interventions in exploited forests are one of the factors that influence the composition of future species in the forest, mainly promoting the abundance of pioneer species, with no significant negative effect on shade-tolerant species [55].

In a certain way, the management of natural forests for wood production is significantly questioned regarding the impact on species composition and richness [14,56], the provision and maintenance of ecosystem services [57], the frequency of silvicultural harvest and post-harvest operations [11,13] and the economic viability [12]. However, research on management in forests that were exploited in the past and that take into account the harvest criteria [58], techniques of harvest optimization and forest densification [11,33] and management compatible with the behavior of each species [59] has revealed great potential for the recovery and rational use of forests.

4.2. Cost–Benefit and Sensitivity Analysis

As a financially viable alternative, the application of the BDq method and the definition of standing wood commercialization criteria have the potential to be applied in exploited tropical forests. This fact is highly promising to keep the forest stand; otherwise, there are serious risks of converting land use to other economic activities [60,61]. This silvicultural management system tends to provide high wood production, while the treated forest maintains its environmental services.

The results found in this study become even more important considering that this region is in the Brazilian Amazon's arc of deforestation [62–64]. In this region, there is great demographic pressure and strong demand for land use in order to develop agricultural and livestock activities.

The deforestation arc is an area of 50 million hectares located in the south and southeast regions of the Amazon and it has high rates of forest degradation due to decades of disorganized and illegal exploitation. The arc's landscapes are composed of a mosaic of lands submitted to agricultural and forestry crops, pastures and exploited forests [65,66].

Financial returns, however, are not the only benefit of applying the BDq method and criteria for the commercialization of standing timber in forests exploited in the Amazon. Fazenda Shet is not part of a public protected area (conservation unit), but of a private area under significant risk of suppression in the face of pressure on alternative land use.

This pressure is due to the advantages of changing the forest cover of the property for land use in more financially competitive activities, such as agriculture and livestock. In this context, the silvicultural management system adopted is considered promising to promote more financially competitive activities in exploited tropical forests in relation to agriculture and livestock.

At this moment, it is necessary to comment on the management of natural forests in the Brazilian Amazon and forests exploited in the past. The forest regulation, according to the current rules, considers a maximum harvest of up to 30 m^3 ha^{-1} in cutting cycles of up to 35 years and a minimum diameter of 50 cm. Within a conservative view, these technical procedures are acceptable for the management of unexploited dense natural forests. However, for the management of already exploited forests or other forest types whenever associated with clearings enrichment, forest regulation should have other criteria that could be similar to the principles of planted forests management such as the use of thinning.

Species that are more commercially known and that have suffered greater pressure due to exploitation, such as *H. courbaril*, *M. huberi* and others with a low growth rate [7], should be harvested with diameters over 50 cm and in cycles of 60 years or more. Intermediate release thinning should occur to reduce competition between trees of different species and improve cash flow, increasing revenues and making management more economically attractive. It is worth mentioning that specific regulations directed to the management of exploited tropical forests are necessary to guarantee their conservation, as well as their productive perpetuity and economic viability.

Finally, it is noteworthy that the identification of a viable solution for the use of degraded forests in eastern Pará is of great importance before they are replaced by other alternatives for using the tarra that are more economically viable. A relevant aspect of our findings in this case study is that there is the possibility of keeping this forest standing productively, through sustainable management and offering wood to local industries. In addition, these forests play an important role in carbon retention in plant biomass and sequestration by faster tree growth, contributing to the mitigation of the effects of climate change.

5. Conclusions

The proposed silvicultural management system has as an object an exploited forest in the Amazon and criteria for harvesting trees with a minimum cut diameter of 25 cm. This management system proved to be viable to generate economic returns, with conservation potential for the continuous supply of forest products and maintenance of the diversity of species forestry. The proposed management system requires normative adjustments in the current legislation for these types of forests, in relation to cutting cycles and technical criteria for harvest planning.

The selection of trees, mainly from market value groups two and three, with a minimum cut diameter of 25 cm in shorter cycles has the potential to promote new commercial species, diversify the income of forest owners and reduce the pressure on more pressed species in the past.

The application of harvesting criteria resulted in positive cost–benefit ratios, being superior to the control treatment in all scenarios.

The set of actions presented is promising to favor the maintenance of biodiversity and expand populations of low density species and the quality of the forest. Besides that, it is more viable for the supply of forest products in a shorter period than that provided for in Brazilian regulations, favoring the viability and the economic sustainability of management.

Exploited tropical forests similar to the one of this study tend to be susceptible to silvicultural interventions in shorter cycles, as long as they are verified to the specificities of each species, as well as the location of the forest and industry. Harvesting must be carried out in areas that are enriched and have a natural regeneration of species with a high tree density, considering the appropriate health conditions of the trees and the biological characteristics of the species.

With the recommendation of solid forest management strategies in the exploited tropical forests worldwide, it is expected to motivate, in a practical way, the maintenance of the forest standing, making it a sustainable financial asset from an environmental, social and economic point of view.

Supplementary Materials: The following are available online at https://www.mdpi.com/article/10.3390/d13110509/s1. Table S1: Species (scientific name and common name) listed by market value group of criteria treatments in the forest census and trees selected for harvest and remaining trees with a dbh ≥25 cm. forest management area (535.6 ha) at Shet Farm. Dom Eliseu. Pará. Brazil. Table S2: Species (scientific name and common name) listed by market value group of control treatments in the forest census and trees selected for harvest and remaining trees with a dbh ≥25 cm. forest management area (535.6 ha) at Shet Farm. Dom Eliseu. Pará. Brazil. Table S3: Species (scientific name and common name) listed description by market value group in the forest at Shet Farm. Dom Eliseu. Pará. Brazil.

Author Contributions: Conceptualization, A.S., M.A.S., J.A.G.Y. and A.R.R.; methodology, A.S., M.A.S., J.A.G.Y. and M.L.d.S.; analysis, A.S., M.A.S., J.A.G.Y., A.R.R. and S.B.V.; resources, A.S. and S.B.V.; data curation A.S. and S.B.V.; writing—original draft preparation, A.S., M.A.S. and S.B.V.; writing—review and editing, A.S. and S.B.V. All authors have read and agreed to the published version of the manuscript.

Funding: This research received no external funding.

Institutional Review Board Statement: Not applicable.

Informed Consent Statement: Not applicable.

Data Availability Statement: Not applicable.

Acknowledgments: To the Rede Biomassa project, Research Support Foundation—FAPESPA, EMBRAPA Amazônia Oriental, Grupo Arboris and Universidade Federal de Viçosa.

Conflicts of Interest: The authors declare no conflict of interest.

References

1. Siviero, M.A.; Lourenço, R.F.; Pacheco, A. *Referência Florestal—Informe Técnico*; Referência Florestal: Curitiba, Brazil, 2007; pp. 45–47. Available online: https://referenciaflorestal.com.br/ (accessed on 28 July 2020).
2. Brasil Resolução No 406 Do Conselho Nacional de Meio Ambiente (CONAMA), 2 de Fevereiro de 2009. Available online: http://http://conama.mma.gov.br/ (accessed on 30 May 2020).
3. Braz, E.M.; Schneider, P.R.; de Mattos, P.P.; Thaines, F.; Selle, G.L.; de Oliveira, M.F.; Oliveira, L.C. Manejo Da Estrutura Diamétrica Remanescente de Florestas Tropicais. *Cienc. Florest.* **2012**, *22*, 787–794. [CrossRef]
4. Sotirov, M.; Arts, B. Integrated Forest Governance in Europe: An Introduction to the Special Issue on Forest Policy Integration and Integrated Forest Management. *Land Use Policy* **2018**, *79*, 960–967. [CrossRef]
5. Sist, P.; Piponiot, C.; Kanashiro, M.; Pena-Claros, M.; Putz, F.E.; Schulze, M.; Verissimo, A.; Vidal, E. Sustainability of Brazilian Forest Concessions. *For. Ecol. Manag.* **2021**, *496*, 119440. [CrossRef]
6. Diário Oficial Da República Federativa Do Brasil. *Brasil Instrução Normativa No 5 Do Ministério Do Meio Ambiente (MMA), de 11 de Dezembro de 2006*; Diário Oficial Da República Federativa Do Brasil: Brasília, Brazil, 2006.
7. Ferreira, T.M.C.; de Carvalho, J.O.P.; Emmert, F.; Ruschel, A.R.; Nascimento, R.G.M. How Long Does the Amazon Rainforest Take to Grow Commercially Sized Trees? An Estimation Methodology for Manilkara Elata (Allemão Ex Miq.) Monach. *For. Ecol. Manag.* **2020**, *473*, 118333. [CrossRef]
8. Gerwing, J.J. Degradation of Forests through Logging and Fire in the Eastern Brazilian Amazon. *For. Ecol. Manag.* **2002**, *157*, 131–141. [CrossRef]
9. Celentano, D.; Rousseau, G.X.; Muniz, F.H.; Varga, I.V.D.; Martinez, C.; Carneiro, M.S.; Miranda, M.V.C.; Barros, M.N.R.; Freitas, L.; Narvaes, I.D.S.; et al. Towards Zero Deforestation and Forest Restoration in the Amazon Region of Maranhão State, Brazil. *Land Use Policy* **2017**, *68*, 692–698. [CrossRef]
10. Bispo, P.D.C.; Pardini, M.; Papathanassiou, K.P.; Kugler, F.; Balzter, H.; Rains, D.; dos Santos, J.R.; Rizaev, I.G.; Tansey, K.; dos Santos, M.N.; et al. Mapping Forest Successional Stages in the Brazilian Amazon Using Forest Heights Derived from TanDEM-X SAR Interferometry. *Remote Sens. Environ.* **2019**, *232*, 111194. [CrossRef]
11. Schwartz, G.; Pereira, P.C.G.; Siviero, M.A.; Pereira, J.F.; Ruschel, A.R.; Yared, J.A.G. Enrichment Planting in Logging Gaps with Schizolobium Parahyba Var. Amazonicum (Huber Ex Ducke) Barneby: A Financially Profitable Alternative for Degraded Tropical Forests in the Amazon. *For. Ecol. Manag.* **2017**, *390*, 166–172. [CrossRef]
12. Siviero, M.A.; Ruschel, A.R.; Yared, J.A.G.; de Aguiar, O.J.R.; Pereira, P.C.G.; Vieira, S.B.; Sales, A. Harvesting Criteria Application as a Technical and Financial Alternative for Management of Degraded Tropical Forests: A Case Study from Brazilian Amazon. *Diversity* **2020**, *12*, 373. [CrossRef]
13. Sist, P.; Ferreira, F.N. Sustainability of Reduced-Impact Logging in the Eastern Amazon. *For. Ecol. Manag.* **2007**, *243*, 199–209. [CrossRef]

14. Poudyal, B.H.; Maraseni, T.; Cockfield, G. Impacts of Forest Management on Tree Species Richness and Composition: Assessment of Forest Management Regimes in Tarai Landscape Nepal. *Appl. Geogr.* **2019**, *111*, 102078. [CrossRef]
15. Bustamante, M. Tropical Forests and Climate Change Mitigation: The Decisive Role of Environmental Governance. *Georget. J. Int. Aff.* 2020. Available online: https://gjia.georgetown.edu/2020/03/20/tropical-forests-climate-change-mitigation-role-of-environmental-governance/ (accessed on 30 August 2021).
16. De Avila, A.L.; Ruschel, A.R.; de Carvalho, J.O.P.; Mazzei, L.; Silva, J.N.M.; Lopes, J.D.C.; Araujo, M.M.; Dormann, C.F.; Bauhus, J. Medium-Term Dynamics of Tree Species Composition in Response to Silvicultural Intervention Intensities in a Tropical Rain Forest. *Biol. Conserv.* **2015**, *191*, 577–586. [CrossRef]
17. Gatica-Saavedra, P.; Echeverría, C.; Nelson, C.R. Ecological Indicators for Assessing Ecological Success of Forest Restoration: A World Review. *Restor. Ecol.* **2017**, *25*, 850–857. [CrossRef]
18. Ciceu, A.; Pitar, D.; Badea, O. Modeling the Diameter Distribution of Mixed Uneven-Aged Stands in the South Western Carpathians in Romania. *Forests* **2021**, *12*, 958. [CrossRef]
19. Meyer, H.A. Structure, Growth, and Drain in Balanced Uneven-Aged Forests. *J. For.* **1952**, *50*, 85–92. [CrossRef]
20. Rangel, M.S.; Calegario, N.; de Mello, A.A.; Lemos, P.C. Melhoria na precisão da prescrição de manejo para floresta natural. *Cerne* **2006**, *12*, 145–156.
21. Glufke, C.; Mainardi, G.L.; Schneider, P.R.; Filho, A.A. Produção de Uma Floresta Natural Em Santa Maria, RS. *Cienc. Florest.* **1994**, *4*, 61–76. [CrossRef]
22. Cunha, U.S. Da Analise da Estrutura Diametrica de Uma Floresta Tropical Umida da Amazonia Brasileira. Ph.D. Thesis, Universidade Federal do Paraná, Curitiba, PR, Brazil, 2013.
23. Pulz, F.A.; Scolforo, J.; Oliveira, A.D.; Mello, J.; Filho, A.T. Acuracidade Da Predição Da Distribuição Diamétrica de Uma Floresta Inequiânea Com a Matriz de Transição. *Cerne* **1999**, *5*, 1–14.
24. Leite, H.G.; Nogueira, G.S.; Campos, J.C.C.; Takizawa, F.H.; Rodrigues, F.L. Um modelo de distribuição diamétrica para povoamentos de Tectona grandis submetidos a desbaste. *Rev. Arvore* **2006**, *30*, 89–98. [CrossRef]
25. Sanquetta, C.R.; Ângelo, H.; Brena, D.A.; Mendes, J.B. Predição da distribuição diamétrica, mortalidade e recrutamento de floresta natural com matriz markoviana de potência. *Floresta* **1994**, *24*. [CrossRef]
26. Campos, J.C.C.; Ribeiro, J.C.; Couto, L. Emprego da distribuição diamétrica na determinação da intensidade de corte em matas naturais submetidas ao sistema de seleção. *Rev. Arvore* **1983**, *7*, 189.
27. De Souza, D.R.; de Souza, A.L. Emprego do método BDq de seleção após a exploração florestal em Floresta Ombrófila Densa de Terra Firme, Amazônia oriental. *Rev. Arvore* **2005**, *29*, 617–625. [CrossRef]
28. Leak, W.B.; Sendak, P.E. Changes in Species, Grade, and Structure over 48 Years in a Managed New England Northern Hardwood Stand. *North J. Appl. For.* **2002**, *19*, 25–28. [CrossRef]
29. Schneider, P.R.; Finger, C.A.G. *Manejo Sustentado de Floresta Inequiânea Heterogênea*; UFSM: Santa Maria, Brazil, 2000.
30. Schwartz, G.; Lopes, J.C.A.; Mohren, G.M.J.; Peña-Claros, M. Post-Harvesting Silvicultural Treatments in Logging Gaps: A Comparison between Enrichment Planting and Tending of Natural Regeneration. *For. Ecol. Manag.* **2013**, *293*, 57–64. [CrossRef]
31. Taffarel, M.; de Carvalho, J.O.P.; Melo, L.D.O.; da Silva, M.G.; Gomes, J.M.; Ferreira, J.E.R. Efeito da silvicultura pós-colheita na população de lecythis lurida (miers) mori em uma floresta de terra firme na amazônia brasileira. *Cienc. Florest.* **2014**, *24*, 889–898. [CrossRef]
32. Sales, A.; Siviero, M.A.; Pereira, P.C.G.; Vieira, S.B.; Berberian, G.A.; Miranda, B.M. Estimation of the Commercial Height of Trees with Laser Meter: A Viable Alternative for Forest Management in the Brazilian Amazon. *Ecol. Evol.* **2020**, *10*, 3578–3583. [CrossRef]
33. Sales, A.; Gonzáles, D.G.E.; Martins, T.G.V.; Silva, G.C.C.; Spletozer, A.G.; Telles, L.A.D.A.; Siviero, M.A.; Lorenzon, A.S. Optimization of Skid Trails and Log Yards on the Amazon Forest. *Forests* **2019**, *10*, 252. [CrossRef]
34. Gomes, J.M.; da Silva, J.C.F.; Vieira, S.B.; de Carvalho, J.O.P.; Oliveira, L.C.L.Q.; de Queiroz, W.T. *Schizolobium parahyba* var. *amazonicum* (Huber ex Ducke) Barneby pode ser utilizada em enriquecimento de clareiras de exploração florestal na Amazônia. *Cienc. Florest.* **2019**, *29*, 417–424. [CrossRef]
35. Vieira, S.B.; de Carvalho, J.O.P.; Gomes, J.M.; da Silva, J.C.F.; Ruschel, A.R. *Cedrela odorata* L. tem potencial para ser utilizada na silvicultura pós-colheita na amazônia brasileira? *Cienc. Florest.* **2018**, *28*, 1230–1238. [CrossRef]
36. Free, C.M.; Landis, R.M.; Grogan, J.; Schulze, M.D.; Lentini, M.; Dunisch, O. NO-VALUE Management Implications of Long-Term Tree Growth and Mortality Rates: A Modeling Study of Big-Leaf Mahogany (Swietenia Macrophylla) in the Brazilian Amazon. *For. Ecol. Manag.* **2014**, *330*, 46–54. [CrossRef]
37. Andrade, C.G.C.; da Silva, M.L.; Torres, C.M.M.E.; Ruschel, A.R.; da Silva, L.F.; de Andrade, D.F.C.; Reis, L.P. Crescimento diamétrico e tempo de passagem de Minquartia guianensis após manejo na Floresta Nacional do Tapajós. *Pesqui. Florest. Bras.* **2017**, *37*, 299–309. [CrossRef]
38. Alvares, C.A.; Stape, J.L.; Sentelhas, P.C.; de Moraes Gonçalves, J.L.; Sparovek, G. Köppen's Climate Classification Map for Brazil. *Meteorol. Z.* **2013**, 711–728. [CrossRef]
39. IDESP Estatística Municipal de Dom Eliseu. O Instituto de Desenvolvimento Econômico, Social e Ambiental do Pará. IDESP: Belém, Brazil, 2014. Available online: http://www.fapespa.pa.gov.br/upload/Arquivo/anexo/626.pdf?id=1499528547 (accessed on 1 March 2020).

40. IBGE. *Manual Técnico Da Vegetação Brasileira: Sistema Fitogeográfico, Inventário Das Formações Florestais e Campestres, Técnicas e Manejo de Coleções Botânicas, Procedimentos Para Mapeamentos*; Instituto Brasileiro de Geografia e Estatística: Rio de Janeiro, Brazil, 2012. Available online: https://biblioteca.ibge.gov.br/index.php/biblioteca-catalogo?view=detalhes&id=263011 (accessed on 14 March 2020).
41. Dos Santos, H.G.; Jacomine, P.K.T.; dos Anjos, L.H.C.; de Oliveira, V.A.; Lumbreras, J.F.; Coelho, M.R.; de Almeida, J.A.; de Araujo Filho, J.C.; de Oliveira, J.B.; Cunha, T.J.F. *Sistema Brasileiro de Classificação de Solos*; Brasília, DF: Embrapa, Brazil, 2018; ISBN 978-85-7035-817-2.
42. Siviero, M.A. *Referência Florestal—Artigo Técnico*; Referência Florestal: Curitiba, Brazil, 2011; pp. 94–96. Available online: https://referenciaflorestal.com.br/ (accessed on 27 July 2020).
43. Sabogal, C.; Silva, J.N.M.; Zweede, J.; Júnior, R.P.; Barreto, P.; Guerreiro, C.A. Diretrizes Técnicas Para a Exploração de Impacto Reduzido em Operações Florestais de Terra Firme na Amazônia Brasileira. 2000. Available online: https://www.embrapa.br/busca-de-publicacoes/-/publicacao/389669/diretrizes-tecnicas-para-a-exploracao-de-impacto-reduzido-em-operacoes-florestais-de-terra-firme-na-amazonia-brasileira (accessed on 22 May 2020).
44. Figueiredo Filho, A.; Dias, A.; Stepka, T.; Sawczuk, A. Crescimento, mortalidade, ingresso e distribuição diamétrica em floresta ombrófila mista. *Floresta* **2010**, *40*. [CrossRef]
45. Sousa, A.L.; Soares, C.P.B. *Florestas Ativas: Estrutura, Dinâmica e Manejo*; Editora UFV: Viçosa, Brazil, 2013.
46. Pára (Estado) IN 05 de 10/09/2015. Publicada No DOE 32969 de 11/09/2015. 2015. Available online: https://www.semas.pa.gov.br/2015/09/11/in-05-de-10092015-publicada-no-doe-32969-de-11092015-paginas-de-37-57/ (accessed on 6 January 2020).
47. Rezende, J.L.P.; Oliveira, A.D. *Análise Econômica e Social de Projetos Florestais*, 3rd ed.; Editora UFV: Viçosa, Brazil, 2013.
48. Das Virgens, A.P.; de Freitas, L.C.; Leite, Â.M.P. Análise Econômica e de Sensibilidade em um Povoamento Implantado no Sudoeste da Bahia. *Floresta E Ambiente* **2016**, *23*, 211–219. [CrossRef]
49. Liocourt, F.D. De l'aménagement Des Sapinières. *Bull. Trimest. Soc. For. Fr. Comte Belfort* **1898**, 396–409. Available online: https://www.worldcat.org/title/de-lamenagement-des-sapinieres-signe-f-de-liocourt/oclc/457477176 (accessed on 23 January 2020).
50. Araujo, H.J.B. de Inventário florestal a 100% em pequenas áreas sob manejo florestal madeireiro. *Acta Amaz.* **2006**, *36*, 447–464. [CrossRef]
51. Cavalcanti, F.J.D.B.; Machado, S.D.A.; Osokawa, R.T.; da Cunha, U.S. Comparação dos valores estimados por amostragem na caracterização da estrutura de uma área de floresta na Amazônia com as informações registradas no censo florestal. *Rev. Arvore* **2011**, *35*, 1061–1068. [CrossRef]
52. Pereira, P.C.G. Potencial Silvicultural Das Espécies Do Gênero Cecropia Na Amazônia. Ph.D. Dissertation, Universidade Federal Rural da Amazônia, Belém, Brazil, 2015. Available online: https://sigaa.ufra.edu.br/sigaa/public/programa/noticias_desc.jsf?lc=pt_BR&id=109¬icia=4973247 (accessed on 23 June 2020).
53. Fontes, C.G. Revelando as Causas e a Distribuição Temporal da Mortalidade Arbórea em Uma Floresta de Terra-Firme na Amazônia Central. Ph.D. Thesis, Instituto Nacional de Pesquisas da Amazônia, Manaus, Brazil, 2012. Available online: http://https://bdtd.inpa.gov.br/handle/tede/2574 (accessed on 23 June 2020).
54. Garcia, L.C.; de Sousa, S.G.A.; Lima, R.M.B. De Seleção de Matrizes, Coleta e Manejo de Sementes Florestais nativas da Amazônia. 2011. Available online: https://www.embrapa.br/busca-de-publicacoes/-/publicacao/906646/selecao-de-matrizes-coleta-e-manejo-de-sementes-florestais-nativas-da-amazonia (accessed on 23 June 2020).
55. Peña-Claros, M.; Peters, E.M.; Justiniano, M.J.; Bongers, F.; Blate, G.M.; Fredericksen, T.S.; Putz, F.E. Regeneration of Commercial Tree Species Following Silvicultural Treatments in a Moist Tropical Forest. *For. Ecol. Manag.* **2008**, *255*, 1283–1293. [CrossRef]
56. Villa, P.M.; Martins, S.V.; de Oliveira Neto, S.N.; Rodrigues, A.C.; Safar, N.V.H.; Monsanto, L.D.; Cancio, N.M.; Ali, A. Woody Species Diversity as an Indicator of the Forest Recovery after Shifting Cultivation Disturbance in the Northern Amazon. *Ecol. Indic.* **2018**, *95*, 687–694. [CrossRef]
57. Schwaiger, F.; Poschenrieder, W.; Biber, P.; Pretzsch, H. Ecosystem Service Trade-Offs for Adaptive Forest Management. *Ecosyst. Serv.* **2019**, *39*, 100993. [CrossRef]
58. Siviero, M.A.; Ruschel, A.R.; Yared, J.A.G.; Pereira, J.F.; de Aguiar, O.J.R.; Junior, S.B.; Pereira, P.C.G.; Vieira, S.B.; Contini, K.P.S.; Sales, A. Manejo de florestas naturais degradadas na Amazônia: Estudo de caso sobre critérios de colheita. *Cienc. Florest.* **2020**, *30*, 43–59. [CrossRef]
59. Vieira, S.B. *Sustentabilidade da Produção Madeireira de Goupia Glabra Aubl. (cupiúba) na Amazônia Brasileira*; Universidade Federal do Pará—Instituto de Geociências: Belém, Brazil, 2019.
60. Paquette, A.; Hawryshyn, J.; Senikas, A.; Potvin, C. Enrichment Planting in Secondary Forests: A Promising Clean Development Mechanism to Increase Terrestrial Carbon Sinks. *Ecol. Soc.* **2009**, *14*. [CrossRef]
61. Keefe, K.; Alavalapati, J.A.A.; Pinheiro, C. Is Enrichment Planting Worth Its Costs? A Financial Cost–Benefit Analysis. *For. Policy Econ.* **2012**, *23*, 10–16. [CrossRef]
62. Fearnside, P.M. Deforestation in Brazilian Amazonia: History, Rates, and Consequences. *Conserv. Biol.* **2005**, *19*, 680–688. [CrossRef]
63. Costa, L.; Miranda, I.; Grimaldi, M.; Silva, M., Jr.; Mitja, D.; Lima, T. Biomass in Different Types of Land Use in the Brazil's 'Arc of Deforestation'. *For. Ecol. Manag.* **2012**, *278*, 101–109. [CrossRef]

64. Schwartz, G.; Ferreira, M.D.S.; Lopes, J.D.C. Silvicultural Intensification and Agroforestry Systems in Secondary Tropical Forests: A Review. *Rev. Cienc. Agrar. Amaz. J. Agric. Environ. Sci.* **2015**, *58*, 319–326. [CrossRef]
65. Lui, G.H.; Molina, S.M.G. Ocupação humana e transformação das paisagens na amazônia brasileira. *Amaz. Rev. Antropol.* **2016**, *1*. [CrossRef]
66. Soares-Filho, B.S.; Nepstad, D.C.; Curran, L.; Cerqueira, G.C.; Garcia, R.A.; Ramos, C.A.; Voll, E.; McDonald, A.; Lefebvre, P.; Schlesinger, P.; et al. Cenários de desmatamento para a Amazônia. *Estud. Av.* **2005**, *19*, 137–152. [CrossRef]

Article

Perceptions of Local Inhabitants towards Land Management Systems Used in the Rainforest Area of Ecuador: An Evaluation Based on Visual Rating of the Main Land Use Types

Alex Vinicio Gavilanes Montoya [1,2], Danny Daniel Castillo Vizuete [1,2,*] and Stelian Alexandru Borz [2,*]

[1] Faculty of Natural Resources, Escuela Superior Politécnica de Chimborazo, Panamericana Sur, Km 1 1/2, Riobamba EC-060155, Ecuador; a_gavilanes@espoch.edu.ec

[2] Department of Forest Engineering, Forest Management Planning and Terrestrial Measurements, Faculty of Silviculture and Forest Engineering, Transilvania University of Brasov, Şirul Beethoven 1, 500123 Brasov, Romania

* Correspondence: danny.castillo@espoch.edu.ec (D.D.C.V.); stelian.borz@unitbv.ro (S.A.B.); Tel.: +593-987-712-497 (D.D.C.V.); +40-742-042-455 (S.A.B.)

Citation: Gavilanes Montoya, A.V.; Castillo Vizuete, D.D.; Borz, S.A. Perceptions of Local Inhabitants towards Land Management Systems Used in the Rainforest Area of Ecuador: An Evaluation Based on Visual Rating of the Main Land Use Types. *Diversity* **2021**, *13*, 592. https://doi.org/10.3390/d13110592

Academic Editors: Michael Wink, Lucian Dinca and Miglena Zhiyanski

Received: 15 September 2021
Accepted: 16 November 2021
Published: 18 November 2021

Publisher's Note: MDPI stays neutral with regard to jurisdictional claims in published maps and institutional affiliations.

Abstract: Land management policy and practice affects a wide segment of stakeholders, including the general population of a given area. This study evaluates the perceptions of local inhabitants towards the land management systems used in the rainforest area of Ecuador—namely, unmanaged (natural) forest, managed forest, croplands, and pasturelands. Data collected as ratings on 12 pictures were used to check the aggregated perceptions by developing the relative frequencies of ratings, in order to see how the perception rating data were associated with the types of land management systems depicted by the pictures, and to see whether the four types of land management could be mathematically represented by a clustering solution. A distinctive result was that the natural forests were the most positively rated, while the managed forests were the least positively rated among the respondents. It seems, however, that human intervention was not the landscape-related factor affecting this perception, since croplands and pasturelands also received high ratings. The ratings generated a clear clustering solution only in the case of forest management, indicating three groups: natural forests, managed forests, and the rest of the land management systems. Based on the results of this study, a combination of the four land use systems would balance the expectations of different stakeholders from the area, while also being consistent to some extent with the current diversity in land management systems. However, a more developed system of information propagation would be beneficial to educate the local population with regards to the benefits and drawbacks of different types of land management systems and their distribution.

Keywords: visual perception; rainforest; land use; aesthetics; Amazon; management implications

1. Introduction

Forests sustain human societies via the provision of a wide range of ecosystem services [1] that are essential for the existence and wellbeing of humans [2]. Forest ecosystems have been always seen by humans as an important source of goods and services, which have supported human existence [3]. However, the sustainable management of forest ecosystems entails a great challenge for all those involved, requiring large spatial and temporal scales [4].

There are nine forest types in Ecuador, one of which is the rainforest [5]. Rainforests are biodiverse, and play an important role in the functioning of the Earth [6,7], accounting for ~36% (14.5 million km^2) of the world's forested areas [8]. The forests of Ecuador are spread over 12.5 million ha, and rainforests account for 42.32% of the country's forested area [9,10]; they provide a wide range of services—such as food, energy, products for medical care, and building materials [11]—for the benefit of their owners and users [12]. In addition, for local communities, forests hold an intrinsic sociocultural value [13], which is why the forest management approach is increasingly considered through the framework of environmental policies [14]. The sustainable management of forest ecosystems requires the legitimacy of key political actors and an integrated support framework in the form of governmental and non-governmental institutions [15,16]. Involving communities in these processes is considered to be one of the key factors for effective forest governance [17]—even more so when the indigenous people are increasingly convinced of their role as guardians of the forest [18]. However, the lack of support policies for land tenure and ownership in local communities prevents the fulfillment of the objectives of sustainable forest management in certain cases [19,20]. For this reason, different forms of forest management have emerged, through which forest authorities work together with local governments and communities to make decisions on forest management, via a participatory approach [21]. The participatory approach to land management planning is seen as a fundamental way to involve local people in the planning process by integrating local knowledge and perspectives [22].

Among the participatory methods for the management of forest ecosystems, and for valuing the landscapes, are those of visual aesthetics. Globally, visual aesthetics is considered an important resource for the assessment of ecosystems, and it is a reliable method used to increase the visual quality of a landscape by design and management activities [23,24]. Visual assessment of landscape quality, or "visual assessment", refers to the procedures implemented to characterize the scenic beauty of landscapes [25]. Accordingly, visual assessment places value on beauty, and identifies key aspects of the landscapes [26]. Aesthetic evaluation is known to be divided into two approaches: the objective approach, which is supported by the physical paradigm, and the subjective approach, which is supported by the psychological paradigm [25,27]. The objective approach considers aesthetic quality as an intrinsic attribute of a landscape, while the subjective approach assumes that aesthetic quality is a subjective value shaped in the evaluator's mind [27,28]. However, many of those who conducted research on the evaluation of aesthetic preferences believe that aesthetic quality is located at the interaction between the physical and psychological characteristics of the landscape and of those who evaluate the landscape (e.g., [28–31]). In addition, taking into account the landscape's quality can increase our understanding of how landscapes change [32]. Today, in environmental planning and management, the approach of preserving the aesthetic diversity of landscapes has become an important part of decision making [33]. On the other hand, the definition of landscape has evolved over the years, so as to focus it on aesthetic values, resources, and the combination of physical, biological, ecological, and human components. The landscape can now be seen as the scene of human activity, and any artificially induced change may affect human perceptions of it. The actions that have generated landscape losses have been mainly anthropogenic, and they have caused a growing and rapid deterioration of the landscape's quality. Therefore, a continuous evaluation of the landscape is required—particularly to measure the state of degradation or improvement of ecosystems. Consequently, data are needed to support the implementation of actions by local governments and competent entities to improve the state of the landscapes.

The aim of this study was to evaluate the perceptions of local inhabitants towards the land management systems used in the rainforest area of Ecuador, based on visual rating of the main land use types. In this sense, the first objective of this study was to evaluate the visual perceptions of local inhabitants with regard to different types of land management systems—namely, natural forests, managed forests, croplands, and pasturelands—by a picture-rating approach. The second objective was to check whether there are important associations of the locals' ratings with the type of land management system, while the third objective of the study was to check whether the locals' ratings shape individualized groups of land management as a collective reflection over the rated land management systems. Potential effects of sociodemographic features, as well as some implications of the study's findings, are also discussed.

2. Materials and Methods

2.1. Study Location

Simón Bolívar Parish (SBP), which is located in the Pastaza Province of the Ecuadorian Amazon Region (Figure 1)—a place with a high biodiversity in the tropical forest—was chosen as the area of study. According to Gavilanes et al. [34], the province is quite abundant in species, as it hosts 540 plants species, of which 507 are native and 12 are included in the endemic group [34]. The choice of this area for study was based on a specific distribution, where the primary (natural, native) forest accounts for 84.31% of the total area of SBP [35], and which is representative for the rest of the rainforest area from Ecuador. The primary (natural) forest is commonly described as a place with a high species richness and without a significant presence of human activities [36]. In addition, the Amazon Region encompasses approximately 75% of the total forested area of Ecuador [37]. The main forest ecosystems in the study area are (1) evergreen forest of the Northern Oriental Mountain Range of the Andes (53.13%); (2) evergreen lowland forest of the "Tigre–Pastaza" Basin (28.56%); (3) floodplain forest of the alluvial plain of the rivers of Andean origin and of the Amazonian mountain ranges (15.72%); (4) floodplain forest with palms from the alluvial plain of the Amazon (0.57%), and (5) flooded forest of the alluvial plain of rivers of Amazonian origin (0.42%) [35].

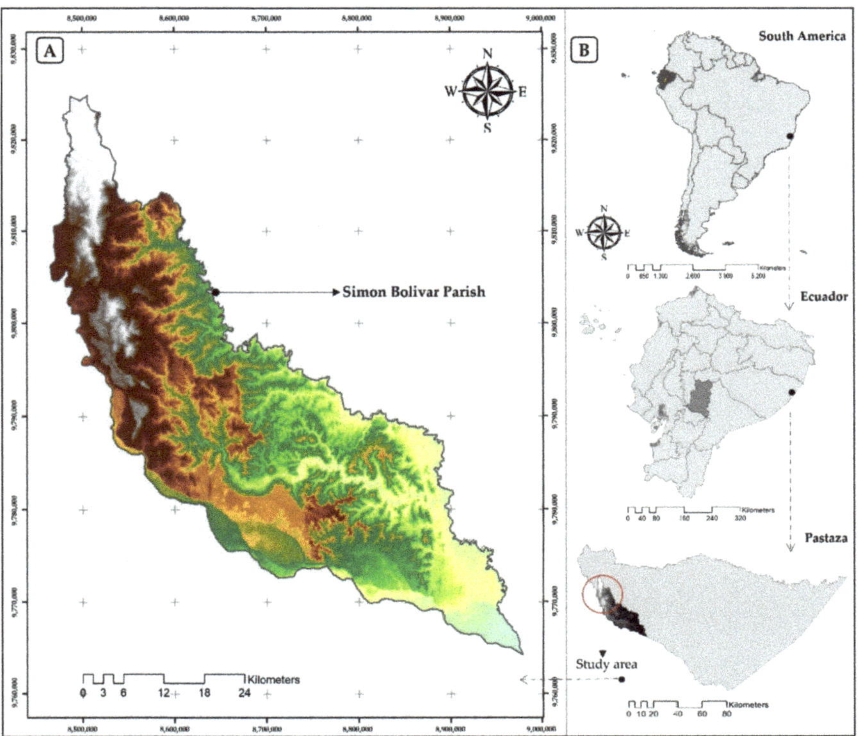

Figure 1. Map of the study area. (**A**) Simon Bolivar Parish location (**B**) location in relation to South America, Ecuador and Pastaza Province.

Ecuador holds natural forests that cover 12,631,198 ha, and which are distributed across 65 locations [38]. The main types of land management systems of SBP are (1) primary (natural) forests, which correspond to rainforest with a coverage of 92,716.38 ha; (2) pasturelands, which include pastures and other systems that are used for cattle breeding (6,831.84 ha); (3) croplands (477.35 ha), as in the study area extensive agriculture is currently used to cultivate cassava, sugar cane, cocoa, and bananas; and (iv) secondary (managed) forest, which is represented by species such as bamboo and palms, with a coverage of 46.44 ha [39].

The majority of SBP's population (67.41%) carries out economic activities related to agriculture, livestock farming, forestry, and fishing. Consequently, the future trend of increasing the categories of land management systems—such as crops, pastures, and inhabited areas—will be maintained at least proportional to the increase in the population and to the development of economic activities, which implies a reduction in the forested area. For instance, in Ecuador, the loss of vegetation cover is frequently associated with anthropogenic activity, population density, dependence on the quality of the land, accessibility, and the level of education [40]. Table 1 describes the main stakeholders in the area of study, their roles related to the land management systems, and their competencies to manage them. These stakeholders evaluated the natural resources of the study area with reference to their knowledge of the local environment, its benefits, and associated activities, as well as the land management systems of the territory.

Table 1. Description of the stakeholders from the study area.

Name	Roles	Decision Making	Use of Resources	Management of Resources
Government of SBP	Activities related to the parish and land planning.	X	X	X
Department of public irrigation of SBP	Controls the use of irrigation water and its rates.		X	X
Public department of potable water	Provision and quality control of water for human use.		X	X
Municipality of Puyo Canton	Plans the categories of land management systems and occupation to regulate the activities to be developed.	X		X
Prefecture of Pastaza	Contributes to the productivity and environment. Develops programs to promote agriculture, conservation, and aquaculture.	X		X
Population of SBP	Consumes the natural resources.		X	X
Farmer organizations	Contribute to decision making in agricultural activities.		X	X
Simón Bolivar Church	Religious components linked to environmental practices.	X		

According to the latest SBP population census, 78.08% of the population was identified in different indigenous ethnic groups, such as the Awa, Achuar, Cofan, Secoya, Shiwiar, Shuar, Waorani, Zapara, Andoa, Kichwa de la Sierra, and Manta. These ethnic groups have occupations mainly related to agriculture, livestock farming, forestry, and fishing [35]. The relationship between biodiversity and the these ethnic groups is based on the values provided by rainforests in biological, ethnobotanical, economic, and cultural terms [41].

2.2. Questionnaire Survey

The research was based on a questionnaire survey implemented through face-to-face interviews with randomly chosen representatives of SBP's local population. The choice of face-to-face interviews was aimed at reducing the incorrect completion of the questionnaires. The original version of the questionnaire was developed by considering five main sections—namely, the (1) sociodemographic features; (2) local context related to the level of importance of rainforests and water resources; (3) socioeconomic component; (4) environmental and cultural components, which included the visual rating of the main land management system types; and (5) willingness to pay for conservation and other attributes. The data on sociodemographic features and visual ratings were used as a baseline for this study. The preliminary version of the questionnaire was tested and refined by the personnel from "Escuela Superior Politecnica de Chimborazo" before being used in the data collection. To estimate the sample size, a probabilistic formula was used at a confidence level of 95%, using as input the population of SBP (8348 inhabitants according to the management plan [35]), which resulted in 368 questionnaires to be implemented. However, a total of 451 questionnaire-based interviews were applied in the field, in order to provide a sufficient pool of respondents, assuming that in some of them the data would be incomplete. The field phase of the study was implemented with the support of 30 researchers who were trained in advance and had an academic background in environmental engineering. The respondents were selected so as to be people over the age of 18 years or the heads of their families.

For the visual perception section, 12 pictures were randomly selected from a larger picture pool and shown to the respondents in order to evaluate how much they liked each of them, using a Likert scale from 1 to 5, where 1 stood for "not at all" and 5 for "very much". The pictures emulated three types of perspective (landscape view, close view, and inside view), and they were shown in three groups and ordered sequentially according to the main land management systems and economic activities of the population described in the management plan [35]: (1) primary (natural) or unmanaged forest, (2) secondary or managed forest, (3) pasturelands, and (4) croplands. Table A1 depicts and describes the pictures shown to the respondents in more detail. The data used in this study were part of a larger study dealing with both subjective and objective views of the inhabitants with regard to the evaluated landscapes [42].

2.3. Data Processing and Statistical Analysis

The data collected via pen-and-paper questionnaires were transferred into a database developed in Microsoft Excel® in the form of binary codes, in order to indicate the presence or absence of given sociodemographic attributes and the ratings given by the respondents. Then, the data were checked for consistency so as to identify the data coming from those questionnaires that were completely filled in. Incomplete data were removed, and only the data from 376 valid questionnaires were retained for further processing, representing ca. 83.4% of the initial sample (451 questionnaires implemented in the field). These data were recoded and reorganized based on sociodemographic features and ratings given at the picture level.

Sociodemographic characteristics were analyzed by the means of absolute and relative frequency of data—a step which was implemented for each sociodemographic feature. For gender, civil status, and age, the number of attributes was kept the same as in the case of those included in the original questionnaire. However, due to a low frequency of responses to some of the attributes, the data on ethnicity and monthly income were reorganized; in addition, for the sake of simplicity, the data on education were reorganized so as to reflect only the main categories, irrespective of the completion of a given education level—a procedure that was also applied to the data on occupation, with the aim being to reflect only the main categories.

Ratings of the respondents were analyzed by the use of relative frequency of responses given on the Likert scale (1 to 5), at two levels of aggregation: The first level was that of the entire respondent cohort, for which the perceptions of the land management systems were characterized by the relative frequencies of ratings from 1 to 5 computed for the 12 pictures shown to the respondents. The second level of aggregation was that characterizing the reorganized attributes of sociodemographic data; as such, relative frequencies of each response given on the Likert scale were computed for the gender (i.e., female vs. male respondents), civil status, age, education level, ethnicity, monthly income, and occupation attributes. For both levels of data aggregation, the results were prepared in the form of radar plots.

Inherently, the used data were available in a multidimensional form. To be able to check the data association and similarity with regard to the instances (pictures) and the ratings (1 to 5), a dimensionality reduction procedure was required. Given the type of the data (categorical), a correspondence analysis was carried out based on a contingency table (5×12), which was developed to contain the frequency of ratings (1 to 5) as rows

(5 rows) and the picture numbers (P1 to P12) as columns (12 columns). The workflow was implemented in R Studio, based on the tutorial available at [43], with the aims of (1) checking whether there were dependencies between the rows and columns (based on the χ^2 statistic), (2) choosing the number of dimensions to characterize the data in a lower dimensional space (based on the inertia and explained variance), (3) characterizing the contribution of row and column data to the developed dimensional solution, and (4) characterizing the data association via symmetrical and asymmetrical biplots.

To answer to the question of whether the respondents' ratings in terms of visual perception would shape individual and cohesive groups of land use management, irrespective of the pictures' scale of view, a hierarchical cluster analysis was implemented by the use of Orange Visual Programming software [44]. For this purpose, the Excel database was fed into a workflow that aimed at computing a distance matrix, a distance map, and a cluster solution, all of which were based on the ratings given by the respondents. Among the parameters used to reach a clustering solution were Spearman's dissimilarity index, which was used as a distance metric, and the complete-linkage algorithm for hierarchical clustering. The choice of Spearman's dissimilarity metric was based on its ability to work with categorical, ranked data, as it represents the square of Euclidian distance applied to given rank vectors. This metric outputs the linear correlation between the rank values remapped as distances in an interval from 0 to 1, where 0 means that there is a perfect match between two given features, while 1 means that the features are dissimilar. The use of the complete-linkage algorithm was chosen as an intermediary solution, mainly to avoid the chaining effect specific to the single-linkage algorithm. The final cluster solution was chosen based on the data grouping so as to reflect the main land use types; to do so, several clustering distances were tested, and the solutions outputted by them were visually evaluated.

3. Results

3.1. Sociodemographic Characteristics

The main statistics of the sample's sociodemographic characteristics are given in Table 2. The share of female (ca. 49%) and male (ca. 51%) respondents was balanced in the sample under study.

More than half of the respondents (ca. 55%) declared to be engaged in some sort of relationship, and most of them (ca. 68%) were aged less than 40 years. Dominant in the sample were those having completed or still studying at the high school level (47%), as well as those belonging to the Metis ethnic group (ca. 75%).

The monthly income was consistent with the wealth of Ecuador [45], and most of the respondents declared that they had a monthly income of less than USD 394. The majority (ca. 86%) of the questioned people had some sort of occupation, such as being employed (ca. 61%) or house care (ca. 25%). According to the statistics available in June 2021, poverty at the national level stood at 32.2%, and extreme poverty at 14.7% of the population, taking into account that a person is considered poor if they have a per capita family income of less than USD 84.71 per month, and extremely poor for an income less than USD 47.74 per month [45]. In addition, according to the National Institute of Statistics and Censuses, the characteristics of employment in the rural sector are described in the following groups: formal sector (21.2%), informal sector (71.6%), domestic employment (1.1%), and not classified by sector (6.1%) [46]. Accordingly, the national data were consistent with the information obtained from the study area, which was representative in terms of sociodemographic statistics.

Table 2. Main statistics of sociodemographic characteristics.

Feature	Item	Absolute Frequency	Relative Frequency (%)
Gender	Female	184	48.9
	Male	192	51.1
Civil status	Single	134	35.6
	Common law	82	21.8
	Married	123	32.7
	Divorced	25	6.6
	Widow(er)	12	3.2
Age	Less than 30 years old	155	41.2
	30–41 years old	103	27.4
	41–50 years old	52	13.8
	51–60 years old	36	9.6
	More than 60 years old	30	8.0
Education	Elementary	128	34.0
	High school	178	47.3
	Bachelor's degree	64	17.0
	Master's degree or more	6	1.6
Ethnicity	Caucasian	6	1.6
	Indigenous	90	23.9
	Metis and Other	280	74.5
Monthly income	Less than USD 394	270	71.8
	USD 395–793	66	17.6
	USD 794–901	21	5.6
	More than USD 901	19	5.1
Occupation	Employed	230	61.2
	House care	92	24.5
	Student	39	10.4
	Unemployed	15	4.0

3.2. Aggregated Frequencies of Ratings

Figure 2 shows the relative frequency of ratings aggregated at the picture level for the sample of respondents under study. For interpretation, in the following section, some of the results are discussed as positive (ratings of 4 and 5), neutral (rating of 3), and negative (ratings of 1 and 2) ratings.

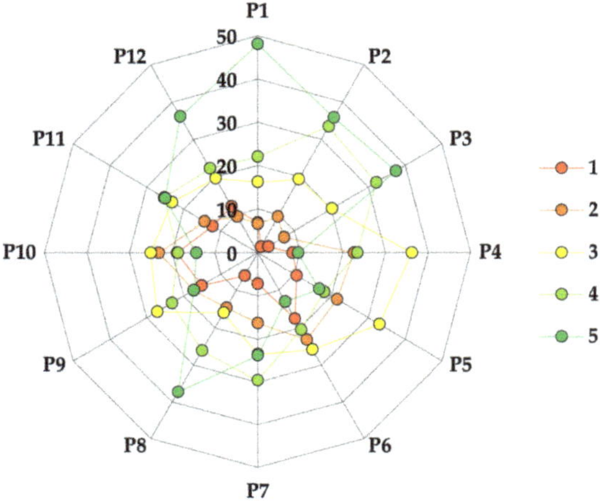

Figure 2. Relative frequencies of ratings at the picture level, aggregated for the sample under study. Legend: 1 to 5 stand for ratings of 1 to 5, P1 to P12 stand for Pictures 1 to 12.

Picture 1, depicting a natural forest landscape, was rated by more than 70% of the respondents as being liked, by ca. 16% as being neutral, and by close to 14% as not being liked. Picture 1 received the most ratings of 5 (ca. 48%), being followed in this regard by Picture 3 (ca. 38%), Picture 8 (ca. 37%), Picture 12 (ca. 36%), and Picture 2 (ca. 36%). At the opposite end of the spectrum, pictures depicting managed forests (P4–P6) received the least positive ratings (by 33, 35, and 34% of the respondents, respectively), and neutral responses dominated the responses with regard to this type of land use management system.

Among the set of pictures describing the continuum in the land use management system, only Pictures 1 to 3 (native forests), 7 and 8 (croplands), and 11 and 12 (pasturelands) were rated positively by more than 50% of the respondents. In the case of pictures depicting native forests, these positive ratings were given by ca. 69–70% of the respondents, while in the case of pictures 7 and 8, depicting croplands, the positive ratings were given by 53 and 63% of the respondents, respectively. Last, but not least, in the case of pictures 11 and 12, positive ratings were given by 50 and 58% of the respondents, respectively.

A rating pattern similar to that shown in Figure 2 was preserved in the case of data that were aggregated at the gender level (Figure A1 in Appendix A), with the difference being that female respondents rated Picture 8 (cropland) better, while male respondents rated the pictures depicting natural forests better. In relation to ethnic groups, the data shown in Figure A2 indicate that Metis and indigenous groups had similar evaluation patterns, with Pictures 1 and 3 (natural forest) being the best rated, while Caucasians gave the highest ratings for Picture 8 (croplands). The ratings associated with marital status are shown in Figure A3, showing that Picture 1 was the best evaluated; however, the group of widowers mostly preferred Picture 8 (croplands) and Picture 12 (pasturelands). Figure A4 shows the results in relation to age category. Those aged less than 30, as well as those aged 41–50 years, gave better ratings to Pictures 1–3 (natural forest) and Picture 8 (croplands). In addition to these ratings, the group of 31–40-year-old respondents positively rated Pictures 11 and 12 (pasturelands). Finally, for the group of those over 60 years old, the best evaluated landscapes were croplands (Picture 8) and pasturelands (Picture 12). Figure A5 shows a relationship between the ratings and education level. The ratings of the primary forest (Pictures 1–3) were correlated with the educational level, being higher as the respondents declared a higher educational level. Figure A6 shows the ratings given by different groups in relation to their occupation. Employees gave the best ratings to Picture 1, while those from the house care group gave the best ratings to the natural forests (Pictures 1–3), croplands (Picture 8), and pasturelands (P12). On the other hand, students highly rated the natural forests (Pictures 1–3) and croplands (Picture 7). Ratings of the groups by monthly income are shown in Figure A7, indicating a trend of highly rating the native forests (Pictures 1–3), with the exception of those with a salary less than USD 394, who also positively rated the agricultural activities (Picture 8).

3.3. Association between Land Use Management Systems and Ratings

The main results of the correspondence analysis are shown in Figures 3 and 4, in the form of symmetrical and asymmetrical biplots complemented with the proportion of explained variance and with the contributions of the row and column data to the dimensions. Finally, a solution with two dimensions was kept, explaining more than 94% of the data variability. Some interesting findings can be seen from a closer look at Figures 3 and 4. Informatively, P1 was the most associated with the rating of 5 (Figure 3a), and it was represented at a considerable distance from its counterparts in the native forest category. P11 was the closest to the average profile of responses and, in general, the grouping of the data on the four types of land management systems was not cohesive.

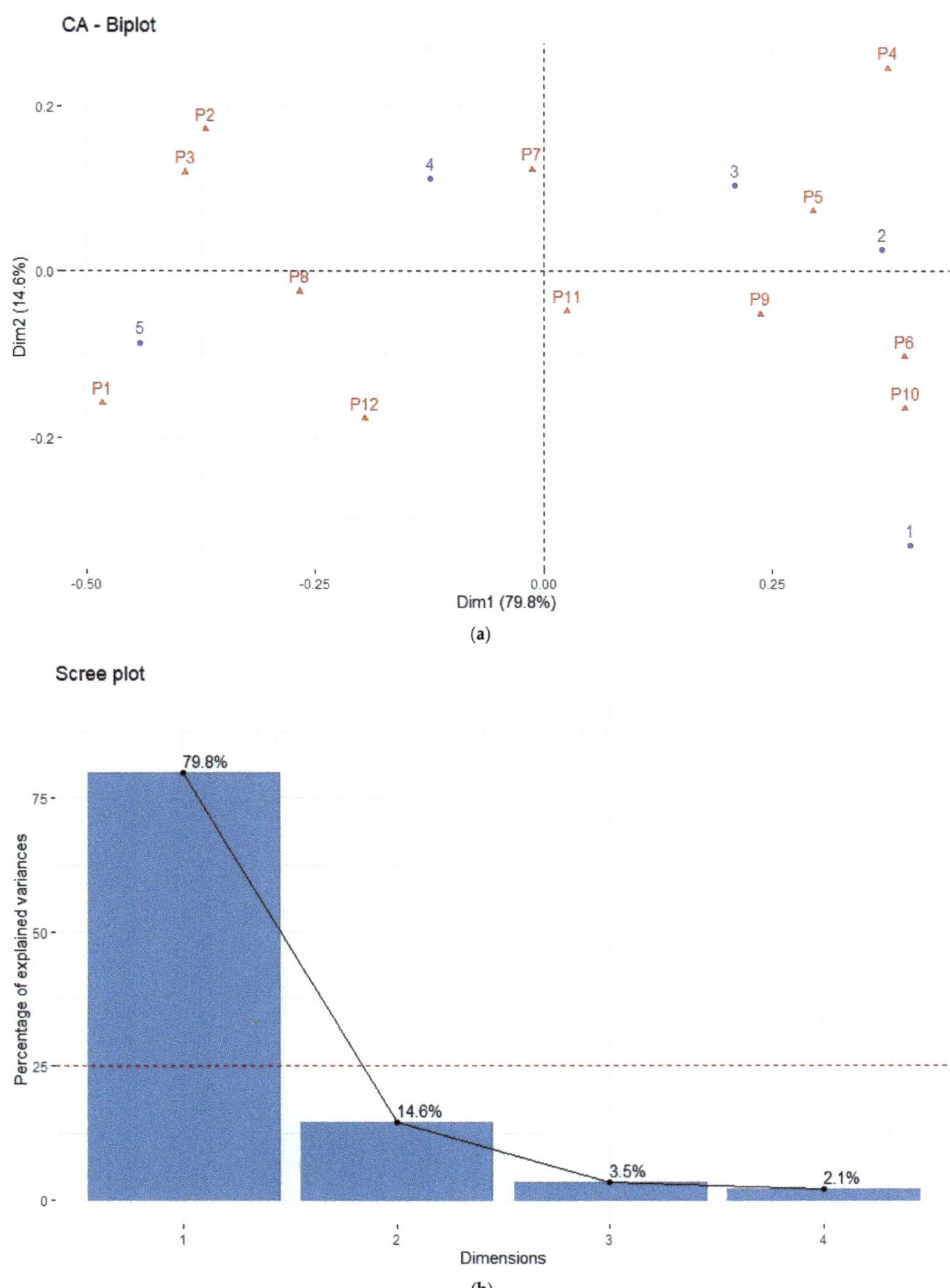

Figure 3. Symmetrical biplot of the correspondence analysis (**a**) and proportion of explained variance (**b**). Legend: blue dots (rows) represent the ratings given, and red triangles (columns) represent the rated pictures.

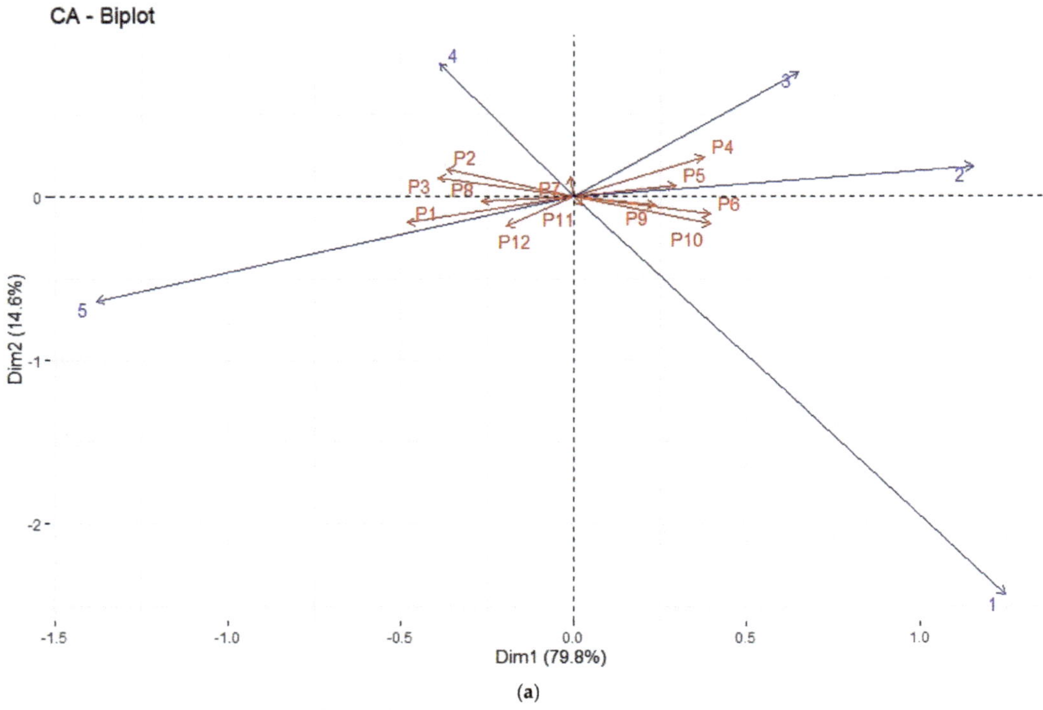

(a)

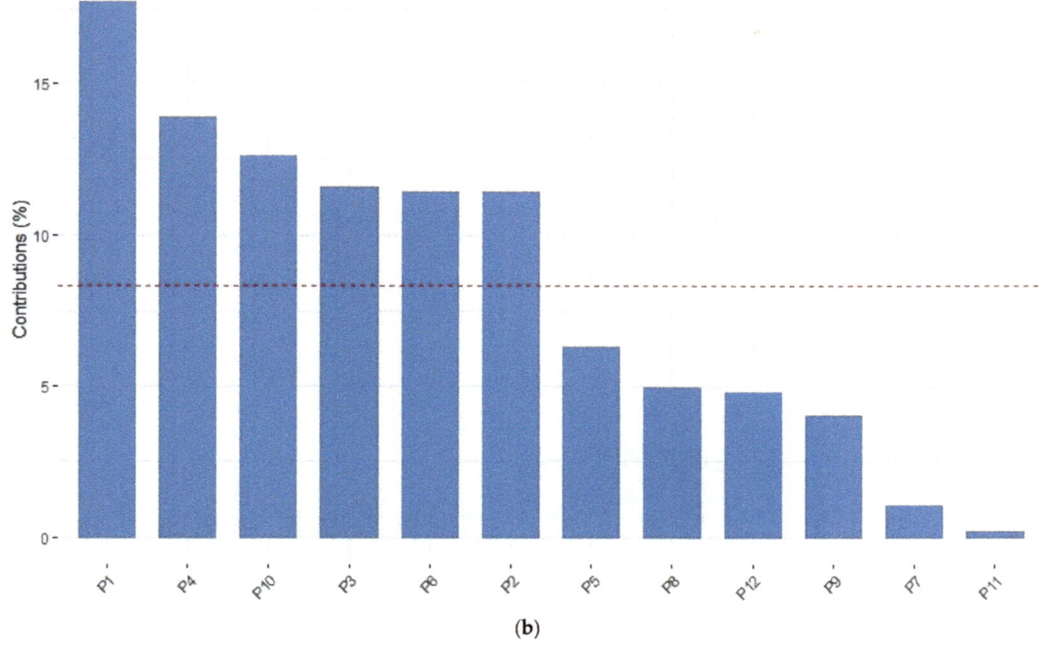

(b)

Figure 4. *Cont.*

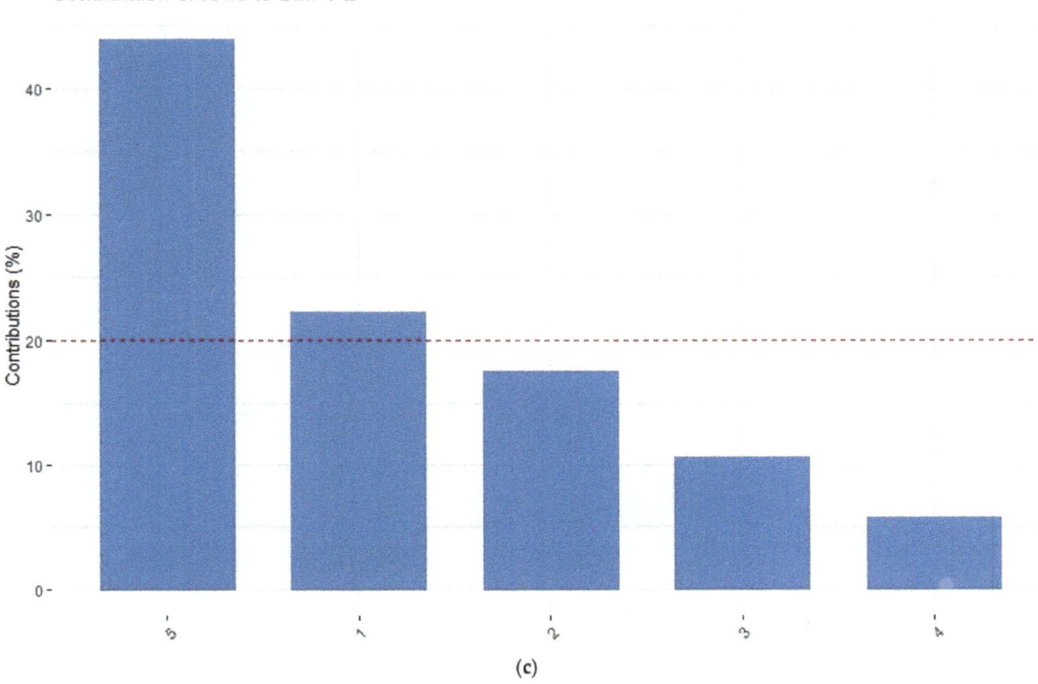

(c)

Figure 4. Asymmetrical plot (**a**), and contributions of columns (**b**) and rows (**c**) to the dimensions.

Figure 4a shows the asymmetrical biplot of the data, by plotting the row (ratings) profiles in the column space (pictures). Interpretation of data on an asymmetrical plot may be done based on the angles made by the arrows and the projection of the data characterizing a given row on the arrows depicting the column profiles [43]. For instance, if the angle between two given arrows is more acute, the association between them is stronger. As such, P1 was more associated with ratings of 5, while at the opposite end was the association between P10 and ratings of 1. Ratings of 1 and 5, and data characterizing P1, P4, P10, P3, P6, and P2, contributed the most to the first two dimensions generated by the correspondence analysis.

In terms of profiles set by the ratings, P1 stood apart from P2 and P3, which were more similar. P9 and P10 seemed to be more similar to P6, although these pictures represented three different land management systems—namely, croplands, pasturelands, and managed forest.

3.4. Grouping of Land Management Systems Based on Collective Visual Ratings

The results of the workflow implemented for the hierarchical cluster analysis are shown in Figures 5 and 6. Figure 5a shows the distance map based on Spearman's dissimilarity metric, while Figure 5b gives the actual distances computed in the range of 0 to 1. Finally, Figure 6 shows the dendrogram built via cluster analysis, in which three clusters were kept in the final solution.

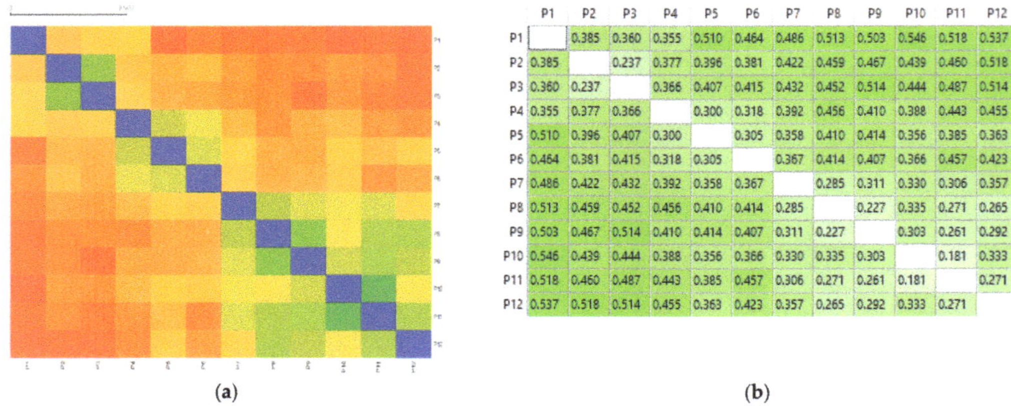

	P1	P2	P3	P4	P5	P6	P7	P8	P9	P10	P11	P12
P1		0.385	0.360	0.355	0.510	0.464	0.486	0.513	0.503	0.546	0.518	0.537
P2	0.385		0.237	0.377	0.396	0.381	0.422	0.459	0.467	0.439	0.460	0.518
P3	0.360	0.237		0.366	0.407	0.415	0.432	0.452	0.514	0.444	0.487	0.514
P4	0.355	0.377	0.366		0.300	0.318	0.392	0.456	0.410	0.388	0.443	0.455
P5	0.510	0.396	0.407	0.300		0.305	0.358	0.410	0.414	0.356	0.385	0.363
P6	0.464	0.381	0.415	0.318	0.305		0.367	0.414	0.407	0.366	0.457	0.423
P7	0.486	0.422	0.432	0.392	0.358	0.367		0.285	0.311	0.330	0.306	0.357
P8	0.513	0.459	0.452	0.456	0.410	0.414	0.285		0.227	0.335	0.271	0.265
P9	0.503	0.467	0.514	0.410	0.414	0.407	0.311	0.227		0.303	0.261	0.292
P10	0.546	0.439	0.444	0.388	0.356	0.366	0.330	0.335	0.303		0.181	0.333
P11	0.518	0.460	0.487	0.443	0.385	0.457	0.306	0.271	0.261	0.181		0.271
P12	0.537	0.518	0.514	0.455	0.363	0.423	0.357	0.265	0.292	0.333	0.271	

(a) (b)

Figure 5. Distance map (**a**) and distance matrix (**b**) of the data used as inputs for clustering.

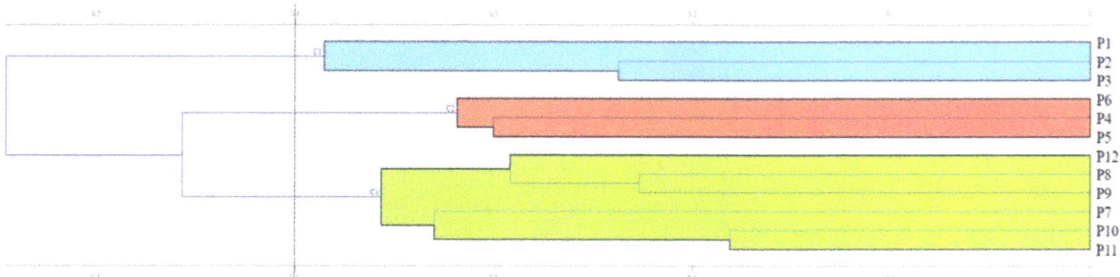

Figure 6. Dendrogram of the land management systems built based on the rating data. Legend: blue—cluster of pictures depicting the natural forest; red—cluster of pictures depicting managed forest; green—cluster of pictures depicting croplands and pasturelands.

The distance set at 0.4 was arbitrarily chosen based on several iterations aiming to make sense of the final data clustering. For instance, P10 and P11 (pasturelands) merged in a cluster at a distance of 0.181, and then they joined a new cluster that included P7 (cropland) at the distance of 0.330 (Figures 5 and 6). The same was found for P8 and P9 (croplands), which merged in a cluster at a distance of 0.227, and then were clustered together with P12 (pasturelands) at a distance of 0.292 (Figures 5 and 6). By moving the distance threshold at which this solution was found (0.4) to 0.335, the data from the cluster of pictures depicting the croplands and pasturelands were split into two new clusters—namely, P7, P10, and P11, and P8, P9, and P1, respectively—which were heterogeneous in terms of land use type (Figure 6). The same solution (distance set at 0.335) has split the first cluster (natural forest) in two new clusters. The data shown in Figure 5 may be used to evaluate the distance at which each two pictures in a given pair are located.

4. Discussion

The fact that people build high consensus in the evaluation of positively perceived land management systems provides a valid argument for the management of valuable landscape scenes in terms of sustainability [33]. This study evaluated the perceptions of the local inhabitants towards the land management systems in the rainforest area of Ecuador, based on 12 pictures depicting the main land management systems in the area—namely, unmanaged (natural) forest, managed forest, croplands, and pasturelands—at three levels of landscape view: far, medium, and inside. A study carried out in Tanzania described the importance of the opinion of the population located near a forest, due to the fact that people carry out economic activities and have an environmental knowledge of the area [47]. In addition, the incidence of the population allows the management of activities that affect the environment and the biophysical aspects of the forest [48]. The importance of accounting for the preferences of local inhabitants with regard to the land management systems was also described by a case study from Botswana where, based on such information, forest management regimes were developed to protect and conserve rainforests via an inclusive, participatory approach [49]. Similar studies have evaluated the importance of forest ecosystems by considering variables such as the ecosystem services, forest area, levels of human intervention, sociodemographic variables [50], ecosystem services categories (provisioning, regulating or cultural) [51], and variables associated with landscape resources [52].

In this study, the respondents' visual perceptions of the types of land management systems were evaluated by considering the typical land use continuum in the area of study. As such, the study included the land use types without human intervention, characterized by density and abundance of vegetation (Pictures 1–3). Previous studies relating the visual perception with the types of scenes shown to respondents have indicated that higher evaluation rates are given to scenes depicting continuous vegetative cover [53], as was the case with the natural forest from this study. Pasturelands, on the other hand, are generally given lower ratings, because they are typically associated with agricultural and livestock practices [54], so respondents associate this type of land management system with human activities and intensive-use practices. Resources such as water, soil, and biodiversity of flora and fauna cause emotive feelings towards a place [55]. It was found in this study that the pictures that showed land management systems from an intermediate perspective (P5—managed forest, and P8—croplands) produced some of the highest ratings within their land management system group. This may be the effect of a differentiated visual perception, owing to recognition of the components of diversity, along with other features such as the shape, density, and position of the landscape features [56].

Thus far, the main sociodemographic variables that may act as drivers of the visual perceptions of the landscapes were found to be the gender, level of education, age, and employment [57,58]. In addition, a study on landscape preferences carried out in Germany reported that female respondents who have completed tertiary education, as well as people with knowledge about the environment, gave positive ratings to different landscapes [54]. However, this study shows that men rated the managed and unmanaged forests higher, which may have been due to the fact that they develop their work in relation to forests, and probably understand the functions and services provided by the forests better [59]. In relation to levels of education, respondents from the bachelor's degree group and above evaluated natural forests with higher scores, but they gave low ratings to croplands (Pictures 7–9); this most likely suggests that their environmental knowledge allowed them to better understand the importance of the rainforest.

The trend of giving higher scores to the natural forest, considering the shares of responses per type of landscape and management system, is possibly related to features such as abundance, structure, and vegetation's diversity. In support of this, a study on the visual quality of rural landscapes indicated that the beauty of a given landscape is linked to its share of flora, and to low homogeneity (color contrast) [23]. In addition, other studies have mentioned that visual perceptions of forests change in relation to the type of ecosystem, stand age, abundance, and diversity [60,61]. For instance, studies on the structural attributes of forests indicate that the population perceives forests under close-to-nature management and low-intensity managed ecosystems positively [62], while the perceptions of the visitors are often related to the functionality and management of the forests [63].

The results of cluster analysis, showing that there is a collective view that can be differentiated into three groups of land management systems, are consistent with past studies on visual perception. For instance, local users liked natural environments, although other groups of people liked scenes depicting various levels of management or human intervention [64]. According to Schmidt et al. [65], people can be characterized as "forest and nature enthusiasts", "traditionalists", and "multi-functionalists". Accordingly, this study indicates that a mathematical differentiation of the main land use systems can be derived based on the collective perception; however, it was more a differentiation between the forest (natural and managed) and the rest of the land management systems. Moreover, in some cases, the results were contradictory, in the sense that some liked both untouched and highly managed land use systems. This contradiction arises from the fact that, in the study area, croplands and pasturelands are commonly created by removing natural forests, thereby changing the land use type. Although the locals' perceptions of intensively managed land uses may be driven by the acknowledgement of the potential economic and professional development, which is not necessarily an unsustainable perspective, public dissemination of knowledge would be useful in order for the local population to understand the benefits and the drawbacks of different land use scenarios, including their share in a given area. This would help not only in shaping, but also in accepting and smoothly implementing local governance policies.

The results of this study highlight the importance of the evaluation of visual perception for the purposes of land use planning and management. To the best of our knowledge, this is the first study of its kind implemented in Ecuador. In addition, a study by De Meo et al. [66] mentioned that the knowledge of communities' perceptions of forests is important for decision makers to develop and implement management strategies [66]; such data are also important for planning or design activities [67]. Finally, we acknowledge that user perceptions may vary according to various factors [68]—for instance, the landscape components and their aesthetic values [69], the backgrounds of different people [70], and lack of knowledge about agricultural practices and environmental attitudes [65]. In this context, our findings provide an overview of the perceptions towards different land management systems, related to the sociodemographic variables. The results of this study can be used to shape a new management approach and objectives taking public perceptions into account, considering that the current management plan of SBP ends in 2021. In addition, some data from this study indicate the importance of educating the local people, in order for them to be more informed about the types of land management systems in the area.

This study evaluated the visual perceptions of the local inhabitants with regard to land management systems via a rather subjective approach which, in addition, implied the use of categorical data in the analytic part of the study. These two important components of the study can also constitute some of its limitations. To what extent other features and mechanisms—such as local and national economics, economic implications, and internal and external trading policies and regulations—would have changed the perceptions of the respondents if brought into the study as quantitative features, remains an open question. In fact, visual assessment is just one component that can be used to evaluate the sustainability of a landscape, and its outcomes need to be balanced with the findings of other quantitative studies. Future studies should add to these findings by extending the assessments over the economic component. By using quantitative data, such studies could elucidate whether the perceptions would remain the same or be shifted based on policy issues and economic implications—particularly those typically brought about by conservation.

5. Conclusions

The results of this research indicate that the natural forest was the most liked by the local people in comparison with the rest of the land use systems. Managed forest was less liked, probably due to the visible impact of human activities. However, one cannot infer that those land management systems with evident human intervention were less liked, as in some cases croplands and pasturelands received high ratings. In addition, there were no significant differences between the scores given to croplands and pasturelands—a fact that was reflected in the clusters formed based on the collective perceptions. Therefore, the results show that locals generally perceived the natural forests positively. Despite that, the perceptions of local people were differentiated, because croplands and pasturelands were probably associated with sources of economic income. However, such land use systems are derived from the removal of natural forest. A combination of the four land use systems would balance the expectations of different stakeholders from the area, while also being consistent to some extent with the current diversity in land management systems. However, a more developed system of information propagation would be beneficial to educate the local population with regards to the benefits and drawbacks of different types of land use and their share.

Author Contributions: Conceptualization, A.V.G.M., D.D.C.V. and S.A.B.; data curation, A.V.G.M., D.D.C.V. and S.A.B.; formal analysis, A.V.G.M. and S.A.B.; investigation, A.V.G.M. and D.D.C.V.; methodology, A.V.G.M., D.D.C.V. and S.A.B.; project administration, S.A.B.; resources, A.V.G.M., D.D.C.V. and S.A.B.; supervision, A.V.G.M., D.D.C.V. and S.A.B.; validation, A.V.G.M., D.D.C.V. and S.A.B.; visualization, A.V.G.M., D.D.C.V. and S.A.B.; writing—original draft, A.V.G.M., D.D.C.V. and S.A.B.; writing—review and editing, A.V.G.M., D.D.C.V. and S.A.B. All authors have read and agreed to the published version of the manuscript.

Funding: This research received no external funding. The APC was funded by the Escuela Superior Politecnica de Chimborazo.

Institutional Review Board Statement: Ethical review and approval were waived for this study due to the fact that the study was carried out based on an informed consent and anonymity of the respondents.

Informed Consent Statement: Each respondent was informed in detail about the objectives of the study and how the data would be used. Each responded agreed verbally to participate in the study under an anonymity clause.

Data Availability Statement: All of the data supporting this study may be made available upon request to the first and second authors of the study.

Acknowledgments: This study reports results that are part of a PhD thesis developed under the supervision of the Doctoral School of the Transilvania University of Brasov. The authors acknowledge the support of Department of Forest Engineering, Forest Management Planning, and Terrestrial Measurements of the Faculty of Silviculture and Forest Engineering, Transilvania University of Brasov. The authors would like to give thanks to the Escuela Superior Politécnica de Chimborazo for technical help during the data collection phase, and to all respondents who participated in this research for supporting our survey.

Conflicts of Interest: The authors declare no conflict of interest.

Appendix A

Table A1. Description of the pictures used for the evaluation of visual preferences.

Photo Content	Name (Abbreviation)	Description
	Picture 1 (P1)	Primary (natural) forest from a far perspective.
	Picture 2 (P2)	Primary (natural) forest from an intermediate position of the observer.

Table A1. *Cont.*

Photo Content	Name (Abbreviation)	Description
	Picture 3 (P3)	Primary (natural) forest from a close perspective of the observer.
	Picture 4 (P4)	Secondary (managed) forest from a far perspective
	Picture 5 (P5)	Secondary (managed) forest from an intermediate position of the observer.
	Picture 6 (P6)	Secondary (managed) forest from a close perspective of the observer.
	Picture 7 (P7)	Croplands from a far perspective.
	Picture 8 (P8)	Croplands from an intermediate position of the observer.
	Picture 9 (P9)	Croplands from a close perspective of the observer.
	Picture 10 (P10)	Pasturelands from a far perspective.
	Picture 11 (P11)	Pasturelands from an intermediate position of the observer.
	Picture 12 (P12)	Pasturelands from a close perspective of the observer.

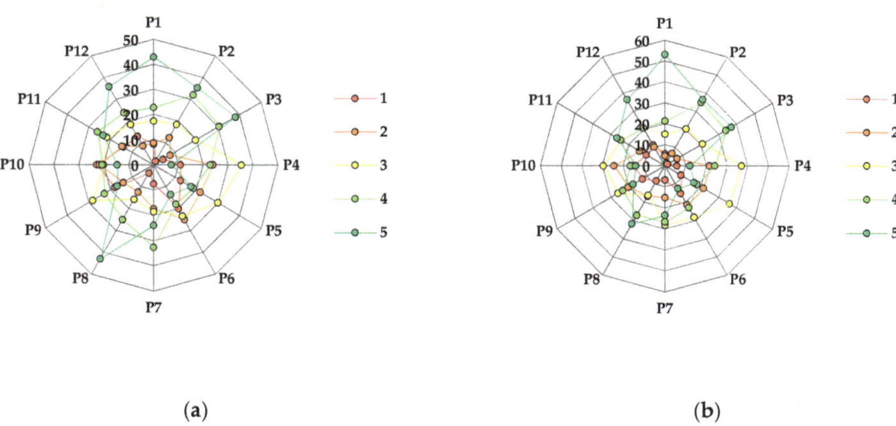

Figure A1. Relative frequency of ratings as a function of gender: (**a**) female respondents; (**b**) male respondents. Legend: P1 to P12—pictures 1 to 12; 1 to 5 stand for ratings of 1 ("not at all") to 5 ("very much").

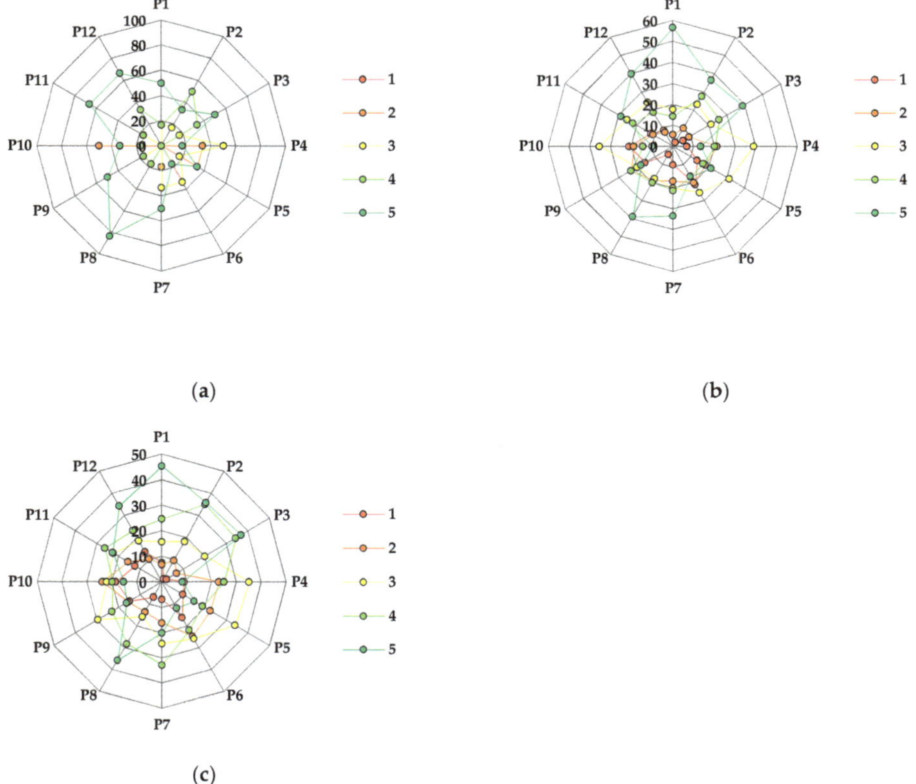

Figure A2. Relative frequency of ratings as a function of the ethnicity: (**a**) Caucasian; (**b**) indigenous; (**c**) Metis and other. Legend: P1 to P12—pictures 1 to 12; 1 to 5 stand for ratings of 1 ("not at all") to 5 ("very much").

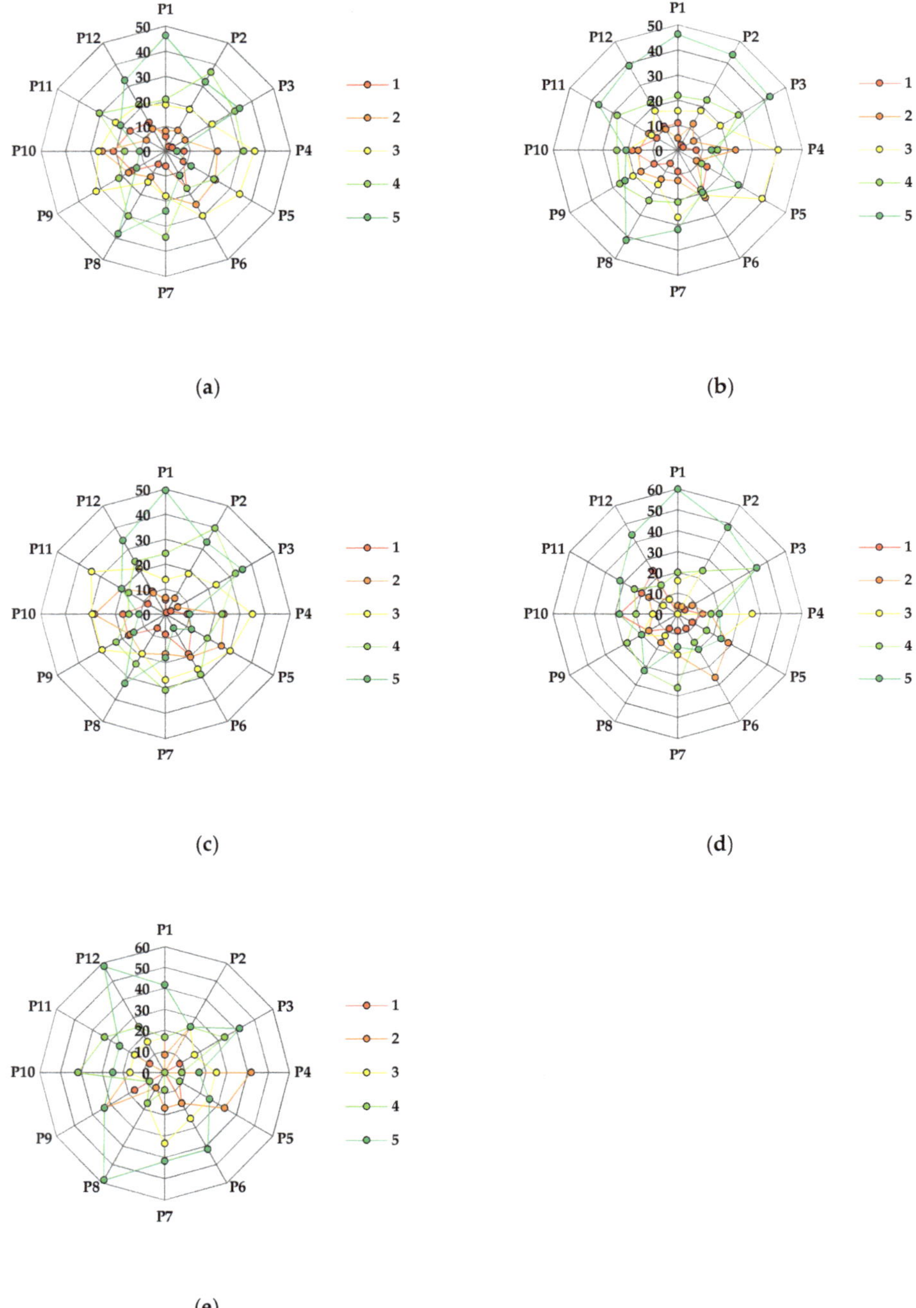

Figure A3. Relative frequency of ratings as a function of marital status: (**a**) single; (**b**) common law; (**c**) married; (**d**) divorced; (**e**) widow(er). Legend: P1 to P12—pictures 1 to 12; 1 to 5 stand for ratings of 1 ("not at all") to 5 ("very much").

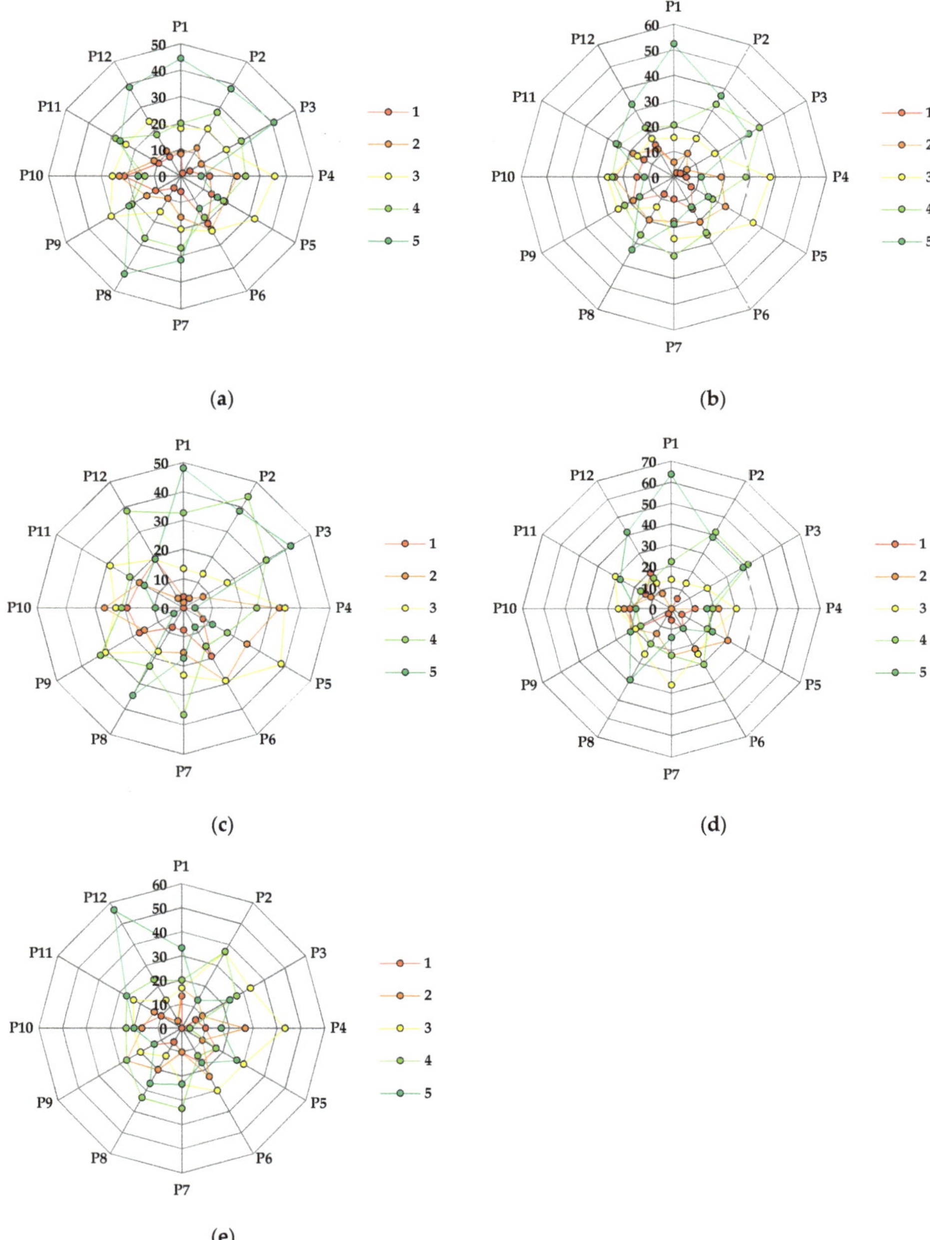

Figure A4. Relative frequency of ratings as a function of the age category: (**a**) less than 30 years old; (**b**) 31 to 40 years old; (**c**) 41 to 50 years old; (**d**) 51 to 60 years old; (**e**) more than 60 years old. Legend: P1 to P12—pictures 1 to 12; 1 to 5 stand for ratings of 1 ("not at all") to 5 ("very much").

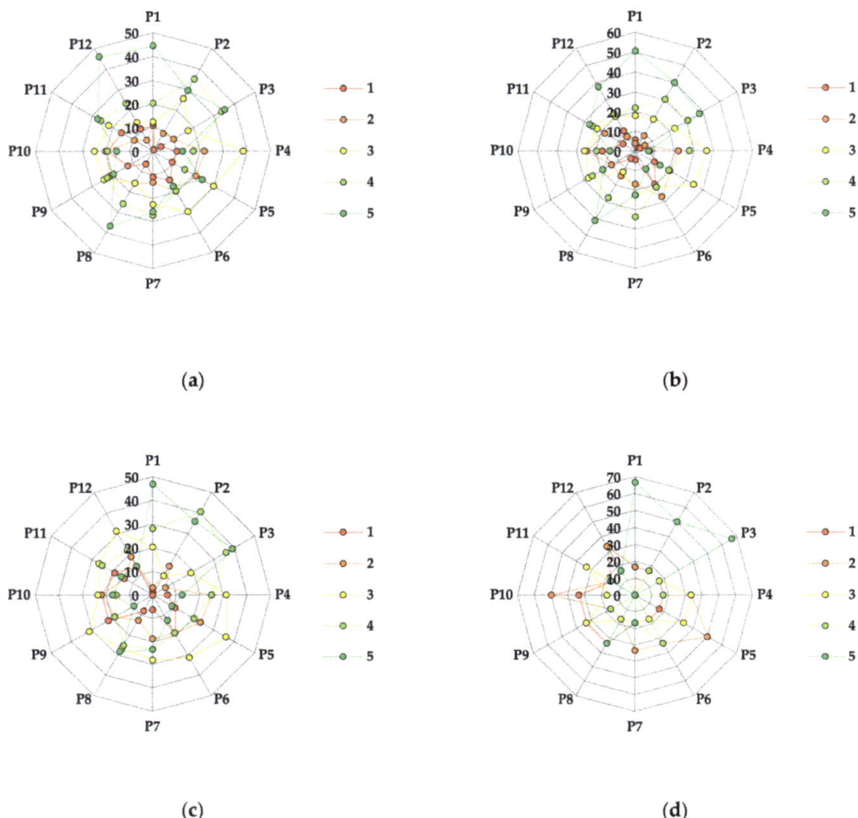

Figure A5. Relative frequency of ratings as a function of the education category: (**a**) elementary school; (**b**) high school; (**c**) bachelor's degree; (**d**) master's degree or more. Legend: P1 to P12—pictures 1 to 12; 1 to 5 stand for ratings of 1 ("not at all") to 5 ("very much").

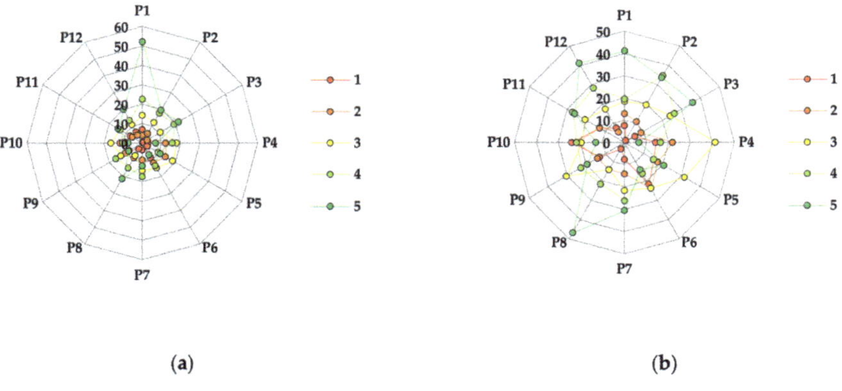

Figure A6. *Cont.*

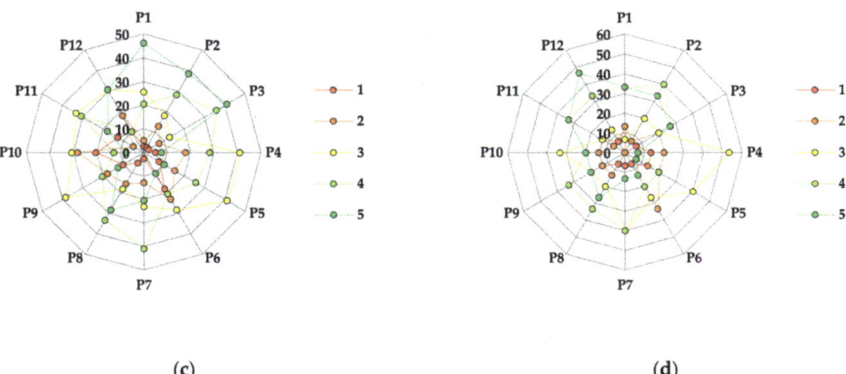

(c) (d)

Figure A6. Relative frequency of ratings as a function of the occupation category: (**a**) employed; (**b**) house care; (**c**) student; (**d**) unemployed. Legend: P1 to P12—pictures 1 to 12; 1 to 5 stand for ratings of 1 ("not at all") to 5 ("very much").

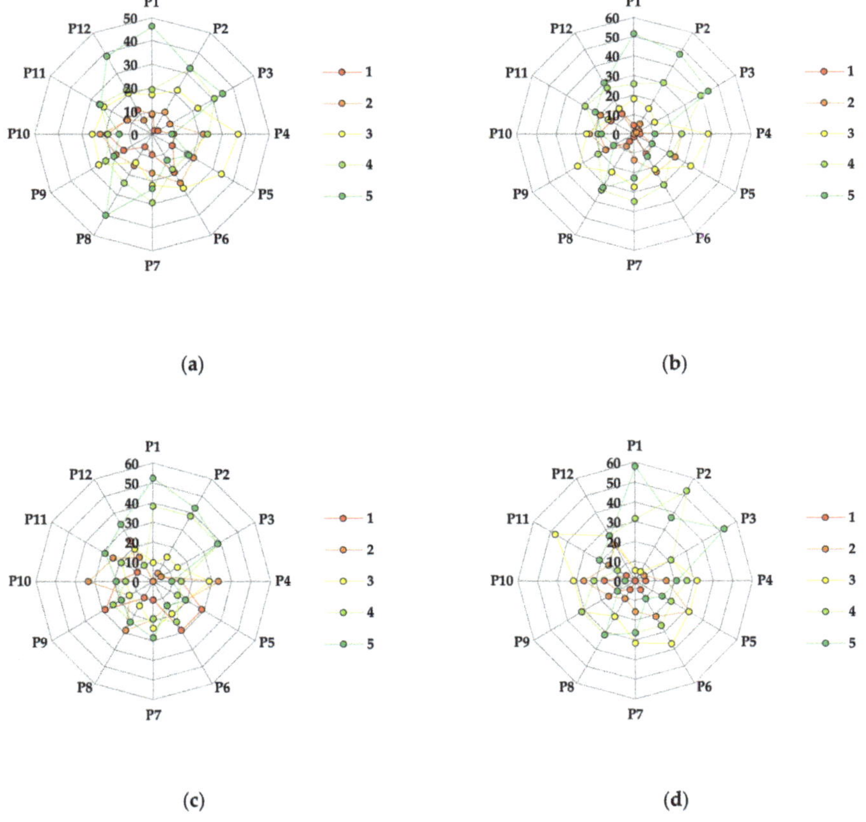

Figure A7. Relative frequency of ratings as a function of the monthly income category: (**a**) less than USD 394; (**b**) USD 395–733; (**c**) USD 734–901; (**d**) more than USD 901. Legend: P1 to P12—pictures 1 to 12; 1 to 5 stand for ratings of 1 ("not at all") to 5 ("very much").

References

1. Juerges, N.; Arts, B.; Masiero, M.; Hoogstra-Klein, M.; Borges, J.G.; Brodrechtova, Y.; Brukas, V.; Canadas, M.J.; Carvalho, P.O.; Corradini, G.; et al. Power analysis as a tool to analyse trade-offs between ecosystem services in forest management: A case study from nine European countries. *Ecosyst. Serv.* **2021**, *49*, 101290. [CrossRef]
2. Taye, F.A.; Folkersen, M.V.; Fleming, C.M.; Buckwell, A.; Mackey, B.; Diwakar, K.; Saint, C. The economic values of global forest ecosystem services: A meta-analysis. *Ecol. Econ.* **2021**, *189*, 107145. [CrossRef]
3. Portela Peñalver, L.; Rivero Galván, A.; Portela Peñalver, L. Valoración económica de bienes y servicios ecosistémicos en montañas de Guamuhaya, Cienfuegos, Cuba. *Rev. Univ. Soc.* **2019**, *11*, 47–55.
4. Fischer, A.P. Forest landscapes as social-ecological systems and implications for management. *Landsc. Urban Plan.* **2018**, *177*, 138–147. [CrossRef]
5. Calva, J.; Ortiz, N.; Calapucha, J.; Changa, G.; Pallo, C. *Los Bosques de Ecuador: Su Importancia y sus Limitaciones*; Universidad Estatal Amazónica: Pastaza, Ecuador, 2020.
6. Malhi, Y.; Gardner, T.A.; Goldsmith, G.R.; Silman, M.R.; Zelazowski, P. Tropical forests in the Anthropocene. *Annu. Rev. Environ. Resour.* **2014**, *39*, 125–159. [CrossRef]
7. Trumbore, S.; Brando, P.; Hartmann, H. Forest health and global change. *Science* **2015**, *349*, 814–818. [CrossRef] [PubMed]
8. Mackey, B.; DellaSala, D.A.; Kormos, C.; Lindenmayer, D.; Kumpel, N.; Zimmerman, B.; Watson, J.E.M.; Hugh, S.; Young, V.; Foley, S.; et al. Policy options for the world's primary forests in multilateral environmental agreements. *Conserv. Lett.* **2015**, *8*, 139–147. [CrossRef]
9. Duffau, M.V.; Guachimboza, P.K.C. Gobernanza ambiental, Buen Vivir y la evolución de la deforestación en Ecuador en las provincias de Tungurahua y Pastaza. *Rev. Derecho* **2020**, *34*, 146–167. [CrossRef]
10. Garí, J. Biodiversity and indigenous agroecology in Amazonia: The indigenous people of Pastaza. *Etnoecologica* **2001**, *5*, 21–37.
11. Ahammad, R.; Stacey, N.; Sunderland, T. Analysis of forest-related policies for supporting ecosystem services-based forest management in Bangladesh. *Ecosyst. Serv.* **2021**, *48*, 101235. [CrossRef]
12. Miura, S.; Amacher, M.; Hofer, T.; San-Miguel-Ayanz, J.; Thackway, R. Protective functions and ecosystem services of global forests in the past quarter-century. *For. Ecol. Manag.* **2015**, *352*, 35–46. [CrossRef]
13. Frick, J.; Bauer, N.; von Lindern, E.; Hunziker, M. What forest is in the light of people's perceptions and values: Socio-cultural forest monitoring in Switzerland. *Geogr. Helv.* **2018**, *73*, 335–345. [CrossRef]
14. Stępniewska, M.; Zwierzchowska, I.; Mizgajski, A. Capability of the Polish legal system to introduce the ecosystem services approach into environmental management. *Ecosyst. Serv.* **2018**, *29*, 271–281. [CrossRef]
15. Loft, L.; Mann, C.; Hansjürgens, B. Challenges in ecosystem services governance: Multi-levels, multi-actors, multi-rationalities. *Ecosyst. Serv.* **2015**, *16*, 150–157. [CrossRef]
16. Keenan, R.J.; Pozza, G.; Fitzsimons, J.A. Ecosystem services in environmental policy: Barriers and opportunities for increased adoption. *Ecosyst. Serv.* **2019**, *38*, 100943. [CrossRef]
17. Gavilanes, A.; Castillo, D.; Morocho, J.; Marcu, V.; Borz, S. Importance and use of ecosystem services provided by the Amazonian landscapes in Ecuador-evaluation and spatial scaling of a representative area. *Bull. Transilv. Univ. Brasov. For. Wood Ind. Agric. Food Eng. Ser. II* **2019**, *12*, 1–26. [CrossRef]
18. Silva, E.; Reardon, H.; Soares, A.; Azeiteiro, U. Identity and environment: Historical trajectories of 'traditional' communities in the protection of the Brazilian Amazon. In *Sustainability in Natural Resources Management and Land Planning*; Springer: Cham, Switzerland, 2021; pp. 233–248. [CrossRef]
19. Catacutan, D.C.; Van Noordwijk, M.; Nguyen, T.H.; Öborn, I.; Mercado, A.R. *Agroforestry: Contribution to Food Security and Climate-Change Adaptation and Mitigation in Southeast Asia*; White Paper; World Agroforestry Centre (ICRAF) Southeast Asia Regional Program: Bogor, Indonesia, 2017.
20. Carrasco, L.R.; Papworth, S.K.; Reed, J.; Symes, W.S.; Ickowitz, A.; Clements, T.; Sunderland, T.; Peh, K.-H. Five challenges to reconcile agricultural land use and forest ecosystem services in Southeast Asia. *Conserv. Biol.* **2016**, *30*, 962–971. [CrossRef]
21. FAO; PROFOR. *Framework for Assessing and Monitoring Forest Governance*; Food and Agriculture Organisation and Program on Forests (PROFOR): Rome, Italy, 2011.
22. Valencia-Sandoval, C.; Flanders, D.N.; Kozak, R.A. Participatory landscape planning and sustainable community development: Methodological observations from a case study in rural Mexico. *Landsc. Urban Plan.* **2010**, *94*, 63–70. [CrossRef]
23. Arriaza, M.; Cañas-Ortega, J.F.; Cañas-Madueño, J.; Ruiz-Aviles, P. Assessing the visual quality of rural landscapes. *Landsc. Urban Plan.* **2004**, *69*, 115–125. [CrossRef]
24. Zhao, J.; Luo, P.; Wang, R.; Cai, Y. Correlations between aesthetic preferences of river and landscape characters. *J. Environ. Eng. Landsc. Manag.* **2013**, *21*, 123–132. [CrossRef]
25. Daniel, T.C.; Meitner, M.M. Representational validity of landscape visualizations: The effects of graphical realism on perceived scenic beauty of forest vistas. *J. Environ. Psychol.* **2001**, *21*, 61–72. [CrossRef]
26. Gobster, P.H.; Ribe, R.G.; Palmer, J.F. Themes and trends in visual assessment research: Introduction to the landscape and urban planning special collection on the visual assessment of landscapes. *Landsc. Urban Plan.* **2019**, *191*, 103635. [CrossRef]
27. Lothian, A. Landscape and the philosophy of aesthetics: Is landscape quality inherent in the landscape or in the eye of the beholder? *Landsc. Urban Plan.* **1999**, *44*, 177–198. [CrossRef]

28. Tveit, M.S. Indicators of visual scale as predictors of landscape preference; a comparison between groups. *J. Environ. Manag.* **2009**, *90*, 2882–2888. [CrossRef]
29. Strumse, E. Demographic differences in the visual preferences for agrarian landscapes in western Norway. *J. Environ. Psychol.* **1996**, *16*, 17–31. [CrossRef]
30. Vouligny, É.; Domon, G.; Ruiz, J. An assessment of ordinary landscapes by an expert and by its residents: Landscape values in areas of intensive agricultural use. *Land Use Policy* **2009**, *26*, 890–900. [CrossRef]
31. Molnarova, K.; Sklenicka, P.; Stiborek, J.; Svobodova, K.; Salek, M.; Brabec, E. Visual preferences for wind turbines: Location, numbers and respondent characteristics. *Appl. Energy* **2012**, *92*, 269–278. [CrossRef]
32. Spielhofer, R.; Hunziker, M.; Kienast, F.; Hayek, U.W.; Grêt-Regamey, A. Does rated visual landscape quality match visual features? An analysis for renewable energy landscapes. *Landsc. Urban Plan.* **2021**, *209*, 104000. [CrossRef]
33. Asur, F.; Deniz Sevimli, S.; Yazici, K. Visual preferences assessment of landscape character types using data mining methods (Apriori algorithm): The case of Altınsaç and Inkoy (Van/Turkey). *J. Agric. Sci. Technol.* **2020**, *22*, 247–260.
34. Gavilanes, A.; Castillo, D.; Ricaurte, C.; Marcu, M.; Borz, S. Known and newly documented uses of 540 rainforest species in the Pastaza region, Ecuador. *Bull. Transilv. Univ. Braşov* **2018**, *11*, 20. [CrossRef]
35. CDTER. Plan de Desarrollo y Ordenamiento Territorial de la Parroquia Simón Bolívar 2015–2019, GADPR Simón Bolivar, Ecuador. 2015. Available online: http://app.sni.gob.ec/sni-link/sni/PORTAL_SNI/data_sigad_plus/sigadplusdocumentofinal/1660011960001_PD%20Y%20OT%20GADPR%20SIM%C3%93N%20BOL%C3%8DVAR_30-10-2015_19-31-05.pdf (accessed on 12 September 2021).
36. Organización de las Naciones Unidas para la Agricultura y la Alimentación (FAO). *Evaluación de los Recursos Forestales Mundiales 2015. ¿Cómo están Cambiando los Bosques del Mundo?* 2a ed.; FAO: Rome, Italy, 2016; pp. 1–54.
37. Bonilla, S.; Estrella, A.; Molina, J.; Herrera, M. Socioecological system and potential deforestation in Western Amazon forest landscapes. *Sci. Total Environ.* **2018**, *644*, 1044–1055. [CrossRef] [PubMed]
38. Ponce, C. Sistema Nacional de Monitoreo de Bosques. 2017. Available online: http://redd.mma.gov.br/images/gttredd/carlosponce_equador.pdf (accessed on 12 September 2021).
39. Ministerio del Ambiente del Ecuador (MAE). *Sistema de Clasificación de Ecosistemas del Ecuador Continental*; Ministerio del Ambiente del Ecuador: Quito, Ecuador, 2013. Available online: http://app.sni.gob.ec/sni-link/sni/PDOT/NIVEL%20NACIONAL/MAE/ECOSISTEMAS/DOCUMENTOS/Sistema.pdf (accessed on 12 September 2021).
40. Mena, C. Deforestación en el Norte de la Amazonía Ecuatoriana: Del patrón al proceso. *Polémika* **2010**, *2*, 1–5.
41. Ríos, M.; Pedersen, H. *Uso y Manejo de Recursos Vegetales. Memorias del Segundo Simposio Ecuatoriano de Etnobotánica y Botánica Económica. En uso y Manejo de Recursos Vegetales: Memorias del Segundo Simposio Ecuatoriano de Etnobotánica y Botánica Económica*; Ediciones Abya-Yala: Quito, Ecuador, 1997.
42. Gavilanes, A. Capacity of the Ecuadorian Amazonian Rainforest to Provide Ecosystem Services: An Evaluation of Plant Uses, Capacity to Provide Products and Services and Perception on the Landscape Management Systems in the View of Local Stakeholders. Ph.D. Thesis, Universitatea Transilvania Brasov, Brasov, Romania, 2020.
43. CA—Correspondence Analysis in R: Essentials. Available online: http://www.sthda.com/english/articles/31-principal-component-methods-in-r-practical-guide/113-ca-correspondence-analysis-in-r-essentials/ (accessed on 10 November 2021).
44. Demsar, J.; Curk, T.; Erjavec, A.; Gorup, C.; Hocevar, T.; Milutinovic, M.; Mozina, M.; Polajnar, M.; Toplak, M.; Staric, A.; et al. Orange: Data Mining Toolbox in Python. *J. Mach. Learn. Res.* **2013**, *14*, 2349–2353.
45. Instituto Nacional de Estadística y Censos (INEC). Encuesta Nacional de Empleo, Desempleo y Subempleo 2021 (ENEMDU). 2021. Available online: https://www.ecuadorencifras.gob.ec/documentos/web-inec/POBREZA/2021/Junio-2021/202106_PobrezayDesigualdad.pdf (accessed on 12 September 2021).
46. Instituto Nacional de Estadística y Censos (INEC). Principales resultados Encuesta Nacional de Empleo, Desempleo y Subempleo. 2021. Available online: https://www.ecuadorencifras.gob.ec/documentos/web-inec/EMPLEO/2021/Julio-2021/202107_Mercado_Laboral.pdf (accessed on 12 September 2021).
47. Degnet, M.; Van der, E.; Ingram, V.; Wesseler, J. Do locals have a say? Community experiences of participation in governing Forest plantations in Tanzania. *Forests* **2020**, *11*, 782. [CrossRef]
48. Payn, T.; Carnus, J.; Freer, P.; Kimberley, M.; Kollert, W.; Liu, S.; Wingfield, M. Changes in planted forests and future global implications. *For. Ecol. Manag.* **2015**, *352*, 57–67. [CrossRef]
49. Garekae, H.; Lepetu, J.; Thakadu, O.; Sebina, V.; Tselaesele, N. Community perspective on state forest management regime and its implication on forest sustainability: A case study of Chobe Forest Reserve, Botswana. *J. Sustain. For.* **2020**, *39*, 692–709. [CrossRef]
50. Carrasco, L.; Nghiem, T.; Sunderland, T.; Koh, L. Economic valuation of ecosystem services fails to capture biodiversity value of tropical forests. *Biol. Conserv.* **2014**, *178*, 163–170. [CrossRef]
51. Quintas, C.; Martín, B.; Santos, F.; Loureiro, M.; Montes, C.; Benayas, J.; García, M. Ecosystem services values in Spain: A metaanalysis. *Environ. Sci. Policy* **2016**, *55*, 186–195. [CrossRef]
52. Zandersen, M.; Tol, R. A meta-analysis of forest recreation values in Europe. *J. For. Econ.* **2009**, *15*, 109–130. [CrossRef]
53. Nahuelhual, L.; Laterra, P.; Jiménez, D.; Báez, A.; Echeverría, C.; Fuentes, R. Do people prefer natural landscapes? An empirical study in Chile. *Bosque (Valdivia)* **2018**, *39*, 205–216. [CrossRef]
54. Häfner, K.; Zasada, I.; van Zanten, B.; Ungaro, F.; Koetse, M.; Piorr, A. Assessing landscape preferences: A visual choice experiment in the agricultural region of Märkische Schweiz, Germany. *Landsc. Res.* **2018**, *43*, 846–861. [CrossRef]

55. Pan, S.; Lee, J.; Tsai, H. Travel photos: Motivations, image dimensions, and affective qualities of places. *Tour. Manag.* **2014**, *40*, 59–69. [CrossRef]
56. Rensink, R. Scene Perception. Oxford University Press. *Encyclopedia of psychology.* **2000**, *7*, 151–155.
57. Kalivoda, O.; Vojar, J.; Skřivanová, Z.; Zahradník, D. Consensus in landscape preference judgments: The effects of landscape visual aesthetic quality and respondents' characteristics. *J. Environ. Manag.* **2014**, *137*, 36–44. [CrossRef] [PubMed]
58. Campbell, C.; Cox, A.; Darby, H.; Roberts, D.; Torrance, L. Landscape Preference and Perception. 2019. Available online: https://macaulay.webarchive.hutton.ac.uk/visualisationlitrev/chap2.html (accessed on 12 September 2021).
59. Affek, A.; Kowalska, A. Ecosystem potentials to provide services in the view of direct users. *J. Ecosyst. Serv.* **2017**, *26*, 183–196. [CrossRef]
60. Nielsen, A.; Gundersen, V.; Jensen, F. The impact of field layer characteristics on forest preference in Southern Scandinavia. *Landsc. Urban Plan.* **2018**, *170*, 221–230. [CrossRef]
61. Sklenicka, P.; Molnarova, K. Visual perception of habitats adopted for post-mining landscape rehabilitation. *Environ. Manag.* **2010**, *46*, 424–435. [CrossRef]
62. Edwards, D.; Jay, M.; Jensen, F.; Lucas, B.; Marzano, M.; Montagné, C.; Peace, A.; Weiss, G. Public preferences for structural attributes of forests: Towards a pan-European perspective. *For. Policy Econ.* **2012**, *19*, 12–19. [CrossRef]
63. Pastorella, F.; Avdagić, A.; Čabaravdić, A.; Mraković, A.; Osmanović, M.; Paletto, A. Tourists' perception of deadwood in mountain forests. *Ann. For. Res.* **2016**, *59*, 311–326. [CrossRef]
64. Schroeder, H. Preference and meaning of arboretum landscapes: Combining quantitative and qualitative data. *J. Environ. Psychol.* **1991**, *11*, 231–248. [CrossRef]
65. Schmidt, K.; Walz, A.; Martín-López, B.; Sachse, R. Testing socio-cultural valuation methods of ecosystem services to explain land use preferences. *Ecosyst. Serv.* **2017**, *26*, 270–288. [CrossRef]
66. De Meo, I.; Paletto, A.; Cantiani, M. The attractiveness of forests: Preferences and perceptions in a mountain community in Italy. *Ann. For. Res.* **2015**, *58*, 145–156. [CrossRef]
67. Palmer, J.; Hoffman, R. Rating reliability and representation validity in scenic landscape assessments. *Landsc. Urban Plan.* **2001**, *54*, 149–161. [CrossRef]
68. Chua, H.; Boland, J.; Nisbett, R. Cultural variation in eye movements during scene perception. *Proc. Natl. Acad. Sci. USA* **2005**, *102*, 12629–12633. [CrossRef] [PubMed]
69. Hagerhall, C. Clustering predictors of landscape preference in the traditional Swedish cultural landscape: Prospect-refuge, mystery, age and management. *J. Environ. Psychol.* **2000**, *20*, 83–90. [CrossRef]
70. Bell, S. Landscape pattern, perception and visualisation in the visual management of forests. *Landsc. Urban. Plan.* **2001**, *54*, 201–211. [CrossRef]

Article

Seed Germination and Seedling Growth of *Robinia pseudoacacia* Depending on the Origin of Different Geographic Provenances

Andrea M. Roman [1,2], Alina M. Truta [1], Oana Viman [1,*], Irina M. Morar [1], Velibor Spalevic [3], Catalina Dan [1,*], Radu E. Sestras [1], Liviu Holonec [1] and Adriana F. Sestras [1]

[1] Faculty of Horticulture, University of Agricultural Sciences and Veterinary Medicine, 3-5 Mănăștur Street, 400372 Cluj-Napoca, Romania; andreearoman34@gmail.com (A.M.R.); alina.truta@usamvcluj.ro (A.M.T.); irina.todea@usamvcluj.ro (I.M.M.); rsestras@usamvcluj.ro (R.E.S.); lholonec@usamvcluj.ro (L.H.); adriana.sestras@usamvcluj.ro (A.F.S.)

[2] Forestry College Transilvania, 425200 Nasaud, Romania

[3] Biotechnical Faculty, University of Montenegro, 81000 Podgorica, Montenegro; velibor.spalevic@gmail.com

* Correspondence: oana.viman@usamvcluj.ro (O.V.); catalina.dan@usamvcluj.ro (C.D.)

Abstract: Black locust (*Robinia pseudoacacia*) is recognised as a forest species of interest due to its multiple uses. The management of forest genetic resources and their efficient conservation suffer from variations in traits and start with seed germination. The aim of the current study was to investigate the germination of seeds obtained from plus trees selected in eight Romanian provenances, as well as to investigate the influence of the origin upon plants' growth and development. Two experiments were undertaken to test seed germination: one treatment involved water-soaked seeds and heat/cold treatment, while the other treatment was based on sulphuric acid, at different concentrations (50, 70, 90%). The results were correlated with the morphological analysis of the seeds. Satu-Mare had the lowest germination rate within both treatments. Sulphuric acid did not improve seed germination as much as the heat treatment. The highest germination rate occurred for the water and temperature treatment on seeds from Bihor provenance (68.2%). The most distant provenance was Bihor, in inverse correlation with Bistrița Năsăud and grouped separately within the hierarchical dendrogram of cluster analysis based on the analysed parameters of the provenances investigated. The results demonstrated that the genotypes and environmental heterogeneity of the seed origin within the provenances may finally result in different performances.

Keywords: black locust; forest; germination; provenances; seed dormancy

Citation: Roman, A.M.; Truta, A.M.; Viman, O.; Morar, I.M.; Spalevic, V.; Dan, C.; Sestras, R.E.; Holonec, L.; Sestras, A.F. Seed Germination and Seedling Growth of *Robinia pseudoacacia* Depending on the Origin of Different Geographic Provenances. *Diversity* **2022**, *14*, 34. https://doi.org/10.3390/d14010034

Academic Editor: Mario A. Pagnotta

Received: 14 December 2021
Accepted: 5 January 2022
Published: 6 January 2022

1. Introduction

The *Robinia* genus includes 19 botanical taxa, eight of which are classified as species, while the others are considered natural varieties or hybrids [1,2]. Black locust (*Robinia pseudoacacia* L.) is native to the eastern part of North America—the Appalachian region [3,4]. In Europe, it was first introduced in France and England around 1600 for ornamental purposes (parks, botanical gardens), from where it spread throughout the continent and some parts of Asia [5]. Nowadays, black locust appears in many European countries, such as Germany, Hungary, Czech Republic, Poland, Slovakia, Austria and Romania [6,7], being added for forestry purposes. *R. pseudoacacia* has been studied because of its economic value rather than its ecology [8,9]. In Romania, it was introduced around 1750 and is considered an important woody species, imposing itself as a forest tree, even though black locust has multiple other uses and exploitation [10], with extensive plantations that cover an area of more than 2,306,000 ha [11].

As a species with rapid growth, *R. pseudoacacia* is also distinguished by the ease of its cultivation and the value of its wood, as it produces durable and rot-resistant wood [12].

Due to the fast vegetative development (production of 14 tons of dried subs/year/ha), it has potential for bio-oil production and fuel ethanol derived from biomass [13–16]. Black locust's rusticity and ecological plasticity (it is the most cultivated exotic species in Europe) are other factors that have determined its spread and wide use. Even more, as part of the Fabaceae family, *R. pseudoacacia* is a nitrogen-fixing species [7], often used on abandoned fields and dry slopes [17]; it also grows on acid, dry and infertile sandy soils, and is also drought-tolerant [18]. Thus, the species is widely planted for revegetation and to control soil erosion. It is also considered a pioneer tree species due to its fast growth, resistance to pollution and strong capacity for improving soil nitrogen content and nutrient availability, the available phosphorus pool and organic carbon sequestration, as well as enhancing erodible sandy soil chemical and microbiological properties [19–22].

Nowadays, *Robinia* is considered as a melliferous species, with the honey collected being highly appreciated [23]; in Romania, there are almost 30 ha of qualified seed orchards, as well as other areas with provenance and melliferous (honey producing) orchards [10]. Other values to be mentioned for black locust are derived from processed products (e.g., wine from black-locust honey), dense wood that can be used for multiple purposes: habitat for insects, birds, fungi and birds and to increase carbon footprint [24]. Some areas lack highly productive native species with wood or growth characteristics suitable for forestry plantation; thus, there is a need to focus largely on exotic species, such as black locust, as it can be easily established on certain sites. Even though this is not necessarily the case for Romania, the species is considered valuable due to its wider physiological adaptability in terms of site conditions, especially on slopes, exploitations and sandy or erodible soils. In Romania, the area occupied by black locust has increased continuously [25,26]. Depending on the favorable ecological conditions, black locust could contribute in the same way as other forest species or by supplementing them, ensuring the multiple functions of the forest at local, zonal, regional, national or global levels [27–34]. The quality of the seeds is the first condition for the afforestation to be successful and for the resulting forests to fulfil their numerous functions; i.e., production, protection, ecological, cultural, educational, etc. [27,35,36]

R. pseudoacacia is considered to be a fast-growing species, as mentioned: in the first season (in the case of sprouts), it can reach heights of 6 m and get to heights of 30 m at the age of 30–35 years, with basal diameters of 30 cm at 15 years, depending on the type of regeneration and site conditions. Several black locust populations cultivated in Romania achieve average increases of 15–17 m^3/year/ha and reach the age of absolute exploitability at 30 years [27,37]. *R. pseudoacacia* reproduces sexually and asexually [17]. The first flowering occurs at the age of 5–6 years old, whereas the fruit is a pod that matures in autumn. A pod contains 4–10 dark brown seeds without endosperm, which are 4–6 mm in length. Every 1 or 2 years, it produces a high quantity of seeds [38]. Black locust seeds can be stored for more than 10 years at a temperature of 0–5 °C [39].

Normally, seed propagation is the easiest and most advantageous method of growing plants, but without previous treatments, black locust seed germination may be low due to the physiological dormancy induced by the impermeable tegument covering the seeds [40,41]. Seed propagation can be an easy and advantageous method of growing, but without previous treatments, seed germination may be low due to the physiological dormancy induced by the impermeable tegument [42], such as within *Robinia* species. The function of the seed coat has multiple benefits, such as protecting the embryo and endosperm from desiccation, mechanical injury, unfavourable temperatures and attacks by bacteria, fungi and insects, but breaking physical dormancy in some forest seeds is a challenge if one is to obtain homogeneous germination for larger seed samples. Both embryo and coat dormancies are components of physiological dormancy, and their interaction determine the degree of "whole-seed" physiological dormancy. The transition from dormancy to germination is a critical control point, leading to the initiation of vegetative growth.

Regarding the germination capacity, various chemical and mechanical treatments are used in forestry to improve the germination rate. Black locust seeds require pre-treatment before sowing in order to break exogenous dormancy, due to the structure of the seed coat, which is impermeable to water and gases [38]. Moreover, the imbibition and course of processes that are essential for germination are inhibited in the dormancy stage [33,43]. The impermeability of the seed coat, understood as physical dormancy, is a very important ecological mechanism for species to ensure that germination occurs only in favorable conditions for seedling growth [44,45]. However, this phenomenon is undesirable when seeds are intended for commercial use and forest practice [46]. Thus, to break the dormancy of seeds, the seed coat must be damaged. Under natural conditions, the coat of black locust seeds can be damaged by either low (frosts) or high temperatures (fires) or by the activity of the soil microflora [47]. For research and practical purposes, three methods are used: mechanical (scarification), thermal and chemical [2,48].

The aim of the present study was to determine the efficiency of two treatments based on sulphuric acid at different concentrations and thermal shock treatment upon seeds to increase the germination rate of black locust seeds as well as to analyse the effect of the treatments and influence of the origin provenances on the development of seedlings in their first stage after emergence.

2. Materials and Methods

2.1. Biological Material

The seeds were collected from mature plus trees from eight Romanian provenances (Figure 1): (1) Galați, (2) Iași, (3) Botoșani, (4) Bihor, (5) Râmnicu Vâlcea, (6) Satu-Mare, (7) Bistrița-Năsăud and (8) Arad, corresponding to the Romanian Gene Reserved Forests and Seed Stands included in National Catalogue of Forest Genetic Resources and Forest Reproductive Materials [49].

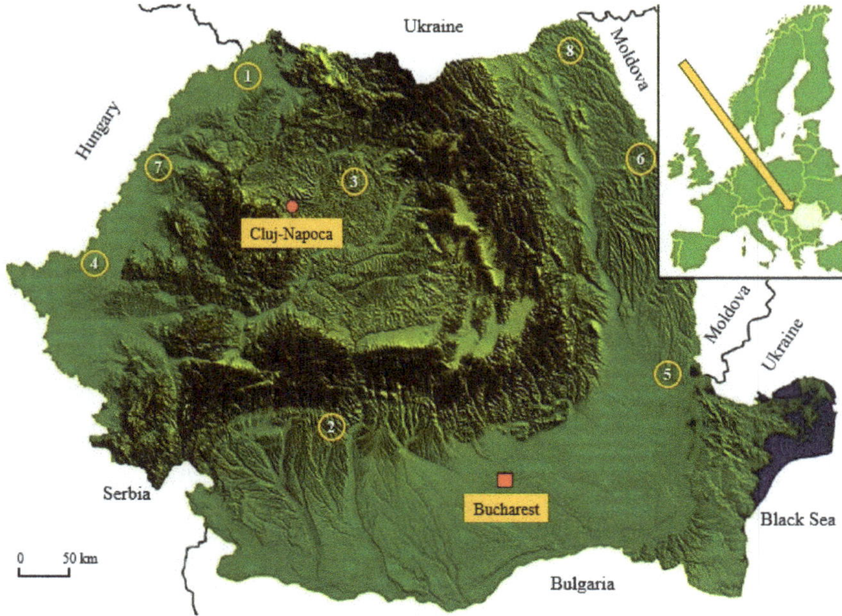

Figure 1. The Romanian provenances of *R. pseudoacacia* seeds.

To characterise the climatic heterogeneity of the different collection sites, mean annual temperature and mean precipitation were considered (Table 1).

Seeds were collected in the autumn of 2020. After harvesting, seeds were labeled for each provenance and deposited in paper bags in a refrigerator (at approximately +4 °C) until further analysis.

Table 1. Location and climatic characteristics of the eight *R. pseudoacacia* seeds provenances in Romania.

Provenance-Population	County	Administrative Location [1]	Latitude/ Longitude	Average Yearly Temperature (°C)	Average Annual Precipitation (mm)	Altitude (m asl)
Foieni	Satu Mare	OS Carei, UPIII, u.a.57L, 58A	47°42′ N/22°24′ E	10.9	619.2	130–130
Budești	Vâlcea	RNP Romsilva, OS Stoiceni UPII, u.a.5L, 5M	45°03′ N/24°26′ E	8.2	813.1	300–350
Pădurea Cetății	Bistrița Năsăud	RNP Romsilva, OSE Lechința UP.7, u.a.80C	47°00′ N/24°20′ E	6.3	802.9	380–400
Moneasa	Arad	Private Orchard Bărzani (Ruben Budău)	46°19′ N/21°40′ E	11.8	575.7	110–150
Drăgănești	Galați	RNP Romsilva, OS Tecuci UP.VI, u.a.49B	45°46′ N/27°30′ E	11.6	480.1	45–45
Borșa	Iași	RNP Romsilva, OS Iași UP. III, u.a.105E	47°25′ N/27°20′ E	10.5	540.9	60–160
Curtuișeni	Bihor	RNP Romsilva, OS Săcuieni UP. IV, u.a.45C	47°32′ N/22°09′ E	10.7	617.6	140–140
Liveni	Botoșani	RNP Romsilva, OS Darabani UP. III, u.a.26D	48°02′ N/27°01′ E	10.1	526.6	200–200

[1] Population description according to the National Catalogue of Forest Reproductive Materials, Bucharest (2012).

Moreover, since water stress is known to induce both morphological and anatomical changes [50–52], seedlings were also studied [53]. The seedlings obtained from treated seeds were further analysed by being kept in the laboratory for three months, then taken outside in natural conditions; at the end of six months, morphological traits were measured: height, stem diameter and number of branches.

2.2. Experimental Procedures Performed in Order to Investigate the Germination Rate

Seed analysis began in March 2021, following the established protocol and steps for the analysis of germination and seedling growth (Figure 2). Before starting the treatments, several measurements of the main phenotypic elements were performed for the seeds: height, diameter and weight. Two treatments were undertaken, each one with three concentrations, at three different temperatures. Four replicates, with 50 seeds per replicate, were used for each combination of population and treatment method. The first treatment involved water-soaked seeds and heat treatment (noted as thermal treatment), as follows: 1. seeds were soaked for 24 h in water at a temperature of 18 °C and then seeded and watered, while kept at room temperature; 2. seeds were stored at an air temperature of 45 °C for two hours (Memmet UF110 oven) and thereafter stored in a refrigerator at −20 °C for two hours and then seeded and watered; 3. seeds were stored at 60 °C for two hours, stored at −20 °C for a further two hours and then seeded and watered. The second treatment was based on sulphuric acid solution at different concentrations (50%, 70% and 90%), within which seeds were soaked for 20 min, at room temperature and light, and then seeded and watered.

For all treatments, watering was performed twice a week with tap water, for a total of four weeks of investigation; all replicates were kept at laboratory temperature (18 °C) and light. To calculate the germination rate, observations were made on days 4, 7, 10, 14, 21 and 28 for all treatments.

Figure 2. Stages of analysis of seed morphological characteristics, seed germination and seedling growth.

2.3. Statistical Analyses

The registered data of the seeds and seedlings were processed as the mean of the traits and standard error of the mean (SEM). An analysis of variance (ANOVA) was applied to the analysed traits, and if the null hypothesis was rejected, a post hoc test was used for the analysis of differences. Among the general class of multiple comparisons procedures, the most suitable for our data and sets of means comparison was considered to be Duncan's Multiple Range Test (Duncan's MRT, $p < 0.05$) [54]. The data were subjected to multivariate statistical analysis, namely principal component analysis (PCA). A multivariate principal components analysis graph for the eight provenances of *R. pseudoacacia* was created using Past software [55]. This software was used also for the construction of a dendrogram, as Euclidean distances among provenances.

3. Results

The treatments investigated to promote black locust germination—respectively, heat treatment and a chemical treatment using diluted sulphuric acid—had different results. Morphological trsaits of the seeds were measured and are presented in Table 2, indicating that their origin was relevant. The longest seeds were found from Bihor provenance, with distinct significative differences, while the shortest seeds were those from Bistrița Năsăud and Galați. In regard to the diameter of block locust seeds, the best value was recorded for Satu-Mare provenance. On the opposite side, the smallest diameter was obtained for seeds from Galați and Botoșani. The heaviest seeds were obtained from plus trees from Satu-Mare and Iași, followed by Bihor seeds.

It is worthwhile to note that the longest seeds were not the same as those that had the largest diameter, nor the heaviest seeds. The only provenance that had high values, meaning that the investigated traits were statistically superior, was Satu-Mare (Table 2).

Traits that have a CV% less than 10% are considered to have a low variability; in this regard, the most uniform length for the seeds was noted within Bihor provenance and the highest value for CV% was 16.9 for Bistrița N., which means the variability was medium for these seeds. Even more, CV% values between 20–30% can illustrate a large variability, and over 30% shows a very high variability [56]. It is to be further analysed if these results correlate with the germination rate after the proposed treatments, so that conclusions can be made with regard to the interaction of seed traits and the capacity of germination. Data are presented in tables that summarise both treatments with all replicates, whereas the results are noted separately and as means, with the significance of differences (Tables 3–6).

Table 2. The main morphological traits of the black locust seeds, as mean * and standard error of the mean (SEM).

Provenance	Seeds Length (mm)		Seeds Width (mm)		Seeds Weight (g)	
	Mean ± SEM	CV%	Mean ± SEM	CV%	Mean ± SEM	CV%
Satu-Mare	$3.13^{ab} \pm 0.27$	15.0	$1.14^{a} \pm 0.04$	6.1	$0.031^{a} \pm 0.002$	10.2
Vâlcea	$3.00^{ab} \pm 0.06$	3.2	$0.63^{cd} \pm 0.05$	14.6	$0.022^{bc} \pm 0.001$	16.0
Bistrița N.	$2.28^{c} \pm 0.22$	16.9	$0.56^{cd} \pm 0.19$	5.7	$0.017^{d} \pm 0.001$	4.5
Arad	$2.77^{bc} \pm 0.15$	9.4	$0.72^{bc} \pm 0.10$	22.9	$0.024^{b} \pm 0.001$	7.7
Galați	$2.36^{c} \pm 0.10$	7.3	$0.40^{d} \pm 0.02$	8.3	$0.019^{cd} \pm 0.000$	3.1
Iași	$2.78^{bc} \pm 0.13$	8.4	$0.89^{b} \pm 0.10$	19.2	$0.031^{a} \pm 0.000$	1.1
Bihor	$3.25^{a} \pm 0.03$	1.4	$0.61^{cd} \pm 0.09$	24.1	$0.028^{a} \pm 0.002$	11.3
Botoșani	$2.71^{bc} \pm 0.15$	9.6	$0.43^{d} \pm 0.06$	23.7	$0.023^{bc} \pm 0.000$	11.5

* The means on the column followed by different letters are significantly different according to Duncan's MRT test ($p < 0.05$).

Table 3. Black locust seed germination (%) depending on the treatments applied.

Provenance	Sulphuric Acid Treatment/Concentration			Mean Provenance **
	50%	70%	90%	
Satu-Mare	5.5^{gh} *	$15.8^{e–g}$	40.0^{ab}	20.4^{CD}
Vâlcea	45.3^{a}	$21.5^{d–f}$	$30.0^{b–d}$	32.3^{A}
Bistrița N.	$30.5^{b–d}$	$25.0^{c–e}$	$34.8^{a–c}$	30.1^{AB}
Arad	39.8^{ab}	$10.3^{f–h}$	$25.5^{c–e}$	$25.2^{A–C}$
Galați	39.8^{ab}	$16.0^{e–g}$	$20.5^{d–f}$	$25.4^{A–C}$
Iași	40.8^{ab}	$16.3^{e–g}$	$26.8^{c–e}$	$27.9^{A–C}$
Bihor	0.0^{h}	$34.8^{a–c}$	0.0^{h}	11.6^{D}
Botoșani	45.0^{a}	4.8^{gh}	$20.5^{d–f}$	23.4^{CD}
Mean treatment ***	30.8^{Z}	18.0^{Y}	24.7^{X}	-
Provenance	**Thermal Treatment/Temperatures**			**Mean Provenance**
	18 °C	**45 °C, −20 °C**	**60 °C, −20 °C**	
Satu-Mare	20.5^{ij}	$49.8^{b–d}$	10.3^{k}	26.8^{EF}
Vâlcea	$30.0^{g–j}$	$37.0^{d–h}$	$43.8^{b–f}$	36.9^{CD}
Bistrița N.	$40.3^{c–g}$	$25.0^{h–j}$	$31.0^{f–j}$	32.1^{DE}
Arad	$25.5^{h–j}$	$25.5^{h–j}$	$20.0^{i–j}$	23.7^{F}
Galați	70.8^{a}	20.8^{ij}	$49.8^{b–d}$	47.1^{B}
Iași	51.8^{bc}	$46.8^{b–e}$	$34.5^{e–i}$	44.3^{BC}
Bihor	55.3^{b}	75.0^{a}	74.3^{a}	68.2^{A}
Botoșani	$30.5^{f–j}$	$50.0^{b–d}$	55.0^{b}	45.2^{BC}
Mean treatment	40.6^{Z}	41.2^{Z}	39.8^{Z}	-

* The means on the column inside the table followed by different small letters are significantly different according to Duncan's MRT test ($p < 0.05$). ** The means on the last column reflect the influence of the provenances, regardless of the treatment applied. The means followed by different capital letters are significantly different according to Duncan's MRT test ($p < 0.05$). *** The means on the last row reflect the influence of the treatments, regardless of the provenances. The means followed by different capital letters are significantly different according to Duncan's MRT test ($p < 0.05$).

Regarding seed germination, the obtained results indicated that the highest germination rate for black locust occurred in the water and temperature variation treatment on seeds from Bihor provenance (68.2%) (Table 3); interestingly, the same location had the smallest values of germination when seeds were treated with sulphuric acid (11.6%). Thus, it can be said that the provenance of seeds and the different treatments had a significant effect on the dynamics of germination.

Table 4. Seedling's height (cm) depending on the treatments applied to black locust seeds.

Provenance	Sulphuric Acid Treatment/Concentration			Mean Provenance **
	50%	70%	90%	
Satu-Mare	24.5 $^{f–j}$ *	28.5 $^{e–h}$	31.5 $^{c–f}$	28.2 ^{BC}
Vâlcea	10.8 ^{l}	8.3 ^{l}	9.3 ^{l}	9.4 ^{E}
Bistrița N.	20.3 $^{h–k}$	19.5 $^{i–k}$	21.5 $^{g–k}$	20.4 ^{D}
Arad	45.0 ^{a}	37.5 $^{a–d}$	40.5 ^{ab}	41.0 ^{A}
Galați	30.0 $^{d–g}$	24.0 $^{f–k}$	27.5 $^{e–i}$	27.2 ^{C}
Iași	39.3 $^{a–c}$	29.8 $^{d–g}$	33.8 $^{b–e}$	34.3 ^{B}
Bihor	0.0 ^{m}	31.8 $^{c–f}$	0.0 ^{m}	10.6 ^{E}
Botoșani	22.0 $^{g–k}$	15.5 $^{k–l}$	18.8 ^{jk}	18.8 ^{D}
Mean treatment ***	24.0 ^{X}	24.3 ^{X}	22.8 ^{X}	-
Provenance	**Thermal Treatment/Temperatures**			**Mean Provenance**
	18 °C	45 °C, −20 °C	60 °C, −20 °C	
Satu-Mare	24.0 $^{h–l}$	31.0 $^{e–g}$	18.8 ^{kl}	24.6 ^{C}
Vâlcea	8.0 ^{m}	8.6 ^{m}	9.5 ^{m}	8.7 ^{F}
Bistrița N.	21.0 $^{j–l}$	17.6 ^{l}	20.0 $^{k–l}$	19.5 ^{E}
Arad	45.0 ^{a}	42.5 ^{ab}	38.5 ^{bc}	42.0 ^{A}
Galați	26.9 $^{f–j}$	18.7 ^{kl}	25.3 $^{g–k}$	23.6 ^{CD}
Iași	38.0 $^{b–d}$	36.0 $^{c–e}$	30.8 $^{e–g}$	34.9 ^{A}
Bihor	28.5 $^{f–i}$	30.0 $^{e–h}$	32.1 $^{d–f}$	30.2 ^{B}
Botoșani	17.5 ^{l}	22.0 $^{i–l}$	23.8 $^{h–l}$	21.1 ^{DE}
Mean treatment	26.1 ^{X}	25.8 ^{X}	24.8 ^{X}	-

* The means on the column inside the table followed by different small letters are significantly different according to Duncan's MRT test ($p < 0.05$). ** The means on the last column reflect the influence of the provenances, regardless of the treatment applied. The means followed by different capital letters are significantly different according to Duncan's MRT test ($p < 0.05$). *** The means on the last row reflect the influence of the treatments, regardless of the provenances. The means followed by different capital letters are significantly different according to Duncan's MRT test ($p < 0.05$).

Even more, the germination rate was significantly higher for the water treatment compared with the sulphuric acid treatment (Table 3). The differences were consistent among provenances, but the strongest impact was found for the applied treatment. The highest percentage of seed germination was obtained for water and 45 °C temperature treatment (41.2%), compared with the other values obtained, which ranged between 18.0% and 30.8% for different concentrations of sulphuric acid treatment.

The results obtained were directly influenced by the concentration of acid (Table 3); in the investigation, the exposure of seeds to 50% sulphuric acid for 20 min resulted in the highest mean for germination for the relevant treatment (30.8% compared with 18% and 24.7%).

Overall, the seeds treated within the thermal variations treatment (45 °C, −20 °C) had the highest germination rate (41.2%). Even so, the water and temperature treatment applied better stimulated the germination of black locust seeds (ranging between 39.8–41.2%) (Table 3).

Depending on the concentration, the doses of 50% and 90% of sulphuric acid determined the continuous increase of the percentage of germinated seeds, after 4, 7, 10 and 14 days from the treatment, after which there was a decrease in the proportion of germinated black locust seeds until the 21 day interval (Figure 3). The regression equation for the 70% concentration did not ensure the same upward trend of the regression line. However, at a concentration of 70%, the coefficient of determination showed the largest contribution of the independent variable (the time interval in which the germination was analysed) of the total variance (37.3%). In addition, it was noted that the correlation coefficient had the highest value (0.611).

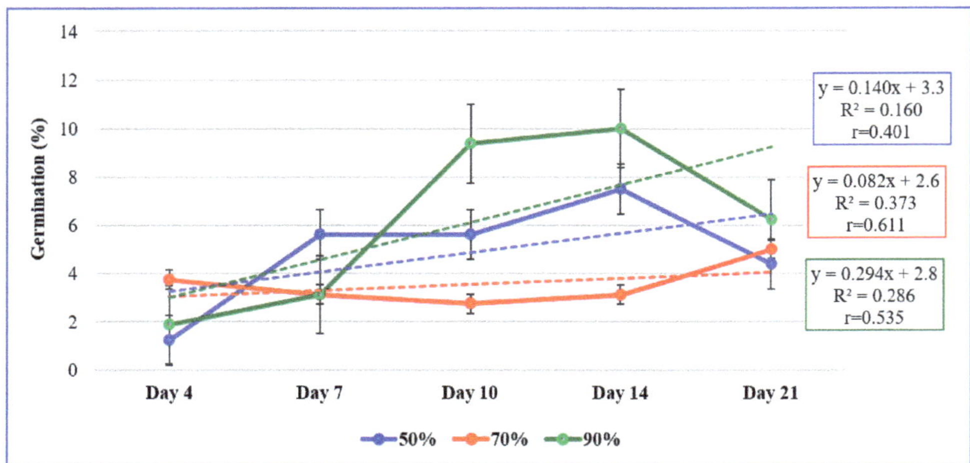

Figure 3. Black locust seed germination (%) evolution in days, depending on the treatments applied with three concentrations of sulphuric acid.

Table 5. Seedling diameter (cm) depending on the treatments applied to black locust seeds.

Provenance	Sulphuric Acid Treatment/Concentration			Mean Provenance **
	50%	70%	90%	
Satu-Mare	3.2 c–g *	4.3 a–d	5.2 a	4.2 A
Vâlcea	5.0 a	3.7 b–f	4.4 a–c	4.3 A
Bistrița N.	2.2 g	2.5 fg	3.1 d–g	2.6 B
Arad	5.3 a	4.0 a–e	4.8 ab	4.7 A
Galați	3.1 d–g	2.2 g	3.0 e–g	2.7 B
Iași	3.5 c–g	2.3 g	2.6 fg	2.8 A
Bihor	0.0 h	3.3 c–g	0.0 h	1.1 C
Botoșani	3.1 c–g	2.3 g	2.5 fg	2.6 B
Mean treatment ***	3.1 X	3.1 X	3.2 X	-
Provenance	**Thermal Treatment/Temperatures**			**Mean Provenance**
	18 °C	**45 °C, −20 °C**	**60 °C, −20 °C**	
Satu-Mare	4.2 a–e	5.3 a	3.7 b–g	4.4 AB
Vâlcea	3.1 c–g	4.0 a–e	4.3 a–d	3.8 BC
Bistrița N.	3.3 c–g	2.8 e–g	3.1 c–g	3.0 CD
Arad	5.3 a	4.7 ab	4.5 a–c	4.8 A
Galați	2.8 e–g	2.3 g	3.0 d–g	2.7 D
Iași	3.8 b–f	3.0 d–g	2.9 e–g	3.2 CD
Bihor	2.9 d–g	3.0 d–g	3.5 b–g	3.1 CD
Botoșani	2.5 f–g	3.4 b–g	3.5 b–g	3.1 CD
Mean treatment	3.5 X	3.5 X	3.7 X	-

* The means on the column inside the table followed by different small letters are significantly different according to Duncan's MRT test ($p < 0.05$). ** The means on the last column reflect the influence of the provenances, regardless of the treatment applied. The means followed by different capital letters are significantly different according to Duncan's MRT test ($p < 0.05$). *** The means on the last row reflect the influence of the treatments, regardless of the provenances. The means followed by different capital letters are significantly different according to Duncan's MRT test ($p < 0.05$).

The three types of thermal variation within the treatments applied to the seeds provided relatively close equations and regression lines (Figure 4). The coefficients of determination had quite low values, whereas the lowest value of the correlation coefficient was registered in the case of the treatment in which the high temperature (+60 °C) was alternated with the low temperature (−20 °C).

Table 6. Seedlings' number of branches per stem depending on the treatments applied to black locust seeds.

Provenance	Sulphuric Acid Treatment/Concentration			Mean Provenance **
	50%	70%	90%	
Satu-Mare	7 $^{d–f}$ *	7 $^{d–f}$	11.9 ^{a}	8.6 ^{AB}
Vâlcea	9 $^{b–d}$	7 $^{d–f}$	7 $^{d–f}$	7.7 ^{BC}
Bistrița N.	8 $^{c–e}$	6 ^{ef}	10 $^{a–c}$	8.0 ^{AB}
Arad	11 ^{ab}	9.1 ^{bd}	8 $^{c–e}$	9.4 ^{A}
Galați	8 $^{c–e}$	5 ^{f}	6 ^{ef}	6.3 ^{CD}
Iași	10 $^{a–c}$	7 $^{d–f}$	8 $^{c–e}$	8.3 ^{AB}
Bihor	0 ^{g}	11 ^{ab}	0 ^{g}	3.7 ^{E}
Botoșani	7 $^{d–f}$	4.9 ^{f}	6 ^{ef}	6.0 ^{D}
Mean treatment ***	7.5 ^{X}	7.1 ^{X}	7.1 ^{X}	-
Provenance	**Thermal Treatment/Temperatures**			**Mean Provenance**
	18 °C	**45 °C, −20 °C**	**60 °C, −20 °C**	
Satu-Mare	9 $^{d–e}$	12 ^{a}	7 $^{d–g}$	9.3 ^{AB}
Vâlcea	6 ^{fg}	7 $^{d–g}$	7 $^{d–g}$	6.7 ^{CD}
Bistrița N.	10 $^{a–c}$	6 ^{fg}	7 $^{d–g}$	7.7 ^{BC}
Arad	9.3 $^{b–d}$	9 $^{d–e}$	7 $^{d–g}$	8.4 ^{B}
Galați	6.3 $^{e–g}$	5 ^{g}	6 ^{fg}	5.8 ^{D}
Iași	9.8 $^{a–d}$	8.3 $^{c–f}$	6.3 $^{e–g}$	8.1 ^{BC}
Bihor	10.0 $^{a–c}$	11.8 ^{ab}	10.8 $^{a–c}$	10.8 ^{A}
Botoșani	7.3 $^{d–g}$	8 $^{c–f}$	8.3 $^{c–f}$	7.8 ^{BC}
Mean treatment	8.4 ^{X}	8.4 ^{X}	7.4 ^{Y}	-

* The means on the column inside the table followed by different small letters are significantly different according to Duncan's MRT test ($p < 0.05$). ** The means on the last column reflect the influence of the provenances, regardless of the treatment applied. The means followed by different capital letters are significantly different according to Duncan's MRT test ($p < 0.05$). *** The means on the last row reflect the influence of the treatments, regardless of the provenances. The means followed by different capital letters are significantly different according to Duncan's MRT test ($p < 0.05$).

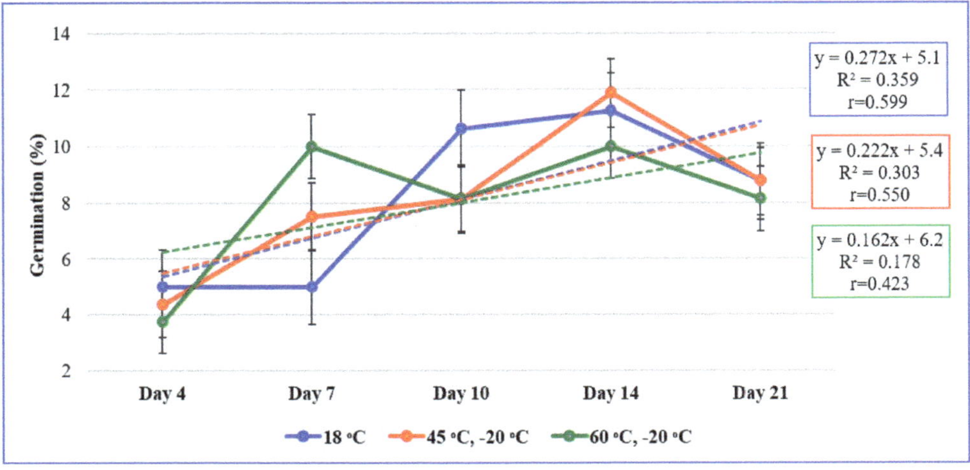

Figure 4. Black locust seed germination (%) evolution in days, depending on the thermal treatments applied with three different levels.

Regarding the seedlings' height, it was noted that Arad provenance seeds had higher values in both treatments (41.0 cm for sulphuric acid treatment and 42.0 cm for thermal treatment). On the other hand, the data obtained for Vâlcea seeds had the smallest values (9.4 cm and 8.7 cm, respectively), with significant differences from the rest of the provenances (Table 4). The smaller concentrations of sulphuric acid tested (50 and 70%) performed better in the current investigation, even though the differences were not statistically assured.

The investigation of the diameter of seedlings confirmed that seeds from Arad provenance are valuable, meaning that the highest values were statistically assured (4.7 cm for acid treatment and 4.8 cm for temperature treatment) (Table 5). Even more, the water imbibition and exposure to heat/low temperature positively influenced the growth of the seedlings after thermal shock, compared with sulphuric acid, both for black locust seedlings' height and diameter.

With reference to the number of branches, it was noted that seeds from Bihor provenance showed different results. Seeds treated with sulphuric acid had the smallest number of ramifications (3.7), while in the water and temperature treatment, the same provenance gave the highest number of branches (10.8) (Table 6).

Different populations of *R. pseudoacacia* show significant differences in terms of seed characteristics and germination capacity, but also variability in terms of the growth and development of seedlings, potentially increasing the ability of this species to adapt to different environmental conditions.

Because the data sets obtained from the combinations of provenances and treatments were large, referring to both seeds and seedlings, it was intended to reduce their size with the help of PCA but maintain all information. The principal component analysis (PCA) illustrated in Figure 5 provides a presentation of the "variability" of the variables, while retaining all the elements analysed in the current study. By identifying new variables as linear functions of those in the original data set, it was found that the first component of the PCA, respectively PC1, accounted for 40.8%, while the second component (PC2) accounted for 26.6% of the total variation observed. The values of these new variables, which successively maximize the variance, and even more are uncorrelated with each other, allowed the reduction of data sets and the summarisation of the eigenvalue problems in a descriptive rather than inferential way. Of the eight Romanian provenances, Satu-Mare and Iași form a relatively homogeneous group, but they are in opposition with Galați and Botoșani, which form another homogeneous group. The most distant provenance is Bihor, which is in a negative correlation with Bistrița Năsăud.

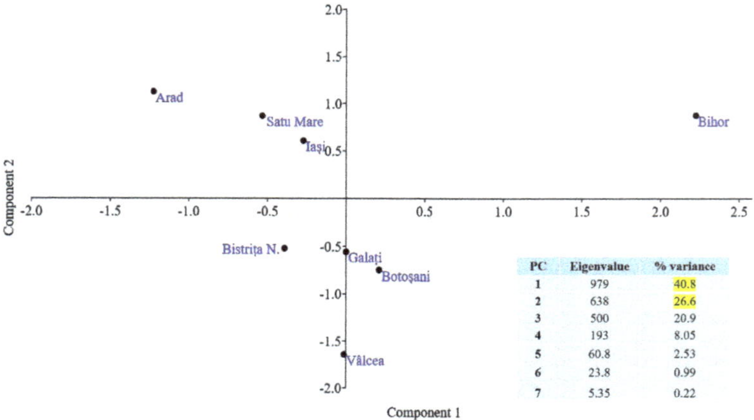

Figure 5. The hierarchical dendrogram of cluster analysis–paired group UPGMA (Unweighted Pair Group Method with Arithmetic Mean)–similarity index (Euclidean), based on the analysed parameters of the *R. pseudoacacia* provenances.

Furthermore, the obtained dendrogram (Figure 6) shows the clusters of the provenances according to the grouping of the observations made and the levels of similarity. The tree diagram confirms the results presented above, both by the pattern by which the clusters were formed and also by the similarity or distance levels of the clusters that contain the analysed *R. pseudoacacia* provenances.

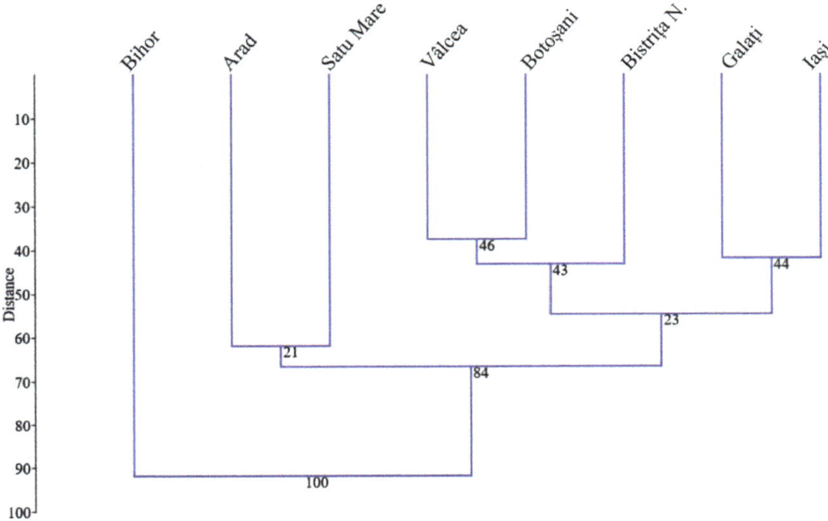

Figure 6. The hierarchical dendrogram of cluster analysis based on the analysed parameters of the *R. pseudoacacia* provenances.

As stated in the discussion about PCA analysis, the distinct cluster of Bihor provenance is obvious. Another cluster consists of Arad and Satu-Mare, separated by a third cluster, the last with two distinct subclusters. Of these subclusters, the one located at the right end of the dendrogram consists of two provenances with certain geographical proximity (Galați and Iași). The other subcluster, which occurs at a close similarity level, contains three provenances—two grouped together at the same level of similarity or distance (Vâlcea and Botoșani) and a separate one (Bistrița Năsăud).

4. Discussion

The management of genetic resources of forest species and their efficient conservation suffer from variations in traits and seed germination [57]. Morphological traits noted for black locust seeds indicated that the origin was relevant for the current investigation, with the data being quite variable. The differences among investigated populations were statistically assured, and the germination was surely influenced by the ecological parameters in the area of provenance, along with the applied treatment.

The environmental heterogeneity of the black locust seed origin within the Romanian provenances may finally result in different germination performances. Indeed, variability in the germination responses of seeds from different collection sites is well documented even in other Fabaceae species [58,59]. According to the specialty literature, larger seeds tend to have a higher germination rate, mainly because it is assumed they contain more resources to support such an intense biological process, which needs a great deal of energy [60] and also leads to healthier seedlings [61]. In the current investigation, Bihor had the largest seeds and also promoted a good germination within the thermal treatment. It is interesting to further analyse the fact that Satu-Mare seeds, which had significant higher values for seed traits, were not found among the provenances with a good germination within all treatments, nor for seedlings with a significant height and diameter. The longest seeds were found from Bihor provenance, with distinct significative differences, which also had the best germination rate for the thermal treatment; interestingly, however, seeds from the same origin also had the poorest germination after being treated with sulphuric acid at the three concentrations tested.

For the germination assay, high germination rates were obtained by immersing seeds in boiled water at temperatures of 60–80 °C, with a soaking time between 20 min to 72 h and several heating water cycles/cooling [62,63]. It can be concluded that the results can be influenced by the treatment that should destroy the seeds' tegument, on which the physiological parameters can interact. Thus, seeds collected in sites characterised by different environmental parameters display a noteworthy difference in the germination dynamics [53], and this aspect should always be considered when procuring biological material and correlated with the final goal.

There are several studies that have shown that the application of sulphuric acid can be more effective compared with other treatments for *R. pseudoacacia* seed germination [48,60,64]. Black locust seeds have a hard and waterproof tegument [65], meaning that scarification is usually applied [61]. Nevertheless [7], seeds soaked for 24 h stimulated a final value of 8% as a germination percentage. By heating the seeds for two hours at 45 °C and 60 °C, followed by two hours at 20 °C as a hot–cold variation, germination reached 23% and 69%, respectively. Thus, the variation among treatments is real, and the best method is still under investigation. Furthermore, in the current investigation, treatments with sulphuric acid were less efficient (the values ranged between 18–30.8%) than the thermic shock (values between 39.8 and 41.2%). Even more, it was observed [7] that the best results on germination capacity were obtained by mechanical scarification [66]. Furthermore, the experiment showed that germination capacity increased with the best results after extreme hot–cold thermal variation (60 °C, −20 °C). It is assumed that after thermal shock, micro-cracks appear in the tegument, meaning that water enters rapidly and light stimuli facilitate the growth processes.

The treatments applied to black locust seeds with both sulphuric acid and thermal differences influenced the germination during the period in which the observations were made. As a whole, the current results confirmed the phenotypic plasticity of black locust as a response to variable water availability and provided evidence for the potential high germination capacity when thermal shock is properly managed. It is important to note the practical results obtained and the resulting grouping of the data, using principal component analysis, for black locust seed germination and the hierarchical dendrogram of cluster analysis; in this context, Bihor provenance stands out, followed by Arad and Satu-Mare, which were at the same level of similarity, as seen in the tree dendogram. It will be interesting to further investigate the possibility to test the seedlings in an experimental field and conduct different studies of similarities (differences) upon their growth.

The geographical area potentially occupied or invaded by *R. pseudoacacia* is expected to increase significantly under the future scenario of global warming [67], despite the germination difficulties. Hence, it is highly recommended to integrate bio-ecological information and to study the species' behaviour and its adaptive capacity [68] under varying environmental factors, with the aim of planning its management and control [69,70]. Conversely, the species tolerates drastic variations of soil water availability in Central Europe and is thus becoming an important tree species to be cultivated on dry, even marginal lands [71]. Nevertheless, the black locust genus is not native to Romania, and some even consider it invasive, but its uses can overcome this inconvenience when proper techniques are applied and the cultivation is made rationally.

5. Conclusions

Germination assays performed in the current study revealed a noteworthy variability in seed germination responses across provenances. The different germination behaviour of *R. pseudoacacia* seeds of the examined Romanian populations may be related to the different environmental parameters of the collection sites. The thermal shock applied to black locust seeds can be concluded to be more efficient with regard to their germination compared with the sulphuric acid treatment. The exposure to heat/cold significantly improved germination but also black locust seedlings' height, diameter and number of branches. As a whole, the current results confirmed that the phenotypic plasticity of black locust may be a response to variable ecological parameters and provided evidence for its potential high germination capacity with proper treatments.

Author Contributions: Data curation, A.M.R.; Formal analysis, A.M.T., C.D., R.E.S. and L.H.; Investigation, A.M.T. and I.M.M.; Methodology, A.M.T., I.M.M. and C.D.; Project administration, R.E.S.; Resources, A.F.S.; Software, A.F.S.; Supervision, V.S. and R.E.S.; Validation, L.H. and R.E.S.; Visualisation, O.V.; Writing—original draft, C.D.; Writing—review and editing, C.D., I.M.M. and A.F.S. All authors have read and agreed to the published version of the manuscript.

Funding: This research was funded by University of Agricultural Sciences and Veterinary Medicine Cluj-Napoca (UASVM) grant number 26011/16.12.2020. The research was partly sustained by Doctoral School from the UASVM, during the Ph.D. study stage granted to A.M.R.

Institutional Review Board Statement: Not applicable.

Data Availability Statement: Not applicable.

Conflicts of Interest: The authors declare no conflict of interest.

References

1. Rédei, K.; Keserű, Z.; Rásó, J. Early evaluation of micropropagated black locust (*Robinia pseudoacacia* L.) clones in Hungary. *For. Sci. Pract.* **2013**, *15*, 81–84. [CrossRef]
2. Jastrzębowski, S.; Ukalska, J.; Kantorowicz, W.; Klisz, M.; Wojda, T.; Sułkowska, M. Effects of thermal-time artificial scarification on the germination dynamics of black locust (*Robinia pseudoacacia* L.) seeds. *Eur. J. For. Res.* **2017**, *136*, 471–479. [CrossRef]
3. Huntley, B. European vegetation history: Palaeovegetation maps from pollen data-13 000 yr BP to present. *J. Quat. Sci.* **1990**, *5*, 103–122. [CrossRef]
4. Pepe, M.; Gratani, L.; Fabrini, G.; Varone, L. Seed germination traits of Ailanthus altissima, Phytolacca americana and *Robinia pseudoacacia* in response to different thermal and light requirements. *Plant Species Biol.* **2020**, *35*, 300–314. [CrossRef]
5. Keresztesi, B. Black Locust: The Tree of Agriculture. *Outlook Agric.* **1988**, *17*, 77–85. [CrossRef]
6. Vítková, M.; Müllerová, J.; Sádlo, J.; Pergl, J.; Pyšek, P. Black locust (*Robinia pseudoacacia*) beloved and despised: A story of an invasive tree in Central Europe. *For. Ecol. Manag.* **2017**, *384*, 287–302. [CrossRef] [PubMed]
7. Carl, C.; Lehmann, J.R.K.; Landgraf, D.; Pretzsch, H. *Robinia pseudoacacia* L. in Short Rotation Coppice: Seed and Stump Shoot Reproduction as well as UAS-based Spreading Analysis. *Forests* **2019**, *10*, 235. [CrossRef]
8. Lukasiewicz, M.; Kowalski, S.; Makarewicz, M. Antimicrobial an antioxidant activity of selected Polish herbhoneys. *LWT* **2015**, *64*, 547–553. [CrossRef]
9. Quinkenstein, A.; Jochheim, H. Assessing the carbon sequestration potential of poplar and black locust short rotation coppices on mine reclamation sites in Eastern Germany—Model development and application. *J. Environ. Manag.* **2016**, *168*, 53–66. [CrossRef] [PubMed]
10. Nicolescu, V.-N.; Hernea, C.; Bakti, B.; Keserű, Z.; Antal, B.; Rédei, K. Black locust (*Robinia pseudoacacia* L.) as a multi-purpose tree species in Hungary and Romania: A review. *J. For. Res.* **2018**, *29*, 1449–1463. [CrossRef]
11. Brus, R. Current Occurrence of Non-Native Tree Species in European Forest. In Proceedings of the Joint WG Meeting of COST Action 402 Non-Native Tree Species for European Forests Experiences, Risks and Opportunities (NNEXT), Lisbon, Portugal, 4 October 2016.
12. Tateno, R.; Tokuchi, N.; Yamanaka, N.; Du, S.; Otsuki, K.; Shimamura, T.; Xue, Z.; Wang, S.; Hou, Q. Comparison of litterfall production and leaf litter decomposition between an exotic black locust plantation and an indigenous oak forest near Yan'an on the Loess Plateau, China. *For. Ecol. Manag.* **2007**, *241*, 84–90. [CrossRef]
13. Grünewald, H.; Böhm, C.; Quinkenstein, A.; Grundmann, P.; Eberts, J.; Von Wühlisch, G. *Robinia pseudoacacia* L.: A Lesser Known Tree Species for Biomass Production. *BioEnergy Res.* **2009**, *2*, 123–133. [CrossRef]
14. Balat, M. Bio-Oil Production from Pyrolysis of Black Locust (*Robinia pseudoacacia*) Wood. *Energy Explor. Exploit.* **2010**, *28*, 173–186. [CrossRef]
15. González-García, S.; Gasol, C.M.; Moreira, M.T.; Gabarrell, X.; i Pons, J.R.; Feijoo, G. Environmental assessment of black locust (*Robinia pseudoacacia* L.)-based ethanol as potential transport fuel. *Int. J. Life Cycle Assess.* **2011**, *16*, 465–477. [CrossRef]
16. Mally, R.; Ward, S.F.; Trombik, J.; Buszko, J.; Medzihorský, V.; Liebhold, A.M. Non-native plant drives the spatial dynamics of its herbivores: The case of black locust (*Robinia pseudoacacia*) in Europe. *NeoBiota* **2021**, *69*, 155–175. [CrossRef]
17. Cierjacks, A.; Kowarik, I.; Joshi, J.; Hempel, S.; Ristow, M.; von der Lippe, M.; Weber, E. Biological Flora of the British Isles: *Robinia pseudoacacia* . *J. Ecol.* **2013**, *101*, 1623–1640. [CrossRef]

18. Wu, W.; Li, H.; Feng, H.; Si, B.; Chen, G.; Meng, T.; Li, Y.; Siddique, K.H. Precipitation dominates the transpiration of both the economic forest (*Malus pumila*) and ecological forest (*Robinia pseudoacacia*) on the Loess Plateau after about 15 years of water depletion in deep soil. *Agric. For. Meteorol.* **2020**, *297*, 108244. [CrossRef]
19. Ussiri, D.A.N.; Lal, R.; Jacinthe, P.A. Soil Properties and Carbon Sequestration of Afforested Pastures in Reclaimed Minesoils of Ohio. *Soil Sci. Soc. Am. J.* **2006**, *70*, 1797–1806. [CrossRef]
20. Qiu, L.; Zhang, X.; Cheng, J.; Yin, X. Effects of black locust (*Robinia pseudoacacia*) on soil properties in the loessial gully region of the Loess Plateau, China. *Plant Soil* **2010**, *332*, 207–217. [CrossRef]
21. Wang, B.; Liu, G.; Xue, S. Effect of black locust (*Robinia pseudoacacia*) on soil chemical and microbiological properties in the eroded hilly area of China's Loess Plateau. *Environ. Earth Sci.* **2011**, *65*, 597–607. [CrossRef]
22. Zhang, H.; Liu, Z.; Chen, H.; Tang, M. Symbiosis of Arbuscular Mycorrhizal Fungi and *Robinia pseudoacacia* L. Improves Root Tensile Strength and Soil Aggregate Stability. *PLoS ONE* **2016**, *11*, e0153378. [CrossRef] [PubMed]
23. Dalby, R. A honey of a tree: Black locust. *Am. Bee J.* **2004**, *144*, 382–384.
24. Spyroglou, G.; Fotelli, M.; Nanos, N.; Radoglou, K. Assessing Black Locust Biomass Accumulation in Restoration Plantations. *Forests* **2021**, *12*, 1477. [CrossRef]
25. Ciuvăț, A.L.; Abrudan, I.V.; Blujdea, V.; Marcu, C.; Dinu, C.; Enescu, C.; Nuță, I.S. Distribution and peculiarities of black locust in Romania. *Rev. Silvic. Cinegetică* **2013**, *32*, 76–85.
26. Sitzia, T.; Cierjacks, A.; de Rigo, D.; Caudullo, G. Robinia pseudoacacia in Europe: Distribution, habitat, usage and threats. In *European Atlas of Forest Tree Species*; San-Miguel-Ayanz, J., de Rigo, D., Caudullo, G., Houston Durrant, T., Mauri, A., Eds.; Publication Office of the European Union: Luxembourg, 2016; pp. 166–167.
27. Nicolescu, V.-N.; Rédei, K.; Mason, W.L.; Vor, T.; Pöetzelsberger, E.; Bastien, J.-C.; Brus, R.; Benčať, T.; Đodan, M.; Cvjetkovic, B.; et al. Ecology, growth and management of black locust (*Robinia pseudoacacia* L.), a non-native species integrated into European forests. *J. For. Res.* **2020**, *31*, 1081–1101. [CrossRef]
28. Matei, I.; Pacurar, I.; Roșca, S.; Bilașco, Ș.; Sestras, P.; Rusu, T.; Jude, E.T.; Tăut, F.D. Land Use Favourability Assessment Based on Soil Characteristics and Anthropic Pollution. Case Study Somesul Mic Valley Corridor, Romania. *Agronomy* **2020**, *10*, 1245. [CrossRef]
29. Curovic, M.; Spalevic, V.; Sestras, P.; Motta, R.; Dan, C.; Garbarino, M.; Vitali, A.; Urbinati, C. Structural and ecological characteristics of mixed broadleaved old-growth forest (Biogradska Gora-Montenegro). *Turk. J. Agric. For.* **2020**, *44*, 428–438. [CrossRef]
30. Kraszkiewicz, A. Productivity of Black Locust (*Robinia pseudoacacia* L.) Grown on a Varying Habitats in Southeastern Poland. *Forests* **2021**, *12*, 470. [CrossRef]
31. Deneau, K.A. The Effects of Black Locust (*Robinia pseudoacacia* L.) on Understory Vegetation and Soils in a Northern Hardwood Forest. Master's Thesis, Michigan Technological University, Houghton, MI, USA, 2013.
32. Annighöfer, P.; Mölder, I.; Zerbe, S.; Kawaletz, H.; Terwei, A.; Ammer, C. Biomass functions for the two alien tree species Prunus serotina Ehrh. and *Robinia pseudoacacia* L. in floodplain forests of Northern Italy. *Eur. J. For. Res.* **2012**, *131*, 1619–1635. [CrossRef]
33. Wei, X.; Liang, W. Multifactor relationships between stand structure and soil and water conservation functions of *Robinia pseudoacacia* L. in the Loess Region. *PLoS ONE* **2019**, *14*, e0219499. [CrossRef]
34. Allam, A.; Borsali, A.H.; Kefifa, A.; Zouidi, M.; DA Silva, A.M.F.; Rébufa, C. Impact of water erosion on the properties of forest soils. *Not. Sci. Biol.* **2021**, *13*, 10921. [CrossRef]
35. Sestras, P.; Bondrea, M.V.; Cetean, H.; Sălăgean, T.; Bilaşco, Ş.; Naș, S.; Spalevic, V.; Fountas, S.; Cîmpeanu, S.M. Ameliorative, Ecological and Landscape Roles of Făget Forest, Cluj-Napoca, Romania, and Possibilities of Avoiding Risks Based on GIS Landslide Susceptibility Map. *Not. Bot. Horti Agrobot. Cluj-Napoca* **2017**, *46*, 292–300. [CrossRef]
36. Hou, G.; Bi, H.; Wei, X.; Wang, N.; Cui, Y.; Zhao, D.; Ma, X.; Wang, S. Optimal configuration of stand structures in a low-efficiency *Robinia pseudoacacia* forest based on a comprehensive index of soil and water conservation ecological benefits. *Ecol. Indic.* **2020**, *114*, 106308. [CrossRef]
37. Ciuvat, A.L.; Abrudan, I.V.; Blujdea, V.; Dutca, I.; Nuta, I.S.; Edu, E. Biomass Equations and Carbon Content of Young Black Locust (*Robinia pseudoacacia* L.) Trees from Plantations and Coppices on Sandy Soils in South-Western Romanian Plain. *Not. Bot. Horti Agrobot. Cluj-Napoca* **2013**, *41*, 590–592. [CrossRef]
38. Olson, D.F.J.; Karrfalt, R.P. *The Woody Plant Seed Manual*; Robinia, L.L., Bonner, F., Karrfalt, R.P., Eds.; Agriculture Handbook; USDA Forest Service: Washington, DC, USA, 2008; Volume 727.
39. DeGomez, T.; Wagner, M.R. Culture and Use of Black Locust. *HortTechnology* **2001**, *11*, 279–288. [CrossRef]
40. Rédei, K.; Csiha, I.; Keserű, Z.; Gál, J. Influence of Regeneration Method on the Yield and Stem Quality of Black Locust (*Robinia pseudoacacia* L.) Stands: A Case Study. *Acta Silv. Lignaria Hung.* **2012**, *8*, 103–112. [CrossRef]
41. Roman, A.M.; Morar, I.M.; Truța, A.M.; Dan, C.; Sestraș, A.F.; Holonec, L.; Ioras, F.; Sestras, R.E. Trees, seeds and seedlings analyses in the process of obtaining a quality planting material for black locust (*Robinia pseudoacacia* L.). *Not. Sci. Biol.* **2020**, *12*, 940–958. [CrossRef]
42. Thapliyal, M.; Kaliyathan, N.N.; Rathore, K. Seed germination response of Indian wild pear (*Pyrus pashia*) to gibberellic acid treatment and cold storage. *Not. Sci. Biol.* **2021**, *13*, 11044. [CrossRef]
43. Lewak, S. Metabolic control of embryonic dormancy in apple seed: Seven decades of research. *Acta Physiol. Plant.* **2011**, *33*, 1–24. [CrossRef]

44. Borkowska, L. Patterns of seedling recruitment in experimental gaps on mosaic vegetation of abandoned meadows. *Acta Soc. Bot. Pol.* **2011**, *73*, 343–350. [CrossRef]
45. Basbag, M.; Aydin, A.; Ayzit, D. The Effect of Different Temperatures and Durations on the Dormancy Breaking of Black Locust (*Robinia pseudoacacia* L.) and Honey Locust (*Gleditsia triacanthos* L.) Seeds. *Not. Sci. Biol.* **2010**, *2*, 125–128. [CrossRef]
46. Mondoni, A.; Tazzari, E.; Zubani, L.; Orsenigo, S.; Rossi, G. Percussion as an effective seed treatment for herbaceous legumes (Fabaceae): Implications for habitat restoration and agriculture. *Seed Sci. Technol.* **2013**, *41*, 175–187. [CrossRef]
47. Zencirkiran, M.; Tümsavaş, Z.; Ünal, H. The Effects of Different Acid Treatment and Stratification Duration on Germination of Cercis siliquastrum L. Seeds. *Not. Bot. Horti Agrobot. Cluj-Napoca* **2010**, *38*, 159–163. [CrossRef]
48. Mirzaei, M.; Moghadam, A.R.L.; Ardebili, Z.O. The induction of seed germination using sulfuric acid, gibberellic acid and hot water in *Robinia pseudoacacia* L. *Int. Res. J. Appl. Basic Sci.* **2013**, *4*, 96–98.
49. Pârnuţă, G.S.E.; Budeanu, M.; Scărlătescu, V.; Marica, F.M.; Lalu, I.; Curtu, A.L. *Catalogul Naţional al Resurselor Genetice 462 Forestiere [National Catalogue of Forest Genetic Resources]*; Editura Silvică: Bucharest, Romania, 2011. (In Romanian)
50. De Micco, V.; Aronne, G. Morpho-Anatomical Traits for Plant Adaptation to Drought. In *Plant Responses to Drought Stress*; Springer: Berlin/Heidelberg, Germany, 2012; pp. 37–61.
51. Belokopytova, L.; Zhirnova, D.; Kostyakova, T.; Babushkina, E. Dynamics of moisture regime and its reconstruction from a tree-ring width chronology of Pinus sylvestris in the downstream basin of the Selenga River, Russia. *J. Arid. Land* **2018**, *10*, 877–891. [CrossRef]
52. Yang, B.; Peng, C.; Harrison, S.P.; Wei, H.; Wang, H.; Zhu, Q.; Wang, M. Allocation Mechanisms of Non-Structural Carbohydrates of *Robinia pseudoacacia* L. Seedlings in Response to Drought and Waterlogging. *Forests* **2018**, *9*, 754. [CrossRef]
53. Giuliani, C.; Lazzaro, L.; Calamassi, R.; Fico, G.; Foggi, B.; Lippi, M.M. Induced water stress affects seed germination response and root anatomy in *Robinia pseudoacacia* (Fabaceae). *Trees* **2019**, *33*, 1627–1638. [CrossRef]
54. Sestras, A.F. *Biostatistica si Tehnica Experimentala Forestiera: Manual Didactic*; Editura Academic Press: Cluj-Napoca, Romania, 2018.
55. Hammer, Ø.; Harper, D.A.; Ryan, P.D. PAST: Paleontological statistics software package for education and data analysis. *Palaeontol. Electron.* **2001**, *4*, 4–9.
56. Sestras, A.F. *Modele Statistice Aplicate în Cercetarea Horticola*; Editura Risoprint: Cluj-Napoca, Romania, 2019.
57. Rajora, O.P.; Mosseler, A. Challenges and opportunities for conservation of forest genetic resources. *Euphytica* **2001**, *118*, 197–212. [CrossRef]
58. Tozer, M.G.; Ooi, M. Humidity-regulated dormancy onset in the Fabaceae: A conceptual model and its ecological implications for the Australian wattle Acacia saligna. *Ann. Bot.* **2014**, *114*, 579–590. [CrossRef]
59. Safdar, M.E.; Wang, X.; Abbas, M.; Ozaslan, C.; Asif, M.; Adnan, M.; Zuan, A.T.K.; Wang, W.; Gasparovic, K.; Nasif, O.; et al. The impact of aqueous and N-hexane extracts of three Fabaceae species on seed germination and seedling growth of some broadleaved weed species. *PLoS ONE* **2021**, *16*, e0258920. [CrossRef] [PubMed]
60. Kheloufi, A.; Mansouri, L.; Aziz, N.; Sahnoune, M.; Boukemiche, S.; Ababsa, B. Breaking seed coat dormancy of six tree species. *Reforesta* **2018**, *5*, 4–14. [CrossRef]
61. Pedrol, N.; Puig, C.G.; López-Nogueira, A.; Pardo-Muras, M.; González, L.; Alonso, P.S. Optimal and synchronized germination of *Robinia pseudoacacia*, Acacia dealbata and other woody Fabaceae using a handheld rotary tool: Concomitant reduction of physical and physiological seed dormancy. *J. For. Res.* **2017**, *29*, 283–290. [CrossRef]
62. Kolbek, J.V.M.; Vetvicka, V. From history of Central European Robinia growths and its communities. *Zprávy České Bot. Společnosti* **2004**, *39*, 287–298.
63. Masaka, K.; Yamada, K. Variation in germination character of *Robinia pseudoacacia* L. (Leguminosae) seeds at individual tree level. *J. For. Res.* **2009**, *14*, 167–177. [CrossRef]
64. Zoghi, Z.; Azadfar, D.; Kooch, Y. The Effect of Different Treatments on Seeds Dormancy Breaking and Germination of Caspian Locust (*Gleditschia caspica*) Tree. *Ann. Biol. Res.* **2011**, *2*, 400–406.
65. Staska, B.; Essl, F.; Samimi, C. Density and age of invasive *Robinia pseudoacacia* modulate its impact on floodplain forests. *Basic Appl. Ecol.* **2014**, *15*, 551–558. [CrossRef]
66. Rédei, K.; Csiha, I.; Keserű, Z.; Végh Ágnes, K.; Gyori, J. The Silviculture of Black Locust (*Robinia pseudoacacia* L.) in Hungary: A Review. *South-East Eur. For.* **2011**, *2*, 101–107. [CrossRef]
67. Kleinbauer, I.; Dullinger, S.; Peterseil, J.; Essl, F. Climate change might drive the invasive tree *Robinia pseudacacia* into nature reserves and endangered habitats. *Biol. Conserv.* **2010**, *143*, 382–390. [CrossRef]
68. Jin, T.T.; Fu, B.J.; Liu, G.H.; Wang, Z. Hydrologic feasibility of artificial forestation in the semi-arid Loess Plateau of China. *Hydrol. Earth Syst. Sci.* **2011**, *15*, 2519–2530. [CrossRef]
69. Bouteiller, X.P.; Porté, A.J.; Mariette, S.; Monty, A. Using automated sanding to homogeneously break seed dormancy in black locust (*Robinia pseudoacacia* L., Fabaceae). *Seed Sci. Res.* **2017**, *27*, 243–250. [CrossRef]
70. Bouteiller, X.P.; Barraquand, F.; Garnier-Géré, P.; Harmand, N.; Laizet, Y.; Raimbault, A.; Segura, R.; Lassois, L.; Monty, A.; Verdu, C.; et al. No evidence for genetic differentiation in juvenile traits between Belgian and French populations of the invasive tree *Robinia pseudoacacia* . *Plant Ecol. Evol.* **2018**, *151*, 5–17. [CrossRef]
71. Mantovani, D.; Veste, M.; Freese, D. Black locust (*Robinia pseudoacacia* L.) ecophysiological and morphological adaptations to drought and their consequence on biomass production and water-use efficiency. *N. Z. J. For. Sci.* **2014**, *44*, 29. [CrossRef]

Article

Conservation of Genetic Diversity of Scots Pine (*Pinus sylvestris* L.) in a Central European National Park Based on cpDNA Studies

Paweł Przybylski [1,*], Anna Tereba [1], Joanna Meger [2], Iwona Szyp-Borowska [1] and Łukasz Tyburski [3]

[1] Forest Research Institute, Braci Leśnej 3, Sękocin Stary, 05-090 Raszyn, Poland; A.Tereba@ibles.waw.pl (A.T.); i.borowska@ibles.waw.pl (I.S.-B.)
[2] Department of Genetics, Kazimierz Wielki University, ul. Chodkiewicza 30, 85-064 Bydgoszcz, Poland; warmbier@ukw.edu.pl
[3] Kampinoski National Park, Tetmajera 38, 05-080 Izabelin, Poland; ltyburski@kampinoski-pn.gov.pl
* Correspondence: p.przybylski@ibles.waw.pl

Abstract: In the old pine stands of national parks, it is possible to observe genetic processes in a state free from disturbance by humans. Studies of this type make it possible to evaluate the effectiveness of the conservation of genetic variation and its transfer between generations. The present study was conducted in the largest national forest park in Poland, located in the Central European pine area. The oldest stands of Kampinos National Park and their natural descendants were selected for detailed analyses. The main objective of the study was to compare the mother pine stand, excluded from forest management, with its progeny generations on the basis of their chloroplast DNA (cpDNA), which was used as a diagnostic tool. The results demonstrate significant genetic difference between the maternal and progeny generations of the studied sites. The degree of variation observed in the maternal generation haplotypes in the present study was found to be reduced in the next generation. A significant proportion of the genetic diversity of the studied stands was also lost in the subsequent progeny generation. The obtained results allow conclusions to be drawn about the genetic processes taking place in valuable old-growth forests.

Keywords: old growth forests; gene flow; molecular markers; legally protected forests

Citation: Przybylski, P.; Tereba, A.; Meger, J.; Szyp-Borowska, I.; Tyburski, Ł. Conservation of Genetic Diversity of Scots Pine (*Pinus sylvestris* L.) in a Central European National Park Based on cpDNA Studies. *Diversity* **2022**, *14*, 93. https://doi.org/10.3390/d14020093

Academic Editors: Mario A. Pagnotta, Lucian Dinca and Miglena Zhiyanski

Received: 27 November 2021
Accepted: 26 January 2022
Published: 28 January 2022

1. Introduction

Scots pine (*Pinus sylvestris* L.), the species investigated in this study, is the second most widespread conifer in the world and has great economic and ecological importance [1]. Pine forests cover 37% of the total global land area and 70% of the land area of the Northern Hemisphere, making pine one of the most important forest-forming trees in the world [1]. Pine is characterised by high morphological and genetic diversity. The maintenance of considerable pine variability is favoured by the transmission of pine genetic information by seeds [2] and pollen [3]. Natural processes affecting the genetic diversity of pine populations can be roughly observed in national parks. In economic stands, forest management leads to the suppression of natural gene flow due to the fragmentation of forest sites, which results from regulation based on forest stand age [4]. The present study was conducted in the largest national forest park in Poland, Kampinos National Park (KNP). Unfortunately, the history of the Kampinos forests indicates periods of the complete deforestation of the national park area influenced by armed conflicts of a global character [5]. On the other hand, the studied ecosystems have been under full legal protection for over 50 years. Therefore, it can be assumed that while the genetic variability of the ancient forests of KNP has been influenced by humans, their regeneration processes are spontaneous. The presented results can be interpreted in the context of adaptation of pine trees to different growth conditions, which follows the results of earlier studies by Przybylski et al. [6].

The results of the presented studies, based on Krzanowska et al. [7] and others, suggest that stands with a high degree of genetic differentiation have a much higher chance of survival and can consequently pass on favourable gene combinations to their offspring. The dynamic changes in climate observed around the world are producing ecosystems that are in a state of dynamic equilibrium in responding to this change. Changes in stands are observed to occur both within and between species [8]. It is undisputed that genetic diversity is important for the long-term survival of species and plays a crucial role in their conservation [9–11]. The genetic diversity of organisms has a major influence on both the adaptive potential of individuals and the response of entire populations to external selection factors [12,13]. However, regulations and policies have long been primarily focused on the more visible elements of biodiversity, such as species or populations, while genetics has been mostly neglected [14]. Ongoing genetic research mainly focuses on stands before and after certain events that have a significant impact on genetic pools, with little work in natural populations under legal protection. The associated analyses usually employ autosomal markers such as microsatellite loci [15,16]. The work presented here replicates the trend of microsatellite DNA sequence analyses, and it should be noted that the study focused on microsatellites in chloroplast DNA (cpDNA). A trend in the study of conifers is the increasing importance of chloroplast DNA analyses (cpSSRs), as observed in, e.g., Fady et al. [17] and Gómez et al. [18]. In Scots pine in Poland, cpDNA has mainly been used to detect hybridisation between *P. sylvestris* × *P. mugo* [19]. Another important scientific study is the description of the complete chloroplast genome of *Pinus uliginosa* (Neumann), which is helpful in explaining the complex taxonomic position of the species [20]. Only a few studies, such as Semerikov et al. [21], have used cpDNA to describe variation in pine stands. The study mentioned above shows a variation between natural populations in Asia and Eastern Europe of 2.1% [21], and slightly lower results were found for populations from Estonia [22]. In Poland, a study of cpDNA diversity between age classes of stocks was conducted [23], which showed a similar level of diversity as between different populations. Wojnicka-Półtorak et al. [23] point to the transfer of foreign pollen to the parent population as one of the reasons for the detected differentiation.

The research presented in this paper focuses on stands excluded from forest management and protected by law in Kampinos National Park, the largest national forest park in Poland. Due to historical events (for example, the First and Second World Wars), it is likely that evolution did not occur by natural regeneration for the stands in the studied areas, in contrast to their progeny. Therefore, the analysis of changes in the gene pool is interesting in terms of the possibility of the intergenerational preservation of genetic diversity and the intensity of gene flow. The formulated hypotheses of the conducted research state that allelic diversity may be preserved in the next generation of a stand. At the same time, it is suspected that pollen from local commercial forest stands influences the formation of the gene pool of the subsequent generation. The main objective of the research presented in this publication is to compare the maternal and progeny generations of pine stands excluded from forest management from the perspective of cpDNA genetic diversity. The obtained results will expand upon basic knowledge regarding the behaviour of pine in natural conditions.

2. Materials and Methods

2.1. Research Area

The study was conducted in Kampinos National Park (52°19′13″ N, 20°47′23″ E), a location dominated by Scots stands in the upper layer and located in strictly protected areas. The characteristics of the study locations are presented in Table 1 and Figure 1.

Table 1. The characteristics of the study locations.

Location	Czerwińskie Góry	Wilków	Granica	Sieraków	Wiersze	Nart	Krzywa Góra
Abbreviation	CG	W	Gr	S	Wi	N	KG
Coordinates	20°23′36.67″ E 52°20′27.693″ N	20°32′34.005″ E 52°21′45.607″ N	20°27′50.019″ E 52°17′21.605″ N	20°46′34.957″ E 52°20′12.144″ N	20°39′45.706″ E 52°18′34.157″ N	20°30′2.718″ E 52°17′49.844″ N	20°25′14.603″ E 52°20′27.744″ N
Age * of the dominant *P. sylvestris*	200–210 (avg.: 205)	180–200 (avg.: 190)	160–170 (avg.: 165)	190–200 (avg.: 195)	app. 160	210–230 (avg.: 220)	app. 108
Plant community	*Querco roboris-Pinetum*	*Querco roboris-Pinetum*	SNFPC/*Querco Carpinetum*	*Querco roboris-Pinetum*	*Querco roboris-Pinetum*	*Tilio-Carpinetum*	*Querco roboris-Pinetum*

* Unpublished data from Kampinoski National Park.

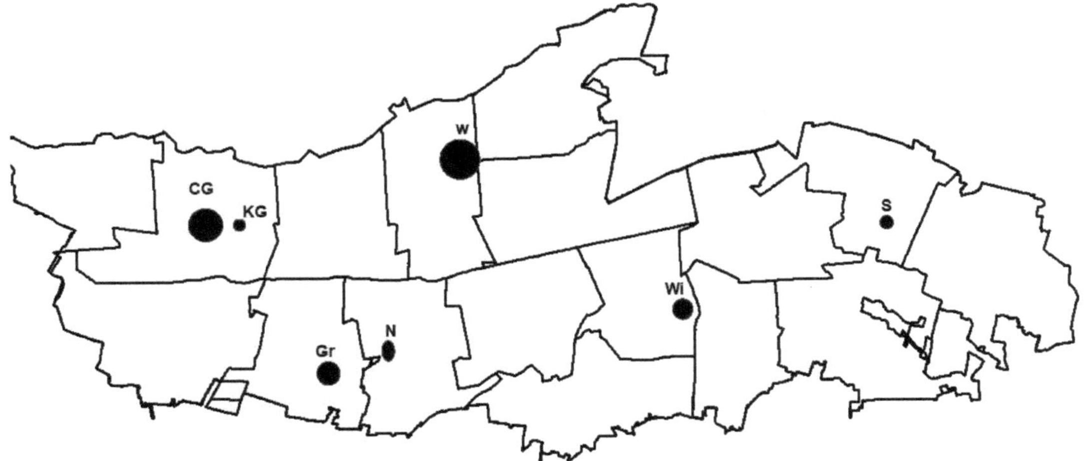

Figure 1. Location of the studied stands in the Kampinos NP area. Points mark survey locations based on their area according to the outermost sampled trees. The size of the marked point is based on the differences in the plots. Description according to the abbreviations used.

2.2. Sample Collection

Plant material (needles) was collected from each location (Table 1) from 50 randomly selected pines constituting the maternal generation (F1) and 50 randomly selected 1-year-old seedlings, referred to as the progeny generation (F2). The condition for selecting F1 was the distance between samples of at least one height of the dominant stand, while F2 was collected in close proximity to F1 (see Tables S1–S7 in the Supplementary Materials). The planned number of individuals of the F2 generation was not collected for most populations because there were no pine seedlings at the study sites (Tables S1–S7, Supplementary Materials). In addition, the vigour of some samples of the F2 generation was poor. The material in the field was placed in 1.5 mL Eppendorf tubes then was stored in the laboratory under cryogenic conditions (−85 °C).

2.3. DNA Extraction and Microsatellites Genotyping

Total genomic DNA was isolated from the collected material using a commercial kit (Macherey–Nagel Gmbh&Co Valenciennes Str. 11; 52,355, Dueren, Germany). The quality of the DNA isolate was controlled using 2% agarose gel and a Quawell (LabX, 334 King street, Midland, ON Canada) spectrophotometer. All samples were diluted to 20–30 ng/uL using deionised water. Molecular analyses were performed using 6 (PCP26106; PCP30277; PCP36567; PCP450712; PCP719872; PCP873142) chloroplast microsatellite markers selected on the basis of other research work [24–26]; the forward primers were fluorescently labelled with the fluorochromes VIC, PET, NED, and 6-FAM. Amplification was performed through two multiplex reactions. Each PCR reaction was performed in a volume of 10 µL, with the following composition: 5 μL Multiplex buffer (Qiagen, Poland), 0.2 μL (10 μM) of each primer, 1 μL of extracted DNA, and PCR-grade water up to a final volume of 10 μL. The PCR thermal profile was as follows: 95 °C for 15 min; followed by 30 cycles at 95 °C for 30 s, 55 °C for 30 s, and 72 ° C for 1 min, with a final extension of 60 °C for 30 min. Genotyping analysis was performed on an ABI 3500 Genetic Analyzer capillary sequencer (Applied Biosystems, Foster City, CA, USA) and allele length analysis was performed using GeneMapper® version 5 (Thermo Fisher Scientific Inc., Carlsbad, CA, USA).

2.4. Data Analysis

Haplotypes were determined as a combination of different microsatellite variants across the cpDNA loci. The chloroplast haplotype variation within populations, i.e., the number of haplotypes (*A*), genetic diversity (*He*), haplotype richness (*Rh*), the number of private haplotypes (*P*), and mean genetic distance between individuals (D^2sh) were calculated using HAPLOTYPE version 1.05 [27].

Genetic structuring of the cpDNA between and within populations was assessed using analysis of molecular variance (AMOVA), implemented in ARLEQUIN version 3.0 [28], with significance tests based on 10,000 permutations. The distance matrix generated by pairwise F_{ST} values between populations was calculated using ARLEQUIN version 3.0 [25]. The statistical significance of F_{ST} values was assessed using 10,000 permutations. The grouping of analysed stands on the basis of the genetic distance of N_{ei} using Principal Coordinates Analisis (PCoA) as implemented in the GenALEx 6.5 [29] program.

The isolation by distance was tested using the Mantel test [30] of association between the F_{ST}-based pairwise genetic distance matrix (i.e., $F_{ij}/(1 - F_{ij})$, where F_{ij} is the F_{ST} for the i-th and the j-th populations) and the matrix of the natural logarithm of geographic distance using GENEPOP version 4.4 software [31]. The significance was assessed with 10,000 random permutations.

3. Results

An average of 4.07 alleles were found in F1 generation populations (Table S8 in the Supplementary Materials), with most in the S location (4.5) and the least in CG (3.7). The average effective number of alleles (Ne) was 2.2, between 2.3 and 2.0. Private allele frequencies were described for S (0.5), Wi (0.33), and KG (0.17), while no private alleles were detected for the other populations. The genetic diversity in the 6 stands of the F1 generation was equal to h = 0.5, except for N, where it was 0.4.

For the F2 generation, the average number of alleles was 3.83 (Table S8 in the Supplementary Materials), with the highest in the KG locality (4.2), while the lowest was in S (3.5). The average effective number of alleles (Ne) in the F2 generation was 2.1, between 2.4 and 1.8. Private allele frequencies were indicated for the locations S (0.16), KG (0.33), and Gr (0.33). The genetic diversity in the 6 stands of the F2 generation was equal to h = 0.5, except for S where it was 0.4. The genetic variability of haplotypes between F1 and F2 of the analysed locations is shown in Figure 2. Note that most of the generated haplotypes are identical for the analysed populations. Genetic diversity arises from differences in allele frequencies between populations and generations and from the occurrence of private alleles.

The genetic diversity parameters of the studied locations are summarised in Table 2. At the studied sites, results were obtained for an average of 47.28 plants of the maternal generation (F1) and an average of 34.42 plants of the progeny generation (F2) (Table 2). The determined number of haplotypes (*A*) ranged from 32–44 for F1 and from 15–34 for F2. The lower number of determined F2 haplotypes is related to the number of samples analysed (N) (Table 2). Private haplotypes (P_h) were indicated for each site, of which there were much more for F1. The (P_h) value was dependent on the haplotype richness (R_h). The results showed no differences in genetic diversity (H_e) among generations and locations (Table 2). The values of mean genetic distance between individuals (D^2sh) were higher for

F1 and particularly high for Sieraków (4.29) and Wiersze (5.52), which were not maintained for F2 (Table 2).

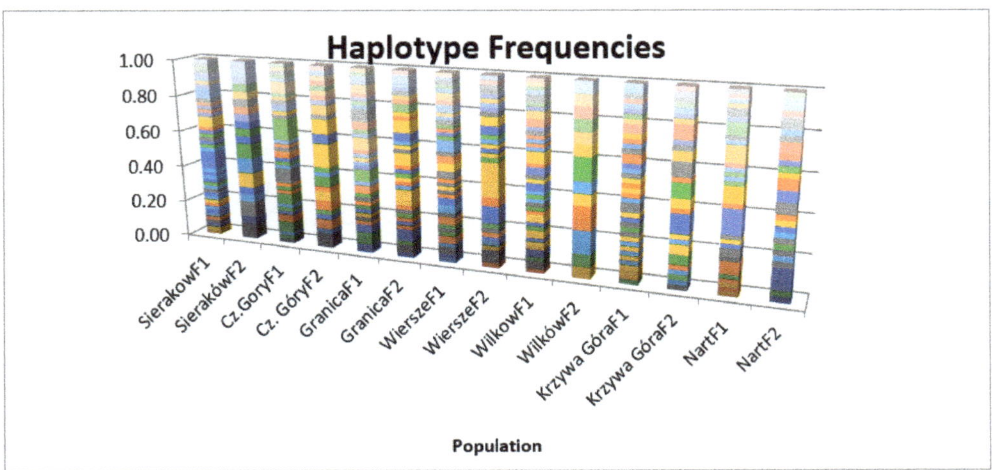

Figure 2. Haplotype frequencies expressed as a percentage (*Y*-axis) for the analysed locations of maternal (F1) and progeny (F2) generations. Each colour on the graph corresponds to one of the 254 identified haplotypes.

Table 2. Genetic diversity within a population of *P. sylvestris* L. based on 6 cpDNA markers. Number of haplotypes (*A*), the number of private haplotypes (*Ph*), haplotype richness (*Rh*), genetic diversity (*He*), and mean genetic distance between individuals (D^2sh).

Population	*N*	*A*	P_h	R_h	H_e	D^2sh
			F1 generation			
Sieraków	50	44	28	35.586	0.992	4.297
Cz. Góry	50	36	19	29.532	0.976	3.625
Granica	46	35	19	31.015	0.986	3.218
Wiersze	49	41	23	34.303	0.992	5.523
Wilków	49	40	18	33.606	0.991	2.754
Krzywa Góra	41	37	17	36.000	0.995	3.594
Nart	46	32	13	28.145	0.971	2.584
Mean	47.286	37.857	19.571	32.598	0.986	3.657
			F2 generation			
Sieraków	24	20	11	13.261	0.986	2.557
Cz. Góry	38	27	14	12.642	0.977	2.893
Granica	46	34	13	13.292	0.985	2.783
Wiersze	43	30	13	12.123	0.961	2.508
Wilków	16	15	8	14.000	0.992	3.950
Krzywa Góra	39	30	13	13.078	0.981	3.620
Nart	35	29	16	13.790	0.990	3.369
Mean	34.429	26.429	12.571	13.170	0.982	3.097

AMOVA analysis of genetic variation demonstrated the presence of significant genetic differentiation between the studied sites for populations in both F1 (p_{value} = 0.000) and F2 (p_{value} = 0.006). The intra-site variation coefficient was significant for F1 and F2 (Table 3, Figure 2) and the percentage of intra-site variation in F1 was more than 0.86, and for F2, it was almost additionally 0.11 higher (Table 3). The Mantel analysis [30] did not demonstrate any relationship between geographical and genetic distances for both F1 (p_{value} = 0.978) and F2 (p_{value} = 0.699) (Table 3, Figures S1 and S2, Supplementary Materials).

Table 3. AMOVA and IBD Mantel [30] results for the two analysed generations of stands.

Source of Variation	d.f.	Sum of Squares	Variance Components	Percentage of Variation
		F1 generation		
Among populations	6	147.358	0.22803	13.17 ***
Among individuals within populations	324	974.007	1.50310	86.83 ***
Within individuals	0	0.000	0.00000	0.00
Total	661	1121.366	1.73112	
IBD Mantel			R	*p*-value
			0.0064	0.978
		F2 generation		
Among populations	6	34.026	0.04251	2.96 **
Among individuals within populations	234	651.633	1.39238	97.04 ***
Within individuals	241	0.000	0.00000	0.00
Total	481	685.660	1.43489	
IBD Mantel			R	*p*-value
			−0.0897	0.6991

Statistical significance of *p*-value ** < 0.01; *** < 0.001.

The value obtained for genetic distance (F_{st}) between F1 locations ranged from 0.00786 to 0.25852 and was considered statistically significant in most cases (Table 4). A lack of genetic distance was detected for two F1 pairs: Granica and Czerwińskie Góry (p_{value} = 0.25586) and Wilków and Sieraków (p_{value} = 0.55176) (Table 4). In F2, the diversity coefficients ranged from 0.0018 to 0.11074 (Table 4), and in 0.47 of cases were not considered genetically diverse in terms of statistical significance. The clustering of analysed locations for F1 and F2 generations allows the visualisation of F2 aggregation into a separate subgroup relative to all locations belonging to the maternal generation (Figure 3).

Table 4. The genetic differentiation (*F*st) between populations analysed (below diagonal) and their statistical significance (*p* values) (above diagonal) results in bold are not statistically significant.

	SierakowF1	SierakówF2	Cz. GoryF1	Cz. GóryF2	GranicaF1	GranicaF2	WierszeF1	WierszeF2	WilkowF1	WilkówF2	Krzywa GóraF1	Krzywa GóraF2	NartF1	NartF2
SierakówF1		0.00098	0.00000	0.01367	0.00000	0.04102	0.00000	0.07227	**0.55176**	0.01953	0.07910	0.00000	0.00098	0.71289
SierakówF2	0.07608		0.00000	0.00684	0.00000	0.00293	0.00000	0.00000	0.00000	0.00586	0.00000	**0.14746**	0.00000	0.00000
Cz. GóryF1	0.23506	0.32715		0.00000	**0.25586**	0.00000	0.00000	0.00000	0.00000	0.00000	0.00000	0.00000	0.00000	0.00000
Cz. GóryF2	0.03988	0.06253	0.33028		0.00000	**0.85645**	0.00000	**0.22168**	0.00879	**0.69531**	0.00000	**0.65137**	0.00000	0.00098
GranicaF1	0.22296	0.32722	0.01388	0.33303		0.00000	0.00000	0.00000	0.00000	0.00000	0.00000	0.00000	0.00000	0.00000
GranicaF2	0.02662	0.06792	0.32038	0.00180	0.31591		0.00000	**0.29883**	0.00781	**0.41797**	0.00000	**0.18848**	0.00000	0.00879
WierszeF1	0.10276	0.23585	0.14030	0.22757	0.12369	0.21324		0.00000	0.00000	0.00000	0.00195	0.00000	0.00000	0.00000
WierszeF2	0.02614	0.09994	0.28641	0.02056	0.28288	0.01585	0.19672		0.00781	**0.19238**	0.00000	0.00195	0.00000	0.00195
WilkówF1	0.00786	0.07290	0.25287	0.04376	0.24788	0.04082	0.11451	0.03907		0.01562	0.03125	0.00000	0.00488	**0.25977**
WilkówF2	0.06101	0.09049	0.35240	0.01116	0.35023	0.02159	0.25149	0.03414	0.06114		0.00000	**0.28613**	0.00000	0.00488
Krzywa GóraF1	0.02506	0.13552	0.22874	0.10931	0.21859	0.09545	0.06186	0.09813	0.03079	0.11282		0.00000	0.00488	**0.09473**
Krzywa GóraF2	0.05783	0.03013	0.34615	0.00709	0.35239	0.01795	0.25523	0.04897	0.05840	0.03121	0.13168		0.00000	0.00000
NartF1	0.05378	0.17311	0.25852	0.15318	0.23628	0.13939	0.09328	0.14344	0.04390	0.16928	0.04168	0.16777		**0.07812**
NartF2	0.00611	0.11074	0.26168	0.06494	0.24697	0.04748	0.11945	0.05946	0.01751	0.08519	0.02894	0.08120	0.02831	

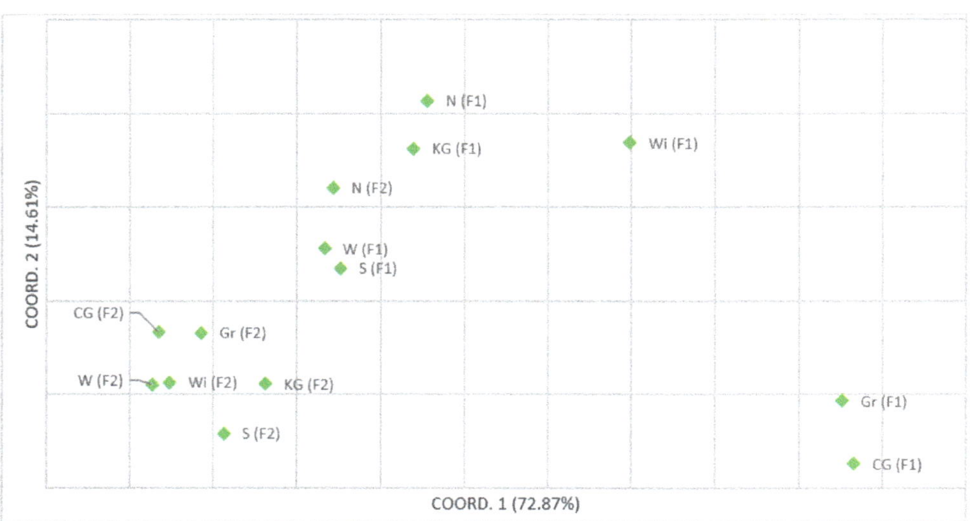

Figure 3. Grouping of studied Scot pine locations using principal coordinates analysis (PCoA) based on Nei genetic distance [32].

4. Discussion

The results of the presented study show that the maternal generation (F1) has a higher number of alleles than the progeny generation (F2). This result is further confirmed by an analysis of the effective number of alleles, the value of which is higher in the F1 generation, and the phenomenon is particularly evident for the S population. Detailed visualisation of the haplotype frequencies for F1 and F2 locations confirmed not only the depletion of the gene pool in F2 but also the presence of haplotypes foreign to F1. This tendency cannot be explained by the presence of private alleles only in F1, which could be due to individual selection during the growth phase of the stand [33], as the mean value of private alleles is identical between generations. The observed differences in the numbers of private alleles are only observed at the level of individual sites, which is evident in the example of stand S, where the value of private alleles in F2 was much lower compared to F1. The observed phenomenon of reductions in the allele numbers for other locations in the F2 generation can be explained by the preference for certain alleles during the reproductive process, which may facilitate population adjustments [34]. In the presented analyses, a more viable cause of changes in the frequency and occurrence of haplotypes may be due to the old forests being contaminated with pollen from younger, surrounding stands. KNP stands younger than those examined in the present study were planted for regeneration after war damage. Their genetic pool is therefore probably poorer in terms of diversity, and intensive pollen production is important for the creation of new generations of KNP. The effect of pollen flow between stands has been described by Lindgren et al. [35], Harju and Nikkanen [36], and Adams and Burczyk [37], among others, who confirm its role in forming the allele pools of the next generation of stands. Detailed analyses of haplotypes confirmed significantly higher values of haplotype richness (R_h) in the F1 generation of the studied sites. The values obtained for both the R_h and D^2sh parameters were almost twice as high in the F1 generation as in F2 generation. On the other hand, it should be noted that there are no clear differences in the values of genetic variability (H_e) between the studied generations. The obtained H_e results were high for both F1 and F2, which is a typical phenomenon in conifers [38]. The results demonstrate significant differentiation at the studied sites, both between and within populations. Typically for conifers, the studied stands show the largest amount of within-population genetic variation, 0.86 for F1 and 0.97 for F2. Notably, the observed variation between populations showed significance of the difference at the level of $p = 0.001$ for F1 and $p = 0.01$ for F2. The IBD–Mantel analysis [30] did not indicate any significant relationship between the values of genetic and geographical distances for the studied sites. Effective pollen transfer between the study sites should therefore be ruled out. The obtained results confirm those discussed above, and we can assume that the emergence of genetic richness was dominated by foreign pollen, probably originating from younger stands in the vicinity. However, this hypothesis requires further investigation, especially in the context of the function of national parks for biodiversity conservation at the genetic level. On 21 January 2000, KNP was recognised by the International Coordinating Council of the UNESCO MaB (Man and Biosphere) Programme as a "Puszcza Kampinoska" Biosphere Reserve with an area of 76 232.57 ha. Since 2004, a fragment of KNP with an area of 37,469.70 ha, which includes the central and buffer zones, has been part of the European Natura 2000 network as a Natura 2000 site with the number PLC 140,001 (https://www.kampinoski-pn.gov.pl/ochrona-przyrody/natura-2000, accessed on 10 April 2021) due to its high bird species richness and unique diversity of plant communities. Ensuring the stability of biodiversity, including at the genetic level, is an important aim for KNP. Considering the results obtained, it is likely that the implementation of total protection for the areas analysed in the project will result in the loss of private alleles. According to some predictions, private alleles may be carriers of genetic variation that allow populations to adapt to a changing environment in the future, even though their current impact on populations may be marginal [39,40]. For the above reasons, we recommend rare genotypes at selected naturally valuable sites be conserved through an ex situ conservation method. Under legally protected natural conditions, it may not be possible to maintain a unique set of alleles for ecologically valuable stands. The genetic distances determined for the studied sites were significant for F1, and the evaluation for F2 indicated that 47% of sites have lost their individual character. The grouping of the studied sites, as presented, illustrates the unification of the genetic variation of F2 in relation to the studied F1. This result confirms the loss, through natural processes, of rare genetic information that determines the individual character of the studied ecosystems and their high ecological value.

5. Conclusions

Significant differences in revealed haplotypes between the analysed generations were shown at the studied sites in the present analyses. Significant genetic differences between sites were also demonstrated, but it should be noted that the genetic distance decreased significantly in the next stand generation. In addition, the studied sites showed higher allelic diversity in F1, but this was not maintained in the subsequent stand generation. It is suggested that this is due to the high influence of site contamination by foreign pollen. However, determining whether this is indeed the cause requires further research. The obtained results indicate the need to maintain, ex situ, the genetic variability of naturally valuable old forest stands.

Supplementary Materials: The following are available online at https://www.mdpi.com/article/10.3390/d14020093/s1, Tables S1–S8: Geographical coordinates of F1 and F2 plant material from the analysed locations; Figures S1 and S2: Correlation of geographical and genetic distance in IBD Mantel [30] for F1 and F2.

Author Contributions: Conceptualization, P.P.; methodology, P.P., A.T. and J.M.; software, P.P.; validation, P.P. and A.T.; formal analysis, P.P., A.T. and J.M.; investigation, P.P.; resources, P.P.; data curation, P.P.; writing—original draft preparation, P.P.; writing—review and editing, P.P., A.T., J.M. and Ł.T.; visualization, P.P. and J.M.; supervision, P.P., I.S.-B; project administration, P.P.; funding acquisition, Ł.T. All authors have read and agreed to the published version of the manuscript.

Funding: This research was funded by Polish State Forest in 2020–2021.

Institutional Review Board Statement: Not applicable.

Informed Consent Statement: Not applicable.

Data Availability Statement: Project data publicly available in the Forest Research Institute library.

Conflicts of Interest: The authors declare no conflict of interest.

References

1. Floran, V.; Sestras, R.E.; Garcia-Gil, M.R. Organelle Genetic Diversity and Phylogeography of Scots Pine (*Pinus sylvestris* L.). *Not. Bot. Horti Agrobot. Cluj-Napoca* **2011**, *39*, 317–322. [CrossRef]
2. Brzeziecki, B.; Zajączkowski, J.; Olszewski, A.; Bolibok, L.; Andrzejczyk, T.; Bielak, K.; Buraczyk, W.; Drozdowski, S.; Gawron, L.; Jastrzębowski, S.; et al. Struktura i dynamika wielogeneracyjnych starodrzewów sosnowych występujących w obszarach ochrony ścisłej Kaliszki i Sieraków w Kampinoskim Parku Narodowym. Część 1. Zróżnicowanie gatunkowe, zagęszczenie i pierśnicowe pole przekroju. *Sylwan* **2020**, *164*, 443–453. [CrossRef]
3. Schuster, W.S.F.; Mitton, J.B. Paternity and gene dispersal in limber pine (*Pinus flexilis* James). *Heredity* **2000**, *84*, 348–361. [CrossRef] [PubMed]
4. Ivetić, V.; Devetaković, J.; Nonić, M.; Stanković, D.; Šijačić-Nikolić, M. Genetic diversity and forest reproductive material—From seed source selection to planting. *iForest* **2016**, *9*, 801–812. [CrossRef]
5. Tyburski, Ł. Kampinos Forest between 1914–1938. *Parki nar. Rez. Przyr.* **2021**, *40*, 85–98.
6. Przybylski, P.; Mohytych, V.; Rutkowski, P.; Tereba, A.; Tyburski, Ł.; Fyalkowska, K. Relationships between Some Biodiversity Indicators and Crown Damage of *Pinus sylvestris* L. in Natural Old Growth Pine Forests. *Sustainability* **2021**, *13*, 1239. [CrossRef]
7. Krzanowska, H.; Łomnicki, A.; Rafiński, J. *Wprowadzenie Do Genetyki Populacji*; Wydawnictwo Naukowe PWN: Warszawa, Poland, 1982.
8. Noss, R.F. Beyond Kyoto: Forest management in a time of rapid climate change. *Conserv. Biol.* **2001**, *15*, 578–590. [CrossRef]
9. Frankham, R. How closely does genetic diversity in finite populations conform to predictions of neutral theory? Large deficits in regions of low recombination. *Heredity* **2012**, *108*, 167–178. [CrossRef]
10. Spielman, D.; Brook, B.W.; Frankham, R. Most species are not driven to extinction before genetic factors impact them. *Proc. Natl. Acad. Sci. USA* **2004**, *101*, 15261–15264. [CrossRef]
11. O'Grady, J.J.; Brook, B.W.; Reed, D.H.; Ballou, J.D.; Tonkyn, D.W.; Frankham, R. Realistic levels of inbreeding depression strongly affect extinction risk in wild populations. *Biol. Conserv.* **2006**, *133*, 42–51. [CrossRef]
12. Hoelzel, A.R.; Bruford, M.W.; Fleischer, R.C. Conservation of adaptive potential and functional diversity. *Conserv. Genet.* **2019**, *20*, 1–5. [CrossRef]
13. Razgour, O.; Forester, B.; Taggart, J.B.; Bekaert, M.; Juste, J.; Carlos Ibáñez, C.; Puechmaille, S.J.; Roberto Novella-Fernandez, R.; et al. Considering adaptive genetic variation in climate change vulnerability assessment reduces species range loss projections. *Proc. Natl. Acad. Sci. USA* **2019**, *116*, 10418–10423. [CrossRef] [PubMed]
14. Laikre, L. Genetic diversity is overlooked in international conservation policy implementation. *Conserv. Genet.* **2010**, *11*, 349–354. [CrossRef]
15. Elsik, C.G.; Minihan, V.T.; Hall, S.E.; Scarpa, A.M.; Williams, C.G. Low-copy microsatellite for *Pinus taeda* L. *Genome* **2000**, *43*, 550–555. [CrossRef]
16. Wang, B.; Chang, R.Z.; Tao, L.; Guan, R.; Yan, L.; Zhang, M. Identification of SSR primer numbers for analyzing genetic diversity of Chinese soybean cultivated soybean. *Mol. Plant Breed.* **2003**, *1*, 82–88.
17. Fady, B.; Lefèvre, F.; Reynaud, M. Gene flow among different taxonomic units: Evidence from nuclear and cytoplasmic markers in Cedrus plantation forests. *Theor. Appl. Genet.* **2003**, *107*, 1132–1138. [CrossRef]
18. Gómez, A.; González-Martínez, S.C.; Collada, C.; Gil, L.; Climent, J. Complex population genetic structure in an endemic Canary Island pine using chloroplast microsatellite markers. *Theor. Appl. Genet.* **2003**, *107*, 1123–1131. [CrossRef]
19. Wachowiak, W.; Lewandowski, A.; Prus-Głowacki, W. Reciprocal controlled crosses between *Pinus sylvestris* and *P. mugo* verified by a species-specific cpDNA marker. *J. Appl. Genet.* **2005**, *46*, 41–43.
20. Celiński, K.; Kijak, H.; Barylski, J.; Grabsztunowicz, M.; Wojnicka-Półtorak, A.; Chudzińska, E. Characterization of the complete chloroplast genome of *Pinus uliginosa* (Neumann) from the *Pinus mugo* complex. *Conserv. Genet. Resour.* **2017**, *9*, 209–212. [CrossRef]
21. Semerikov, V.L.; Semerikova, S.A.; Dymshakova, O.S.; Zatsepina, K.G.; Tarakanov, V.V.; Tikhonova, I.V.; Ekart, A.K.; Vidyakin, A.I.; Jamiyansuren, S.; Rogovtsev, R.V.; et al. Microsatellite loci polymorphism ofchloroplast DNA of the pine tree (*Pinus sylvestris* L.) in Asia and Eastern Europe. *Genetika* **2014**, *50*, 660–669.
22. Pazouki, L.; Shanjani, P.S.; Fields, P.D.; Martins, K.; Suhhorutsenko, M.; Viinalass, H.; Niinemets, U. Large within-population genetic diversity of the widespread conifer *Pinus sylvestrisat* its soil fertilitylimit characterized by nuclear and chloroplast microsatellite markers. *Eur. J. For. Res.* **2016**, *135*, 161–177. [CrossRef]
23. Wojnicka-Półtorak, A.; Celiński, K.; Chudzińska, E. Genetic Diversity among Age Classes of a *Pinus sylvestris* (L.) Population from the Białowieża Primeval Forest, Poland. *Forests* **2017**, *8*, 227. [CrossRef]
24. Vendramin, G.G.; Lelli, L.; Rossi, P.; Morgante, M. A set of primers for the amplificationof chloroplast microsatellites in Pinaceae. *Mol. Ecol.* **1996**, *5*, 595–598. [CrossRef] [PubMed]
25. Provan, J.; Soranzo, N.; Wilson, N.J.; McNicol, J.W.; Forrest, G.I.; Cottrell, J.; Powell, W. Gene-poolvariation in Caledonian and European Scots pine (*Pinus sylvestris* L.) revealed by chloroplastsimple-sequence repeats. *Proc. R. Soc. Lond. Ser. B Biol. Sci.* **1998**, *265*, 1697–1705. [CrossRef] [PubMed]

26. Nowakowska, J.A.; Oszako, T.; Tereba, A.; Konecka, A. Forest tree species traced with a DNA-based proof for illegal logging case in Poland. In *Evolutionary Biology: Biodiversification from Genotype to Phenotype*, 1st ed.; Springer: Berlin/Heidelberg, Germany, 2015.
27. Eliades, N.-G.; Eliades, D.G. *Haplotype Analysis: Software for Analysis of Haplotypes Data. Distributed by the Authors*; Forest Genetics and Forest Tree Breeding; Georg-Augst University: Goettingen, Germany, 2009.
28. Excoffier, L.; Lischer, H.E.L. Arlequin suite ver. 3.5: A new series of programs to perform population genetics analyses under Linux and Windows. *Mol. Ecol. Resour.* **2010**, *10*, 564–567. [CrossRef]
29. Peakall, R.; Smouse, P. Genealex 6.5: Genetic Analysis in Excel. Population genetic software for teaching and research. *Mol. Ecol. Notes* **2006**, *6*, 288–295. [CrossRef]
30. Mantel, N. The detection of disease clustering and a generalized regression approach. *Cancer Res.* **1967**, *27*, 209–220.
31. Nei, M. Estimation of average heterozygosity and genetic distance from a small number of individuals. *Genetics* **1978**, *89*, 583–590. [CrossRef]
32. Rousset, F. Genepop'007: A complete re-implementation of the genepop software for Windows and Linux. *Mol. Ecol. Resour.* **2008**, *8*, 103–106. [CrossRef]
33. Kosińska, J.; Lewandowski, A.; Chalupka, W. Genetic Variability of Scots Pine Maternal Populations and Their Progenies. *Silva Fenn.* **2007**, *41*, 5. [CrossRef]
34. Hu, X.S.; Li, B. Linking evolutionary quantative genetics to the conservation of genetic resources in natural forest populations. *Silvae Genet.* **2001**, *51*, 177–183.
35. Lindgren, D.; Paule, L.; Shen, X.H.; Yazdani, R.; Segerström, U.; Wallin, J.E.; Lejdebro, M.L. Can viable pollen carry Scots pine genes over long distances? *Grana* **1995**, *34*, 64–69. [CrossRef]
36. Harju, A.; Nikkanen, T. Reproductive success of orchard and non-orchard pollens during different stages of pollen shedding in a Scots pine seed orchard. *Scand. J. For. Res.* **1996**, *26*, 1096–1102.
37. Adams, T.; Burczyk, J. Magnitude and implications of gene flow in gene conservation reserves. In *Forest Conservation Genetics: Principles and Practice*; Young, A., Boshier, D., Boyle, T., Eds.; CSIRO Publishing & CABI Publishing: Collingwood, Australia, 2000; pp. 215–224. ISBN 0 643 06260 2.
38. Burczyk, J.; Lewandowski, A.; Chałupka, W. Local pollen dispersal and distant gene flow in Norwey spruce. *For. Ecol. Manag.* **2004**, *197*, 39–48. [CrossRef]
39. Müller-Starck, G. Protection of genetic variability in forest trees. *For. Genet.* **1995**, *2*, 121–124.
40. Klisz, M.; Ukalski, K.; Ukalska, J.; Jastrzębowski, S.; Puchałka, R.; Przybylski, P.; Mionskowski, M.; Matras, M. What can we learn an erly test on the adaptation of silver fir populations to marginal envirments. *Forest* **2018**, *9*, 441. [CrossRef]

 diversity

Article

Damage and Tolerability Thresholds for Remaining Trees after Timber Harvesting: A Case Study from Southwest Romania

Ilie-Cosmin Cântar [1,*], Cătălin-Ionel Ciontu [1], Lucian Dincă [2], Gheorghe Florian Borlea [3] and Vlad Emil Crişan [2]

1 "Marin Dracea" National Research and Development Institute in Forestry, 8 Padurea Verde Street, 300310 Timisoara, Romania; ciontu_catalin@yahoo.com
2 "Marin Dracea" National Research and Development Institute in Forestry, 13 Cloşca Street, 500040 Braşov, Romania; dinka.lucian@gmail.com (L.D.); vlad_crsn@yahoo.com (V.E.C.)
3 Faculty of Agriculture, Banat's University of Agricultural Sciences and Veterinary Medicine, "King Michael I of Romania" from Timisoara, Calea Aradului No. 119, 300645 Timisoara, Romania; fborlea@yahoo.com
* Correspondence: ilie.cantar@icas.ro

Abstract: The present study analyses the damage of remaining trees after timber harvesting from 24 logging sites from southwest Romania. The purpose was to establish tolerability thresholds within which damaged trees recover in a short amount of time, reducing the possibility of further rot apparition and tree health deterioration. Observations were resumed after the growing season had passed. Healed damage was analysed in regard to damage type, width, orientation and tree circumference. By using the ratio between the width of healed damage and the circumference of trees as experimental variants, equations were elaborated to determine the tolerance threshold of trees in logging. This is expressed as a maximum value between the damage width and the damaged tree circumference for which the damage is curable. The correlation between the circumference and the abovementioned relation was analysed, and differences between the values of the analysed relation for different cardinal orientations of the damage were statistically tested. The value of this ratio, which can be considered a tolerance threshold for trees in logging, records values of 0.09 (for thinnings, for cuttings to increase the light availability for regeneration and for final cuttings from shelterwood systems) and 0.10 (for first-intervention cuttings, as well as preparatory and seed cutting from shelterwood systems or selections systems).

Keywords: residual trees; logging technology; silvicultural works; tree healed wound

Citation: Cântar, I.-C.; Ciontu, C.-I.; Dincă, L.; Borlea, G.F.; Crişan, V.E. Damage and Tolerability Thresholds for Remaining Trees after Timber Harvesting: A Case Study from Southwest Romania. *Diversity* **2022**, *14*, 193. https://doi.org/10.3390/d14030193

Academic Editor: Tibor Magura

Received: 30 January 2022
Accepted: 4 March 2022
Published: 6 March 2022

1. Introduction

Timber extraction from forests must be realized in profitable economic conditions, with expenses accepted by society at a given moment. This means using machines with a high productivity while remaining in compliance with the objectives of sustainable forest management. For logging, this involves activities with a damage level that does not exceed the tolerance threshold of the forest ecosystem.

Timber logging affects all the forest ecosystem's components: residual trees, soil [1,2] and seedlings [3–5]. Logs are usually extracted from the forest site to the landing areas by machines on skid trails [6]. As they are moved, these transported loads harm the abovementioned components of forest ecosystems, especially trees, in which case injury can lead to death. Different management practices in timber harvesting can lead to different tree mortality rates, even within the same forest type [7]. Silvicultural treatments are sometimes applied to maintain forest health and productivity. However, the necessary interventions vary from one forest to another [8], leading to differences in the number and intensity of injuries caused to trees. Damage from harvesting using different technologies has been studied by different authors and has shown a proportion of 22–44% wounded stems for conventional logging using chainsaw and skidder and 20–31% for mechanized operations using a harvester [9]. Regarding logging technologies with reduced impact on forest ecosystems, such as cable yarding systems, higher first-year mortality rates have been observed in less severely damaged trees located in conventional logging areas than in those from logging areas with a reduced impact [10]. Wound area and the ratio of the maximum wound's width-to-tree circumference at breast height, as well as the percentage of dead crown and the growth rate, have been tested as variables in some models that determine tree mortality risks [11].

Some studies have compared the influence of timber harvesting with that of forest fires on forest biodiversity and productivity. The resulting perturbation did not decrease productivity or plant diversity when compared with fires, which represent the main natural disturbance in these areas [12]. Other researchers have shown that anthropogenic disturbances, such as timber harvesting, can promote the stability of certain non-native and invasive plants [13].

Silvicultural and harvesting activities are commonly believed to primarily affect forest populations and species content [14]. The negative impact of logging extends not only to trees, soil and seedlings but also to wildlife and the entire regional landscape [15,16]. Regardless of this, some biodiversity benefits from reduced-impact logging may accrue over longer periods after logging ends [17]. Selective logging poses a lesser threat, largely because most—although not all—of the species found in primary forests appear to be able to persist in logged forests [18]. Windstorms can improve biodiversity indicators in these areas, but salvage logging may reduce these positive impacts across most indicators [19,20]. Similarly to logging, extreme phenomena, such as storms, can cause torrential leaks, with a negative impact on forests, leading to extensive damage to trees, seedling and soil [21]. Furthermore, besides the damage caused by logging and extreme phenomena, a major impact on exploited stands can also be caused by climatic changes [22–24].

The scope and severity of residual tree damage depends on, among other things, the harvest intensities and the layout of the skid trail network. Most damage occurs due to construction of main roads, but with an increase in harvest intensity, damage resulting from tree felling and skid trails dominates [25]. Additionally, different curvatures of skid trails and different site conditions (soil moisture, soil type, terrain slope) contribute to the amount and intensity of residual tree damage [26]. A high damage rate within mechanized harvesting was reported in stands impacted by the handling of wood parts and trees. Thus, the experience of machine operators can correlate with a lower handling rate of wood in forests, resulting in a low damage rate of residual trees. The actions of operators and the harvested tree size can influence an important part of residual tree damage [27]. The choice of mechanized harvesting machines used in logging should be based on their impact on the forest ecosystem because mechanized harvesting has a longer-lasting effect for residual trees compared to non-mechanized harvesting [28]. Residual tree damage produced by forwarders is approximately half the damage produced by harvesters [29].

Another factor that influences the number and intensity of residual tree damage is the size of the harvested trees and the size of the skidded wood piece. The amount of residual tree damage significantly increases when the harvested diameter of harvested trees at breast height increases [30]. The length of wood pieces is dictated by the logging method used. Some research has shown that using the cut-to-length method can lead to residual trees being more damaged than when using the tree-length method [31]. The most widely used harvesting methods in Romania are tree-length, usually used in the mountainous and hilly area, and the cut-to-length method, used in lowland forest areas or in mountainous areas in the process of installing cable yarding capabilities [32]. The most damaging method to the forest ecosystem is the whole-tree method, which is forbidden in Romania [32].

In order to avoid tree mortality after harvesting and the previously mentioned connected negative effects on biodiversity, it is necessary to coordinate logging technologies with treatments that are appropriate for the essential characteristics of stands in order to preserve the protection potential of forests. This practice is necessary in order to harvest wood material and fulfil the necessary conditions for natural regeneration and the creation of healthy and economically valuable stands. In this way, causing damage that exceed the tolerance thresholds of trees is avoided so that the damaged samples recover in a short period of time.

Silvicultural and functional requirements can be satisfied by establishing damage thresholds (limits) for the remaining standing trees, seedlings and soil. These thresholds can be tolerated by the forest, avoiding derangement of the production and protection functions of the forest ecosystems [33]. If a wound closes quickly, the tree is less prone to decay at the stem level [34]. Based on tree DBH, as well as hierarchical and geographical positions within the stand, and based on position, size and depth of wound, some authors calculated the average synthetic index, establishing values for "tolerable" damage [35]. Knowing the tolerance thresholds of trees and the impact of logging within the limit where trees heal without developing rot can lead to management solutions that will increase the quality of the wood that will be harvested in the following periods.

Based on field observations realized before the present study, our hypothesis was that there is a link between healed damage, damage width, tree circumference and the damage cardinal orientation. Our observations have shown that only a certain percentage of tree damage heals in a short period of time after logging.

The purpose of this study was to establish the tolerance thresholds for trees under timber harvesting actions from the southwest of Romania. This purpose can be achieved by attaining objectives based on the study's hypothesis, which are detailed below.

The first objective is to identify and evaluate tree damage from harvesting sites located in southwestern Romania from all relief forms and including a wide range of work. The healed damage was identified after the growing season had passed.

The second objective consists in establishing a link between healed damage and damage width, tree circumference and damage orientation.

The third objective consists in establishing tree tolerance thresholds against the action posed by timber extraction based on an existent relationship between healed damage and different variables.

Based on the results of this research, a series of good practices for the management of timber harvesting were brought forward in order to minimize the negative impact on both the forest ecosystem and biodiversity.

Based on the abovementioned scope and objectives, the goal of the research is to establish the tolerability thresholds for trees in logging at harvesting sites in southwestern Romania.

2. Materials and Methods

The research for this study was conducted at 24 harvesting sites in southwestern Romania in forests located in the plains, mountains and hill areas. The distribution of research variants was realized by taking into account the harvesting sites, depending on silvicultural work, for variants and, within them, depending on the relief by repeating the observations twice at harvesting sites from each relief form.

In this study, the harvesting sites were chosen as variants depending on applied silvicultural work as follows: thinnings (variant V1), shelterwood systems (variants V2, V3 and V4) and selection systems (variant V2). For variant V2, we studied harvesting sites with first-intervention cuttings, as well as preparatory and seed cuttings in shelterwood systems or selection systems.

A regular shelterwood system is a silvicultural system wherein regeneration is initiated and supported by the removal of the harvestable trees in two or more successive steps of cutting [36]. In this paper, three steps were used, which correspond to preparatory and seed cutting (V2), as well as several successive cuttings, in order to increase the light availability for regeneration (V3) and final cutting (V4). The temporarily remaining old trees provide seeds and protect the natural regeneration from climatic extremes. The higher amount of light available due to these cuttings also promotes the growth of the remaining old tress [36]. Shelterwood cutting and later thinning produce an evenly aged stand with a homogenous vertical and horizontal structure. Only at the regeneration stage, when the shelter of mature trees covers seedlings and saplings, is the shelterwood system characterized by two clear canopies [36].

A selection system (V2) is a silvicultural system that results in unevenly aged stands. Individual trees or small groups of trees are cut periodically to obtain a yield in order to improve the forest structure and growth and to support the regeneration at the same time and in the same area. There are no defined cutting areas that are managed or harvested at a specific time [36].

Considering the above descriptions of applied treatments, their structure with respect considered variants is as follows: variant V1: thinnings; variant V2: first-intervention cuttings, preparatory and seed-cutting (as part of shelterwood system) or selections system; variant V3: cuttings to increase light availability for regeneration (as part of the shelterwood system); variant V4: final cuttings (as part of the shelterwood system).

Harvesting operations on harvesting sites considered in the study were finished during the year 2018. The period between the finishing time of the harvesting operation and the assessment was one vegetation season. Tree damage was identified and evaluated during the vegetative resting period between November 2019 and March 2020. The re-evaluation of damage was conducted between September 2021 and November 2021. The research was conducted at harvesting sites in Banat region managed by the Caransebeş Experimental Basis of INCDS "Marin Drăcea" and the forest departments of the Caraş-Severin forest directorate that belong to ROMSILVA (Băile Herculane, Bocşa Montană, Bocşa Română, Moldova Nouă and Văliug) (Figure 1).

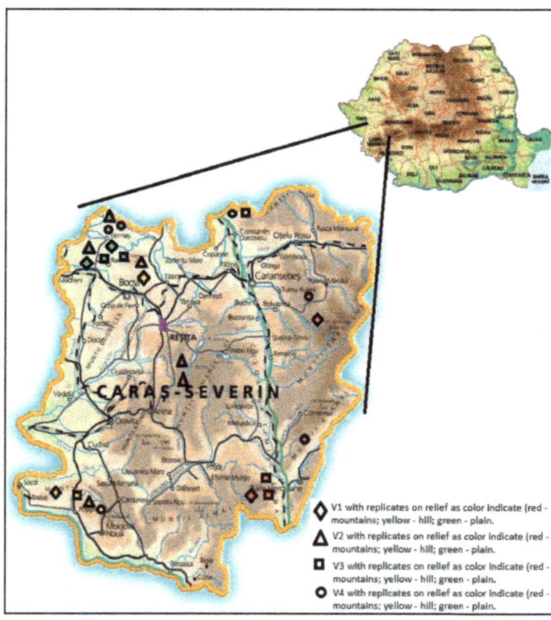

Figure 1. Location of the analysed harvesting sites [37,38].

Data collection was performed in the established sample plots with FieldMap equipment, Vertex, a compass, a riglet, and writing and labelling instruments. These were used to determine the tree position within the sample plots, to mark trees, to determine their biometric characteristics, to measure and determine the locations of the identified damage and to note the gathered data. The data were processed with specific software for table calculation and graphics processing.

The distribution of harvesting sites in terms of variants and replicates in the study is as follows (see Appendix A):

- Variant V1: harvesting sites with thinnings; two harvesting sites on each relief form;
- Variant V2: harvesting sites with first-intervention cuttings; preparatory and seed-cutting from shelterwood systems or selection systems; two harvesting sites on each relief form;
- Variant V3: harvesting sites with cuttings to increase light availability for regeneration from shelterwood systems; two harvesting sites on each relief form;
- Variant V4: harvesting sites with final cuttings from the shelterwood system; two harvesting sites on each relief form.

Tree damage was identified, evaluated and re-evaluated along skid trails in three sample plots with a length of 100 metres, measured along the driving direction of forest machines on the main skid trail for each logging site. This was done for each harvesting site on the ascent, at the middle and on the descent (Figure 2a).

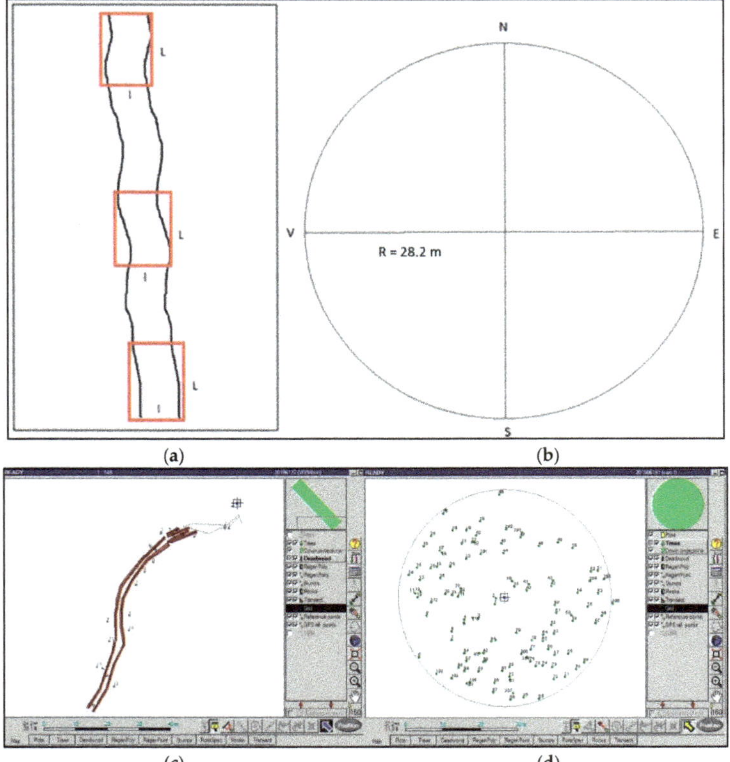

Figure 2. Location of sample plots: (**a**) along skid trails (schematic view); (**b**) inside the harvesting site (schematic view); (**c**) FieldMap representation of a sample plot from the skid trail; (**d**) FieldMap representation of a surface from inside the harvesting site.

The same tree observations were made in a circular sample plot of 2500 square metres in the middle of the harvesting site, avoiding skid trails (Figure 2b). This prevented the sample plot from overlapping with the skid trails. Random distribution of sample plots was not used because the harvesting operations took place only in certain parcels covered with regeneration or thinning works and only in certain accessible areas in which regeneration meshes were opened according to the applied treatment. In this case, inherent bias was avoided by placing sample plots according to a similar previous plan for all studied plots, regardless of the field situation, using three criteria: (1) to be as far away as possible from the border of the harvesting site; (2) to be as far away as possible from the skid trails; and (3) to be in an area where harvesting operations had been applied.

Injuries caused after timber extraction were identified and evaluated in the sections located along the skid trails. On the other hand, injuries caused during the collection and harvesting were identified and evaluated in the circular sample plot.

The classification of the identified types of tree damage was adapted from specialized literature classifications [39,40] as follows:

- Galling: partial removal of bark or rind without affecting the cambial area;
- Barking: removing parts of the bark up to the wood;
- Splintering: removing parts of bark and wood;
- Breaking branches or the trunk;
- Partial or total uprooting.

In addition to the identified types of damage, other characteristics were noted and stored, presented below in the FieldMap system.

During the initial evaluation of trees with injuries, several aspects were recorded: the species, damage type, age of damage based on previous forestry work (new/old), position of the damaged tree (FieldMap or polar coordinates), tree height (measured with VERTEX), tree circumference (at 1.3 m high), damage measurements (length, width and depth), position on the tree (trunk, crown or root insertion), height of the damage, cardinal position (exposition) of the damage and phase of the timber-harvesting technological process. When re-evaluating trees, we aimed to identify the tree, observe the status of the tree (dead/alive), identify the previously determined damage based on its type (galling, barking, splintering, breaking or uprooting) [39,40], remeasure the width of the damage, frame a new type of damage if it had changed and identify repeated damage.

The width of the damage that was taken into account represents the maximum distance between the edges of the damage found on the trunk of the tree, measured horizontally along the circumference of the tree (Figure 3).

We established a link between the healed damage width, tree circumference and the damage cardinal orientation by analysing the considered variants; the minimum and maximum value of the ratio between the healed damage's width; and the tree's circumference, amplitude, average, standard deviation and variation coefficient for ratio values.

The average value of cardinal orientation for the ratio between the healed width of the damage and the circumference of the damaged tree was identified, together with the standard deviation and the variation coefficient of the obtained value. An *ANOVA* test was applied in order to test the hypothesis that significant differences exist between the values of the ratio between the healed damage widths and the tree circumference for different damage orientations. The cardinal orientation of the damage was considered the cardinal orientation from the middle of the width of the damage, measured in the horizontal direction. The correlation between the studied relation and the circumference of the damaged tree was analysed with the help of the *Pearson* correlation coefficients.

Assuming that tree-growth reduction and the possible death of trees is mainly caused by reduced sap flow, some authors have used the ratio between the width of the damage and circumference of the damaged trees to describe the severity of damage [41,42]. In the case of similar investigations, because the normality of wound size and data regarding wound–stem size was not satisfying in the initial assumption, wound size and wound–stem ratio values were analysed [43].

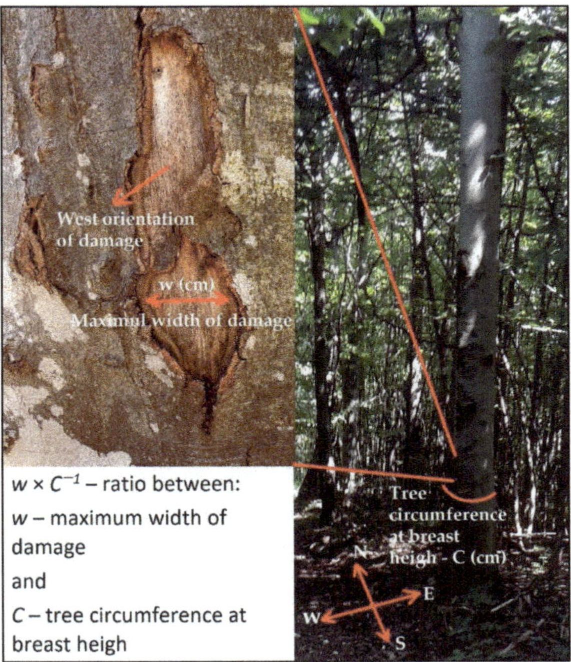

Figure 3. Diagram of measuring damage width and tree circumference and determining damage orientation.

We establishing the tree tolerance thresholds for the remaining trees after timber harvesting based on the relationship between the tree circumference (C) and the ratio of healed damage width and tree circumference ($w \times C^{-1}$). We assumed that damage that does not heal is more likely to develop rot, which can cause wood and tree health to depreciate. Tolerance thresholds were established for damage, considering silvicultural works from the harvesting sites and the cardinal orientation of the damage. This fact was obtained by analysing the value of the ratio between the damage width and the tree circumference for the damage that healed during the analysed period, modelling the statistical connections between the considered variables (w and C) using power functions.

Equations that lead to tree tolerance thresholds towards harvesting operations were elaborated. In order to see whether the observed data corresponded sufficiently to the expected values, χ^2 for goodness of fit was used to test the obtained models. The tolerance threshold was expressed as a maximum value of the relation between the damaged width and the tree circumference, for which the damage is curable. The correlation between the circumference and the abovementioned relation was analysed with the *Pearson* correlation coefficient. A model for the tolerance threshold concerning the main species was also elaborated.

For each silvicultural works, according to the considered variants, we identified the minimum and the maximum value of the ratio between the width of the healed damage and the circumference of the tree, the amplitude and average values of this ratio, the standard deviation and the coefficient of variation of values obtained from this ratio. As for tolerance threshold, expressed as the value of the ratio between the width of the healed damage and the circumference of the tree, we used the averages of the data sets with the lowest spread (lowest coefficient of variation) among the studied variants.

The experimental design of the conducted research, as well as the workflow of field work and data processing, according to the method presented above, is schematically represented in Figure 4 [44].

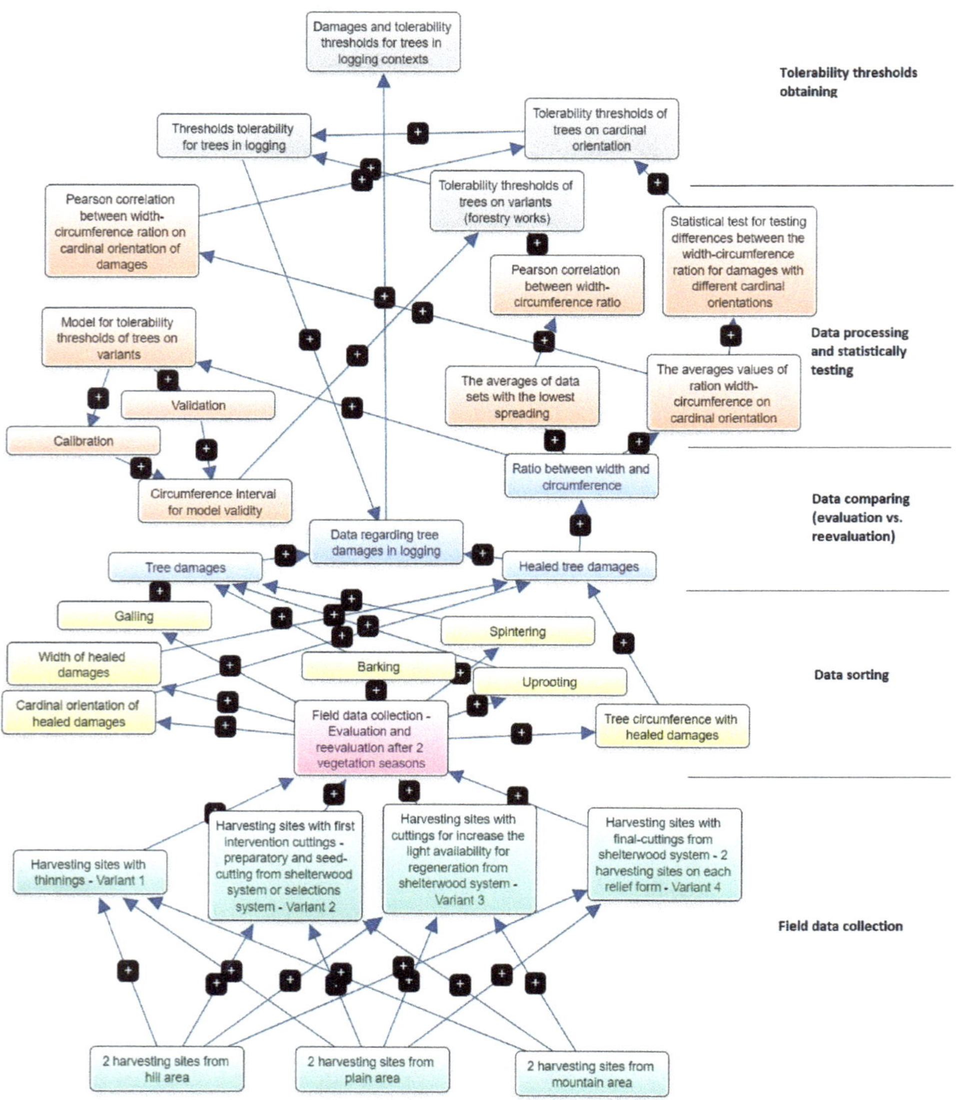

Figure 4. Design of conducted research and the workflow of field work and data processing [44].

The specific work conditions from the 24 harvesting sites (Table 1) have a direct impact on the development of the harvesting operations, as they define the spaces in which this work occurs. The cut-to-length-method was the harvesting method used in 23 of the 24 harvesting sites. The tree-length-method was applied only in one harvesting site. These are most often used harvesting methods in Romania, but depending on the operational conditions and the equipment used, many intermediary adaptations are in practice [32]. This is also the case of western Romania, which is the reason for taking into consideration the last harvesting site with a different harvesting method than the others, respecting the percentage of usage of this method in the studied area. Considering the above, the obtained results below are representative from this point of view.

Table 1. Characteristics of case-study areas in terms of stand, terrain and harvesting operations *.

Variant	Harvesting Site	Forest District, Production Unit, Management Unit	Average Tree Volume (m^3)	Total Number of Extracted Trees/ha (pcs.)	Trunk Volume of the Harvested Trees (m^3)	Extracted Wood Volume per ha (m^3)	Number of Trees per ha (pcs.)	Slope in the Sample Surface (%)
V1—Thinnings	h1	Bocșa Română, III, 49	0.52	35	166.6	18.4	554	2
	h2	Bocșa Română, II, 58A	0.13	133	22.8	17.5	762	5
	h3	Bocșa Montană, VI, 95A	0.19	500	34.4	93	1246	22
	h4	Moldova Nouă, III, 15B	0.10	28	176.6	28.1	1440	30
	h5	Băile Herculane, II, 23	0.32	286	466.1	91	903	23
	h6	Caransebeș, VI, 99A	0.29	101	324.7	30	1390	30
V2—First-intervention cuttings	h1	Bocșa Română, II, 55	0.6	161	160.7	100	525	5
	h2	Bocșa Română, I, 11C	0.7	137	136.9	88.9	319	13
	h3	Bocșa Montană, IV, 62B	0.9	139	139.2	126	306	26
	h4	Moldova Nouă, III, 212A	1.1	105	105.1	116	308	22
	h5	Văliug, VI, 15A	1.19	15	379.9	18	263	25
	h6	Văliug, VI, 16A	1.39	29	741.6	40	336	25
V3—Cuttings to increase light availability for regeneration	h1	Bocșa Română, III, 76A	0.9	137	136.5	121.4	429	6
	h2	Bocșa Română, III, 28B	1	141	140.8	135.7	282	0
	h3	Moldova Nouă, III, 176A	1	79	79.3	75.4	143	30
	h4	Caransebeș, II, 30B	0.8	114	113.7	90.4	235	23
	h5	Băile Herculane, II, 99	1.19	107	606.8	127	224	30
	h6	Băile Herculane, II, 100A	2.24	44	1638.4	99	123	29

Table 1. *Cont.*

Variant	Harvesting Site	Forest District, Production Unit, Management Unit	Average Tree Volume (m^3)	Total Number of Extracted Trees/ha (pcs.)	Trunk Volume of the Harvested Trees (m^3)	Extracted Wood Volume per ha (m^3)	Number of Trees per ha (pcs.)	Slope in the Sample Surface (%)
V4—Final cuttings	h1	Bocșa Română, I, 1E	1.2	192	191.6	230.1	163	14
	h2	Bocșa Română, I, 14A	0.6	167	166.5	92.5	150	14
	h3	Moldova Nouă, III, 162B	1.7	74	74.1	122.8	36	25
	h4	Caransebeș, I, 46D	0.9	234	234.1	218.6	238	30
	h5	Băile Herculane, IV, 98A	3	40	1122.3	120	25	28
	h6	Caransebeș, V, 16A	3.27	114	3134.8	371	58	32

* Data were processed using the sources of technical documentation from the studied harvesting sites, as well as from forest management plans, within the forest management headquarters.

3. Results

3.1. Tree Damage and Healed Damage

The research allowed us to compare data from re-evaluations to those from initial evaluations and to emphasize the damage dynamics regarding dimension, migration towards other types of damage (e.g., transforming galling in barkings by slicing bark under the pressure of cambial growth), the apparition or evolution of rot, or, on the contrary, the healing of damage.

The tree tolerance threshold was established in relation to the size of the damage identified in the initial evaluation but considered healed during re-evaluations. The other evaluated variables were also taken into account.

A total of 1237 damaged trees were identified in the sample plots from analysed harvesting sites in the initial evaluation. These were distributed as follows: 537 trees in variant V1, 254 trees in variant V2, 215 trees in variant V3 and 231 trees in variant V4. We identified 1945 injuries, with many trees presenting multiple injuries. Barking represented the majority of injuries (78.9%), followed by splintering (11%), galling (7.7%), broken trees (1.8%) and uprooting (0.6%).

Most cases of damage were identified as thinnings (V1), where stand density was the highest (Figure 5).

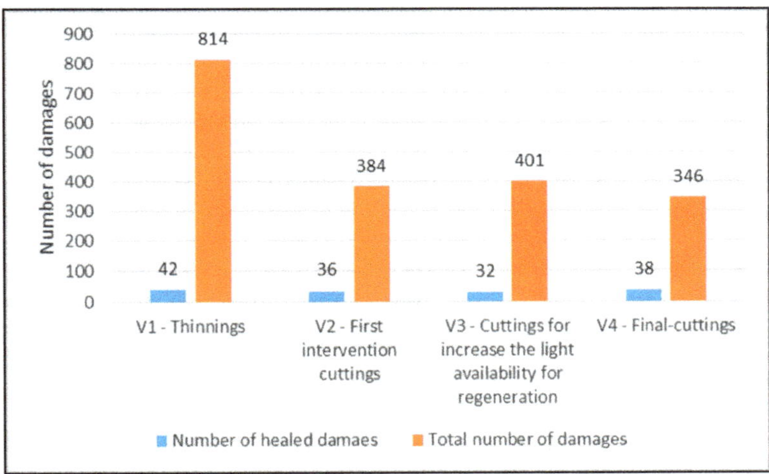

Figure 5. Distribution of healed and unhealed damage by variant.

The synthesis shown in Table A1 was created based on the field data gathered from the sample plots.

Taking into account the fact that a significant amount of the studied damage healed during the analysed period, it is important to find a dimensional damage threshold up to which healing was possible. In the analysed sample plots, 64 cases of damage were initially identified as healed, whereas 108 cases of damage healed during the two years in which the research was realized. From a percentage perspective, healed damage represents 8.8% of the total amount of damage.

The research was carried out mostly in beech stands in the mountains and hills and in mixed stands in the plains, with resinous species found in a small percentage of the analysed areas. However, the percentage of injuries healed in the specimens of sampled conifers was 28% (17 cases of damage healed out of a total of 60 cases found). In deciduous species, the healing rate of damage was only 8.2%.

3.2. Tolerance Thresholds concerning Types of Forestry Work and Species

If the growing type of damaged bark and field observations are taken into consideration, it can be said that the most significant influence on damage healing is represented by damage width in relation with the entire circumference of the damaged tree. As such, in the analysed sample plots, the average of the ratio between the healed damaged widths and the tree circumference is approximately 0.10 (0.09 for damage healed during the analysed period and 0.12 for damage identified as healed in the initial evaluation).

The value of the ratio between the damage width and the tree circumference for the damage healed in the studied variants, as well as the statistical indicators that characterize the value of this relation, is presented in Table 2.

Table 2. Values of the ratio between the healed damage width (w) and the tree circumference (C) in the studied variants.

			Galling		
Statistical indicator	**Minimum value of $w \times C^{-1}$**	**Maximum value of $w \times C^{-1}$**	**Average value of $w \times C^{-1}$**	**Standard deviation**	**Variation coefficient**
V1—Thinnings	0.0164	0.2128	0.0696	0.0493	70.8320
V2—First-intervention cuttings	0.0192	0.2963	**0.1047**	0.0799	**76.2434**
V3—Cuttings to increase light availability for regeneration	0.0195	0.2286	**0.0868**	0.0595	**68.5909**
V4—Final cuttings	0.0323	0.2429	**0.0914**	0.0676	**73.9230**
TOTAL	0.0164	0.2963	0.0896	0.0650	72.5425
			Barking		
Statistical indicators	**Minimum value of $w \times C^{-1}$**	**Maximum value of $w \times C^{-1}$**	**Average value of $w \times C^{-1}$**	**Standard deviation**	**Variation coefficient**
V1—Thinnings	0.0088	0.2222	0.0889	0.0529	59.5267
V2—First-intervention cuttings	0.0244	0.3684	0.1265	0.1002	79.1722
V3—Cuttings to increase light availability for regeneration	0.0137	0.1852	0.0734	0.0547	74.5544
V4—Final cuttings	0.0072	0.2653	0.0954	0.0797	83.6130
TOTAL	0.0072	0.3684	0.0969	0.0729	75.2008
			Splintering		
Statistical indicators	**Minimum value of $w \times C^{-1}$**	**Maximum value of $w \times C^{-1}$**	**Average value of $w \times C^{-1}$**	**Standard deviation**	**Variation coefficient**
V1—Thinning	0.0781	0.3214	0.1792	0.1268	70.7651
V4—Final cuttings	0.0606	0.3846	0.2226	0.2291	10.9195
TOTAL	0.0606	0.3846	0.1965	0.1474	74.9960

Table 2. *Cont.*

Amount of total damage					
Statistical indicators	**Minimum value of $w \times C^{-1}$**	**Maximum value of $w \times C^{-1}$**	**Average value of $w \times C^{-1}$**	**Standard deviation**	**Variation coefficient**
V1—Thinnings	0.0088	0.3214	0.0886	0.0592	66.7758
V2—First-intervention cuttings	0.0192	0.3684	0.1186	0.0908	76.6220
V3—Cuttings to increase light availability for regeneration	0.0137	0.2286	0.0827	0.0564	68.2538
V4—Final cuttings	0.0072	0.3846	0.0994	0.0869	87.4290
TOTAL	0.0072	0.3846	0.0971	0.0744	76.6439

The variation coefficient of the ratio between the damage width and the circumference had values over 50% in all analysed cases, namely for all three damage types identified as healed and for all studied variants. The highest values for the variation coefficient were recorded in splintering, for which the amount of healed damage was very small. Generally speaking, splintering is a type of width damage that affects not only the bark but also the wood. In the current research, the average diameter of splintered and healed trees was 11.5 cm during the studied period. Most of these trees were young and identified in thinnings and final cuttings. They had some of the lowest values of the ratio between the width of the healed damage and the circumference of the tree when compared with other splinterings.

A high variation coefficient signals a spreading of the studied relation in all variants. This means that the damage with a small width reported to the tree circumference is most likely to heal. However, over time, the amplitude value of this relation increases.

As can be seen in Table 2, the amplitude value of the ratio between the healed damage width and the tree circumference has the lowest values for variant V3.

The low amplitude of variant V3 indicates that the healing process occurs only at a low value within this relation. The widest healed damage reported in relation the tree circumference was observed in the variant where forestry work opening regeneration areas was applied, namely V2. Here, the amplitude of the studied relation is the highest, with the exception of splintering. However, the maximum value of the studied relation for most cases is higher than 0.2.

Taking this information into account, in order to determine the tolerance threshold (expressed as the value of the ratio between the healed damage width and tree circumference), one can use the average of data sets with the lowest spreading rounded to decimals (the lowest variation coefficient) from the studied variants (emphasized in the last column of Table 2).

As such, the following tolerance thresholds for timber harvesting were proposed. They are expressed as a ratio between the damage width (*l*) and the circumference of the damaged tree (C):

- V1—Thinnings: $w \times C^{-1} = 0.09$;
- V2—First-intervention cuttings: $w \times C^{-1} = 0.10$;
- V3—Cuttings to increase light availability for regeneration: $w \times C^{-1} = 0.09$;
- V4—Final cuttings: $w \times C^{-1} = 0.09$.

As can be seen, the value of the ratio between the damage width and the tree circumference, which can be considered a tolerance threshold for trees from timber harvesting operations, has similar values for all studied variants. The value of this ratio is of 0.09 for thinnings, cuttings to increase light availability for regeneration and final cuttings and 0.10 for first-intervention cuttings.

If the ratio of damage width to circumference is higher than 0.10 in first-intervention cuttings or higher than 0.09 for the other cuttings, the damage is not likely to heal within three years of the end of harvesting operations.

Although the variation coefficient of $w \times C^{-1}$ is big, a connection was observed to exist between tree circumference and the ratio between damage width and circumference (Figure 6).

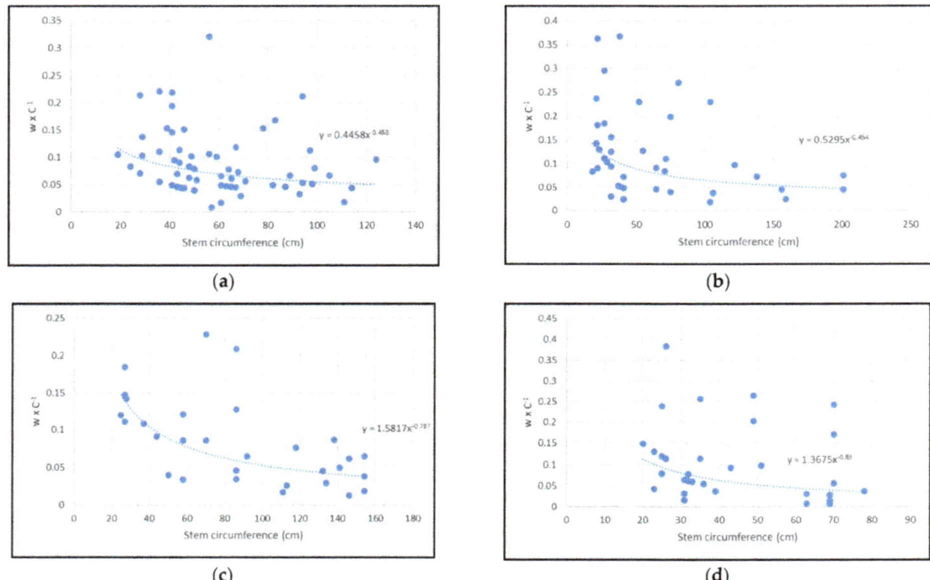

Figure 6. The relation between tree circumference and the ratio between healed damage width and tree circumference: (**a**) V1 –thinnings; (**b**) V2—first-intervention cuttings; (**c**) V3—cuttings to increase light availability for regeneration; (**d**) V4—final cuttings.

In the above figure, the average values for $w \times C^{-1}$ (y axis) are as shown in Table 2. The average values for circumference (C; x axis) are as follows: 60 for V1, 64 for V2, 88 for V3 and 42 for V4.

For all variants, the functions that define the abovementioned relation are exponential monotonous decreasing functions of data intervals of studied circumferences in the following form:

$y = 0.4458x^{-0.453}$ for V1—Thinnings;

$y = 0.5295x^{-0.454}$ for V2—First-intervention cuttings;

$y = 1.5817x^{-0.737}$ for V3—Cuttings to increase light availability for regeneration;

$y = 1.3675x^{-0.830}$ for V4—Final cuttings.

where $y = w \times C^{-1}$; $x = C$, in which:

w—the width of the healed damage;

C—the circumference of the damaged tree.

The equations for models that establish the link between circumference and $w \times C^{-1}$ for the healed damage are presented in Figure 6. By calibrating the models presented above using the average values obtained for the studied ratio of 0.10 in V2 and 0.09 in the other variants, it can be observed that the $w \times C^{-1}$ value can exceed these values for some small circumferences ($C < 35$ in V1; $C < 40$ in V2; $C < 49$ in V3; $C < 27$ in V4; values determined with the obtained equations). By calibrating the models for the minimum values of the ratio between width and circumference for which healed damage was identified as respecting the equations from the figure above, namely solving the equations, we obtained the maximum circumferences for proposed models of 124 cm in V1 (thinnings), 201 cm in V2 (first-intervention cuttings), 154 cm in V3 (cuttings to increase light availability for regeneration) and 78 cm in V4 (final cuttings).

Therefore, the validity of the models presented above may be applied to damage in trees with circumferences within the above values but larger than 35 cm in thinnings, 40 cm in first-intervention cuttings, 49 cm in cuttings to increase light availability for regeneration and larger than 27 cm in final cuttings.

The fitting of models was tested using χ^2 for goodness of fit between observed values for $w \times C^{-1}$ and expected values of this ratio obtained using the above models for studied variants (Table 3).

Table 3. Statistical data used to test models for goodness of fit.

Variant	χ^2 Values	Confidence Level	Degree of Freedom	Critical Values of χ^2
V1—Thinnings	3.109	95%	64	63.335
V2—First-intervention cuttings	3.451	95%	41	56.943
V3—Cuttings to increase light availability for regeneration	1.107	95%	30	43.773
V4—Final cuttings	4.777	95%	37	52.192

As can be observed for all variants, the proposed models fit the obtained experimental data; in all cases χ^2 obtained by testing was lower than the critical value for χ^2 for a specific degree of freedom. With 95% confidence, we conclude that the observed data follow the distribution of the proposed models.

By calculating the *Pearson* correlation coefficient between circumference and $w \times C^{-1}$ for healed damage, negative correlation coefficients were obtained. This marks an inverted correlation of different degrees, which are presented below, together with the value of the correlation coefficient for each variant:

- In V1, $r = -0.21$—weak correlation;
- In V2, $r = -0.33$—weak correlation;
- In V3, $r = -0.57$—reasonable correlation;
- In V4, $r = -0.20$—weak correlation.

Neither variant showed an inexistent correlation ($r > -0.2$). A linear Person correlation (not only exponential, as shown above) is also present in all variants, although this correlation is between weak and reasonable.

For each variant (type of silvicultural work), if we replace x from the equations in Figure 6 with the circumference of the damaged tree, we obtain the tolerance threshold (y), expressed as a maximum value of the ratio between damage width and tree circumference from which the damage can heal.

The relationship between the circumference of the trees and the ratio between the width of the healed injury and the circumference for groups of species and for the main species is shown in Figure 7.

As can be seen, the small amount of damage in conifers and various deciduous species and the implicitly low percentage of participation of these species in the studied stands led to a poor statistical fitting of the above relationship. (Figure 7a,b).

For European beech, the coefficient of determination, R^2, takes the highest values in the case of variant V3 (cuttings to increase light availability for regeneration). In this variant, the damage was not exposed to sunstroke as in the case of sparse stands in variant V4; the trees were mature, unlike variant V1 (thinnings) and were not diseased with rot or other defects because they were extracted at the first cutting of the shelterwood system.

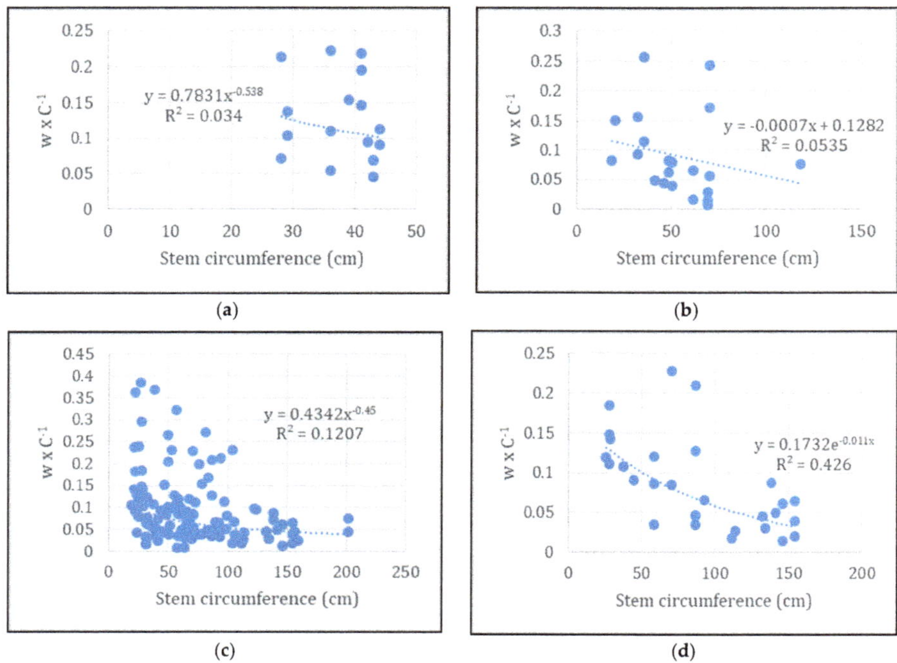

Figure 7. The relation between the tree circumference and the ratio between healed damage width and tree circumference: (**a**) coniferous species, all variants; (**b**) various deciduous species, all variants; (**c**) European beech, all variants; (**d**) European beech in V2 (first-intervention cuttings).

In view of the above, the maximum value of the relationship between the damage width and the circumference described by the model below can be proposed as a threshold of tolerability for beech:

$$y = 0.1732\ e^{-0.011x}$$

where $y = w \times C^{-1}$; $x = C$, in which:

w—the width of the healed damage;

C—the circumference of the damaged tree.

The value of the correlation coefficient of $r = -0.61$ is obtained by calculating the *Pearson* correlation coefficient between the circumference and $w \times C^{-1}$ of the healed damage in European beech in V3. There is therefore a high inverse correlation between the values considered in the proposed model ($r < -0.6$).

3.3. Tolerance Threshold and the Orientation of Damage

A factor with an important influence on the healing of damage in trees is represented by damage orientation. This is expressed through the cardinal position towards which the damage is oriented. It can be observed that healed damage is oriented towards all cardinal orientations, with N as the most common orientation (Table 4).

Analysing Table 4 and Figure 8 regarding the percentage of healed damage in relation to the total amount of damage (without breaks and uprooting) in each cardinal orientations (N, NE, E, SE, S, SW, W, NW), it can be seen that the most important percentage of healed damage is oriented in the N orientation, with 13.5% of the observed damage already healed. High percentages of healed damage are also observed in the neighbouring cardinal orientations, namely NW (11.3%) and NE (10.3%).

Table 4. Healed tree damage in relation to the orientation of damage and total damage.

Damage Orientation	Amount of Healed Damage	Total Amount of Damage	Percentage of Healed Damage Based on Orientation (%)
N	35	260	13.5
NE	19	184	10.3
E	26	310	8.4
SE	17	188	9.0
S	19	226	8.4
SW	12	191	6.3
W	21	337	6.2
NW	23	203	11.3

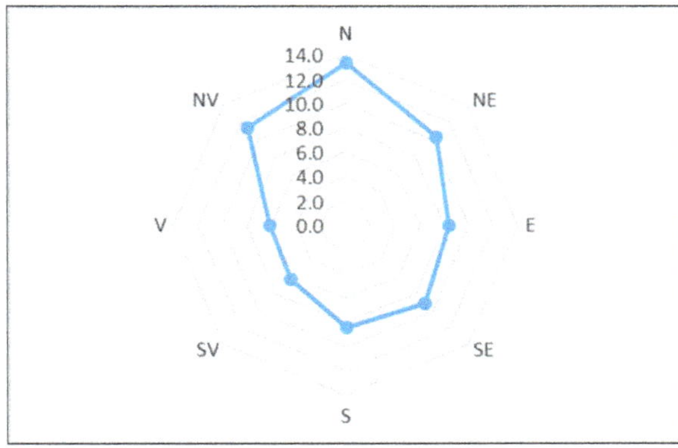

Figure 8. Percentage expression of the amount of healed damage based on orientation.

The lowest percentage of healing is in the W (6.2%) and SW orientations (6.3%).

The differences regarding the healing rate depending on the orientation of the damage may be due to the fact that the sunstrike is stronger in the case of south-facing injuries. These are exposed to drying, and bark growth is inhibited, which leads to a lower healing rate compared to that of north-facing damage. In the latter case, at the level of injury, the humidity is higher and does not negatively influence the healing rate but favours the long-term development of rot for the injuries that fail to heal.

In regard to the tolerance threshold of damaged trees according to cardinal orientation, as can be seen in Table 5, the $w \times C^{-1}$ average, which can be considered a tolerance threshold, varies based on the orientation. The values range between 0.076 in NE and 0.138 in SW orientation.

Although the percentage of healed damage is the highest for damage with a NW-N-NE orientation, the widest damage reported in relation the damaged tree circumference was found in the SW orientation (Table 5). Here, the variation coefficient is relatively small when compared with the other cases (Table 5). Furthermore, the average of the studied relation is higher than 0.1 for healed damage oriented to the N, NW and W.

An *ANOVA* test was applied in order to test the hypothesis that significant differences exist between the $w \times C^{-1}$ values between different damage orientations. These types of significant differences ($p^* < 5\%$) can be observed only for two pairs of samples, namely between SW and S orientations and between SW and NE orientations (Table 6).

Table 5. Values of the ratio between healed damage width and damaged tree circumference based on cardinal orientation.

Damage Orientation	Average	Standard Deviation	Variation Coefficient
N	0.101	0.101	100.000
NE	0.076	0.045	59.278
E	0.091	0.081	88.792
SE	0.087	0.058	66.173
S	0.075	0.053	70.198
SW	0.138	0.094	67.941
W	0.110	0.082	74.216
NW	0.106	0.076	71.802

Table 6. p values (*ANOVA* test) to test different $w \times C^{-1}$ values between different damage orientations (significant (p < 5%) = *).

Damage Orientation	N	NE	E	SE	S	SW	W	NW
N	1	0.207	0.635	0.532	0.221	0.192	0.705	0.836
NE	0.207	1	0.440	0.519	0.952	0.016 *	0.111	0.129
E	0.636	0.440	1	0.845	0.446	0.114	0.446	0.519
SE	0.532	0.519	0.845	1	0.527	0.084	0.349	0.407
S	0.221	0.952	0.446	0.527	1	0.024 *	0.130	0.147
SW	0.192	0.016 *	0.114	0.084	0.024 *	1	0.384	0.261
W	0.705	0.111	0.446	0.349	0.130	0.384	1	0.851
NW	0.836	0.129	0.519	0.407	0.147	0.261	0.851	1

As such, differences between samples with different damage orientations are mainly insignificant, so representative $w \times C^{-1}$ values could not be determined to be considered tolerance thresholds based on cardinal orientation.

The *Pearson* correlation coefficients between the studied relation and the tree circumference indicate a reverse correlation for some orientations, as follows:

- N, $r = -0.23$—weak correlation;
- NE, $r = -0.13$—very weak correlation;
- E, $r = -0.21$—weak correlation;
- SE, $r = -0.29$—weak correlation;
- S, $r = -0.54$—reasonable correlation;
- SW, $r = -0.44$—reasonable correlation;
- On V, $r = -0.31$—weak correlation;
- On NV, $r = -0.32$—weak correlation.

A very weak correlation ($r > -0.2$) is present in the NE orientation between $w \times C^{-1}$ and the circumference, whereas the correlation is reasonable in the S and SW orientations ($-0.6 < r < -0.4$). All the other cases present a weak correlation ($-0.4 < r < -0.2$).

If all this is considered, the lack of significant differences from a statistical perspective regarding the $w \times C^{-1}$ value between different orientations, as well as the existence of a linear *Pearson* correlation between the relation's value and the circumference, the previously obtained average values (Table 5) are proposed as tolerance thresholds with respect to circumference intervals obtained by calibrating and validating the models presented above.

4. Discussion

4.1. Tree Damage and Tolerance Thresholds

Similar results regarding tree damage were obtained in studies focused on animal logging in which barking was identified as the most common damage (61.5%), a lower percentage than that obtained in the present study (78.9%) [39]. Unlike the current study, the higher percentage of minor damage was due to the use of animal logging, a low-impact logging technique. Thus, animal logging is correlated with a high percentage of light damage, such as squashed bark (23.1%), compared to the percentage obtained in the current study (7.7%), where conventional logging was used. Other studies have shown that, on average, logging damage affected 40% of residual trees, with 21% injured and 19% killed trees [45].

Related investigations have shown similar values for healed damage, representing 12% of the total amount of identified damage but in smaller samples (10 healed damage) [34]. The present study comprises 172 cases of healed damage during the research period, representing 13.5% of the total amount of identified damage.

In young trees, some authors have mentioned that the cambial tissue never survives exposure. Thus, even if only small bark pieces are removed, the xylem is open to an invasion of pathogenic agents [8]. In the studied variants in the present study, the amount of healed damage was superior in variants where thinnings were applied, where the trees' healing power is higher than in old stands. The healing of bark lesions varies by the quantity of removed bark, as well as by vigour and species, whereas all damaged trees maintain rot pockets even after minor injury, regardless of age. Heavier damage results in interior trunk rot over the following decades [8]. Healing tree damage to the greatest extent possible in stands where thinnings were applied reduced the risk of obtaining depreciated wood with root at harvesting age. In the present study, most of the damage was found in thinnings. This is due to the space between the residual trees because when spacing is narrow, there is a higher probability of increased residual tree damage when logs are skidded [46].

Regarding the ratio between damage width and tree circumference, some research has shown that in the case of using skidders, the ratio between damage width and the stem circumference has values of 0.093 (9.3%) for the most frequent damage and 0.12 (12%) for the most severe damage [42]. These data are similar to those obtained in the current study, where the average of the ratio between the width of healed damage and tree circumference was approximately 0.10 in the analysed sample plots (0.09 for damage healed during the analysed period and 0.12 for damage identified as healed in the initial evaluation). The wound healing rate is related to DBH (circumference in the present study), and the rate decreases with increasing wound width [47].

The results of other investigations regarding the ratio between stem and wound for scrapes were similar among treatments, but the same ratios for gouges and scuffs were larger under high-intensity treatment [43]. In the present study, there was an increase in the value of this ratio from 0.09 for thinning to 0.10 for first-intervention cuttings.

Observations in beech stands show that all wounds with an initial width of less than 5 cm were healed [48]. Results of the present study reveal a high inverse correlation between the width and the circumference of damage in beech trees.

Similar investigations studying the healing rate of damage from poplars have found similar healing rates. This is especially true in the N orientation, where the highest percentage of healed damage is recorded [49]. Similarly, in the present study, the highest percentage of healed damage was recorded for damage with N orientation (13.5%), although significant percentages of healed damage were also maintained in the neighbouring cardinal orientations, namely NW (11,3%) and NE (10.3%).

4.2. Discussion and Recommendations Regarding Good Practices for Logging Management and Respecting Tolerance Thresholds

To protect the forest ecosystem, some studies have shown that applying specific measures can reduce the damage to residual trees by 25–33% [50]. One such measure is equipping forests with a network of roads to reduce the average distances for collecting wood [51].

A basic necessity for the ecological harvesting of forests is the use of a method with a low impact on the forest ecosystem. This includes the cut-to-length harvesting method [52]. This method was used in 23 out of the 24 studied harvesting sites. As previously mentioned, this harvesting method causes the fewest seedling and soil injuries, as well as using collecting methods at capacity. Using the shortwood system in the final cutting is the dominant practice in Nordic countries [53]. Studies from Poland have shown that the shortwood system caused the fewest tree injuries in stands of all ages [54].

Another basic condition for the long-lasting development of forests that is not widely used in Romania (a fact also observed in the studied variants) is represented by linking a system of improved machines with the forest regeneration regimes and treatments. For example, modern machines were used in the thinning work and in certain field conditions that were developed and improved along with other adaptations for agricultural tractors [55]. Different adaptations and improvements of forest equipment are necessary, as the damage potential increases in some cases due to the difficulty of manipulating machines in very dense stands [56].

By studying the ratio between the damage width and tree circumference, research comparing crawlers with cable yarders for collecting timber has shown a smaller surface affected by damage in the latter case [42]. A series of additional measures are recommended to ensure favorable effects after harvesting operations. These measures must be technically feasible and economically acceptable [57].

These measures were not typically seen in the analysed parcels; however, some of the practical measures that can decrease the damage caused by timber harvesting will be mentioned. The measures that can be applied include using sustaining coils for load cables to transport pieces by semi-suspension and using direction coils to form the load; using tractors equipped with roller chains in fields with a low carrying capacity; proper assignments; protecting the soil and trees in places where timber is stocked; protecting trees that border skid trails; protecting seedlings by placing paths and collecting tracks outside seedling loci; protecting seedlings with wood ramparts; and using harvest remains to reduce erosion in certain areas.

The results of these studies suggest that low-impact timber harvesting operations should be accompanied by a close surveillance of field personnel, by a financial motivation to encourage—or, based on the case, to discourage—negative activities and by post-harvest inspections to verify proper implementation [58].

5. Conclusions

Based on the data obtained from sample plots, we established that the value of the ratio between the damage width and the tree circumference can be considered a tolerance threshold for trees in logging. The value of this ratio is of 0.09 for thinnings, cuttings to increase light availability for regeneration and final cuttings and 0.10 for first-intervention cuttings.

Equations for each variant and for the main species have been elaborated. The equations were used to obtain the tolerance threshold expressed as the maximum value of the relation between the damage width and the tree circumference, for which the damage is curable.

The *ANOVA* test showed significant differences ($p^* < 5\%$) between the damage width and the tree circumference ratio for two pairs of samples, namely SW and S orientations, as well as SW and NE. As such, differences between samples based on damage orientation are mainly insignificant, so it is not possible to determine representative values that can be considered tolerance thresholds according to cardinal orientation. Therefore, we adopted values obtained previously in silvicultural work as tolerance thresholds concerning the limit to which tree damage in different cardinal orientations is healed in a certain time period.

Author Contributions: Conceptualization, I.-C.C. and L.D.; methodology, I.-C.C.; software, V.E.C.; validation, L.D., G.F.B. and V.E.C.; formal analysis, I.-C.C.; investigation, I.-C.C. and C.-I.C.; resources, I.-C.C.; data curation, C.-I.C.; writing—original draft preparation, I.-C.C. and L.D.; writing—review and editing, I.-C.C., L.D. and V.E.C.; visualization, C.-I.C. and G.F.B.; supervision, L.D.; project administration, I.-C.C.; funding acquisition, I.-C.C. All authors have read and agreed to the published version of the manuscript.

Funding: This paper was carried out with the support of the Ministry of Research, Innovation and Digitization (MCID) through the "Nucleu" program—project no. 19070507 and through "Programul 1—Dezvoltarea sistemului naţional de cercetare—dezvoltare, Subprogram 1.2—Performanţă instituţională—Proiecte de finanţare a excelenţei în CDI"—project "Creșterea capacității și performanței instituționale a INCDS "MarinDrăcea") în activitatea de CDI—CresPerfInst" (Contract No. 34PFE./30.12.2021).

Institutional Review Board Statement: Not applicable.

Data Availability Statement: The data presented in this study are available from the corresponding author upon request. The data are not publicly available becuase the project through which the research was funded is not yet completed.

Acknowledgments: We are thankful to all colleagues for constructive advice and especially to Ion Chisăliţă, former director of Timişoara Research Station.

Conflicts of Interest: The authors declare no conflict of interest.

Appendix A

The arrangement of the harvesting sites in variants and relief forms is the following:

- V1—harvesting sites with thinnings from:
  - Plains: Forest department (OS) Bocşa Română—Production unit (UP) II, management unit (u.a.) 58A and UP III, u.a. 49;
  - Hill: OS Bocşa Montană—UP VI, u.a. 95A and OS Moldova Nouă—UP III, u.a. 15B;
  - Mountain: OS Băile Herculane—UP II, u.a. 23 and Caransebeş experimental basis (BE)—UP VI, u.a. 99A.
- V2—harvesting sites with first-intervention cuttings—preparatory and seed-cutting from shelterwood system or selections system, from:
  - Plains: OS Bocşa Română—UP I, u.a. 11C and UP II, u.a. 55;
  - Hill: OS Bocşa Montană—UP IV, u.a. 62B and OS Moldova Nouă—UP III, u.a. 212A;
  - Mountain: OS Văliug—UP VI, u.a. 15A and u.a. 16A.
- V3—harvesting sites with cuttings to increase light availability for regeneration from shelterwood system, from:
  - Plains: OS Bocşa Română—UP III, u.a. 28B and 76A;
  - Hill: BE Caransebeş—UP II, u.a. 30B and OS Moldova Nouă—UP III, u.a. 176A;
  - Mountain: OS BăileHerculane—UP II, u.a. 99A and 100A.
- V4—harvesting sites with final cuttings from shelterwood system, from:
  - Plains: OS Bocşa Română—UPI, u.a. 1E and 14A;
  - Hill: BECaransebeş—UP I, u.a 46D şi OS Moldova Nouă—UP III, u.a. 15B;
  - Mountain: OS BăileHerculane—UP IV, u.a. 98A şi BE Caransebeş—UP V, u.a. 16A.

Appendix B

Table A1. Synthesis of data regarding damage and characteristics.

Variant	Harvesting Site—Forest District, Production Unit, Management Unit	Healed Damage				Total number of Injuries in Sample Plots	from Which:				
		Initial Assessment		Revaluation							
		Number of Healed Injuries	Ratio between the Initial Damage Width and Tree Circumference	Number of Healed Injuries	Ratio between the Initial Damage Width and Tree Circumference		Galling	Barking	Splintering	Breaking	Uprooting
V1	h1—Bocșa Română, III, 49	0	0	1	0.07	38	2	33	3	0	0
	h2—Bocș aRomână, II, 58A	0	0	1	0.08	18	2	15	0	1	0
	h3—Bocșa Montană, VI, 95A	0	0	1	0.03	92	18	70	1	1	2
	h4—Moldova Nouă, III, 15B	0	0	6	0.04	229	28	190	11	0	0
	h5—Băile Herculane, II, 23	6	0.08	9	0.08	230	2	189	39	0	0
	h6—Caransebeș, VI, 99A	28	0.12	14	0.08	207	8	162	32	4	1
Total plain		0	0.00	2	0.08	56	4	48	3	1	0
Total hill		0	0.00	7	0.04	321	46	260	12	1	2
Total mountain		34	0.11	23	0.08	437	10	351	71	4	1
Total thinnings		34	0.11	32	0.07	814	60	659	86	6	3

Table A1. *Cont.*

Variant	Harvesting Site—Forest District, Production Unit, Management Unit	Healed Damage				Total number of Injuries in Sample Plots	from Which:				
		Initial Assessment		Revaluation							
		Number of Healed Injuries	Ratio between the Initial Damage Width and Tree Circumference	Number of Healed Injuries	Ratio between the Initial Damage Width and Tree Circumference		Galling	Barking	Splintering	Breaking	Uprooting
V2	h1—Bocșa Română, II, 55	0	0	0	0	14	0	12	2	0	0
	h2—Bocșa Română, I, 11C	0	0	0	0	36	0	33	2	0	1
	h3—Bocșa Montană, IV, 62B	0	0	0	0	40	4	33	3	0	0
	h4—Moldova Nouă, III, 212A	1	0.04	5	0.08	87	20	57	10	0	0
	h5—Văliug, VI, 15A	3	0.22	9	0.11	118	5	90	23	0	0
	h6—Văliug, VI, 16A	8	0.17	10	0.12	89	7	67	12	3	0
Total plain		0	0.00	0	0.00	50	0	45	4	0	1
Total hill		1	0.04	5	0.08	127	24	90	13	0	0
Total mountain		11	0.18	19	0.12	207	12	157	35	3	0
Total first-intervention cuttings		12	0.17	24	0.11	384	36	292	52	3	1

Table A1. *Cont.*

Variant	Harvesting Site—Forest District, Production Unit, Management Unit	Healed Damage				Total number of Injuries in Sample Plots	from Which:				
		Initial Assessment		Revaluation							
		Number of Healed Injuries	Ratio between the Initial Damage Width and Tree Circum-ference	Number of Healed Injuries	Ratio between the Initial Damage Width and Tree Circum-ference		Galling	Barking	Splintering	Breaking	Uprooting
V3	h1—Bocșa Română, III, 76A	0	0	0	0	53	0	39	8	4	2
	h2—Bocșa Română, III, 28B	0	0	1	0.08	18	1	13	1	3	0
	h3—Moldova Nouă, III, 176A	0	0	13	0.09	105	15	79	10	1	0
	h4—Caransebeș, II, 30B	0	0	0	0	43	1	38	2	2	0
	h5—Băile Herculane, II, 99	2	0.05	3	0.08	76	2	60	12	2	0
	h6—Băile Herculane, II, 100A	2	0.04	11	0.10	106	12	78	9	4	3
Total plain		0	0.00	1	0.08	71	1	52	9	7	2
Total hill		0	0.00	13	0.09	148	16	117	12	3	0
Total mountain		4	0.05	14	0.10	182	14	138	21	6	3
Total cuttings to increase light availability for regeneration		4	0.05	28	0.09	401	31	307	42	16	5

Table A1. *Cont.*

Variant	Harvesting Site—Forest District, Production Unit, Management Unit	Healed Damage				Total number of Injuries in Sample Plots	from Which:				
		Initial Assessment		Revaluation							
		Number of Healed Injuries	Ratio between the Initial Damage Width and Tree Circumference	Number of Healed Injuries	Ratio between the Initial Damage Width and Tree Circumference		Galling	Barking	Splintering	Breaking	Uprooting
V4	h1—Bocșa Română, I, 1E	0	0	0	0	32	2	28	1	1	0
	h2—Bocșa Română, I, 14A	0	0	0	0	27	1	23	2	1	0
	h3—Moldova Nouă, III, 162B	1	0.26	6	0.17	75	9	56	8	1	1
	h4—Caransebeș, I, 46D	5	0.09	8	0.05	91	7	68	13	3	0
	h5—Băile Herculane, IV, 98A	2	0.15	5	0.04	65	1	58	4	2	0
	h6—Caransebeș, V, 16A	6	0.05	5	0.17	56	4	43	6	2	1
Total plain		0	0.00	0	0.00	59	3	51	3	2	0
Total hill		6	0.12	14	0.10	166	16	124	21	4	1
Total mountain		8	0.08	10	0.11	121	5	101	10	4	1
Total final cuttings		14	0.09	24	0.10	346	24	276	34	10	2
TOTAL		64	0.12	108	0.09	1945	151	1534	214	35	11

References

1. Dincă, L.; Badea, O.; Guiman, G.; Bragă, C.; Crișan, V.; Greavu, V.; Murariu, G.; Georgescu, L. Monitoring of soil moisture in Long-Term Ecological Research (LTER) sites of Romanian Carpathians. *Ann. For. Res.* **2018**, *61*, 171–188. [CrossRef]
2. Murariu, G.; Dinca, L.; Tudose, N.; Crisan, V.; Georgescu, L.; Munteanu, D.; Dragu, M.D.; Rosu, B.; Mocanu, G.D. Structural Characteristics of the Main Resinous Stands from Southern Carpathians, Romania. *Forests* **2021**, *12*, 1029. [CrossRef]
3. Meadows, J.S. Logging damage to residual trees following partial cutting in a green ash-sugarberry stand in the Mississippi Delta. In Proceedings of the 9th Central Hardwood Forest Conference, West Lafayette, IN, USA, 8–10 March 1993; USDA Forest Service, Southern Forest Experiment Station: Washington, DC, USA, 1993; pp. 248–260.
4. Pinard, M.A.; Barker, M.G.; Tay, J. Soil disturbance and post-logging forest recovery on bulldozer paths in Sabah, Malaysia. *For. Ecol. Manag.* **2000**, *130*, 213–225. [CrossRef]
5. Nystrand, O.; Granström, A. Predation on Pinus sylvestris seeds and juvenile seedlings in Swedish boreal forest in relation to stand disturbance by logging. *J. Appl. Ecol.* **2000**, *37*, 449–463. [CrossRef]
6. Hawthorne, W.D.; Marshall, C.A.M.; Juam, M.A.; Agyeman, V.K. *The Impact of Logging Damage on Tropical Rainforest, Their Recovery and Regeneration: An Annotated Bibliography*; Oxford Forestry Institute, Department of Plant Sciences: Oxford, UK, 2011; p. 123.
7. Kasenene, J.M.; Murphy, P.G. Post-logging tree mortality and major branch losses in Kibale Forest, Uganda. *For. Ecol. Manag.* **1991**, *46*, 295–307. [CrossRef]
8. Putz, F.E.; Sist, P.; Fredericksen, T.; Dykstra, D. Reduced-impact logging: Challenges and opportunities. *For. Ecol. Manag.* **2008**, *256*, 1427–1433. [CrossRef]
9. Nichols, M.T.; Lemin, R.C., Jr.; Ostrofsky, W.D. The impact of two harvesting systems on residual stems in a partially cut stand of northern hardwoods. *Can. J. For. Res.* **1994**, *24*, 350–357. [CrossRef]
10. Pinard, M.A.; Putz, F.E. Retaining forest biomass by reducing logging damage. *Biotropica* **1996**, *28*, 278–295. [CrossRef]
11. Guillemette, F.; Bédard, S.; Fortin, M. Evaluation of a tree classification system in relation to mortality risk in Québec northern hardwoods. *For. Chron.* **2008**, *84*, 886–899. [CrossRef]
12. Reich, P.B.; Bakken, P.; Carlson, D.; Frelich, L.E.; Friedman, S.K.; Grigal, D.F. Influence of logging, fire, and forest type on biodiversity and productivity in southern boreal forests. *Ecology* **2001**, *82*, 2731–2748. [CrossRef]
13. Brown, K.A.; Gurevitch, J. Long-term impacts of logging on forest diversity in Madagascar. *Proc. Natl. Acad. Sci. USA* **2004**, *101*, 6045–6049. [CrossRef] [PubMed]
14. Putz, F.E.; Redford, K.H.; Robinson, J.G.; Fimbel, R.; Blate, G.M. *Biodiversity Conservation in the Context of Tropical Forest Management*; The International Bank for Reconstruction and Development/THE WORLD BANK: Washington, DC, USA, 2000; p. 80.
15. Fedorca, A.; Popa, M.; Jurj, R.; Ionescu, G.; Ionescu, O.; Fedorca, M. Assessing the regional landscape connectivity for multispecies to coordinate on-the-ground needs for mitigating linear infrastructure impact in Brasov—Prahova region. *J. Nat. Conserv.* **2020**, *58*, 125903. [CrossRef]
16. Fedorca, A.; Fedorca, M.; Ionescu, O.; Jurj, R.; Ionescu, G.; Popa, M. Sustainable landscape planning to mitigate wildlife-vehicle collisions. *Land.* **2021**, *10*, 737. [CrossRef]
17. Edwards, D.P.; Woodcock, P.; Edwards, F.A.; Larsen, T.H.; Hsu, W.W.; Benedick, S.; Wilcove, D.S. Reduced-impact logging and biodiversity conservation: A case study from Borneo. *Ecol. Appl.* **2012**, *22*, 561–571. [CrossRef] [PubMed]
18. Wilcove, D.S.; Giam, X.; Edwards, D.P.; Fisher, B.; Koh, L.P. Navjot's nightmare revisited: Logging, agriculture, and biodiversity in Southeast Asia. *Trends Ecol. Evol.* **2013**, *28*, 531–540. [CrossRef] [PubMed]
19. Augustynczik, A.L.; Asbeck, T.; Basile, M.; Jonker, M.; Knuff, A.; Yousefpour, R.; Hanewinkel, M. Reconciling forest profitability and biodiversity conservation under disturbance risk: The role of forest management and salvage logging. *Environ. Res. Lett.* **2020**, *15*, 0940a3. [CrossRef]
20. Crișan, V.E.; Dincă, L.C.; Oneț, A.; Bragă, C.I.; Enescu, R.A.; Teușdea, A.C.; Oneț, C. Impact of windthrows disturbance on chemical and biological properties of the forest soils from Romania. *Environ. Eng. Manag. J.* **2021**, *20*, 1163–1172.
21. Tudose, N.C.; Ungurean, C.; Davidescu, Ș.; Clinciu, I.; Marin, M.; Nita, M.D.; Adorjani, A.; Davidescu, A. Torrential flood risk assessment and environmentally friendly solutions for small catchments located in the Romania Natura 2000 sites Ciucas, Postavaru and Piatra Mare. *Sci. Total Environ.* **2020**, *698*, 134271. [CrossRef]
22. Ducci, F.; De Rogatis, A.; Proietti, R.; Curtu, L.A.; Marchi, M.; Belletti, P. Establishing a baseline to monitor future climate-change-effects on peripheral populations of Abies alba in central Apennines. *Ann. For. Res.* **2021**, *64*, 33–66. [CrossRef]
23. Kutnar, L.; Kermavnar, K.; Pintar, A.M. Climate change and disturbances will shape future temperate forests in the transition zone between Central and SE Europe. *Ann. For. Res.* **2021**, *64*, 67–86. [CrossRef]
24. Vlad, R.; Constandache, C.; Dincă, L.; Tudose, N.C.; Sidor, C.G.; Popovici, L.; Ispravnic, A. Influence of climatic, site and stand characteristics on some structural parameters of scots pine (*Pinus sylvestris*) forests situated on degraded lands from east Romania. *Range Manag. Agrofor.* **2019**, *40*, 40–48.
25. Gullison, R.E.; Hardner, J.J. The effects of road design and harvest intensity on forest damage caused by selective logging: Empirical results and a simulation model from the Bosque Chimanes, Bolivia. *For. Ecol. Manag.* **1993**, *59*, 1–14. [CrossRef]
26. Naghdi, R.; Solgi, A.; Zenner, E.K.; Tsioras, P.A. Effect of skid trail curvature on residual tree damage. *Aust. For.* **2019**, *82*, 1–8. [CrossRef]
27. Hwang, K.; Han, H.S.; Marshall, S.E.; Page-Dumroese, D.S. Amount and location of damage to residual trees from cut-to-length thinning operations in a young redwood forest in Northern California. *Forests* **2018**, *9*, 352. [CrossRef]

28. Akay, A.E.; Yilmaz, M.; Tonguc, F. Impact of mechanized harvesting machines on forest ecosystem: Residual stand damage. *J. Appl. Sci.* **2006**, *6*, 2414–2419.
29. Han, H.S.; Kellogg, L.D. Damage characteristics in young Douglas-fir stands from commercial thinning with four timber harvesting systems. *West. J. Appl. For.* **2000**, *15*, 27–33. [CrossRef]
30. Danilović, M.; Kosovski, M.; Gačić, D.; Stojnić, D.; Antonić, S. Damage to residual trees and regeneration during felling and timber extraction in mixed and pure beech stands. *Šumar. List* **2015**, *139*, 253–262.
31. Naghdi, R.; Raafatnia, N.; Sobhany, H.; Jalali, G.; Hossieni, M. Evaluation of tree length and assortment logging methods with respect to residual damage and productivity in Caspian forest (north of Iran). *Nauk. Visn. NLTU Ukr.* **2004**, *14*, 296–302.
32. Moskalik, T.; Borz, S.A.; Dvořák, J.; Ferencik, M.; Glushkov, S.; Muiste, P.; Lazdiņš, A.; Styranivsky, O. Timber harvesting methods in Eastern European countries: A review. *Croat. J. For. Eng.* **2017**, *38*, 231–241.
33. Dămăceanu, C.; Gava, M. *Cercetări Privind Stabilirea Pragurilor de Vătămare a Arborilor, Semințișului și Solului Prin Lucrările de Exploatări Forestiere*; ROMSILVA, Institutul de Cercetări și Amenajări Silvice, Seria a II-a, Centrul de Material Didactic și Propaganda Agricolă, Redacția de Propaganda Tehnică Agricolă: București, Romania, 1991; p. 99.
34. Tavankar, F.; Picchio, R.; Lo Monaco, A.; Nikooy, M.; Venanzi, R.; Bonyad, A.E. Wound healing rate in oriental beech trees following logging damage. *Drewno* **2019**, *62*, 5–22.
35. Picchio, R.; Magagnotti, N.; Sirna, A.; Spinelli, R. Improved winching technique to reduce logging damage. *Ecol. Eng.* **2012**, *47*, 83–86. [CrossRef]
36. Mund, M.; Schulze, E. Impacts of forest management on the carbon budget of European beech (*Fagus sylvatica*) forests. *Allg. Forst Jagdztg.* **2006**, *177*, 47–63.
37. Agenţia Pentru Fianţarea Investiţiilor Rurale. Available online: https://portal.afir.info/informatii_institutionale_structuri_teritoriale_oficii_judetene_ojfir_caras_severin (accessed on 21 September 2021).
38. Interferențe ... Sursa ta de Cunoaștere. Available online: http://www.interferente.ro/harta-romaniei-pe-regiuni-judete-si-orase.html (accessed on 21 September 2021).
39. Knežević, J.; Gurda, S.; Musić, J.; Halilović, V.; Sokolović, D.; Bajrić, M. The Impact of Animal Logging on Residual Trees in Mixed Fir and Spruce Stands. *South-East Eur. For.* **2018**, *9*, 107–114. [CrossRef]
40. Ciubotaru, A. *Exploatareapădurilor*; Editura Lux Libris: Braşov, Romania, 1998; p. 351.
41. Gill, R.M.A. A review of damage by mammals in north temperate forests: 3. Impact on trees and forests. *Forestry* **1992**, *65*, 363–388. [CrossRef]
42. Spinelli, R.; Magagnotti, N.; Nati, C. Benchmarking the impact of traditional small-scale logging systems used in Mediterranean forestry. *For. Ecol. Manag.* **2010**, *260*, 1997–2001. [CrossRef]
43. Behjou, F.K. Effects of wheeled cable skidding on residual trees in selective logging in Caspian forests. *Small-Scale For.* **2014**, *13*, 367–376. [CrossRef]
44. Mental Modeller. Available online: https://www.mentalmodeller.com (accessed on 15 December 2021).
45. Bertault, J.G.; Sist, P. An experimental comparison of different harvesting intensities with reduced-impact and conventional logging in East Kalimantan, Indonesia. *For. Ecol. Manag.* **1997**, *94*, 209–218. [CrossRef]
46. Camp, A. Damage to residual trees by four mechanized harvest systems operating in small-diameter, mixed-conifer forests on steep slopes in Northeastern Washington: A case study. *West. J. Appl. For.* **2002**, *17*, 14–22. [CrossRef]
47. Tavankar, F.; Nikooy, M.; Lo Monaco, A.; Latterini, F.; Venanzi, R.; Picchio, R. Short-term recovery of residual tree damage during successive thinning operations. *Forests* **2020**, *11*, 731. [CrossRef]
48. Vasiliauskas, R. Damage to trees due to forestry operations and its pathological significance in temperate forests: A literature review. *Forestry* **2001**, *74*, 319–336. [CrossRef]
49. David, E.C.; Enache, L.N. Research concerning the frequency of the external defects for poplar (*Populus nigra* L.) from alignments. *Bul. Trans. Univers. Bras. Ser. II For. Wood Ind. Agric. Food Eng.* **2011**, *4*, 21.
50. Johns, J.S.; Barreto, P.; Uhl, C. Logging damage during planned and unplanned logging operations in the eastern Amazon. *For. Ecol. Manag.* **1996**, *89*, 59–77. [CrossRef]
51. Chisăliţă, I. *Tehnologii Ecoproductive în Exploatările Forestiere*; Eurobit: Timişoara, Romania, 2011; p. 172.
52. Horodnic, S. *Sisteme Tehnologice Forestiere cu Impact Ecologic Redus*; Editura Universităţii Suceava: Suceava, Romania, 2014; p. 174.
53. Fjeld, D.; Granhus, A. Injuries after selection harvesting in multi-stored spruce stands–the influence of operating systems and harvest intensity. *J. For. Eng.* **1998**, *9*, 33–40.
54. Karaszewski, Z.; Giefing, D.F.; Mederski, P.S.; Bembenek, M.; Dobek, A.; Stergiadou, A. Stand damage when harvesting timber using a tractor for extraction. *For. Res. Pap.* **2013**, *74*, 27–34. [CrossRef]
55. Marchi, E.; Picchio, R.; Spinelli, R.; Verani, S.; Venanzi, R.; Certini, G. Environmental impact assessment of different logging methods in pine forests thinning. *Ecol. Eng.* **2014**, *70*, 429–436. [CrossRef]
56. Picchio, R.; Neri, F.; Maesano, M.; Savelli, S.; Sirna, A.; Blasi, S.; Baldini, S.; Marchi, E. Growth effects of thinning damage in a Corsican pine (*Pinus laricio* Poiret) stand in central Italy. *For. Ecol. Manag.* **2011**, *262*, 237–243. [CrossRef]
57. Jonkers, W.B.J. *Logging Damage and Efficiency: A Study on the Feasibility of Reduced Impact Logging in Cameroon*; Wageningen University: Wageningen, The Netherlands, 2000; p. 57.
58. Jackson, S.M.; Fredericksen, T.S.; Malcolm, J.R. Area disturbed and residual stand damage following logging in a Bolivian tropical forest. *For. Ecol. Manag.* **2002**, *166*, 271–283. [CrossRef]

 diversity

Article

Plant Species Turnover on Forest Gaps after Natural Disturbances in the Dinaric Fir Beech Forests (*Omphalodo-Fagetum sylvaticae*)

Blanka Ravnjak [1,*], Jože Bavcon [1] and Andraž Čarni [2]

1 Biotechnical Faculty, University Botanic Gardens Ljubljana, 1000 Ljubljana, Slovenia; joze.bavcon@bf.uni-lj.si
2 Research Centre of the Slovenian Academy of Science and Arts, Jovan Hadži Institute of Biology, 1000 Ljubljana, Slovenia; carni@zrc-sazu.si
* Correspondence: blanka.ravnjak@bf.uni-lj.si; Tel.: +386-31336507

Abstract: We studied species turnover and changes of ecological conditions and plant strategies on forest gaps created by natural disturbances (sleet, windthrow). We studied five forest gaps and a control plot within in the Dinaric silver fir-beech forest in the southern part of Slovenia. Forest gaps varied in age and size. The total number of recorded species in gaps was 184, with the highest number (106) at the largest forest gap and with the 58 species at the control locality in a juvenile beech forest. Forest gaps were predominantly colonised mostly by species of understory, forest margins, and forest clearings. The species presented in all forest gaps are representatives of the understory of beech forests. Species colonising forest gaps prefer habitats with more sunlight, medium wet to dry soil, and are tolerant to high daily and seasonal temperature fluctuations. In gaps, the community of plant species has a competitive strategy, which is also complemented with a stress-tolerator strategy. We determined that a forest gap represents a significant habitat patch, especially for those plant species which were not present there before.

Keywords: forest gaps; plant colonisation; community strategy; Dinaric fir-beech forests; Ellenberg values

Citation: Ravnjak, B.; Bavcon, J.; Čarni, A. Plant Species Turnover on Forest Gaps after Natural Disturbances in the Dinaric Fir Beech Forests (*Omphalodo-Fagetum sylvaticae*). *Diversity* **2022**, *14*, 209. https://doi.org/10.3390/d14030209

Academic Editors: Lucian Dinca, Miglena Zhiyanski and Michael Wink

Received: 15 February 2022
Accepted: 8 March 2022
Published: 11 March 2022

1. Introduction

Forest gaps are areas in a forest stand that have been created by the death of individual or multiple trees [1]. Forest gaps are part of forests and a typical stage of forest dynamics [2]. Every newly formatted gap results in changes to certain environmental factors, also affecting the flora and fauna. The created forest gaps can vary in size and shape. Depending on the size and gap shape, there are also changes in environmental factors [3]. When a forest gap forms, the light availability and precipitation regime change (increased soil moisture) in the gap area [4,5]. Illumination of the forest gap largely depends on the size of the gap and the size of the surrounding tree stand. Large trees shade the edges of the forest gap and, if the gap is small, can shade it almost in full [3]. The amount of light that reaches the floor of the forest gap also depends on the latitude at which the gap is located, the slope of the terrain, and its position on the sunny or shady side. All three items affect the angle of incidence of sunlight. The amount of incoming light also varies depending on the season and daily rhythm [5–8]. The greater the gap, the more surface is exposed to light. Air temperature largely depends on location exposure to sunlight and wind. Kermavnar et al. [5] showed that the microclimate in Dinaric fir-beech forests also depends on topographic factors (i.e., within-sinkhole position). Temperature fluctuations in forest gaps are greater than those in a compact forest [9]. Higher daily and seasonal fluctuations in air temperatures were also shown by Kermavnar et al. [5]. A forest gap has higher soil humidity than the forest, as it is directly exposed to precipitation [3]. Humidity depends on the amount of litterfall remaining in the forest gap after its formation. Woody debris (trunks, branches, leaves) retains moisture [3,8,10].

The forest dynamics in forest gaps are similar in all forest types and progress in the direction of overgrowing with tree species [2], but can also differ among gaps according to pre-disturbance biotic and abiotic conditions [11]. However, there is a difference in the rate of overgrowing and the species composition of the plants. At the beginning of a newly formatted forest gap, the trees growing on the edges of the forest gap start spreading their canopies towards the gap, i.e., where they have sufficient space. This is noticeable particularly in deciduous trees [12]. This is an edge effect that, in addition to the effect on light, also has an effect on the expansion of plants from the edge to the forest gap [8]. In forest gaps, the tree species that stagnated in the understory before the formation of the gap, where they did not have enough light to grow, begin growing more quickly. Specimens of some plant species adapted to a dense forest stand (growing there before gap formation) either die or adapt to new conditions [13], like *Cyclamen purpurascens*, in which the synthesis of anthocyanins is increased at forest gaps compared to a compact forest stand [14]. In deciduous forests of temperate climates, the first colonisers of forest gaps are spring flowering plant species (like *Omphalodes verna*) of the herbaceous layer from a nearby forest stand [15]. They begin their vegetation period when trees do not develop leaves yet and, of course, they retain this trait at forest gaps. In forest gaps near roads with high traffic, some invasive plant species can also occur [16].

For some seeds, a forest gap also represents a favourable environment for germination. The soil seed banks at the gaps as a form of post-disturbance regeneration are especially important for some herbaceous plant species. A forest gap represents a sink for seeds from near and far surroundings, carried there by winds, animals, or man [12]. The more successful colonisers of forest gaps include species that have long-lived seeds in a seed bank and species whose seeds can disperse over long distances [17]. Generally, the species whose seeds are spread by animals are more successful.

Shortly after the gap formation, the species richness is increasing and continues to rise over time, until the last stages of succession when the gap is again overgrown. Constantly changing abiotic and biotic factors, as well as interspecific relation over the years, cause species turnover. Some species disappear or decrease, others appear or increase, but at the beginning of the succession process on newly formed forest gaps, the species richness increases [17]. Forest gaps, as a result of disturbance, affect changes in the community structure which depend on the type [18] and intensity of the disturbance and interactions of various disturbances in time and space [19–22].

In relation to the occurrence of limiting stress factors (altered water and light regime, suboptimal temperature, change in the amount of nutrients) and the disturbance itself (partial or complete destruction of plant biomass), the strategy of the whole community in the forest gap changes, from the time of formation to overgrowing along the succession line [23]. Understanding the changes resulting from the formation of the forest gap in the forest ecosystem and its effects on local biodiversity can be very important for forest management [24].

Based on findings of other studies [2,9,16,25] describing the change of abiotic and biotic factors in forest gaps and general patterns of colonisation of plant species in forest gaps, our study examined the occurrence of plant species in the initial stages (immediately after disturbance) of newly formed forest gaps and of those older ones (after four years). However, we selected only forest gaps that were formed as a result of disasters and were, therefore, of different sizes. Disturbances, e.g., ice storms and windthrow, are increasingly common in forests due to climate changes, resulting in greater likelihood of forest gap formation [26]. In forest management, the question arises as to whether such newly formed areas should be actively managed or left to natural processes. By studying colonisation of such areas by plant species, we can obtain important data that will help in forest management and thus preserve the great biodiversity of the forest as an ecosystem.

The goal of our study was to determine the plant species population dynamics at forest gaps in Dinaric silver fir-beech forests (*Omphalodo-Fagetum*). We studied the plant species turnover at forest gaps, which is a result of colonised and resident plant species. Our study focused on species diversity, composition of plant community, and its ecological strategy in relation to the size and age of the forest gaps. We predicted:

- that plant species diversity would vary between specific forest gaps, depending on their age;
- that mostly plant species from surrounding the forest will spread to the forest gaps;
- there would be no significant appearance of invasive plants;
- that beech would be the most common tree species in forest gaps.

With the research, we want to enlighten about the natural succession and forest dynamics at naturally formed forest gaps. Our research could help forest management to make decisions about whether newly formed gaps should be planted with tree species or should be left to natural succession with higher plant diversity at the first stage of forest gap formation.

2. Materials and Methods

2.1. Study Sites

The study was conducted in Dinaric silver fir-beech forests (*Omphalodo-Fagetum*) in the southern part of Slovenia (Kočevje-Ribnica area). The geological substrate of the area is diverse soils changing in a small scale due to local topography and is formed predominantly by Jurassic and Cretaceous limestone and, to a lesser extent, dolomite. The surface in higher areas is drier because of shallow humus limestone soil (rendzina), which is formed on parent material, while more or less shallow brown soil formed in the lowlands.

The average annual rainfall in the lowlands is between 1400 and 1500 mm and, at higher altitudes, from 1600 mm to 2000 mm [27]. The two peak rainfall times usually occur in June and from September to October (www.meteo.arso.gov.si/met/sl/climate/maps/monthly-mean-precipitation-maps, accessed on 14 February 2022). The average annual temperature of the Kočevje-Ribnica area ranges between 7 °C and 9 °C in the lowlands and between 6 °C and 7 °C in the mountain range [27]. Weather phenomena include frequent sleet and wet snow, which cause trees to break and fall. Windthrow can occur in summer months due to a strong south-western wind [27,28]. Thunderstorm winds cause the most severe damage, but blowdown patches are typically limited to stand-scales (e.g., 10 s of ha). Ice storms and heavy snow typically cause intermediate severity damage and affect much larger areas [29]. Because of these weather phenomena and their consequences, the formation of forest gaps in this area is very common. Dinaric silver fir-beech forests of the studied area are part of the Dinaric phytogeographic region [30]. The area is characterised by Central European flora with a more or less strong influence of the Illyrian-Balkan flora and Alpine floral element [30–32].

For the purposes of our study, we selected six survey localities (L) within the research area (Figure 1), of which five were forest gaps that were formed at different times and of different sizes (other features are listed in Table 1) and one was in the stand of juvenile beech trees (with trunk diameter $\leq$ 10 cm) (Stone Wall). Namely in the rejuvenation phase of the stand, beech predominates, and only later the fir saplings develop under the beech [33]. The localities were chosen according to their position in the same region, (Kočevje-Ribnica area), similarities in gap formation, altitude, and exposition. We could not obtain exact data on their formation, so we estimated the age of forest gaps on the basis of the state of bigger organic decomposing material (trunks and branches) and the annual growth of spruce saplings (faster growth on sunlight and habitus of a single plant) [34–36]. All forest gaps formed as a result of natural disturbances (sleet or windthrow), which was determined on the basis of uprooted and decaying trees, the condition of tree trunks, and the size of the forest gap [37,38].

Figure 1. Locations of 6 sampling sites on map (1: Stone Wall, 2: Below Barnik, 3: Above Barnik, 4: Goteniški Snežnik, 5: Goteniška gora, 6: Above Draga). (Source: http://gis.arso.gov.si/atlasokolja/profile.aspx?id=Atlas_Okolja_AXL@Arso, accessed on 12 January 2022).

Table 1. Forest gaps and their features (coordinates, UTM quadrant, altitude, aspect, substrate, soil, year of origin). The location marked * is control site with juvenile phase of beech forest.

Locality/Forest Gap	Coordinates	Altit.	Ex.	Surface	Substrate	Year
L1 (Stone wall) *	Y: 5479575 X: 5052223	1068 m	SE	175 m^2	limestone	2000
L2 (Below Barnik)	Y: 5478774 X: 5051499	1132 m	SW	600 m^2	dolomite	2007
L3 (Above Barnik)	Y: 5478904 X: 5051210	1161 m	SE	1400 m^2	limestone, dolomite	2007
L4 (Goteniški Snežnik)	Y: 5480085 X: 5049403	1205 m	E	1300 m^2	limestone, dolomite	2013
L5 (Goteniška gora)	X: 5055948 Y: 5476338	1100 m	SE	3200 m^2	limestone, dolomite	2011
L6 (Above Draga)	Y: 5473400 X: 5052341	954 m	E	1480 m^2	dolomite	2014

2.2. Floristic Survey and Sampling

In 2016 and 2017, vegetation surveys were performed a total of nine times on all five forest gaps and in the beech stand in order to (Table 2) get a total species pool at localities (hereafter they will be named with the locality number and year of creation, e.g., L1-2000). We conducted 9 surveys over three years because we wanted to obtain, during different seasons, a record of plant species as comprehensive as possible at each forest gap [39]. Floristic surveys at total forest gap area were done for a control, including how many and which species can we expect at a single gap. In addition, we estimated the spatial distribution of plant species at a single forest gap.

Table 2. Floristic surveys and sampling scheme.

Locality	Floristic Survey Dates	Sampling Dates and No. of Sampling Plots
L1 (Stone wall) *	24 August 2015 21 April, 7 June, 8 July, 3 August, 27 September 2016 29 March., 25 April, 24 May, 11 July 2017	11 July 2017/8 plots
L2 (Below Barnik)	24 August 2015 21 April, 7 June, 8 July, 3 August, 27 September 2016 29 March, 25 April, 24 May, 11 July 2017	11 July 2017/8 plots
L3 (Above Barnik)	24 August 2015 21 April, 7 June, 8 July, 3 August, 27 September 2016 29 March, 25 April, 24 May, 11 July 2017	11 July 2017/8 plots
L4 (Goteniški Snežnik)	24 August 2015 21 April, 7 June, 8 July, 3 August, 27 September 2016 29 March, 25 April, 24 May, 20 July 2017	20 July 2017/14 plots
L5 (Goteniška gora)	24 August 2015 21 April, 7 June, 8 July, 3 August, 27 September 2016 29 March, 25 April, 24 May,12 July 2017	12 July 2017/20 plots
L6 (Above Draga)	24 August 2015 21 April, 7 June, 8 July, 3 August, 27 September 2016 29 March, 25 April, 24 May, 21 July 2017	21 July 2017/20 plots

During the last vegetation survey in 2017 (in July), we divided the forest gaps into 3-by-3-metre sample plots, which were evenly (systematically) distributed over the surface of each locality. The number of sample plots per locality/forest gap varied between individual gaps, as the forest gaps were of different sizes (Tables 1 and 2). We selected 8 sample plots at L1-2000 (beech stand), L2-2007, and L3-2007 forest gaps, 14 at the L4-2013 gap, and 20 at the L5-2011 and L6-2014 gaps. In each sample plot, we conducted a sampling using the Braun-Blanquet method [40] to get more objective sampling for statistical analyses (Table 2). We used the Braun-Blanquet cover score, which was then transformed in cover percentage in JUICE 7.0 software [40]. Only herb [41] and shrub layer were included in the recording. The sampling was done during the last floristic survey in July because, at that time, there is a peak of vegetation season.

2.3. Data Analysis

For the data analyses, we used data collected in sample plots. Analyses of plant species composition in sample plots at individual forest gaps (location) were conducted using the JUICE 7.0 software [42]. The calculation of parameters was based on averages calculated from all sample plots at a single locality (forest gap). Since we wanted to determine whether the coverage of the species present in a single forest gap is changing during succession, we calculated the minimum, maximum, and average cover of each species at each forest gap and the frequency of their occurrence in forest gaps. For species that occurred in at least five or six forest gaps (locations), and whose cover differed between locations, we used the Kruskal-Wallis test with Bonferoni correction for multiple comparisons in the Statistica 8.0 software [43] to calculate whether the difference in cover between individual forest gaps was statistically significant for these species ($p < 0.05$). The calculation was also based on data obtained in sample plots.

We attributed geoelements according to Pignatti et al. [44] to all species and calculated the share of species belonging to a specific geoelement. For each forest gap, we identified unique species, as well as diagnostic, constant, and dominant plant species [45]. For the fidelity threshold of diagnostic species, we chose the occurrence of the species in at least 50% of the sample plots at each locality compared to other localities (at phi between −1 and 1). A species was constant if it occurred in at least 60% of sample plots, and dominant when it had a cover of more than 30% on each sample plot. Incidental species were those that were present at only one sample plot [42]. The Fischer test in the JUICE 7.0 software was used to identify rare species [46]. To determine similarities in plant community (species composition) between forest gaps, we used the Jaccard similarity coefficients [39,47] and nearest neighbour amalgamation procedure, which were calculated using JUICE 7.0 (as Cluster Analysis–PC-ORD) [48] and plotted a dendrogram.

Using the BIOLFLOR database [49] and the C-S-R Signature Calculator 1.2 software [50], we determined the life strategy of plant communities at individual locality and produced a C-S-R diagram [23]. With the help of Ellenberg indicator values [51], we analysed the environmental factors of individual locality and consequently determined the ecological requirements of plant species that inhabit the studied forest gaps.

3. Results

3.1. Species Richness

A total of 184 different plant species were recorded in the herb layer together on all localities (forest gaps and beech stand) and a total of 140 species on all sample plots (78 plots of size 3 × 3). Comparatively among forest gaps, the highest number of species was found in the largest 5-year-old forest gap, L5-2011, namely, in the survey of entire locality, 106 and, in the surveys within single sample plot, 74 species. The fewest (84) were found on the smallest 9-year-old forest gap, L2-2007 (Table 3).

Table 3. Number of plant species at single locality (L1–2000; L2–2007; L3–2007; L4–2013; L5–2011; L6–2014).

	L1	L2	L3	L4	L5	L6
No. of all species per locality/forest gap	58	84	92	96	106	93
Only on individual locality	2	9	9	8	13	23
Only on sample plots	0	3	4	10	7	14
Only outside sample plots but at the locality	27	27	26	28	32	20

Of all the species, only one non-native invasive species (*Erigeron annuus*) was recorded. It was present at two forest gaps, with one or two specimens. The highest number of species present only at an individual forest gap was found at the youngest forest gap, L6-2014 (23), and the largest forest gap, L5-2011 (13) (Table 3).

The locations/forest gaps differed from each other also in the diagnostic, constant, and dominant species and in its number. The youngest forest gap, L6-2014, had the largest number of diagnostic species (7). The highest number of constant species was recorded on the forest gap L3-2007 (14) and the highest of dominant species on the L5–2011 forest gap. The dominant species on all forest gaps was *F. sylvatica.* All other diagnostic, constant, and dominant species at single locations are shown in Table 4.

According to the Jaccard and coefficients, the locations L3-2007 and L4-2013 were the most similar, and the least similar were the locations L3-2007 and L6-2014 (Figure 2). The spatial distribution of investigated forest gaps is similar, which indicates that the similarities and differences between them (according to the plant community) are due to conditions at the given location.

Table 4. Diagnostic, constant, and dominant plant species on the studied forest gaps (1: L6-2014, 2: L4-2013, 3: L5-2011, 4: L2-2007, 5: L3-2007, 6: L1-2000 (control)). The presence of species in studied forest gaps is marked with *.

Species	Diagnostic						Constant						Dominant					
	1	2	3	4	5	6	1	2	3	4	5	6	1	2	3	4	5	6
Abies alba								*										
Acer pseudoplatanus							*	*	*	*	*	*						
Adenostyles glabra				*														
Aegopodium podagraria	*																	
Agrimonia eupatoria									*	*								
Ajuga reptans			*												*			
Anemone nemorosa										*								
Asarum europaeum	*																	
Athyrium filix-femina		*																
Brachypodium sylvaticum								*	*		*	*		*	*	*	*	
Calamagrostis epigejos														*			*	
Calamintha grandiflora					*													
Cardamine trifolia										*		*						
Carex sylvatica									*	*	*							
Circaea lutetiana		*																
Cyclamen purpurascens	*																	
Digitalis ambigua	*																	
Epimedium alpinum													*					
Euphorbia amygdaloides							*			*	*							
Fagus sylvatica							*	*	*	*	*	*	*	*	*	*	*	*
Fragaria vesca								*	*		*			*	*		*	
Galeobdolon flavidum								*		*	*	*						
Galium odoratum								*	*	*	*	*				*		
Hacquetia epipactis	*												*					
Hedera helix	*																	
Helleborus niger				*														
Heracleum sphondylium				*														
Hypericum perforatum													*					
Laserpitium krapfii	*																	
Maianthemum bifolium													*					
Mercurialis perennis							*		*				*					
Mycelis muralis							*	*	*				*					
Omphalodes verna										*	*	*	*		*	*		
Oxalis acetosella		*																
Petasites albus													*				*	
Picea abies									*		*				*			
Polygonatum verticillatum											*	*						
Polystichum aculeatum		*																
Prunella vulgaris															*			
Ranunculus platanifolius				*														
Rosa pendulina				*														

Table 4. *Cont.*

Species	Diagnostic						Constant						Dominant					
	1	2	3	4	5	6	1	2	3	4	5	6	1	2	3	4	5	6
Rubus idaeus								*	*	*	*		*			*		
Salvia glutinosa		*									*			*				
Sanicula europaea							*		*			*			*			
Scrophularia nodosa		*												*				
Senecio ovatus							*	*	*	*	*					*		
Tussilago farfara															*	*		
Urtica dioica															*			
Veronica officinalis															*			
Vicia oroboides			*															
Viola canina								*										

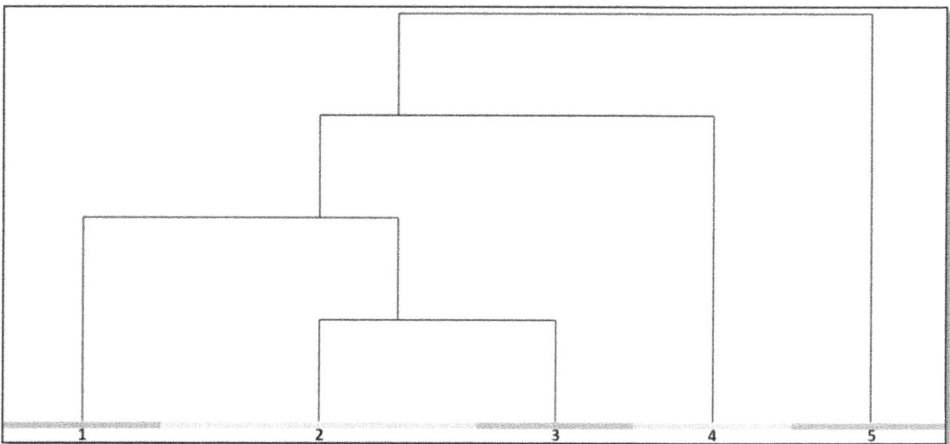

Figure 2. Dendrogram of similarities between individual locations based on Jaccard coefficient (1: control L1-2000; 2: L3-2007 and L4-2013; 3: L5-2011; 4: L2-2007; 5: L6-2014).

3.2. Analysis of Species Composition

At individual forest gaps, the cover of most plant species was less than 50% in the sample plots. In terms of forest gap age, at the youngest forest gap, L6-2014, and oldest, L5-2011, less than a quarter of species in the sample plots had average cover below 2% and at second youngest, L4-2013, even less. At L6-2014, *Epimedium alpinum* L. had the highest average coverage (21%) and at L4-2013, species *Fragaria vesca* L. (26%), *F. sylvatica* (25%), and *S. glutinosa* (22%). At the forest gap L5-2011, species *Tussilago farfara* L. (36%), *B. sylvaticum* (27%), and *F. vesca* (24%) had the highest average cover. This is followed by the oldest forest gaps, with species *T. farfara* (63%), *F. sylvatica* (38%), *B. sylvaticum* (22%), and *S. ovatus* (21%) having the highest percentage of average cover in the sample plots at the forest gap L2-2007. Comparing to other forest gaps, this forest gap had many more species with a cover of less than 2%. At the forest gap L3-2007, as many as one third of the species had an average cover of less than 2% in the sample plots, with species *Petasites albus* (L.) Gaertn. (55%) having high average cover in addition to species *F. sylvatica* (32%) and *B. sylvaticum* (44%). The highest average (at least 20% or more) and maximum cover (more than 60%) of species in total at all forest gaps are shown in Table 5. In terms of frequency of occurrence of species on sample plots, *F. sylvatica* (100%) and *A. pseudoplatanus* (above 60%) had the highest frequency at all localities, *Galium odoratum* (L.) Scop. (above 70%) and *Galeobdolon flavidum* (F. Herm.) (above 50%) at five localities and *B. sylvaticum* (over 70%) and *Carex sylvatica* Huds. and *F. vesca* at four localities with more than 50%.

Table 5. Plant species with highest average (at least 20% or more) and maximum (more than 60%) cover at all sample plots of studied forest gaps.

Average Cover in %	Maximum Cover in %
Fagus sylvatica (29%)	*Fagus sylvatica* (88%)
Brachypodium sylvaticum (24%)	*Brachypodium sylvaticum* (63%)
Tussilago farfara (20%)	*Tusilago farfara* (63%)
Epimedium alpinum (21%)	*Fragaria vesca* (63%)
	Petasites albus (63%)
	Salvia glutinosa (63%)
	Senecio ovatus (63%)

For 15 species that were present in the sample plots on most forest gaps, statistically significant differences in cover were observed between individual forest gaps (Figure 3a,b). Species *B. sylvaticum* had a statistically higher average cover on the forest gap L3-2007 compared to the forest gap L6-2014, where this species was not present at all, and from the beech stand L1-2000 and the forest gap L2-2007. For species *C. grandiflora*, there were statistically significant differences in the average cover between the site L3-2007 and the sites L5-2011, L4-2013, beech stand L1-2000, and L6-2014, but at L3-2007, this species had greater cover than at other sites. At L2-2007, species *Cardamine trifolia* L., *G. odoratum*, *Omphalodes verna* Moench and *S. ovatus* had a statistically significant highest cover of all in comparison to other five sites. At the site L6-2014, however, *Mercurialis perennis* L., *Mycelis muralis* (L.) Dumort. and *Euphorbia amygdaloides* L. had a statistically significant higher average cover than at the other five sites. While species *F. vesca* had a statistically significant higher average cover at the location L4-2013 compared to the locations L6-2014, beech stand L1-2000 and L2-2007, species *S. glutinosa* had, in addition to the same statistical differences as the previously mentioned species, significant differences in cover compared to locations L3-2007 and L5-2011. Species *Sanicula europaea* L. had characteristically the highest average cover at the site L5-2011 compared to the beech stand L1-2000, L3-2007, L4-2013, and L2-2007. For species *A. reptans*, there were statistically significant differences only between the sites L5-2011 and L4-2013, where its average cover was significantly lower or was absent. Species *C. sylvatica* and *G. flavidum* also had a significantly different cover on L5-2011. The first one had the highest cover and the second one, the lowest cover on L5-2011. The first one had a significantly higher average cover there than at the location L6-2014, and the latter, the lowest compared to the location L2-2007.

3.3. Analysis of Environmental Factors and Ecological Strategy of Community

Based on Ellenberg phytoindication estimates of environmental variables, we found that the Ellenberg light index at all studied forest gaps was between 4 and 5, temperature index between 4.8 and 5.2, and humidity index between 5 and 5.4. In the Ellenberg index that describes the pH of the soil, there were differences between the forest gap L6-2014 and others. At the forest gap L6-2014, it was 7, while in the other four forest gaps, it was between 6 and 7. In terms of the amount of nitrogen and other nutrients in the soil, the average Ellenberg index was between 5.5 and 6 for four locations and between 6 and 6.5 for two locations. According to the Ellenberg continentality index, which was between 3 and 4 on forest gaps, forest gaps in the studied area are mainly inhabited by plant species widespread in Central Europe.

By analysing the ecological strategy of plant communities, we found that plant communities have a stress tolerator–competitor/competitor–stress tolerator–ruderal (SC/CSR) strategy in most of the studied forest gaps. Only the plant community at the forest gap L2-2007 differs slightly, as it has a competitor/competitor–stress tolerator–ruderal (C/CSR) strategy. In all communities of studied forest gaps, the largest component is represented by strategy C (competitor), which has a share of over 45% in all of them (Figure 4).

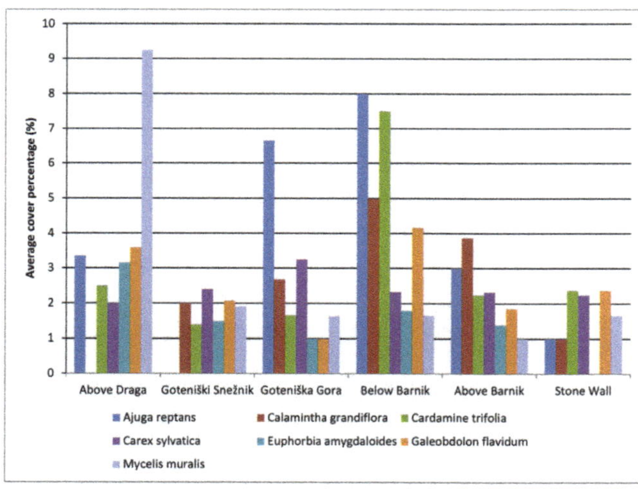

(a)

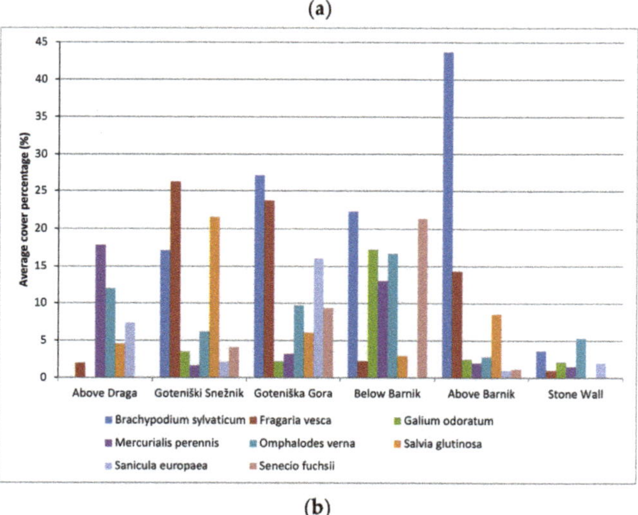

(b)

Figure 3. (**a**) Cover percentage change diagram between single sampling sites for first set of 15 chosen plant species (Above Draga/L6-2014, Goteniški Snežnik/L4-2013, Goteniška gora/L5-2011, Below Barnik/L2-2007, Above Barnik/L3-2007, control Stone Wall/L1-2000). (**b**) Cover percentage change diagram between single sampling sites for second set of 15 chosen plant species (Above Draga/L6-2014, Goteniški Snežnik/L4-2013, Goteniška gora/L5-2011, Below Barnik/L2-2007, Above Barnik/L3-2007, control Stone Wall/L1- 2000).

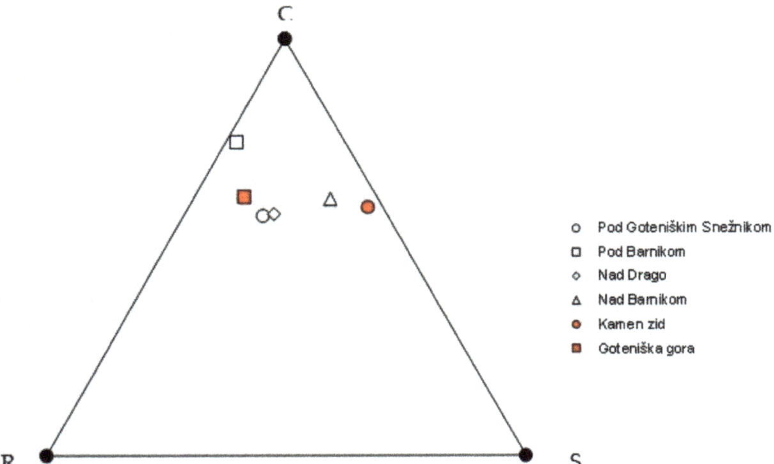

Figure 4. C-S-R strategy of plant communities at single sampling sites (L4-2013 (Pod Goteniškim Snežnikom), L2-2007 (Pod Barnikom), L6-2014 (Nad DRago), L3-2007 (Nad Barnikom), L1-2000 (Kamen zid), L5-2011 (Gotaniška gora)).

4. Discussion

In our study, we recorded a total of 186 species in forest gaps, which is much more than in a similar study in Hungary, where the largest number of species was 61 [2,52]. We can, therefore, conclude that the plant species diversity in forest gaps included in our study is rich and contributes to the impact on the biodiversity of the entire area [53]. This can already be explained by the high α-diversity (local diversity) of individual forest gaps, which in turn leads to high β-diversity (ratio between regional and local diversity) [47]. Therefore, forest gaps L5-2011 and L2-2007 provide support for the fact that the size of the forest gap affects its biodiversity [7,14,53]. At the first one, we recorded the largest number of species, which is due to the fact that it is the largest of all gaps and its mosaic structure, which is a result of a natural disturbance, provides many microhabitats for the growth of plant species with different ecological niches [14]. Some shade species (e.g., *O. verna, M. perennis, Maianthemum bifolium* (L.) F. W. Schmidt) were still present at this gap, representing the remnants of forest understory species [54], while species of sunny habitats were predominant on shallow soils and locations exposed to the sun. Furthermore, next to decaying organic material, we observed species that thrive on nutrient-rich soils (e.g., *Urtica dioica* L., *R. idaeus*) [55] and representatives of species with high stalk plants, characteristic of forest gaps (e.g., *S. glutinosa, S. nodosa, S. ovatus*) [54]. A higher degree of mineralisation occurs in the initial stages of forest gaps, as the disturbance causes more decomposing organic material on the ground [56]. The opposite was found out for the smallest forest gap, L2-2007. Lower species number is the influence of the canopy of edge trees. Specifically, in this forest gap, there appear no plants of sunny habitats because the canopies of edge trees form shade over most of the forest gap [2,7,16].

The species present in our studied forest gaps were mostly species of forest understory, forest edge, and clear-fells [14,53]. There were almost no meadow species, which is logical since the studied forest gaps are located in the middle of forests, and there are no such meadow surfaces nearby that could be a source of meadow plant species colonisation. Some meadow species, e.g., *Bellis perennis* L., *Plantago major* L., *Medicago lupulina L., Crepis biennis* L., and *Taraxacum officinale* Weber in Wiggers, probably spread to forest gaps through forest roads and trails, which are used for wood harvesting and are connected to non-forest surfaces [57]. These species were predominantly located along forest trails. Colonisation of non-forest species along roads and trails to forest gaps also represents a potentially dangerous possibility for transfer of invasive plant species, so biodiversity along paths should be regularly monitored in forest management to promptly prevent the introduction of invasive species. However, at each of the forest gaps, we recorded a few species that appeared only there and nowhere else. We surveyed most of such species (23) at the forest gap L6-2014. The reason for this could be that this forest gap is the farthest from all the others. Therefore, the floristic composition of the surroundings (as a source of colonising species) of the forest gap L6-2014 may be different from those surrounding areas of other forest gaps [58]. The other reason is also that the gap was formed the most recently. This forest gap also had a larger share of annuals and biennials, which are mainly pioneer species–this is characteristic mainly of the initial phase of succession [12].

The species that were present at all forest gaps can be classified into two groups, specifically those that are typical representatives of beech forest understory (e.g., *O. verna, C. trifolia, C. sylvatica, G. flavidum, E. amygdaloides, M. muralis, S. europaea*) [59,60] and those that are common representatives of forest edges and clearings (e.g., *F. vesca, S. glutinosa, Scrophularia nodosa, R. idaeus*) [61]. Differences in their cover between individual forest gaps, however, indicate that the age and size of the forest gaps have an impact on their occurrence and spread. Namely, the greater cover of species *M. perennis, H. epipactis*, and *E. alpinum* at the forest gap L6-2014 again indicates that this forest gap was still in formation phase at the time of the study, as the forest understory species had the largest cover. These again represent the remnants of species that grew under canopy before the felling of trees and forest gap formation [16,53,62]. This is also confirmed by the statistically significant higher cover of forest understory species *M. muralis* and *E. amygdaloides* (in addition to *M. perennis*) at the site L6-2014. The same was found in the study by Kermavnar et al. [12]. At the forest gap L2-2007, shade-loving species (forest understory species) had a statistically significant higher cover because this forest gap is the smallest and longest and, consequently, shadier. These species of forest understory in forest gaps are, in a way, remnants of the former beech forest [12], which still thrive primarily on the shadier parts of forest gaps, and are, therefore, characteristic of newly formed forest gaps. However, they reappear in the final stages of succession overgrowing, when tree vegetation begins overgrowing the forest gap [16]. For species of forest edges, the forest gaps represent a new favourable environment, to which they colonise from already existing forest edges. These species thus tolerate more open surfaces with more light, as well as partial shade. These are species that are present on forest gaps of medium age [2,6,53,63].

However, the low cover of plant species (at least one third of the species on forest gaps did not exceed cover of 2%) indicates that the plant species on forest gaps did not appear in larger closed populations. Small cover of species allows the coexistence of a larger number of species, increasing the biodiversity of the site [14,53]. Species, which appeared with the highest cover on all forest gaps, except the youngest one, L6-2014, were *F. sylvatica, T. farfara, B. sylvaticum, F. vesca, S. glutinosa, P. albus*, and *S. ovatus*. The gap L6-2014 was at the beginning of succession and species did not yet develop bigger populations. The high cover of *F. sylvatica* is expected, as it represents the predominant tree species in all locations. Species *B. sylvaticum* is also a common species in forests. The greater cover of species *T. farfara* and *P. albus* indicates their pioneer character and morphological adaptations, which allow them to propagate rapidly (production of large quantity of seeds and propagation by rhizomes) primarily on moist limestone soils with rock debris [63].

The frequency of occurrence on the sample plots and the constancy of specific species confirmed their uniform even distribution on the entire surface of the forest gaps, as the sample plots were evenly distributed on the forest gaps. The uniform representation of a species on the surface can again be either the result of environmental factors (uniform influence of the environmental factor on the entire surface) or the adaptability of a specific species to various environmental factors. One third of the species were present in more than 50% of the sample plots of an individual forest gap, i.e., they were more or less evenly distributed on the forest gaps. Their share varied slightly between forest gaps. On the forest gap L2-2007 thrive most such species, probably because of its small size, which resulted in lower fluctuations of environmental factors. Species *S. ovatus* stood out with its 100%

frequency of occurrence at three forest gaps. We can conclude that this species has a wide ecological niche. It appeared on all sample plots, both on the smallest and shadiest forest gap, L2-2007, as well as on the largest forest gap with a very mosaic structure, L5-2011, and on the nutrient-rich forest gap, L4-2013. Its broad ecological valence is also cited in the literature. It is even said to be a typical species of forest gaps formed after fires, snow damage, and blowdowns [54,64].

According to Ellenberg indices, we found that forest gaps are colonised primarily by semi-shady plant species that grow in places with more than 10% light intensity, but rarely thrive in full light and are characteristic of moderately warm and submontane areas [51]. These are also species that thrive on moderately moist to dry soils, as forest gaps are open surfaces where fluctuations in humidity are greater. The surface of all forest gaps is also structured into pit and mound, with humidity being higher in pit with accumulated organic material than in raised, exposed parts [65–67]. The mosaic structure thus allows those species that need slightly more moisture to thrive as well as those that grow in dry habitats. According to Ellenberg indices, differences between forest gaps became apparent in the amount of nutrients in the soil, which is the result of the gap formatting time and the decaying plant material left there after the disturbance. On L5-2011 and L4-2013, there were several species that grow in soils richer in nutrients, which is the result of a larger amount of litterfall (branches, trunks, stumps) that is in the last stage of decomposition at both forest gaps. As decomposition increases, the pH of the soil decreases [68,69], which we determined by the Ellenberg index. Therefore, plant species growing on soils with medium to weak acidity are present at all forest gaps except the youngest (L6-2014). At the youngest forest gap, Ellenberg soil pH index determined the presence of plant species growing on moderately acidic to weakly alkaline soils. The reason for this is that decomposition of organic material after ice storms and salvage harvesting is only in its initial phase.

Every newly formed forest gap can represent a new area either for colonisation by plant species or for increasing the population of species already present there. Competitively more successful plant species occupy the surface with sufficient resources faster or their population there increases more rapidly [47]. Due to the larger amounts of decomposing plant material in the soil, there are enough nutrients and also enough light for photosynthesis in the studied forest gaps. Therefore, when studying the ecological strategy of communities at all forest gaps, we found the most pronounced competitor component of the C-S-R diagram that characterises the community of plant species with a competitor strategy [23]. At the same time, it is classified as a community that has a stress-tolerator strategy [70]. The opposite was found by Eller et. al. [71], where the C-S-R strategy shifted from stress-tolerators in pre-logging conditions to a more ruderal component in post-logging stands (when the forest gap was formed). One reason for differences in research could be that, in our research, the gaps were in the middle of dense forest stands and there were no donor populations of ruderal plant species around, and the second reason is that our forest gaps, except one Above Draga, were not newly formed gaps and were not at the beginning of succession. Additionally, in our research, we chose naturally formed forest gaps, and in aforementioned research of Eller et al. [71], they were formed by logging. The difference between gap forming is that, with logging, process machines damage the understory vegetation and forest floor. The 'opened' floor patches are then suitable environments for ruderal species. The same happens on forest gaps where high digging activity of bigger forest mammals is present [15]. With natural gap formatting and leaving the trunks and fallen wood on the floor, there are fewer 'opened' floor patches and a less suitable environment for ruderal species.

The study confirmed the complexity of colonisation of forest gaps by plant species. The characteristics of plant species colonisation in gaps is that, in the initial stage of a gap (after formation of a gap), the most common species are sciophytic plants, which represent the remaining forest understory and ruderal species. With time, an older gap starts to be colonised by heliophytic plant species (representative of species with tall stems), with greater populations of grasses and sedges. Tree species start growing on gap edges, primarily beech in the Dinaric fir-beech forests. The species that grow in forest gaps prefer habitats with more sunlight, medium wet to dry soil, and are tolerant to high daily and seasonal temperature fluctuations, but at same time, forest gaps are a habitat patch for some species that did not exist there before. Colonisation, as well as species turnover, occurs very quickly with the change of environment in the process of overgrowing. Forest gaps also represent important windows of increased biodiversity and the chance for forest regeneration. Regrowth is fast enough, so there is no need for additional planting on formed forest gaps.

5. Conclusions

In the forest ecosystem, forest gaps represent an important developmental stage of the forest and a habitat for certain plant species. In our research, we found that the species composition and distribution of plants at forest gaps are significantly influenced by the size of forest gaps, their greater or lesser structure in microenvironment, amount of woody debris, abiotic factors, plant species composition of nearby forest gaps, and colonisation pathways.

In our study, we found greater biodiversity of plant species at forest gaps compared to the juvenile beech stand. The number of species was highest at the largest of the forest gaps. We found that the investigated forest gaps inhabit mostly species of forest understory, forest edges, and felling. The species present at all forest gaps are typical representatives of the beech forest understory, as well as common representatives of forest edges. Forest understory species are only remnants of former compacted forest before the forest gaps formatting, which thrive mainly on the shadier parts of forest gaps.

Plant species typical of forest understory have a greater cover on newly emerging forest gaps. In older forest gaps (4 years and older), there is a species turnover, where species typical of more open areas have a greater coverage. Populations of forest understory species are beginning to decrease, while populations of grasses, sedges, and other plant species typical of open areas are beginning to increase.

The plant species composition of an individual forest gap largely depends on the proximity of donor populations. Our studied forest gaps were inhabited by plant species characteristic of temperate and submontane areas. Characteristics of species that colonise forest gaps are preference for sites with more light, with moderately moist to dry soils, and tolerance to large daytime, night, and seasonal temperature fluctuations. Those forest gaps, on which more decaying plant material is present, are inhabited by species that need more nutrients in the soil to grow. Due to this, forest gaps colonisation with plant species is also significantly influenced by the amount of decaying plant material.

Author Contributions: Conceptualization, B.R. and J.B.; methodology, B.R. and J.B.; formal analysis, B.R. and A.Č.; investigation, B.R. and J.B.; writing—original draft preparation, B.R.; writing—review and editing, J.B. and A.Č.; visualization, B.R. and A.Č. All authors have read and agreed to the published version of the manuscript.

Funding: This research received no external funding.

Informed Consent Statement: Not applicable.

Data Availability Statement: Not applicable.

Conflicts of Interest: The authors declare no conflict of interest.

References

1. Helms, J.A. *The Dictionary of Forestry;* Society of American Foresters and CABI Publishing: Bethesda, MD, USA, 1998; 210p.
2. Frelich, L.E. *Forest Dynamics and Disturbance Regimes;* Cambridge University Press: Cambridge, UK, 2002; 90p.
3. Gálhidy, L.; Mihok, B.; Hagyo, A.; Rajkai, K.; Standovár, T. Effects of gap size and associated changes in light and soil moisture on the understorey vegetation of a Hungarian beech forest. *Plant Ecol.* **2006**, *183*, 133–145. [CrossRef]
4. Vilhar, U.; Roženbergar, D.; Simončič, P.; Diaci, J. Variation in irradiance, soil features and regeneration patterns in experimental forest canopy gaps. *Ann. For. Sci.* **2015**, *72*, 253–266. [CrossRef]
5. Kermavnar, J.; Ferlan, M.; Marinšek, A.; Eler, K.; Kobler, A.; Kutnar, L. Effects of various cutting treatments and topographic factors on microclimatic conditions in Dinaric fir-beech forests. *Agric. For. Meteorol.* **2020**, *295*, 108–186. [CrossRef]

6. Latif, Z.A.; Blackburn, G.A. The effects of gap size on some microclimate variables during late summer and autumn in a temperate broadleaved deciduous forest. *Int. J. Biometeorolgy* **2010**, *54*, 119–129. [CrossRef] [PubMed]
7. Seyednasrollah, B.; Kumar, M. Net radiation in a snow-covered discontinuous forest gap for a range of gap sizes and topographic configurations. *J. Geophys. Res. Atmos.* **2014**, *10*, 323–342. [CrossRef]
8. O'Hara, K. *Multiaged Silviculture: Managing for Complex Forest Stand Structures*; Published to Oxford Scholarship Online: Oxford, UK, 2014; 262p.
9. Duguid, M.C.; Frey, B.R.; Ellum, D.S.; Kelty, M.; Ashton, M.S. The influence of ground disturbance and gap position on understory plant diversity in upland forests of southern New England. *For. Ecol. Manag.* **2013**, *303*, 148–159. [CrossRef]
10. Kollár, T. Light Conditions, Soil Moisture, and Vegetation Cover in Artificial Forest Gaps in Western Hungary. *Acta Silv. Et. Lignaria Hung.* **2017**, *13*, 25–40. [CrossRef]
11. Kermavnar, J.; Eler, K.; Marinšek, A.; Kutnar, L. Post-harvest forest herb layer demography: General patterns are driven by pre-disturbance conditions. *For. Ecol. Manag.* **2021**, *491*, 119–121. [CrossRef]
12. Feldmann, E.; Drößler, L.; Hauck, M.; Kucbel, S.; Pichler, V.; Leuschner, C. Canopy gap dynamics and tree understory release in a virgin beech forest, Slovakian Carpathians. *For. Ecol. Manag.* **2018**, *415–416*, 38–46. [CrossRef]
13. Kermavnar, J.; Eler, K.; Marinšek, A.; Kutnar, L. Initial understorey vegetation responses following different forest management intensities in Illyrian beech forest. *Appl. Veg. Sci.* **2019**, *22*, 48–60. [CrossRef]
14. Ravnjak, B.; Bavcon, J.; Osterc, G. Physiological response of local populations of species *Cyclamen purpurascens* Mill. To forest gaps. *Appl. Ecol. Environ. Res. Int. Sci. J.* **2019**, *17*, 11489–11508. [CrossRef]
15. Fahey, R.T.; Puettmann, K.J. Ground-layer disturbance and initial conditions influence gap partitioning of understorey vegetation. *J. Ecol.* **2007**, *95*, 1098–1109. [CrossRef]
16. Collins, B.S.; Dunne, K.P.; Pickett, S.T.A. Reponses of Forest Herbs to Canopy Gaps. In *The Ecology of Natural Disturbance and patch Dynamics*; Picket, S.T.A., White, P.S., Eds.; Academic Press: London, UK, 1985; pp. 217–234.
17. Kelemen, K.; Mihók, B.; Gálhidy, L. Dynamic response of herbaceous vegetation to gap opening in a central European beech stand. *Silva Fenn.* **2012**, *46*, 53–65. [CrossRef]
18. Vukelić, J.; Mikac, S.; Baričević, D.; Šapić, I.; Bakšić, D. Vegetation and structural features of Norway spruce stands (Picea abies Karst.) in the virgin forest of Smrčeve doline in northern Velebit. *Croat. J. For. Eng. J. Theory Appl. For. Eng.* **2011**, *32*, 85–86.
19. Schoennagel, T.; Smithwick, E.A.H.; Turner, M.G. Landscape heterogeneity following large fires: Insights from Yellowstone National Park, USA. *Int. J. Wildland Fire* **2008**, *17*, 742–753. [CrossRef]
20. Bottero, A.; Garbarino, M.; Dukic, V.; Goveda, Z.; Lingu, E.; Nagel, T.A.; Motta, R. Gap-phase dynamics in the old-growth forest of Lom, Bosnia and Herzegovina. *Silva Fenn.* **2011**, *45*, 875–887. [CrossRef]
21. Muscolo, M.; Bagnato, S.; Sidari, M.; Mercurio, R. A review of the roles of forest canopy gaps. *J. For. Res.* **2014**, *25*, 725–736. [CrossRef]
22. Hilmers, T.; Friess, N.; Bässler, C.; Heurich, M.; Brandl, R.; Pretzsch, H.; Seidl, R.; Müller, J. Biodiversity along temperate forest succession. *J. Appl. Ecol.* **2018**, *55*, 2756–2766. [CrossRef]
23. Grime, J.P. Evidence for the existence of three primary strategies in plants and its relevance to ecological and evolutionary theory. *Am. Nat.* **1997**, *3*, 1169–1194. [CrossRef]
24. Meterc, G.; Skudnik, M.; Jurc, M. Vpliv gospodarjenja na biotsko raznovrstnost saproksilnih hroščev [The impact of forest management to the biodiversity of saproxylic beetles]. *Gozdarski Vestn.* **2015**, *73*, 3–18.
25. Honnay, O.; Hermy, M.; Coppin, P. Impact of habitat quality on forest plant species colonization. *For. Ecol. Manag.* **1999**, *115*, 157–170. [CrossRef]
26. Dale, V.H.; Joyce, L.A.; McNulty, S.; Neilson, R.P.; Ayres, M.P.; Flanningan, M.D.; Hanson, P.J.; Irland, L.C.; Lugo, A.E.; Peterson, C.J.; et al. Climate Change and Forest Disturbance: Climate change can effect forests by altering the frequency, intensity, duration, and timing of fire, drought, introduced species, insects and pathogen outbreaks, hurricanes, windstorms, ice storms or landslides. *BioScience* **2001**, *51*, 723–734. [CrossRef]
27. Slovenian Environment Agency—ARSO. *Podnebne Razmere v Sloveniji (Obdobje 1985–2010) [Weather Conditions in Slovenia (Period 1985–2010)]*; Slovenian Environment Agency: Ljubljana, Slovenija, 2010; 27p.
28. Kordiš, F. *Dinarski Jelovo-Bukovi Gozdovi v Sloveniji [Dinaric Silver Fir-Beech Forests in Slovenia]*; Biotechnical Faculty, Department of Forestry: Ljubljana, Slovenija, 1993; 139p.
29. Nagel, T.A.; Mikac, S.; Dolinar, M.; Klopčič, M. The natural disturbance regime in forests of the Dinaric Mountains: A synthesis of evidence. *For. Ecol. Manag.* **2017**, *388*, 29–42. [CrossRef]
30. Wraber, M. Pflanzengeographische Stellung und Gliederung Sloweniens. *Vegetatio* **1969**, *17*, 176–199. [CrossRef]
31. Wraber, M. Vegetacija slovenskega bukovega gozda v luči ekologije in palinologije. *Biološki Vestn.* **1964**, *12*, 77–95.
32. Praprotnik, N. *Ilirski Florni Element v Sloveniji. Doktorska Disertacija*; Univerza Edvarda Kardelja v Ljubljani, Biotehniška fakulteta, VTOZD za biologijo: Ljubljana, Slovenija, 1987; 234p.
33. Marinček, L. *Bukovi Gozdovi na Slovenskem [Beech Forests in Slovenia]*; Delavska Enotnost: Ljubljana, Slovenija, 1987; 153p.
34. Fraver, S.; Wagner, R.G.; Day, M. Dynamics od coarse woody debris following gap harvesting in the Acadian forests of central Maine, U.S.A. *Can. J. For. Res.* **2002**, *32*, 2094–2105. [CrossRef]
35. Müller-Using, S.; Bartsch, N. Decay dynamic of coarse and fine woody debris of a beech (*Fagus sylvatica* L.) forest in Central Germany. *Eur. J. For. Res.* **2009**, *128*, 287–296. [CrossRef]

36. Sefidi, K.; Marvie Mohadjer, M.R. 2010: Characteristics of coarse woody debris in successional stages of natural beech forests (*Fagus orientalis*) of Northern Iran. *J. For. Sci.* **2010**, *56*, 7–17. [CrossRef]
37. Jakša, J. Natural disasters in Slovenian forest. *Gozdraski Vestn.* **2007**, *65*, 177–192.
38. Jakša, J. Natural disasters in Slovenian forest. *Gozdraski Vestn.* **2007**, *65*, 241–256.
39. Tarman, K. *Osnove Ekologije in Ekologija Živali*; Državna Založba Slovenije: Ljubljana, Slovenija, 1992; 547p.
40. Braun-Blanquet, J. *Pflanzesoziologie. Grundzüge der Vegetationskunde*; Springer: Berlin/Heidelberg, Germany, 1964; 330p.
41. Gilliam, F.S. The Ecological Significance of the Herbaceus Layer in Temperate Forest Ecosystems. *BioScience* **2007**, *57*, 845–858. [CrossRef]
42. Tichý, L. JUICE, software for vegetation classification. *J. Veg. Sci.* **2002**, *13*, 451–453. [CrossRef]
43. Zur, A.F.; Ieno, E.N.; Smith, G.M. *Analysing Ecological Data*; Springer: Berlin/Heidelberg, Germany, 2007; 672p.
44. Pignatti, S.; Menegoni, P.; Pietrosanti, S. Valori di bioindicazione delle piante vascolari della flora d'Italia [Bioindicator values of vascular plants of the Flora of Italy]. *Braun-Blanquetia* **2005**, *39*, 1–97.
45. Jarolímek, I.; Šibik, J. *Diagnostic, Constant and Dominant Species of the Higher Vegetation Units in Slovakia*; VEDA: Bratislava, Slovakia, 2008; 329p.
46. Sokal, R.R.; Rohlf, F.J. *Biometry*, 3rd ed.; W.H. Freeman: New York, NY, USA, 1995.
47. Smith, R.L.; Smith, T.M. *Ecology & Field Biology*; Benjamin Cummings: San Francisco, CA, USA, 2001; 771p.
48. Tichy, L.; Holt, J. *JUICE Program for Management Analysis and Classification of Ecological Data*; Vegetation Science Group, Masaryk University Brno: Brno, Czech Republic, 2006; 98p.
49. Klotz, S.; Kühn, I.; Durka, W. *BIOLFLOR—Eine Datenbank zu Biologisch-Ökologischen Merkmalen der Gefässpflanzen in Deutschland—Schriftenreihe für Vegetationskunde 38*; Bundesamt für Naturschutz: Bonn, Germany, 2002.
50. Hunt, R.; Hodgson, J.G.; Thompson, K.; Bungener, P.; Dunnett, N.P.; Askew, A.P. A new practical tool for deriving a functional signature for herbaceous vegetation. *Appl. Veg. Sci.* **2004**, *7*, 163–170. [CrossRef]
51. Ellenberg, H.; Weber, H.E.; Düll, R.; Wirth, V.; Werner, W.; Paulissen, D. *Zeigerwerte von Pflanzen in Mitteleuropa*; Erich Goltze: Göttingen, Germany, 1992; 262p.
52. Večeřa, M.; Divíšek, J.; Lenoir, J.; Jiménez-Alfaro, B.; Biurrun, I.; Knollova, I.; Agrillo, E.; Campos, J.A.; Čarni, A.; Crespo, G.; et al. Alpha diversity of vascular plants in European forests. *J. Biogeogr.* **2019**, *46*, 1919–1935. [CrossRef]
53. Degen, T.; Devillez, F.; Jacquemart, A.-L. Gaps promote plant diversity in beech forests (Luzulo-Fagetum), North Vosges, France. *Ann. For. Sci.* **2005**, *62*, 429–440. [CrossRef]
54. Šilić, Č. *Šumske Zeljaste Biljke*; Zavod za Udžbenike i Nastavna Sredstva: Beograd, Serbia, 1988; 271p.
55. Taylor, K. Biological Flora of the British Isles: *Urtica dioica* L. *J. Ecol.* **2009**, *97*, 1436–1458. [CrossRef]
56. Mayer, M.; Matthews, B.; Rosinger, C.; Douglas, H.S.; Godbold, L.; Katzensteiner, K. Tree regeneration retards decomposition in a temperate mountain soil after forest gap disturbance. *Soil Biol. Biochem.* **2017**, *115*, 490–498. [CrossRef]
57. Čarni, A. Vegetation of trampled habitats in the Prekmurje region (NE Slovenia). *Hacquetia* **2005**, *4*, 151–159.
58. Burton, J.I.; Mladenoff, D.J.; Clayton, M.K.; Jodi, A.; Forrester, J.A. The roles of environmental filtering and colonization in the fine-scale spatial patterning of ground-layer plant communities in north temperate deciduous forests. *J. Ecol.* **2011**, *99*, 764–776. [CrossRef]
59. Willner, W.; Jimenez-Alfaro, B.; Agrillo, E.; Biurrun, I.; Campos, J.A.; Čarni, A.; Cassela, L.; Csiky, J.; Ćušterevska, R.; Didukh, Y.P.; et al. Classification of European beech forests: A Gordian Knot? *Appl. Veg. Sci.* **2017**, *20*, 494–512. [CrossRef]
60. Zupančič, M.; Vreš, B. Phytogeographic analysis of Slovenia. *Folia Biol. Geol.* **2018**, *59*, 160–211.
61. Mucina, L.; Bueltmann, H.; Dierßen, K.; Theurillat, J.P.; Raus, T.; Čarni, A.; Šumberová, K.; Willner, W.; Dengler, J.; Gavilán, R.; et al. Vegetation of Europe: Hierarchical floristic classification system of vascular plant, bryophyte, lichen, and algal communities. *Appl. Veg. Sci.* **2016**, *19*, 3–264. [CrossRef]
62. Kermavnar, J. Vplivi Gospodarjenja na Funkcionalne Lastnosti Gozdne Vegetacije in Ekološke Razmere v Dinarskih Jelovo-Bukovih Gozdovih [Impacts of Forest Management on Functional Properties of Vegetation and Ecological Conditions in the Dinaric Fir-Beech Forests]. Ph.D. Thesis, Biotechnical Faculty, University of Ljubljana, Ljubljana, Slovenia, 2021; 225p.
63. Myerscough, P.J.; Whitehead, F.H. Comparative biology of *Tussilago farfara* L.; *Chamaenerion angustifolium* (L.) Scop., *Epilobium montanum* L. and *Epilobium adenocaulon* Hausskn. *Gen. Biol. Germination* **1965**, *1*, 192–210.
64. Rahmonov, O.; Jędrzejko, K.; Majgier, L. The secondary succession in the area of abandoned cemeteries in northern Poland. In *Landscape Ecology—Methods, Applications and Interdisciplinary Approach*; Barančoková, M., Krajčí, J., Kollár, J., Belčáková, J., Eds.; Institute of Landscape Ecology, Slovac Academy of Sciences: Bratislava, Slovakia, 2010; pp. 647–657.
65. Ulanova, N. The effects of windthrow on forest at different spatial scales: A review. *For. Ecol. Manag.* **2000**, *135*, 155–167. [CrossRef]
66. Mollaei Darabi, S.; Kooch, Y.; Hosseini, S. Dynamic of Plant Composition and Regeneration following Windthrow in a Temperate Beech Forest. *Int. Sch. Res. Not.* **2014**, *2014*, 1–9. [CrossRef]
67. Marinšek, A.; Celarc, B.; Grah, A.; Kokalj, Ž.; Nagel, T.A.; Ogris, N.; Planinšek, Š.; Roženbergar, D.; Veljanovski, T.; Vochl, S.; et al. Žledolom in njegove posledice na razvoj gozdov—Pregled dosedanjih znanj [Impacts of Ice Storms on Forest Development—A Review]. *Gozdarski Vestn.* **2015**, *73*, 392–405.
68. Kraigher, H.; Jurc, D.; Kalan, P.; Kutnar, L.; Levanič, T.; Rupel, M.; Smolej, I. Beech coarse woody debris characteristic in two virgin forests reserves in southern Slovenia. *Zb. Gozdarstva Lesar.* **2002**, *69*, 91–134.

69. Spears, J.D.H.; Lajtha, K. The imprint of coarse woody debris on soil chemistry in the western Oregon Cascades. *Biogeochemistry* **2004**, *71*, 163–175. [CrossRef]
70. Bavcon, J. *Belo Cvetoče Različice v Slovenski Flori [White-Flowered Varieties in Slovenian Flora]*; University Botanic Gardens Ljubljana: Ljubljana, Slovenia, 2014; 349p.
71. Eler, K.; Kermavnar, J.; Marinšek, A.; Kutnar, L. 2018: Short-term changes in plant functional traits and understory functional diversity after logging of different intensities: A temperate fir-beech forest experiment. *Ann. For. Res.* **2018**, *61*, 223–242. [CrossRef]

Article

Whole System Data Integration for Condition Assessments of Climate Change Impacts: An Example in High-Mountain Ecosystems in Rila (Bulgaria)

Kostadin Katrandzhiev *, Kremena Gocheva and Svetla Bratanova-Doncheva

Institute of Biodiversity and Ecosystem Research, Bulgarian Academy of Sciences, 2 Yurij Gagarin St., 1113 Sofia, Bulgaria; kremena.gocheva@iber.bas.bg (K.G.); sbrat@abv.bg (S.B.-D.)
* Correspondence: kmkatrandjiev@gmail.com; Tel.: +359-899169870

Abstract: To study climate impacts, data integration from heterogeneous sources is imperative for long-term monitoring in data sparse areas such as the High Mountain Ecosystems in the Rila Mountain, Bulgaria—difficult to both access and observe remotely due to frequent clouds. This task is especially challenging because discerning trends in vegetation location, condition and functioning requires observing over decades. To integrate the existing sparse data, we apply the Whole System framework adapted nationally in the Bulgarian Methodological Framework for Mapping and Assessment of ecosystem services. As the framework mainly relies on field data, we complement it with remote sensing vegetation indices (NDVI, NDWI and NDGI) for 42 years, together with Copernicus High Resolution Layer products and climate change reanalysis data for 40 years. We confirmed that the Whole System framework is extensible and semantically, ontologically and methodologically well suited for heterogeneous data fusion, co-analysis, reanalysis and joint interpretation. We found trends in ecosystem extent and functioning, in particular species composition, in line with climate change trends since around 1990 and exclusively attributable to climate change since 2015. Furthermore, we specified a data crosswalk between habitats and ecosystems at Level 3 (ecosystem subtype), and define new candidate indicators suitable for remotely monitoring climate change's effects on the ecosystems' extent and condition, as candidates for inclusion in the methodological framework.

Keywords: whole system approach; data fusion and integration; semantic and ontological compatibility; mountain ecosystems; ecosystem condition; remote sensing; vegetation indices; climate change reanalysis; dominant species level forest ecosystem classification

Citation: Katrandzhiev, K.; Gocheva, K.; Bratanova-Doncheva, S. Whole System Data Integration for Condition Assessments of Climate Change Impacts: An Example in High-Mountain Ecosystems in Rila (Bulgaria). *Diversity* **2022**, *14*, 240. https://doi.org/10.3390/d14040240

Academic Editors: Michael Wink, Lucian Dinca and Miglena Zhiyanski

Received: 7 October 2021
Accepted: 18 March 2022
Published: 25 March 2022

1. Introduction

The ecologically complex high mountain ecosystems (HMEs) are important study areas for climate change impact on ecosystem structure and functions. There is a wide range of concepts about ecosystem functioning, reviewed by Pettorelli et al. (2017) [1].

Scientists studying HMEs attempt to explain the influence of climate change on the vegetation structure [2,3], causation processes [4], spatial distribution in HMEs [5–7] and the adaptive capacity of HMEs to the influence of natural and/or anthropogenic changes [8]. In doing so, they invariably come across the overwhelming complexity of the interrelated study objects of HMEs and climate change.

As a means to reduce complexity, research often focuses on specific aspects such as the ecotone rather than the entire ecosystem [5,9–15], abiotic factors [16], species composition and dynamics [17,18]. However, a truly holistic assessment of both condition and functioning trends for these highly sensitive and vulnerable zones in their entirety requires an integrated interdisciplinary approach. Different types of data are collected at different points in time and space, different scale, for different purposes and using different—in the cases of remote sensing and climate modelling significantly improved—instruments and methods over time. These objective challenges include:

- Combining continuous and discrete data. Verifying ecosystem type typically relates to one-off observations of ground data; remote sensing imagery is continuous in space but discrete in time, while climate models have spatial and temporal continuity;
- Data heterogeneity. Even within the same ecosystem, the scope, field measurement methods and even the underlying indicators, classifications and other conceptual elements may vary significantly, making the results difficult to reconcile and reuse. Climate models render differently accurate results and are periodically corrected through reanalysis, resulting in different variables for the new data series;
- Data imbalances—biases in the scientific and management interest with data collection (predominantly about the forest ecosystem), and technological imbalances due to the different quality of remote sensing equipment over time and the increasing number of missions in the last years. As a result, there is a much better availability and quality of remote sensing scenes in the last five years than at any time before, and this development is accelerating;
- In addition to the usual difficulties of data integration listed above, the study of HMEs meets specific difficulties that lead to a data scarcity:
    - HMEs are often difficult to access physically for surveys, species inventories or installing/maintaining monitoring equipment for long-term observations;
    - Automated measurements meet the challenges of weak or lacking internet connectivity and natural hazards such as avalanches that damage or annihilate the equipment;
    - During the vegetation season HMEs in good condition act as a biotic pump that increases air humidity [19,20] and hence their remote sensing often reveals clouds. Even where there are no visible clouds, there often are variable patterns of mist or fog [21,22] that may remain undetected in cloud masking. These atmospheric conditions distort the sensor readings and cause errors in the values of vegetation indices; their precise detection typically requires another data source, i.e., ground data validation or accurate meteorological data. This makes remote phenological observations virtually unfeasible;
    - Scale discrepancies: Microclimatic influences largely shape developments on the ground, while climate models such as ECMWF Re-Analysis (ERA) we use in this study are much coarser. Satellite imagery is constantly improving, with the pixel size reduced by orders of magnitude between early products and current very high-resolution ones, making it commensurate with the size of field sites. Such great differences in scale make data fusion imperative for the use of ERA Interim and future use of ERA 5 together with other available data since no sufficient downscaling is possible in such a discrepant scenario. While ERA Interim uncertainty is relatively low at a large scale for the key parameters relevant to our study (precipitation dry bias of -1% for Europe according to [23], temperature uncertainty of $\pm 2\%$ according to [24]), regional (micro-climatic) variation is not captured well due to their coarse granularity [25] and downscaling them is challenging (as evidenced by the modelling effort behind the Vito dataset "Climate variables for cities in Europe from 2008 to 2017" documented at https://cds.climate.copernicus.eu/cdsapp#!/dataset/sis-urban-climate-cities?tab=overview (accessed on 17 March 2022) which is not maintained beyond the contract and did not manage significant downscaling even in areas with a high density of meteorological observation points). This great difference in scales of datasets is the reason to consider climate models only for exploring HMEs qualitatively at this stage, as they are far from sufficiently detailed at the scales commensurate with most monitoring needs. Their use remains indispensable despite the large scale since they are the only available approximation of weather information in data poor areas.

- Finally yet importantly, protected areas containing HMEs are not so interesting commercially which exacerbates the bias of data collection and monitoring towards inhabited or managed territories such as cities or agricultural areas.

Despite these challenges, complex assessments mandate going beyond the sole use of field observation methods that mainly inform on ecosystem structure, not functions [1]. This is especially true for Bulgaria where field observations are still the preferred method of ecosystem research (also reflected in the parameter measurement methods in [26]). Climate models are increasingly accessible and precise, and the verification in different regions shows their suitability for observing trends even in the presence of biases in some parameters [27]. They therefore provide useful guidelines on climate change over longer periods in areas with no direct observations available. Remote sensing methods are applicable to ecosystem monitoring [28] and widely used for landscape (ecosystem) change detection and land cover classification, monitoring condition and functioning trends and disturbance factors [29], changes in growth conditions, growing season length, shifts in vegetation [30] and natural capital accounting [31,32]. The latest observations benefit from the improved data quality due to the technological (sensor) innovation in the past decades and increasingly allow for observing ecosystem services [33].

Combining data from sources as diverse as field surveys, remote sensing and constrained climate reanalysis models bring their own set of technological and cognitive challenges. Different data require different sets of expertise to process and analyze. No single researcher is equally proficient in all these methods and structuring teamwork for interdisciplinary data integration requires each member of the interdisciplinary team to be sufficiently aware of all data and the processing steps everyone else performs on it. Therefore, the optimal and efficient reuse of the efforts of large teams in producing assessments and models on the national or continental scale becomes essential for smaller research teams. Data processing disciplines like ecoinformatics [34] rely upon solutions that follow the Findable, Accessible, Interoperable and Reusable (FAIR) data principles [35] and use standardized technology [36–39]. However, attempting to bridge the data silos between research disciplines for interdisciplinary ecosystem research meet their cognitive limits, as abundantly discussed by Villa et al. [40] who note the domain dependence of any formalized system of terms and their relations (ontology). Each scientific discipline has its own semantic structure, data and metadata collection models, and nomenclatures and indicator systems (a few examples: (1) Mapping ecosystem to habitat types is a difficult task which becomes ambiguous if such mapping is sought across climatic zones; (2) taxonomical standards for Animalia and Plantae obey similar but not identical principles; (3) the way anthropogenic pressures are described varies between the Driver-Pressure-State-Impact-Response framework (DPSIR) and the reporting guidelines to the EU Habitats Directive; (4) even the same indicator such as biomass may be observed differently depending on the ecosystem type, e.g., using NDVI for terrestrial ecosystems and Chlorophyll A for water bodies, and intercalibrating such different methods to obtain comparable results may be very difficult); some researchers argue that assumptions not contained in the sematic annotation (priors, as termed by Chollet [41]) will always implicitly influence such semantic structures and models. At the end, due to the very diverse and complex nature of their research object, ecosystem researchers have the task to unify conceptual bases in a convergent manner as a mean to integrate the corresponding data initially collected for other purposes (divergent exploration [42]).

In effect, not resolving these issues prevents researchers from efficiently analyzing all available data. Therefore, our first hypothesis is based on the assumption that, by design, a unifying conceptual framework at the ecosystem scale allows for data integration that

(1) is more ecologically meaningful,
(2) is more reliable,
(3) is extensible in terms of indicators and methods, especially in cases of sparse and biased data which can significantly reduce the accuracy of many automated approaches such as machine learning; and

(4) allows for using all available data from multiple sources across space and time to the extent possible.

In the context of this hypothesis, ecologically meaningful data integration refers to the explainability, through known ecological processes, of data with different types and provenances measured/modelled either at the same time or within ecologically meaningful time intervals. For example, change trends in climate parameters (such as temperature and precipitation) modeled over time, if correlating with field data indicating change of species composition, are consistent with the ecologically meaningful process of upward shift of deciduous species in coniferous forests. Similarly, climate change datasets together with remote sensing data on changing ecosystem extent in the absence of land management are consistent with succession. Of course, having a single ecologically meaningful hypothesis consistent with the data is seldom the case, and in practice it is often necessary to estimate the influence of different co-occurring natural or anthropogenic environmental factors to avoid double counting [43,44]. More reliable data integration refers to the formulation within the same conceptual framework of reproducible new indicators for new policy or research purposes [45]. Such extension must by necessity be founded on the use of different data sources to crosscheck data and exclude incorrect data instances and outliers. Extensible data integration refers to integration methods that allow for adding more data, updating data to perform reanalysis, upscaling or downscaling of data or other necessary changes in the types and properties of data used in the analysis, while at the same time preserving the ecologically meaningful interpretation. To test this hypothesis, in this study we apply the Whole System approach [46,47] as adapted in the Methodological Framework for assessment and mapping of ecosystem condition and ecosystem services in Bulgaria [26], and use all available heterogeneous data. For more on the framework, see Section 2.2.

To our knowledge, to date no studies have been performed in the Whole System context for the long timeframe of several decades on HMEs in Bulgaria or elsewhere.

Based on earlier work [2,5,7], we formulated a second hypothesis, namely: We expected that the species in our study areas' HME develop in a changing climate, and assess the response of the HME to changes in climate parameters.

The objective of this study is testing these two closely linked hypotheses—the first pertaining to the general methodological principles of data fusion in the Whole System context, and the second to a specific application concerning studying effects of climate change in a representative ecologically interesting but data poor HME area. Testing the second hypothesis using the framework of the first hypothesis allows for simultaneously testing both the method and the ecological hypothesis at once.

To the extent possible, quantitative evaluation criteria are a focus of this assessment—both measurable changes in ecosystems' spatial distribution at the landscape level (e.g., succession), and observable ecosystem change parameters such as change in species composition [48,49] traceable over the past decades.

2. Materials and Methods

2.1. Study Area

The selected study area is located in the southwestern part of Rila Mountain, Bulgaria (Figure 1). It has a total area of 14,334 ha and includes parts of the communal lands of several populated settlements in Blagoevgrad District, as well as the Parangalitsa Reserve which has been assigned as a reserve since 1933 and was later included in the National Ecological Network Natura 2000 as a protected area for birds and habitats. The highest point of the study area is Dzherman peak (also called Ezernik, 2485 m). Its main ecosystems are woodland and forests, shrubs and grasslands as characterized by [50–54]. Small water bodies including Lake Dzherman and the upper stream of Draglishka river (water body BG4ME800R089 as per the River Basin Management Plan of the West Aegean River Basin Directorate, Blagoevgrad) belong to "rivers and lakes" ecosystem type [53]. The steep rocky surfaces near the top contain sparsely vegetated areas [54]. Thus, five of the nine ecosystem types found in Bulgaria are present in the study area. Due to its inaccessibility, harsh climate, the lack of long series of ground data and the limited number of usable remote sensing images, the study area is representative in terms of data sparsity in the context of being rich in biotic and abiotic diversity—traits it shares with many of Bulgaria's remote wild areas.

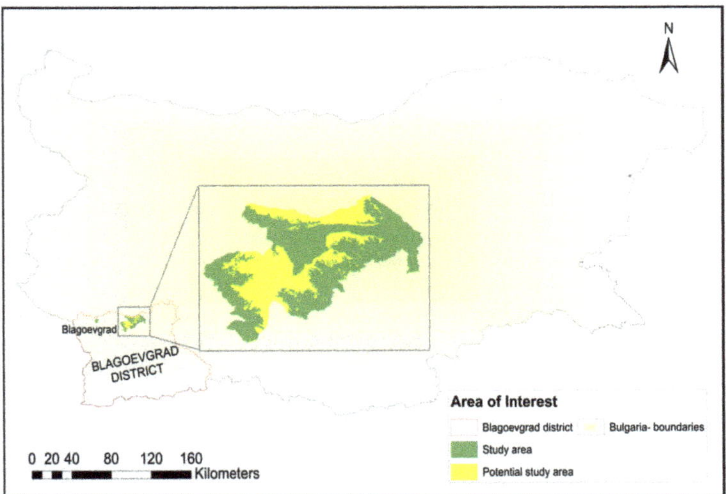

Figure 1. Map of the area of interest: Potential study area (yellow) and parts of it studied in the current paper (green).

Land use is mainly limited to pastoral habitat. Protection measures under the Common Agricultural Policy include subsidy schemes for managed extensive grazing in protected areas within the Natura 2000 network (we verified this using the Copernicus Ploughing layers for 2015 and 2018). While no logging permits are issued in the study area itself, the surrounding forests are logged actively, in some cases beyond the permitted scope—as visualized in the forest website maintained by the Executive Forestry Agency and WWF Bulgaria, which also contains crowdsourced information on irregularities (https://gis.wwf.bg/mobilz/#/23.38057/42.01433/12 (accessed on 17 March 2022).

This study area, although too small for regional or other large-scale assessments, is commensurate with the characteristic scale [55] of the objects of our observation—high mountain ecosystems and ecotones, and their growth patterns and dynamics. Having in mind the availability of ground data on the location of ecosystems, we were able to use the fact-based basic entity approach ([56]) instead of resorting to statistical methods that may increase uncertainty. The study area satisfies three key criteria:

(i) its scale corresponds to the study objects;
(ii) it contains the widest possible variety of HME ecosystems and representative forest-grassland and forest-shrub ecotones; and
(iii) it only covers protected areas so that the impact of other factors, in particular abandoning the grazing or transition from intensive to extensive grazing, would not overlap on the effects caused by climate change and distort the observations.

The selected study area is furthermore a good example of a place with biased and sparse available data. Large scale field verification is limited due to the inaccessibility of the alpine landscape; moreover, field verification through digitalization or combining datasets is difficult to perform and often impossible for older data. Testing our first hypothesis in such an environment is important since in such places it typically is much more difficult to ensure the reliability and reach desired lower uncertainty levels when including them in national or European scale wall-to-wall maps and models.

In addition, the study area illustrates the great variety of scales in available data sources. ERA Interim pixel size is 12 km and the size of our study area is covered by approximately 2 pixels of that model while it also contains over 22,000,000 pixels of very high resolution satellite products. The lack of meteorological data time series makes climate models the only source of information on temperature and precipitation which we cannot derive from other remote sensing data. At the same time, the large pixel size of ERA Interim causes this dataset to be very imprecise at the smaller scale of the actual ecological processes, which in turn increases the uncertainty of any statistical processing (as discussed by Jelinski and Wu [56] in their exploration of the scale problem's dependence on grain size in the data).

2.2. The Whole System Framework—A Versatile Tool for Data Fusion and Co-Analysis

The Whole System approach regards the ecosystem as a complex of five subsystems—structural (biodiversity, abiotic heterogeneity) and functional (balance of energy, matter and water). As such, it is integrative to the extent of unifying the research subjects of ecosystem functioning (as part of the biodiversity subsystem) and climate change (as part of the abiotic heterogeneity subsystem). Being dynamic, the Whole System framework is furthermore conductive to working with time series by observing the manifestations of the matter balance and water balance subsystems through vegetation indices and the energy balance—indirectly through the changes in temperature and the related evapotranspiration intensity. Therefore, it is possible to use both climate models and remote sensing methods for observing the system level structural and functional parameters within this framework, as presented in Table S1 of the Supplementary Materials.

Our first hypothesis is based on the understanding that, by its design, the Methodological Framework for assessment and mapping of ecosystem condition and ecosystem services in Bulgaria further facilitates wall-to-wall landscape, regional and national data integration on a number of levels in a manner supportive of data fusion and, ultimately, data integration and co-analysis. The layers of compatibility built into the framework are:

- Semantic compatibility: This applies the same indicator system to all ecosystems, as detailed in Figure 2. In this manner, indicators vary little across ecosystems, creating semantic links, whereas the huge diversity of observable ecosystem manifestations is mostly contained at the parameter level but clearly linked to the indicators and, through them, to the ecosystem structure or functioning.
- Ontological compatibility (for the purpose of this article, we understand ontology not in the philosophical sense but as commonly defined in information technology and semantic web applications as a means for users to create their own set of definitions—in our case, definitions of ecosystem types/subtypes, habitats, etc. Pan (2006) [58] and Serafini and Bogrida (2005) [59] derive mathematical formalism of ontological compatibility and reasoning in the context of decentralized systems such as a multi-ecosystem assessment): Creating links (crosswalks) is another systematic feature of the Methodological Framework. Each Level 2 ecosystem type contains more differentiated Level 3 subtypes (Figure 3). This creates an unambiguous basis for cross-referencing of indicators and parameters collected under reference frameworks as different as ecosystem or habitat classification, plant or animal taxonomies or genetic sequences specific to each subtype. Thus, if lacking ground truth observations, field data collected in another context (e.g., forest inventory including habitat data) about a subset of parameters can be sufficient to find the ecosystem subtype (Level 3), as we demonstrate in this study. Cross-referencing the classification of Level 3 to finer grained ecological concepts such as habitats, while not replacing a detailed assessment, may allow to narrow down the expected habitat types, species composition and other ecosystem traits even with sparse ground data. Moreover, the Methodological Framework includes an in situ verification guide [60] that provides a mechanism for resolving inconsistencies between observations in a landscape, thus addressing the concerns raised by Pan [58].

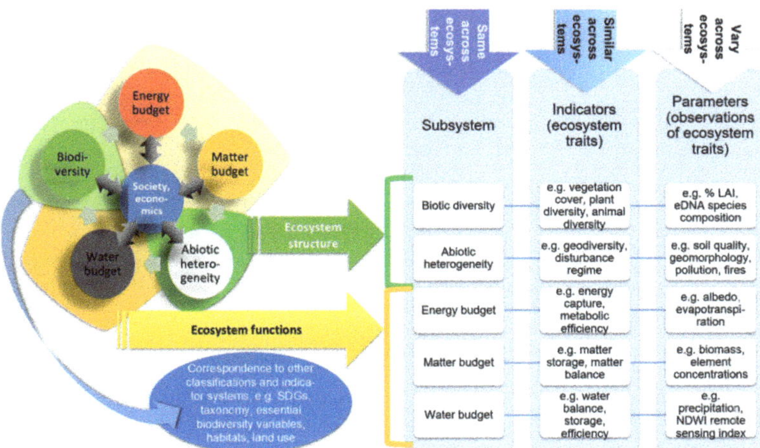

Figure 2. Semantic compatibility of the Methodological Framework: Translating highly variable ecosystem structure and functions to a coherent set of indicators and parameters (general principle and hierarchy of ecosystem subdivision). Adapted from [26,57].

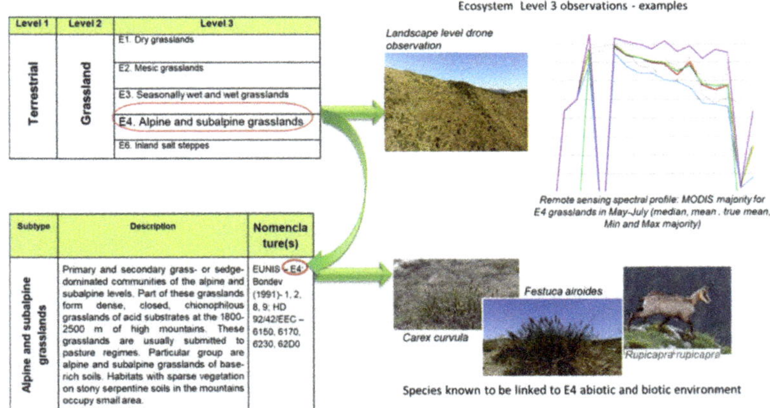

Figure 3. Crosswalk links between different hierarchies of the Whole System and different observations at the appropriate scale: Above—system level; below—species and populations level.

We note that obtaining ontological compatibility across knowledge domains comes at the cost of losing precision and growing uncertainty [61]. Therefore, automated fusion and processing of datasets matched on the ontological level will be subject to using appropriate mathematical apparatus and methods such as fuzzy logic [62] and fuzzy graphs [63] in the inference of semantic and ontological links and processing of big data.

- Methodological compatibility: One key step of the Methodological Framework is to assign non-dimensional numeric values between 1 and 5 to the ecological parameters and 0 to 5 for ecosystem services (0 being assigned when a service is not provided by this ecosystem). These scores are based on initial expert assessment and later field verification (Table 1). This step ensures the possibility to assess semi-quantitatively different parameters measured using different methods and expressed in diverse measurement units, against the reference values established for different ecosystem condition. Furthermore, the Framework foresees aggregating these semi-qualitative reference values to form single indices for assessing the ecosystem integrity, overall condition and service provision capacity (IP index). Akin to the carbon equivalent in climate change, the IP index both enables a numeric expression of the ecosystem condition/service provisioning capacity, and allows for an overall cross-ecosystem comparison at the indicator or ecosystem level, which facilitates the compilation of wall-to-wall assessments and analyses by IP index or indicator across all ecosystems present on the landscape level.

Table 1. Example of cross-ecosystem compatibility of parameter assessments—(**a**) assessment scale for a grassland condition parameter; (**b**) assessment scale for grassland ecosystem services; and (**c**) adjustment of reference values for the same parameter after field verification with reference values determined by ecosystem subtype (red indicates subtype present in our study area). These reference values may be candidates for incorporating an update of the Methodological Framework.

(a)

Parameter	Unit	Methodology	Assessment scale				
			Score 1 (bad)	Score 2 (poor)	Score 3 (moderate)	Score 4 (good)	Score 5 (excellent)
Vegetation cover	%	Estimation	0–10	11–30	31–50	51–70	71–100

(b)

			Score 0 (Not applicable)	Score 1 (bad)	Score 2 (poor)	Score 3 (moderate)	Score 4 (good)	Score 5 (excellent)
P1 Reared Animals and their output	Livestock units/ha	Statistics, EC condition assessment	Not relevant	0.01–0.5	0.51–0.75	0.76–0.9	0.91–1	>1.01

(c)

Indicator group	Indicator	Parameter	Unit	Apply to ES subtype	Score 1 (bad)	Score 2 (poor)	Score 3 (good)	Score 4 (very good)	Score 5 (excellent)
Biotic diversity	**Plant diversity**	**Vegetation cover**	%	E1	30 or less	31–40	41–60	61–80	81 or more
				E2, E3	60 or less	61–70	71–80	81–89	90 or more
				E4	50 or less	61–70	71–80	81–89	90 or more
				E5	10 or less	11–20	21–40	41–60	61 or more

As with the ontological compatibility, the selected approach to methodological compatibility also introduces a degree of uncertainty and therefore any future automated workflow for data processing may also require the use of fuzzy methods when analyzing and modelling data across ecosystems.

- Information compatibility: Data reuse is possible by utilizing parameter observations from existing sources, e.g., for forest ecosystems—data from forest inventories; for water ecosystems—monitoring data collected while implementing the Water Framework Directive or Marine Strategy Framework Directive (Figure 4a). In addition, the Methodological Framework prescribes the same database structure and processing workflow in the specific methodologies for assessing each ecosystem type (Figure 4b), hence allowing for a meaningful data fusion [64] from the semantical down to the instrumental level.

a)

Ecosystem type	Key plant diversity indicators	Parameters and units	Score 1 (bad)	Score 2 (poor)	Score 3 (moderate)	Score 4 (good)	Score 5 (very good)
Woodland and Forest	Plant diversity	species composition (% mixed tree species growing stock)	0–20%	21–40	41–60	61–80	81–100
		stand dynamic phase age (years) estimated FMP	81–100% young	61–80% young	41–60% initial to optimal subphase	21–40% initial to optimal subphase	0–20% initial to optimal subphase
		grass cover %	0–10	11–20	21–30	31–40	41–60
		grass cover age	91–100	81–90	71–80	61–70	<61
Grassland	Plant diversity	Vegetation cover, %	<10	11–30	31–50	51–70	70
		Plant species richness (Nr. species per sample plot)	<5	6–10	11–20	21–30	>30
Sparcely vegetated areas	Vegetation cover	Vegetation cover, %	<1	2–5	6–10	11–50	>50
	Plant species richness	Number of species per sample plot	0	1	2–4	5–7	>7
Heathland and shrubs	Shrub layer cover	% cover of shrubs	≤30	31–40	41–50	51–70	>70
	Plant diversity	Plant species richness (Nr. species per sample plot)	≤5	6–10	11–20	21–30	>30
Rivers and lakes	Plant diversity (both phytoplankton/phytobentos and macrophytes)	diatom index (IPS) number	>17.5	>13.5	>9.5	>5.5	<5.5
		diatom index (IPS) EQR	>0.85	>0.64	>0.43	>0.22	<0.22
		macrophytes ref. index number	100 to 3	2 to −45	−46 to −69	−70 to −100	None
		macrophytes ref. index EQR	1-0.52	0.51-0.28	0.27-0.16	0.15-0	n/a

Figure 4. *Cont.*

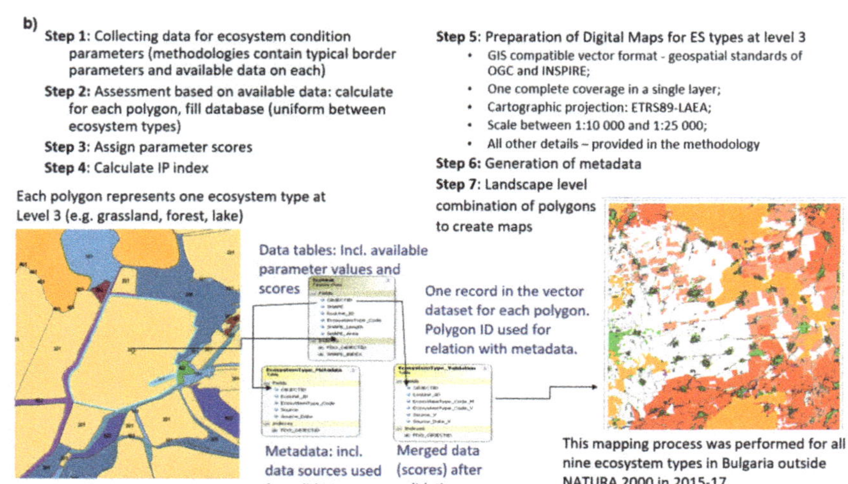

Figure 4. Data fusion, processing and integration into workflows: (**a**) Example of cross-ecosystem comparison of indicators by semi-quantitative parameter assessment (here, the key biodiversity indicators for plants for the ecosystem types presented in our study area); (**b**) workflow of a landscape level assessment, standard database structure and results for mapping and assessment.

- Extensibility: Incremental improvement within the same framework is key for accommodating changes in research objectives and methods of observation. This is especially important in a time of rapidly emerging new or improved technologies, and is a key test of our first hypothesis. To enable data cross-checking, a necessary first step is to match the available datasets to ecosystem parameters and, from there, to condition indicators in the Methodological Framework. In this process, we also assess the existing indicators on their fitness for purpose. The extent and condition indicators for the ecosystems grassland, forest and heathland and shrubs (based on vegetation cover and species composition) are static and therefore by themselves insufficiently informative for assessing the ecosystem extent dynamics over time. This dynamic must, however, be captured in order to cross-analyze it with time series on the climate change parameters and other possible processes influencing the ecosystems such as land management practices or protected-area management plans. This makes the formulation of new indicators necessary. Since establishing these indicators at the national scale is a process beyond the scope of this study, the testing of the extensibility as part of our first hypothesis is limited to formulating candidate indicators based on observations in a single study area.

The sparsity of existing field data is an additional challenge that renders difficult to impossible the application of many of the existing approaches to fusing heterogeneous data to a complete and consistent dataset, as proposed by [58]:

- Yager [65] implies that there is a single, well-defined variable that can be inferred from different data sources, which is not the case in established literature either with ecosystem integrity or with climate change. While such a variable is defined in the Methodological Framework through the use of IP index, it has not yet been used to characterize NATURA 2000 protected areas (including our study area).
- The too large scale of data derived from climate models (effectively the entire study area is covered by a small number of pixels) makes them too coarse for automated processing until reliable downscaling is developed—a problem faced even in data rich areas like the cities that need more detailed projections to tackle urban heat islands (the difficulties of downscaling climate models are apparent from the dataset Climate

Variables for Cities in Europe from 2008 to 2017—a project that required standalone modelling, has higher local uncertainty and is apparently no longer maintained. The dataset is available online at https://cds.climate.copernicus.eu/cdsapp#!/dataset/sis-urban-climate-cities?tab=overview (accessed on 17 March 2022)).

- Automated data fusion through machine learning or AI typically uses high-resolution images or consistent single-ecosystem data series [66,67], or performs experiments with ground truth knowledge in curated datasets [68,69]. The latter also perpetuate bias in algorithmic data fusion by being highly anthropocentric (such as [70], and the datasets quoted therein, US Merced (http://weegee.vision.ucmerced.edu/datasets/landuse.html (accessed on 17 March 2022))), focused on ground truth labeling a single ecosystem [71,72] or generally labeling a single variable for a given task ([73], the Brazilian coffee dataset (www.patreo.dcc.ufmg.br/downloads/brazilian-coffee-dataset/ (accessed on 17 March 2022), etc.). In contrast, our remote sensing data is sparse, of different resolution over time, heavily biased and does not form a regular data series that covers the annual vegetation growth cycles. Its ancillary ground data is also sparse and biased. Furthermore, landscape level classification including more than one ecosystem type is more difficult algorithmically due to additional bias caused by the different scale or area of landscape features representing different ecosystems and their representation in a limited number of satellite bands. For example, rivers and small landscape forms, like small woody features occupying in some cases less than one pixel, need specific extraction different from the processing of remote sensing data for vegetation massives like forest and grassland; these are in turn less homogeneous than cropland monocultures. Finally yet importantly, its land cover (and the seasonal land cover dynamics) is quite different from the existing labeled datasets, so transfer learning would be difficult to impossible.

Due to these limitations, in this study we resort to joint semi-quantitative analysis of existing data sources as detailed in the following sections. Automating the method through a consequent implementation of fuzzy graph analysis is the subject of ongoing research, and therefore outside the scope of this study.

The Methodological Framework focuses on fieldwork or other existing terrestrial data while it mainly uses remote sensing information such as ortho-photo for visual inspection and study object identification at the preparation stage prior to fieldwork. This approach, however, causes a number of qualified scientists to become a bottleneck for large-scale data collection, leading to data deficits, particularly in inaccessible or difficult study areas like ours. In the currently accepted parameter system and its respective protocols, there are no parameters based on vegetation indices or modeling (including climate change reanalysis). Thus, the existing indicator and parameter system of the Methodological Framework is well suited for mapping and assessment but not for continuous remote monitoring. This poses another significant challenge in data sparse and/or inaccessible areas. Therefore, our current study contributes to extending the Methodological Framework by proposing a means to complement it additively with back-up methods for continuous remote monitoring in sparse data environments (as detailed in sections Results and Discussion).

The Methodological Framework was developed in 2015–2017 before some significant developments in the mapping and assessment of ecosystem services at the global and EU level (such as the 2020 revision of the System of Environmental Economic Accounting (https://seea.un.org/content/seea-experimental-ecosystem-accounting-revision (accessed on 17 March 2022)) and the development of a systematic supply and demand spatial analysis for ecosystem services led by the EU's Joint Research Centre [74,75]). It therefore lacks the conceptual basis for ecosystem extent and condition accounts and the supply side of ecosystem services is insufficiently linked with the demand side. Against this background, we emphasize the extent and condition assessment as two important elements of extensibility when testing the first hypothesis in this study.

2.3. Data and Its Processing

Based on the Whole System approach outlined above, we test our first hypothesis by attempting to use all available data to the extent possible. Therefore, we collected heterogeneous data outlined below, and did not exclude any data available to us based on metadata only (e.g., due to insufficient spatial or temporal resolution). This scope used data wider than in studies that focus on a smaller number of datasets. Our approach is to seek ways to ensure, for each available dataset, as a minimum semantic, but ideally also ontological, methodological and/or information compatibility. The data selection process also ensures that we aim at achieving at least partial data availability for the ecosystems' biodiversity, abiotic structure and functions across all five subsystems of the Whole System. We deem data to be similar in the temporal scale if the observed ecosystems' phenomena remain stable between two data points. For example when exploring phenology, similar data are the data collected within the same vegetation stage, whereas, when studying succession, the similar data may include data points collected within 2 or even 5 years. We derive such timeframes for the different ecosystem types based on the Monitoring Guide [76] of the Methodological Framework. For the purpose of our study, data from Copernicus products for base year 2018 and data on the most dynamic meadow ecosystems from the forestry database (2016) are less than 3 years apart and therefore we consider the two datasets essentially similar across all ecosystems. If considering only slowly changing ecosystems like forests, this interval may be even larger. Grounds for excluding a data item or dataset may be its redundancy to more detailed or reliable data, or its proven high uncertainty when cross analyzing with other data. Still, the exclusion of data is rather the exception and not the rule in our data fusion method. Such an all-inclusive approach to data collection and processing is important for uncovering synergies between datasets; there is also the important emergent effect of ecologically meaningful reanalysis of multiple datasets to improve the overall analytical quality or spatial/temporal scale.

2.3.1. Field Data

As a main source of information, we used official partial forestry inventory data in GIS format (Executive Forestry Agency, 2016) to distinguish the main ecosystems on the territory of the study area. This data includes information on non-forest formations such as meadows, and was therefore suitable to delineate spatially the ecosystem types. Since mapping and assessment of ecosystems within Natura 2000 protected areas has not been performed yet in Bulgaria, for the ecosystems other than forest we used a variety of other available data sources:

- data from the official agricultural subsidy layers for High Natural Value pasture grasslands eligible for funding and actually funded (since 2015) (State Agriculture Fund, access rules at https://www.dfz.bg/bg/selskostopanski-pazarni-mehanizmi/school_milk/doc-up-uml-3-2/ (accessed on 17 March 2022))
- data from the surface water monitoring under the Water Framework Directive as contained in the River Basin Management Plan 2013–2015 (A central database is maintained by the Executive Environment Agency, more at http://eea.government.bg/bg/nsmos/water (accessed on 17 March 2022) (in Bulgarian only))
- field photography data obtained by the lead author using a drone in the period 2016–2017
- publications (Bondev, 1991) [77] as digitized in 2013; other scientific publications on ecosystem condition for the study area
- soil map (Executive Environment Agency) (the metadata for this dataset is available at http://eea.government.bg/bg/nsmos/spravki/Spravka_2020/soil (accessed on 17 March 2022) (in Bulgarian only)
- real-time forest information system maintained by the Executive Forestry Agency and WWF-Bulgaria and containing data on old forests, land use mode, logging permits and crowdsourced information on irregularities reported by volunteers (view at https://gis.wwf.bg/mobilz/, accessed on 17 March 2022).

We found a strong bias both in the research and administrative efforts made to inventory the different ecosystem types. Forest ecosystems are subject to routine data collection, and their growth, disturbances and pests are much better studied and mapped with a relatively high level of detail on species composition (dominant species and up to 7 other species are being entered). In contrast, much sparser data was available in the forestry database on ecosystems other than forest, except for the heathland and shrub ecosystems also classified under the EU Habitats Directive as habitat 4070—bushes with *Pinus mugo* and *Rhododendron hirsutum*. In contrast, the only description of grasslands as "meadow" is very broad. Data on the availability of alpine and subalpine grasslands outside the forest management area are insufficiently georeferenced (either due to missing coordinates or due to coarse scale mapping). In the sense of data integration, this limited the scope of much of the present analysis—determining Level 3 ecosystem subtype had to be constrained to the spatial–temporal trends in the parts of the study area inventoried in the forest management databases (colored in green in Figure 1). At this time, the only practical way to classify ecosystems in the remaining part of the potential study area (outlined in yellow in Figure 1) is to resort to Level 2 classification as the use of remote sensing and climate data requires additional ground truth validation by a larger team.

The study area is subject to vivid scientific interest with over 100 publications. However, there is a strong disparity of study objects on the ecosystem level, with prevailing forestry studies and scarcely any publications on the grassland, sparsely vegetated or river/lake ecosystems. Forestry publications taken into consideration include forest condition assessments, climate studies fieldwork, water balance and pollution studies; we discarded species level papers focusing on single species of flora, fauna or fungi, which form the bulk of the publications. Particularly relevant to our study are publications on the condition of forest ecosystems, including publications on large-scale disturbances causing changes in condition or functioning that may influence the vegetation index values [78–85]. These sources show that windthrows seem to be the major cause of disturbance. To further specify the classification of ecosystems across the study area, we restored to the crosswalks defined by Kostov et al. (2017) [50], Velev et al. (2017) [51] and Apostolova et al. (2017) [52].

2.3.2. Remote Sensing Data and Products

In this study, we use three types of remote sensing data: Ortho-photo data (2013) (Data upon request, GIS viewer at http://gis.mrrb.government.bg/ (Bulgarian only) (accessed on 17 March 2022)), satellite derived expert products to determine the current ecosystem extent and a long time series of remote sensing scenes as means to derive vegetation indices to assess ecosystem condition. This section lists and describes the data while details of data processing to overcome the shortcomings in each product type are provided in Supplementary Materials, Section S5.

Expert products are better suited for static features such as the positioning and characteristics of specific vegetation types. They are released in a sparse data series (for Copernicus—typically every 3 or every 6 years, depending on the product) and their production involves significant processing and quality control that does not need to be repeated by the authors. A shortcoming of expert products is that they do not provide complete (wall-to-wall) coverage but focus on single ecosystem types and therefore their joint use leaves gaps when pixels do not belong to any of the ecosystem types in the respective products.

For ecosystem extent, we chose the Copernicus High Resolution Layers (HRL) expert products for grassland (grassland extent layers) and forest (tree cover density and dominant leaf type), the two products that, taken together, provide the most coverage in our study area's landscape, while also distinguishing between dominant forest type and detailing the tree density. We chose reference years 2015 and 2018 since these are the only two reference years available for both the grassland and forest HRLs.

In contrast, remote sensing scenes are processed to a much lesser extent (e.g., orthorectification, cloud removal, etc.) but are available on average once in two to three weeks, which makes them suitable for forming longer, seasonal data series to perform functional observations or confirm vegetation location to explore spatial extent.

On ecosystem condition, we performed a selection of vegetation indices (VIs). Essentially, VIs are based on the reflectance properties of the vegetation and are designed to measure vegetation quantity and vitality in different aspects—canopy area and structure, concentration of chlorophyll [86]. For the long-term condition and functional assessment of the selected HME in Bulgaria's Rila Mountain, we applied an empirical method based on two well documented vegetation indices (the Normalized Difference Vegetation Index (NDVI) and Normalized Difference Water Index (NDWI) [87]) as well as the newly developed Normalized Difference Greenness Index (NDGI) [88]. Research on the relationships between NDVI and climatic variables already enables predictions and trend definitions [89], albeit at much larger scale.

We selected for use of multispectral satellite data from MultiSpectral Instrument (MSI) sensors of Landsat from the United States Geological Survey (USGS) database covering 42 years (1977–2019) and Sentinel 2 data with high spatial, spectral and radiometric resolution [90].

2.3.3. Climate Data

In our study, we use climate data for the parameters 't2m' (temperature 2 m above ground level), 'tp' (total precipitations), 'evpt' (evaporation) and 'v10' (10 m V wind component) extracted from the public daily subset of the ECMWF Re-Analysis (ERA Interim) dataset of the European Centre for Medium-Range Weather Forecasts (ECMWF) database.

The multidimensional files in NC format contain information for each parameter within the vegetation periods (May–September) for most of the studied period—40 years from 1979 (the starting year of ERA Interim reanalysis time series), to the time series' end in August 2019. We prepared these NC files for further use by ArcMap 10.3 processing. We produced raster files of each parameter for every month within the vegetation period for the 40 years of climate simulations and converted them into shapefiles to facilitate data extraction. We extracted the attribute data and transferred it to SigmaPlot 11.0 where we performed the regression and line plot analyses.

The retrieval and processing of climate data also has some limitations. ECMWF discontinued ERA Interim—the dataset containing relevant variables for our study in August 2019. It provides the follow-up reanalysis dataset ERA 5 [91] in stages and currently ERA 5 is only available starting 1981. In addition, it had some errata and is still being verified across the globe [91–93] with no verification yet for Europe on all climate variables needed for replicating our study using ERA 5 data.

2.3.4. Data Processing Workflow

Processing of Single Datasets

The remote sensing data preprocessing steps include:

- Review and selection of scenes from both sensors. While satellite missions regularly fly over our study area, it often is cloudy or misty, and in some years there is snow in parts of it well into May. Therefore, we had to handpick suitable scenes (listed in Table S4 in Supplementary Materials). Unfortunately, these scenes do not adequately cover all vegetation seasons with satellite imagery. Even in the handpicked scenes that we analyze in this study, there are still some clouded areas (visible in the false color renderings in Figures S7e, S9d, S10a,d, S12a,h, S13d and S16d,g,j of the Supplementary Materials and Figure 5). Based on cloud coverage masks over the study area and additional visual review, we selected a set of 33 satellite scenes with minimal to no cloud coverage, taken within a time frame of 42 years;
- Creating composite images containing spectral bands of the images. In this step we used ERDAS 14.0 software.
- Creating a raster file of each scene containing the Area of Interest (AOI), i.e., study area boundary.

- Calculation of VI from the selected scenes. We calculated Normalized Difference Vegetation Index (NDVI) for all 33 scenes and produced raster files. Older Landsat sensors with fewer bands do not allow calculating NDWI and therefore in 12 of the Landsat MSI satellite scenes we could not obtain the values of this index.

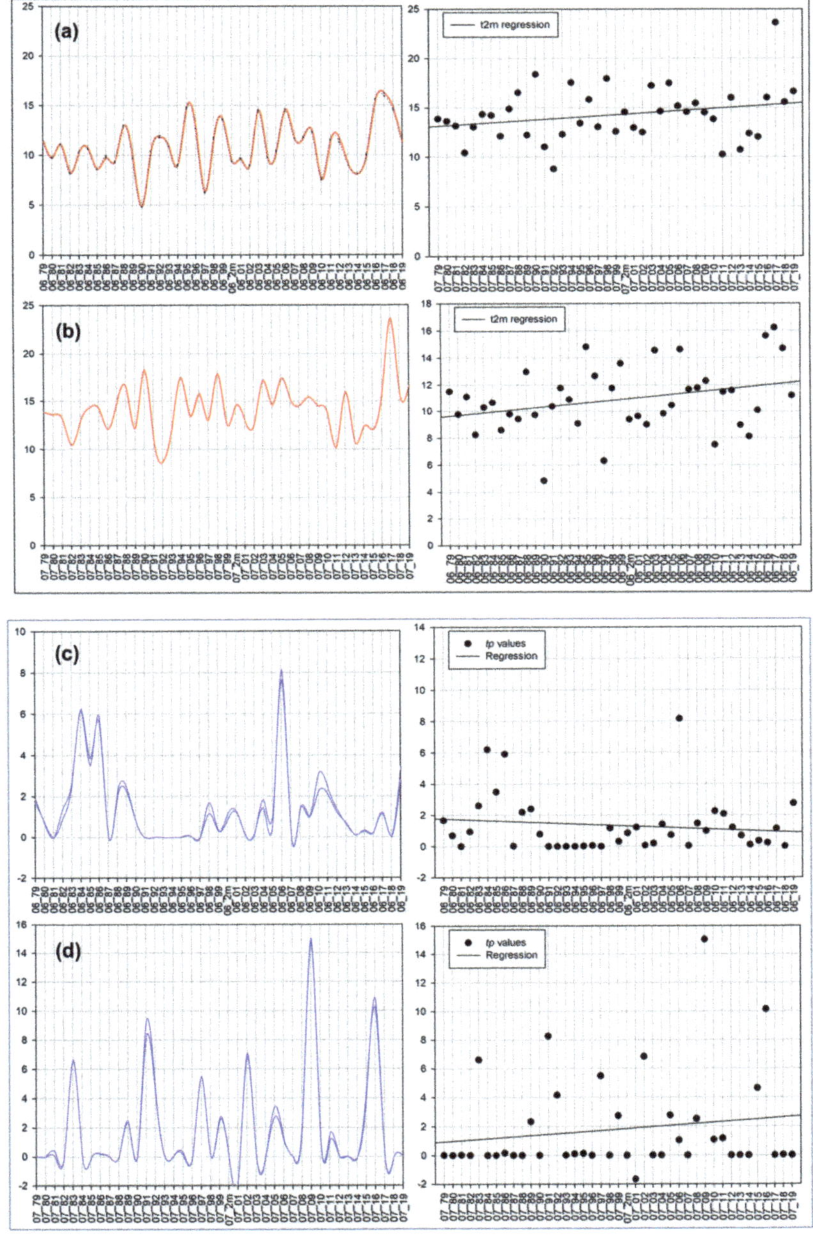

Figure 5. Processing of climate data series: t2m (**a**,**b**) and tp (**c**,**d**) plots and regression results for the months of June and July in 1979–2019.

Table S4 of the Supplementary Materials contains a summary of the data provenance of the satellite scenes used in the study.

NDVI is the most widespread index for vegetation condition assessment indicating its energy absorption and reflectance capabilities, photosynthetic capacity and biomass concentration [94–97]. According to Pettorelli et al., 2005 [89], "it could be used to predict the ecological effects of environmental change on ecosystems functioning".

To clarify the relationship between climate parameters and NDVI values and define their influence on NDVI values within the studied period (1977–2019), we analyze the VIs of single scenes within a vegetation season together with the available climate data. We aim at inferring trends of condition changes for the HME in the study period.

We visualize the distribution of NDVI through 3D graphics including the index values for the forest within the HME as a whole. Because of the large data volume, the elaboration of a single 3D graph for this period is challenging, both computationally and visually. Therefore, we constructed six 3D graphs (Figure S6 of the Supplementary Materials) containing NDVI data from dates in the beginning (Figure S6a,b), in the middle (Figure S6c,d) and in the end of the period (Figure S6e,f) to present the distribution of NDVI values. We based the data grouping by dates on calculation of correlation coefficients between NDVI values in the satellite scenes.

After the calculation of the VIs, we prepared 33 thematic maps (TM) of NDVI, 21 TM of NDWI and 5 TM of NDGI using ArcMap 10.3, as well as six 3D graphic models using SigmaPlot 11.0.

Since our satellite time series is composed of images from several satellite missions with different sensor characteristics that influence the respective NDVI values, we cross-verified the results using the sensor-invariant Normalized Differential Greenness Index (NDGI) [88]. NDGI is a dynamic index calculated for pairs of satellite scenes in remote sensing time series and therefore suitable for change detection and verifying the changes in vegetation conditions in situations where NDVI from different sensors may not be fully compatible between scenes. It is very sensitive to even small increments of vegetation development; furthermore, it is designed to detect changes in vegetation cover for a specific time interval, which makes it useful for vegetation processes (photosynthetic capacity), i.e., vegetation functional assessment. Based on the spectral reflectance characteristics of the vegetation, this index is defined through the greenness component obtained via orthogonalization of the satellite images. Using the greenness component causes NDGI to be sensor and time invariant. Furthermore, the error stemming from external factor influences during satellite image capturing is significantly diminished [88]. The index values range from -1 to $+1$, where the negative values ($NDGI < 0$) correspond to a decrease of the vegetation process (loss of vegetation) and the positive values ($NDGI > 0$) indicate an increase of the vegetation process, i.e., photosynthesis takes place (appearance, development of vegetation). When $NDGI = 0$, biomass production remains unchanged for the studied period.

Due to the limited number of available cloud free scenes, we only generated five NDGI TMs (Figure 8). Four of them have a full set of correspondent climate parameters for the years in which the starting and final scenes of the NDVI comparison were taken. For the period 1984–1977, we lack the climate parameters for 1977 to compare with NDGI. We assessed the suitability of scene pairs for calculating NDGI based on correlation analysis.

For Copernicus HRL products, we performed GIS-based analysis and identified areas of interest based on the different attributive information in pixels with the same location in 2015 and 2018 to identify changes over time. We used this data to localize changes in extent for grassland and forest ecosystems, as well as changes in species composition in forests.

For the retrieved climate data, we extracted the single variables of interest for each month and performed regression analysis on them. Figure 5 presents a sample of the results, for the variables t2m and tp in the months of June and July during the period 1979–2019, while the full range of climate data time series plots and regression analyses for tp is available in the Supplementary Materials (Figures S1–S5).

Cross-Validation of All Available Data

We use the Whole System paradigm to organize data and jointly analyze different high mountain ecosystems in our study area and the different datasets available from their measurements. To this end, we re-compiled a table of combined data sources complementing the parameters in the Methodological Framework (for a list of indicators common across the ecosystems see Table S1 and for their availability in the ecosystems in our study area Table S2 of the Supplementary Materials). We furthermore perform the following analyses:

- At the semantic level, we explore relationships between potential new parameters derived from remote sensing and climate models, and the indicator they describe. This approach allows for identifying and proposing alternative observation techniques for the same indicator (see Table S3 of the Supplementary Materials). Cross-calibrating such data sources to identify their accuracy, measuring ecosystem parameters can help to establish them as alternatives to field observation data (the standardization of observations for different ecosystem types is part of ongoing work in the European Long-Term Ecosystem Research Network and globally in several initiatives. As such, performing it is currently outside of this paper's scope). Furthermore, complementing the Methodological Framework with Earth observation and climate modelling allows us to look at a larger scale landscape mosaic and easily locate the most dynamically changing areas of interest such as the forest ecotone or areas with rapid changes in species distribution. In this manner data co-analysis at the semantic level allows for using the synergies between available data from all sources to both observe changes in the whole area of interest and identify focus areas for observation efficiency, both within the same conceptual framework. The semantic linking also allows us to combine data mostly related to ecosystem extent and data typically used to study ecosystem conditions by applying ecological knowledge to infer the links between extent and condition.
- At the ontology level, we verify/specify the crosswalks between habitat type data in the forestry database and ecosystem types, as applicable to this specific study area. The crosswalks are established as a one-to-many relationship in the respective ecosystem mapping methodologies—parts of the Methodological Framework [50–54]; see Figure 6a. Verification consists of a georeferenced view of available data for habitats and an on-the-spot check of respective ecosystem type/subtype having in mind the overall ecosystem structure. The level of detail of the crosswalk depends on the relations of different labels of available data according to different classifications in the crosswalk. For example, the finer grained habitat labelling of a polygon allows for deducing the coarser ecosystem subtype or type classification for the same polygon while data on the ecosystem type does not automatically translate to ecosystem subtypes or habitats and typically requires further field validation and/or co-analysis with other available data. Heathland and shrub ecosystems are specified in the forestry database to have Pinus mugo as their dominant species and were therefore assumed to be of the "Arctic, alpine and subalpine shrub" ecosystem subtype. To overcome the limitations of existing data on grasslands, we analyzed the forestry database's information on elevation, slope, exposition and soil types (the latter being also verified using the classification in the Executive Environment Agency's soil type layer). We then defined the possible ecosystem type as Alpine and subalpine grasslands based on Apostolova et al. (2017) [52] and an unpublished interpretation key produced during the mapping and assessment of grassland ecosystems in 2015–2017 by the same authors.
- At the methodological level, we fuse semantically and ontologically compatible datasets to verify the hypothesis that climate change causes changes in ecosystem spatial distribution within the landscape mosaic and ecosystem condition/functioning over time, including change in species composition. We use the spatial distribution of ecosystem types in this crosswalk derived from ortho-photo and forest database data, together with Copernicus HRL products for grassland distribution and forest

tree cover density for 2015 and 2018 to geolocate areas of recent location and dynamically changing composition of ecosystems in the landscape. In the resulting mosaic (Figure 6b, bottom), the data fusion of remote sensing and climate data conforms to the ground truth observations concerning the dominance and location of forest ecosystems. The highest parts of the study area are covered with shrub vegetation and grasslands, which are unevenly distributed, covering sub-alpine and alpine territories. Following the crosswalk validation, we combined data from the different available sources to form time series in the sequence depicted in Figure 6b. We performed a correlation analysis of the NDVI TMs generated for the satellite images to analyze the spatial correlation of NDVI across satellite scenes and find pairs of scenes of analytical interest (Figure 7). This data is complemented with ecosystem location and species composition towards the end of the period, as derived from ground data and Copernicus HRL.

- Having in mind the need to update the Methodological Framework with developments that occurred after its publication, we furthermore used the co-analysis of climate and spatial distribution data from satellite products to derive reference values for candidate indicators based on remote sensing and thus test the framework's extensibility.

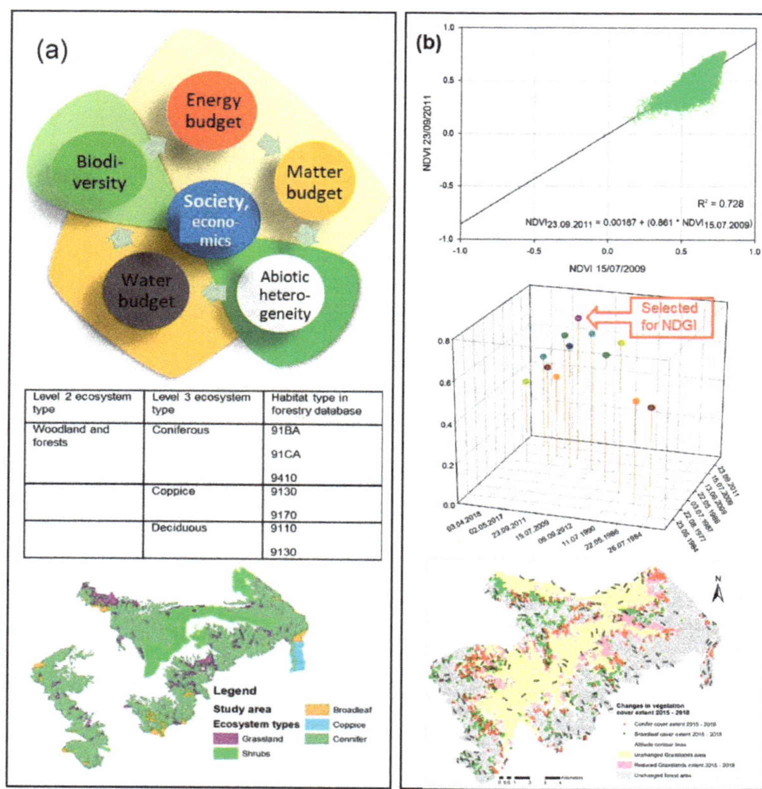

Level 2 ecosystem type	Level 3 ecosystem type	Habitat type in forestry database
Woodland and forests	Coniferous	91BA 91CA 9410
	Coppice	9130 9170
	Deciduous	9110 9130

Figure 6. (**a**) Whole System approach: Conceptual view, data crosswalking and ecosystem type (level 2 or level 3) distribution on the ground; (**b**) Selection of pairs of satellite scenes by correlation analysis of NDVI between scenes; 3D graphic of the correlation coefficient values obtained via Pearson correlation analyses of NDVI between scenes, and georeferenced supplementary information on ecosystem types derived earlier and used for reference along with the results of vegetation change detection in Copernicus HRL products.

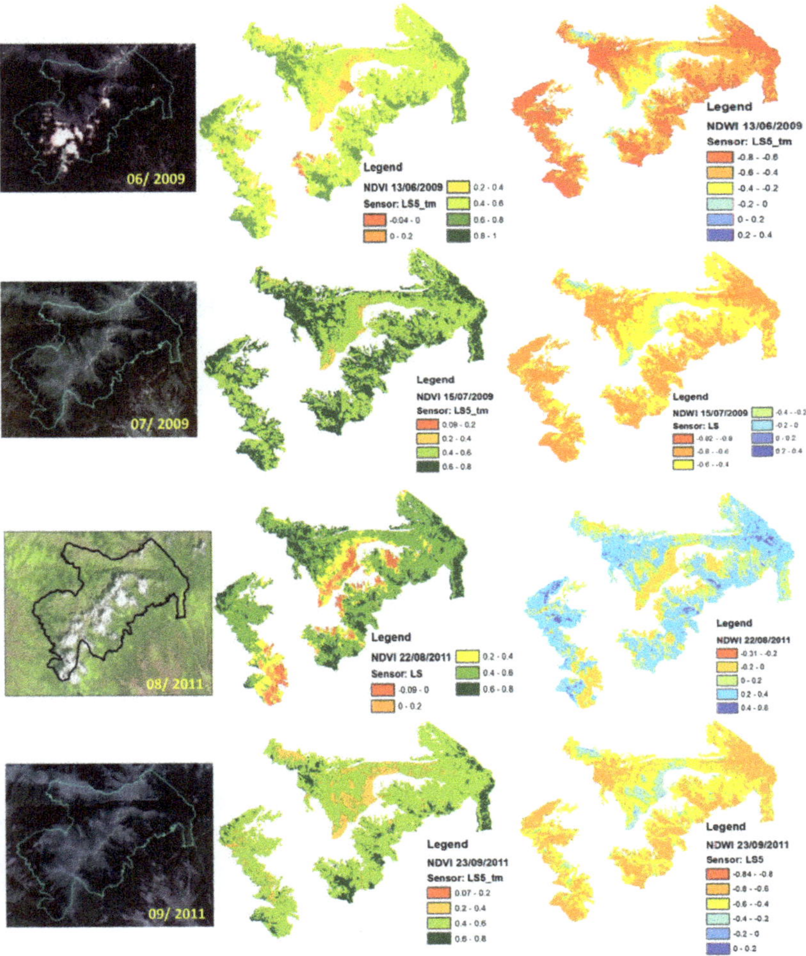

Figure 7. Cross-verification of single satellite images' time series using different vegetation indices. Anomalies in NDVI and NDWI can help locate the residual clouds in the TM (such is the case over the mountain ridge in the scene from August 2011).

Using all available data on both extent and condition proved necessary for the reanalysis of data in the earlier part of the study period and therefore complements the missing or insufficiently detailed ground data on the beginning of succession in the forest–grassland ecotone, as well as the upward crawl of deciduous species, causing the change in species composition in forests. To verify the data fusion, we defined that the datasets for spatial extent must contain georeferenced data on ecosystem existence at least at Level 3 (ecosystem type) but preferably at ecosystem subtype or habitat level. Data on ecosystem condition must relate to condition and functional parameters over time, such as biomass production observed via the proxy of vegetation indices. The stages of cross-validation at the methodological level are detailed below:

(1) We performed a combined analysis of NDVI and where available—also NDWI to verify the observed vegetation growth in each scene. A sample of such results is presented in Figure 7, while the full set of processed images is available in the Supplementary Materials (Figures S7–S16 and S18). This approach is necessary due to the chosen scale of observation—in contrast to expert products that cover all of a given ecosystem type but do not contain its dynamic characteristics, factors like clouds render parts of the few available scenes unusable, which only allows partial processing of these scenes.

(2) We analyze the derived vegetation indices (NDVI and where available—NDWI) for each year together with climate data for the respective vegetation season to observe the ecosystem functioning and its relations to the climate parameters in the respective year. Since the climate reanalysis does not yield detailed projections, the only way to observe the influence of local parameters determining the microclimate, such as elevation, slope, aspect, is by comparing the geolocated vegetation indices in scenes from different stages of the vegetation period with the data on abiotic factors derived from the forestry database. At this stage, we perform this analysis in a qualitative manner.

(3) To improve the precision of change detection in the early Landsat images, we restore to the sensitive NDGI index [88]. NDGI is a valuable part of the remote sensing indices portfolio for different purposes, including crop monitoring [98,99], disturbance detection and response [100,101], flood detection [102], ecosystem risk assessment [28], wetland ecosystem services [103], etc. Earlier work [7,90] proves its usefulness for evaluating shorter time series of remote sensing images in this same study area. In this study, we use NDGI to cope with the lack of georeferenced historical data on ecosystem species composition. The earliest available spatial atlas [77] has very low resolution (1:600,000) and a description of communities that is very different from the current EU level and global classifications. Therefore, its usefulness concerning inferring the semantic and ontological compatibility is only limited to ecosystem type and does not allow for tracking subtler changes in ecosystem conditions or species composition. In a first analytical step, we analyzed the NDGI TMs together with climate data (Figure 8) to identify the climatic constraints to vegetation growth (such as extreme temperatures or insufficient precipitation). The resulting scenes are useful both for observing the ecosystem conditions' dependence on climate and local environmental factors influencing microclimate (in particular elevation), and for determining changes in spatial distribution/exploring species composition, as detailed below.

As an additional analytical step, the generated NDGI TMs also proved useful for detecting outliers and eliminating problematic satellite scenes such as the image derived on 13 September 2019, which we identified through co-analyzing the NDGI TM between 13 June 2009 and 16 September 2019 and climate data. The reasoning behind this process is detailed in the Supplementary Materials, Section S5.

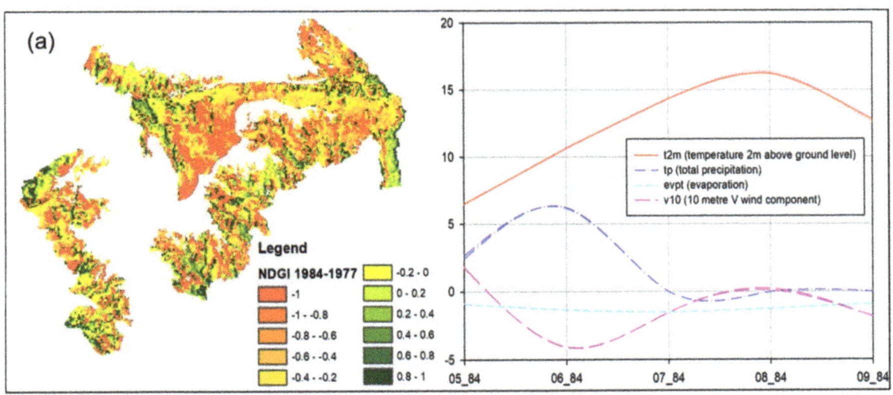

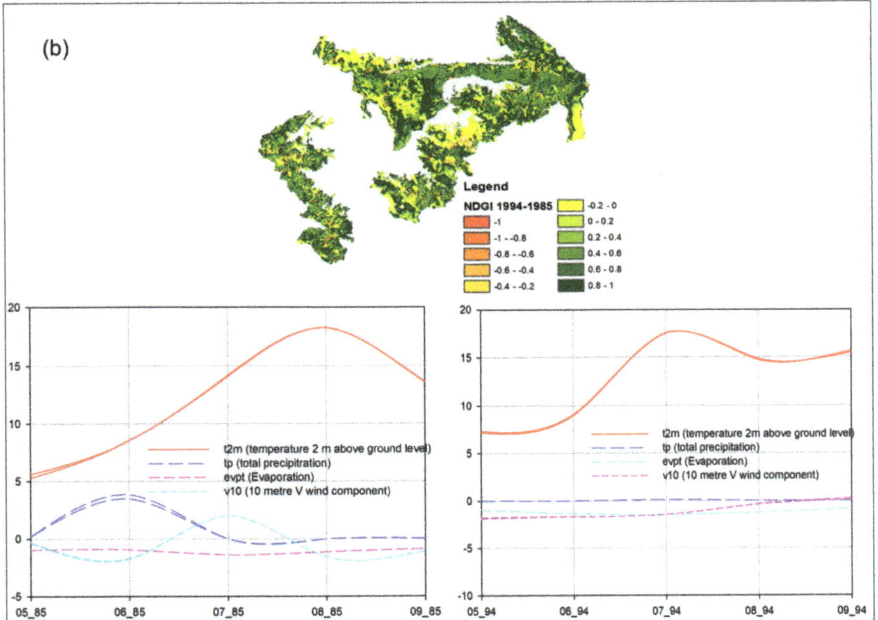

Figure 8. *Cont.*

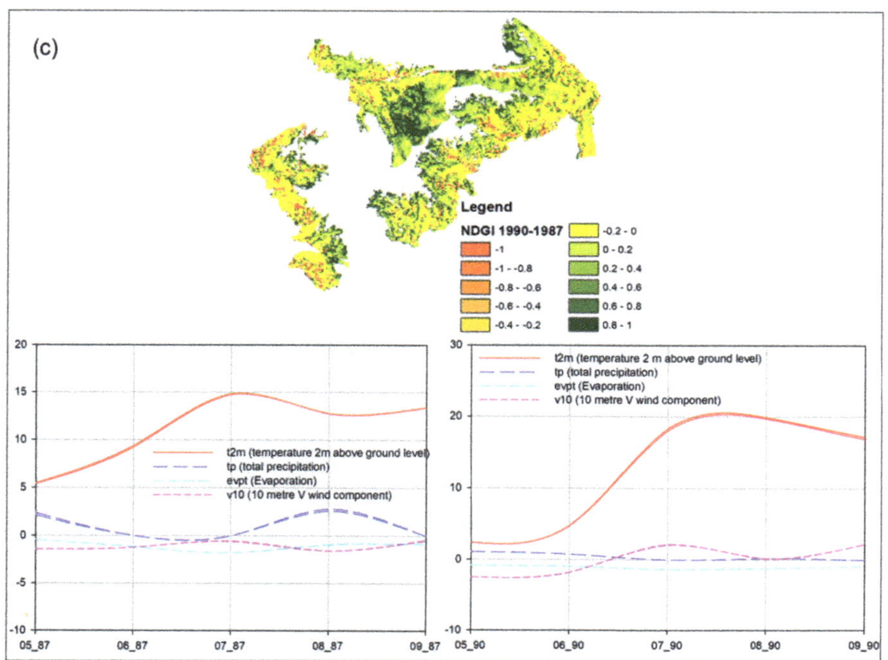

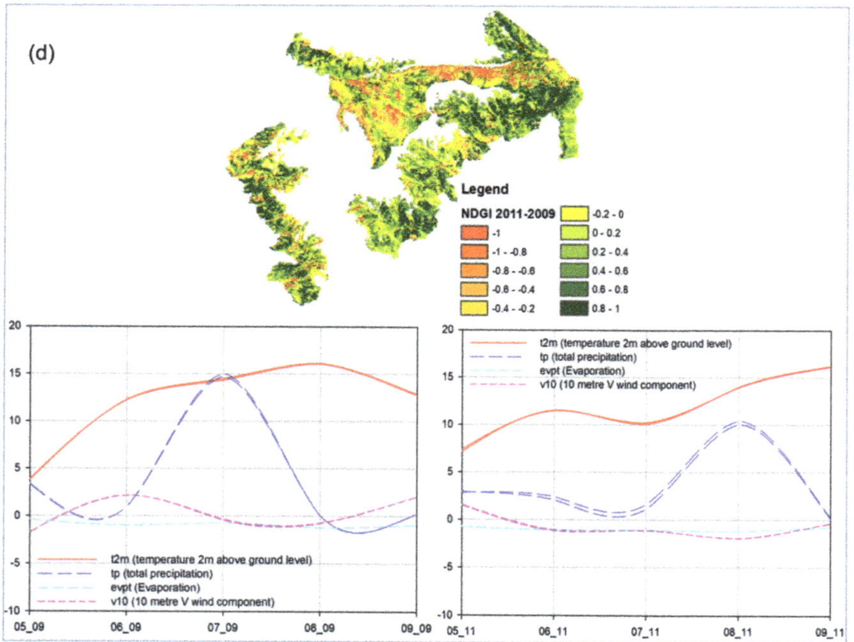

Figure 8. *Cont.*

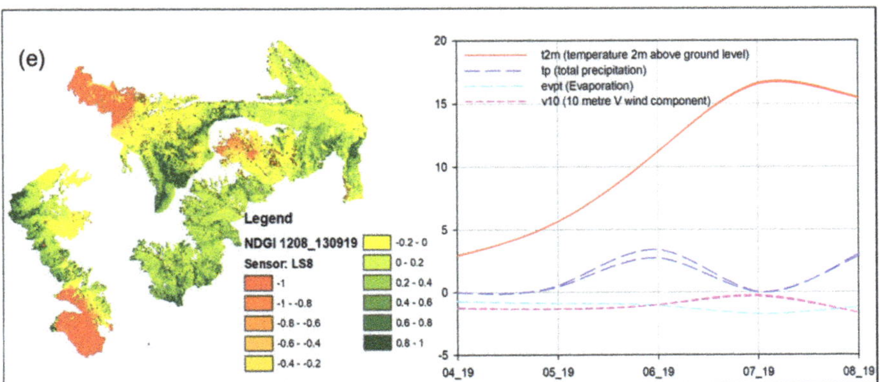

Figure 8. Relation between NDGI and the respective climate conditions: (**a**) TM of NDGI between 22 August 1977 and 23 May 1984 and diagram of climate parameter dynamics for 1984; (**b**) TM of NDGI between 27 June 1985 and 29 June 1994 and diagrams of climate parameter dynamics for 1985 and 1994; (**c**) TM of NDGI between 3 July 1987 and 11 July 1990 and diagrams of climate parameter dynamics for 1987 and 1990; (**d**) TM of NDGI between 13 June 2009 and 23 September 2011 and diagrams of climate parameter dynamics for 2009 and 2011; (**e**) TM of NDGI between 12 August 2019 and 13 September 2019 and diagram of climate parameter dynamics for August and September 2019.

(4) The lack of a long time series for tracking changes in spatial extent of the ecosystems within the landscape is a particularly challenging case for data fusion. This is so partially due to the coarser resolution of early satellite imagery that increases the uncertainty in detecting succession, and partly due to the changing signal within the same ecosystem type caused by changes in species composition, local microclimate, etc., which prevents geospatial reasoning on ecosystem extent based on single satellite scenes. At the same time, our data does not cover full vegetation seasons to enable reliable location of vegetation types by phenology. The most convenient data sources on ecosystem extent among the available data are the Copernicus High Resolution Layer (HRL)—new products currently only available with the information needed for our analysis in two baseline years, 2015 and 2018. Such a short "timeline" of two points is by itself insufficient for exploring trends in the change of ecosystem extent, being at most sufficient to specify the current extent and approximate species composition of ecosystems. Observing long-term changes in extent therefore requires creating a semantic link to data not directly attributable to extent. Such semantic linking allows, in effect, for performing reanalysis of multiple, various scale data sources for extent and condition. In the case of succession, the semantic inference that enables reanalysis back in time includes detecting the earliest signals of active growth in locations with proven later changes in ecosystem extent detected at the end of the period. In this manner we can use, information on the ecosystem type, location and first detection of the vigorous new growth to approximately date the beginning of the succession. Such an approach also has its limitations since it only applies to clearly defined and spatially stable ecosystems (deciduous and coniferous forests, heathland and shrubs). It is therefore not applicable to ecosystems with features less discernible through remote sensing (in our study area: Grasslands, water ecosystems). The data reanalysis consists of locating stable features detected at the end of the period and locating the earliest available signals for the forming of these features. To cross-check the changes in ecosystem extent, we compared the resulting change maps from the Copernicus Grassland HRL to the corresponding decline in grassland extent, and cross-validated the findings with earlier ortho-photo, our own drone imagery and the forestry database. Figure 9 shows representative spots of cross-

checking the spatial extent. Targeted collection of dendrochronological information through ground studies in the identified spots could provide for further reduction of uncertainties, better dating and ground-truthing to form datasets for machine learning or AI applications.

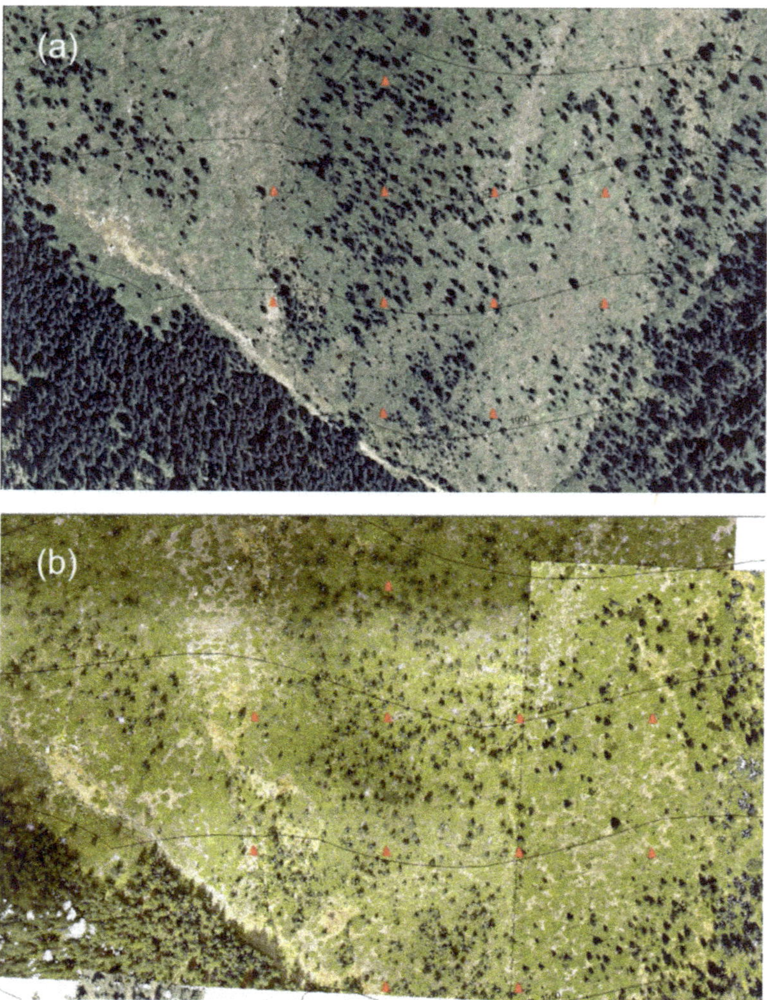

Figure 9. Cross-verification of ecosystem extent change detection in the upper ecotone—succession from grasslands through shrubs to coniferous forests. Verification of Copernicus HRL change in tree cover density (red pixel centroids represent increase in tree cover density between the base years 2015 and 2018 at 100 m pixel) as compared to the actual vegetation cover in (**a**) ortho-photography (2011), and (**b**) our georeferenced drone imagery (2016). The northernmost and southernmost of these centroids are outside the scope of the 2016 forestry database, whereas the remaining are within a polygon of the coniferous ecosystem type.

For the purpose of species composition data reanalysis, semantic inference on the change in species composition includes locating places of known changes in the ecosystem composition in Copernicus HRL between 2015 and 2018 (in our case—spread of broadleaved species within coniferous forests) and observing the vegetation indices back in time to date the beginning of this process. The process is illustrated for representative parts of our study area in Figure 10.

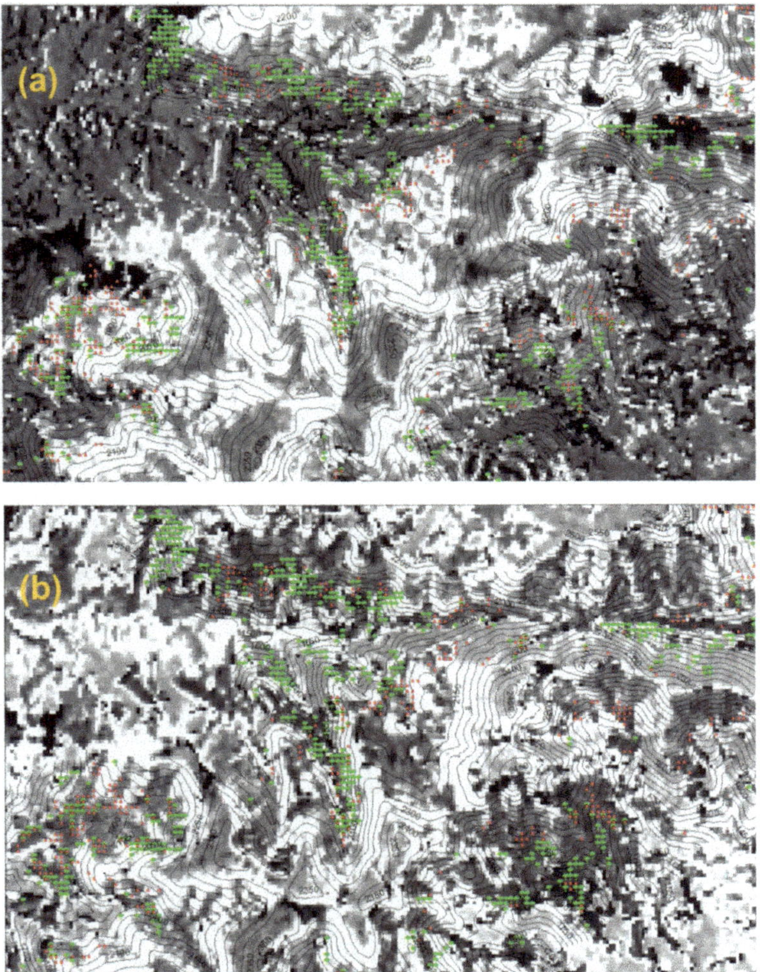

Figure 10. Data reanalysis: Spatial patterns of change in species composition at locations with known long-term modifications. Copernicus HRL change in tree cover density (red pixel centroids represent increase in coniferous and green pixel centroids represent increase in the broadleaved tree cover density between the base years 2015 and 2018 at 100 m pixel) is the baseline to locate changes in NDGI values for earlier periods: (**a**) 07.1987 to 07.1990; (**b**) 06.1985 to 06.1994. Similar values of NDGI correspond to similar changes in vegetation growth intensity between the pairs of years used to generate NDGI. A comparison between the two NDGI TMs and the HRL spatial overlay suggests that active change in tree species composition started around 1990. The higher precipitation in June 1985 strengthens the baseline signal and therefore accounts for smaller differences found in NDGI and vegetation growth in the second scene.

In this manner, the fusion of different data sources can form a longer quasi time series of heterogeneous data sources on extent and condition. In our case, the composite data series covers the period 1987–2018 and includes the earliest Landsat 1–3 missions at the beginning of the period (images of 80 m pixel in two spectral bands); orto-photo and terrestrial data for the 2011–2016 period; and Copernicus HRL (100 m pixel) grassland and tree cover density products for the base years 2015 and 2018.

3. Results

3.1. Results of the Remote Sensing Data Processing

3.1.1. Remote Sensing of Spatial Extent and Distribution Trends

The cross-verification of the expert products together with forest database data shows a high degree of correlation between the datasets (see Figure 11). The available HRL products for grassland and forest ecosystems (which comprise most of the landscape mosaic) are presently insufficient to form a timeline of only two data points for three years. However, as detailed in Section 2.3.4, data fusion and re-analysis allows for forming a longer timeline of different but semantically compatible datasets. In particular, the co-analysis with NDGI for earlier satellite scenes and climate data (correlations can be seen in Table S5 of the Supplementary Materials) supports our second hypothesis about climate change-induced ecosystem dynamics. This is so since the remaining abiotic parameters are largely unchanged, while anthropogenic activity was traditionally weaker and extensive in most of the inaccessible parts of the study area and has ceased completely since its inclusion in the Natura 2000 protected areas network. The comparison of spatial trends shows upward expansion of both broadleaved and coniferous forests—with broadleaved expansion occurring as high as 2000–2100 m—and decline in grassland ecosystems' extent in the lower elevations bordering the forest. Figure 11 presents the results of data reanalysis and fusion across the entire study area. The upward extension found by joint analysis of Copernicus HRL and the forest database is confirmed and dated more precisely by extending the data series back in time through NDGI between selected satellite scenes. The fused data further reveal that southern slopes are more conductive to accelerated growth of broadleaved species, whereas accelerated upward expansion of coniferous species is more frequent on the northern slopes. This finding is important for focusing future field monitoring and dendrochronological verification.

The available ground data also support the observation on the transition from coniferous to deciduous forests due to climate change. In addition to our spatial analysis, all 24 polygons assigned to coppices in the forestry database are marked as "forest in transition" in the "forest type" field, some having been earlier classified as "rock" or "non-afforestable". After disturbances, four polygons underwent transition to habitat 9130—Asperulo-Fagetum beech forests.

We observe the same general dynamics across the study area when analyzing the earlier vegetation growth through NDGI (see Figure S18 in the Supplementary Materials). The only negative NDGI values we observed are in the comparison of August 1977 and May 1984. They are attributable to the earlier stage of seasonal vegetation growth in the second scene.

The greatest difference in the NDGI TMS comparing scenes taken in the same stage of vegetation growth (between July 1987 and July 1990, and between June 1985 and June 1994) is noticeable in higher altitudes, both in coniferous forests and shrubs. There are areas showing growth trends despite the seasonal difference in temperature and precipitation in the two years (coppices, broadleaf deciduous forests, coniferous forests in lower altitudes). They contain faster growing vegetation that has net biomass gain and builds up carbon depots in the seven years between scenes, whereas grasslands' small positive and negative NDGI numbers point to a relatively stable state or decline.

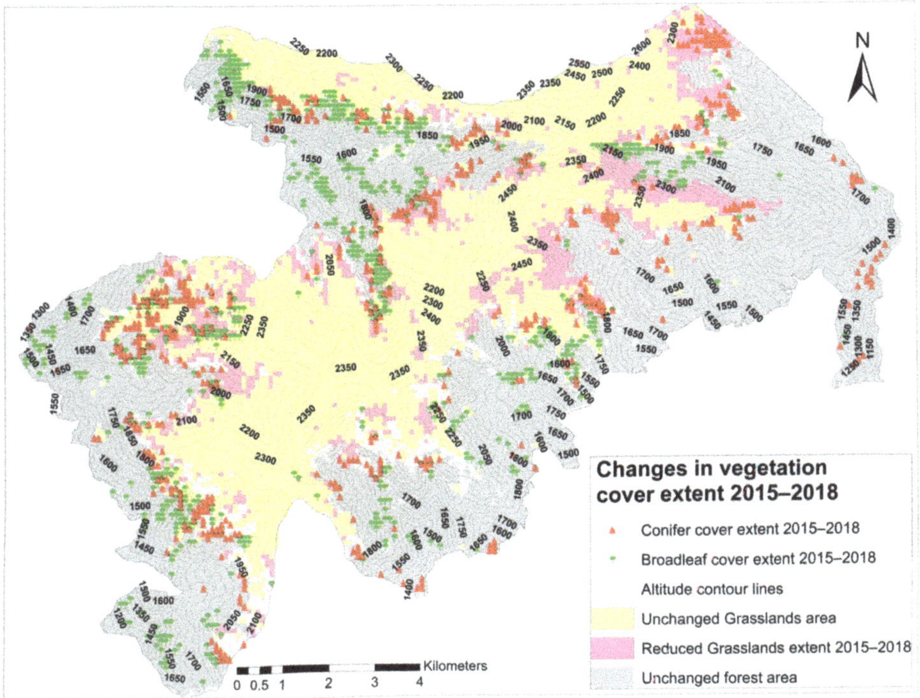

Figure 11. Changes in spatial distribution of grassland and forest ecosystems at pixel size of 100 m between base years 2015 and 2018. The extent of areas with higher tree cover density in 2018 is marked with coniferous and deciduous symbols at the centroids of the respective pixels. The changes are confirmed by NDGI in satellite scenes as early as 1990.

Both NDGI TMs between July 1987 and July 1990, and between June 1985 and June 1994 demonstrate the importance of temperature differences for the vegetation growth, especially in higher-altitude coniferous forests and shrubs containing coniferous trees. With most of the parameters being similar, both the rise in temperature between 1987 and 1990 and the earlier onset of higher temperatures seem to have a positive influence on all ecosystem types in the higher altitudes. Almost lacking precipitation appears to be the limiting factor on growth for all ecosystem types in the lower altitudes. The earlier onset of higher temperatures has an even more marked effect in June (comparison between 1985 and 1994) even though the temperature differences themselves are minimal and 1985 was a year with higher precipitation at the beginning of the vegetation season.

3.1.2. Condition and Functional Dynamics of the HME—Vegetation Indices (VI) and Climate Data Co-Analysis

Following the reanalysis of combined climate and vegetation indices data, we found correlations between NDGI that indicate patterns of active vegetation growth between two satellite scenes to be much stronger when calculated at the observation object's characteristic scale. This is illustrated in Figure 12, which compares potential determinants of growth for each characteristic area of rapid coniferous and broadleaf growth (as identified by the changes in extent towards the end of the period) with the much more heterogeneous an large scale study area as a whole. In both focal areas, the correlation between ecosyst type and overall seasonal growth indicated by the NDGI signal is stronger than the sa type of correlation in the entire study area (Figure 12b,c).

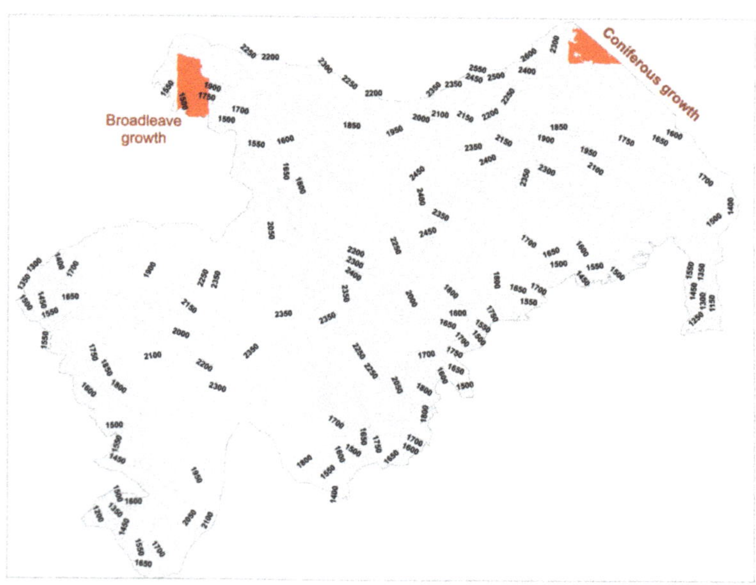

(a)

Correlation scope: Seasonal comparison	Ecosystem type	Elevation	Aspect	Slope	Relief
Coniferous active growth					
22 August 1977 to 23 May 1984	−0.42	0.02	0.44	−0.45	0.00
13 June 2009 to 23 September 2011	0.66	−0.46	0.34	0.28	0.00
Broadleaf active growth					
22 August 1977 to 23 May 1984	0.23	−0.38	0.02	−0.06	0.63
13 June 2009 to 23 September 2011	0.25	−0.58	−0.49	0.26	0.35
Entire forest area					
22 August 1977 to 23 May 1984	−0.12	−0.30	−0.05	−0.01	0.26
13 June 2009 to 23 September 2011	0.30	−0.06	0.25	0.08	−0.11

(b)

Figure 12. *Cont.*

Correlation scope: Year-to-year comparison	Ecosystem type	Elevation	Aspect	Slope	Relief
Coniferous active growth					
27 June 1985 to 29 June 1994	0.02	−0.64	−0.72	−0.07	0.00
03 July 1987 to 11 July 1990	0.27	0.80	0.00	0.51	0.00
Broadleaf active growth					
27 June 1985 to 29 June 1994	0.20	0.17	0.54	−0.36	0.27
03 July 1987 to 11 July 1990	−0.28	0.28	0.48	−0.32	−0.21
Entire forest area					
27 June 1985 to 29 June 1994	0.21	0.10	−0.03	0.10	0.02
03 July 1987 to 11 July 1990	−0.09	0.38	−0.13	0.05	−0.11

(**c**)

Figure 12. (**a**) Location of smaller scale focal areas with observed predominant coniferous and predominant broadleaf active growth within the study area; (**b**) Correlation coefficients between NDGI mean and other environmental factors within a vegetation season (note that correlations in the first scene are inverse since the 1977 scene is later in the vegetation season than the 1984 scene); (**c**) Correlation coefficients between NDGI mean and other environment factors as year-to-year comparison in the same part of the vegetation season.

In addition, both in areas with broadleaf and coniferous expansion, the seasonal growth strengthens between earlier and later NDGI scenes as the expanding species grow in their new location, indicating their role as driver of overall growth (Figure 12b).

At the same time, the strong positive correlation of coniferous growth and elevation in coniferous expansion areas on a year-to-year basis between 1987 and 1990 (Figure 12c) is in line with the observed upward shift in ecosystem extent and emergence of succession in former grassland areas. The relatively strong negative correlation between dominant coniferous species and the expanding broadleaf species in the same year is in line with the change in species composition from coniferous towards mixed forest.

Annual climate differences, however, can significantly influence the year-to-year growth patterns, as shown in the NDGI comparison between early seasons of 1985 and 1994 (Figure 12c). While ERA Interim's scale is much greater than the two focus areas, it informs the understanding of observed NDGI correlation anomalies. The climate modelling shows that 1994 had virtually no precipitation in June, whereas in 1985 the peak precipitation was in June (Figure S18c–e). As a result, the much weaker growth for 1994 in the more water-deprived higher plots of coniferous forest translates to a negative correlation between year-on-year growth and elevation (and to smaller extent, aspect). A slower upward expansion of broadleaf species is also observed in the less water deprived broadleaf focus area where the dominant coniferous species have grown much stronger than the expanding broadleaf species (as evidenced by the positive correlation between the coniferous dominated ecosystem type and NDGI in the 1985–1994 scene).

Positive NDVI values prevailed throughout the study period, which indicates good conditions with the HMEs. NDVI values vary according to the phenological stage within the vegetation period; we found that they correspond to the changes of t2m parameter, which confirms the correlation between t2m and NDVI established by a number of authors—Wang et al. (2003) [97], Pavlova and Nedkov (2005) [94], Katrandzhiev (2018) [90] and Katrandzhiev and Bratanova-Doncheva (2019) [7]. The seasonal dynamics of NDVI is most obvious in the last 3D graphic (Figure S6f) which presents the dynamics of NDVI changes in 2019. At the beginning of the VP (24 April 2019) when t2m values varied within 4 °C (see Figure 9) NDVI values were negative and close to 0 due to the presence of snow cover (Figure S16a). In the middle of the VP (03 July 2019) when the t2m varied within 16.5 °C high values of the NDVI were observed (0.5–1), i.e., the amount of the leaf biomass increased and the functioning of the vegetation cover became more active. At the end of the VP (Figure S17b,e,h,j) we observe decreased NDVI values, suggesting decreased values of the t2m parameter. This confirms the expected influence of temperature on NDVI.

We assessed the selected HME condition by means of NDVI calculation based on a set of satellite data and the influence of the t2m parameter on the NDVI. We verified the results via NDWI and NDGI calculations (Figures S7–S18)

Snow cover is one of the major distorting factors in analyzing time series of vegetation indices from satellite imagery during the vegetation period, appearing as late as May/June and as early as October (Figures S7c, S8d, S14a,i, S15a and S16a in the Supplementary Materials).

Close to the beginning of the VP, the seasonal fluctuations in temperature and precipitation seem to have a strong influence on the intensity of vegetation growth. The snow in April and early May in Figures S14b,j, S15b and S16b had less impact on vegetation growth farther down from the ridge than the snow in late May at the beginning of the time series (Figures S7d and S8e). This finding corresponds to the trend of rising temperatures over the period, as well as a steepening of the temperature curves in the 1990s. It suggests that the lengthening of the vegetation season in September contributes to increasing the HME's resilience to cold spells occurring at the beginning of the vegetation season, with overall conditions remaining good despite the seasonal shift. This proves that resilient ecosystems can better mitigate climate stress.

The presence of snow in remote sensing imagery corresponds to low values of t2m from the climate model, thus proving the value of fusing these two types of data despite the lack of sufficient spatial resolution in the climate model. Temperatures vary between 6 °C in May 1984 and up to 8 °C in May 2017, and the NDVI values are correspondingly low (0–0.2 for the alpine shrub and grass vegetation, 0.2–0.4 in conifer forests and above 0.5 in the broadleaf and coppice stands in May). As a reference, NDVI is almost uniform 0.2–0.4 in most ecosystems and 0–0.2 in the shrubs closest to the snow cover. This finding confirms the usefulness of cross-referencing between the remote sensing data and climate model.

There is strong seasonal dynamics during the growing season, with similar growth patterns in June and July for all ecosystem types except the broadleaf deciduous forests: Growth has not reached its maximum and its speed distribution appears to be mostly dependent on the altitude and not so much on the ecosystem type. The leaf biomass of broadleaved deciduous forests grows faster than all other ecosystem types. At the beginning of the available observations, NDVI is in the range 0–0.2 for the alpine shrub and grass vegetation, 0.2–0.4 in conifer forests and above 0.5 in the broadleaf and coppice stands. In this period, whenever precipitation is not a constraining factor, the intensity of leaf biomass growth appears to best correlate with temperature (and grow as the temperature rises and the vegetation season possibly lengthens). This progression is visible in Figures S9b,e, S10b, S11b,f and S13e,h,k. At the end of the period, the NDVI values increased (0.6–0.8) in the dominant conifer forests in unison with the rise in temperature (~12 °C in June 2019, Figure S16g in the Supplementary materials).

The vegetation growth peaks in August and September, as seen in the higher NDVI values in the late summer thematic maps. Apart from 1977 which has low NDVI values in most ecosystems, the index varies between 0.4 and 0.6 in most territories in the study area in 1985 (Figure S8b) and between 0.4 and 0.6 for shrubs, 0.6 and 0.8 for the dominant forests and up to 1 for deciduous forests in 2019, confirming the increase in forest biomass production and carbon sequestration with warming climate.

At the end of the VP, the prevailing NDVI values vary from 0.2 to 0.4 in the higher parts and 0.4 to 0.6 in the lower part of the study area in the first available scene from October 1986 (Figure S8e,h). With growing temperatures, vegetation is more active in

October with slowdown mostly in the shrubs at the highest elevation, as evidenced in the two scenes of Figure S15h,k.

Apart from these general trends, the second part of the study period displays a number of anomalies. In 1994, the climate model shows the lowest precipitation in the entire time series, causing a break in the early season growth pattern. There is a relation between the climate model showing almost no precipitation at the beginning of the vegetation season (Figure S10f), and the NDVI readings of a slow growth in the NDVI thematic map (Figure S10e). In comparison, low precipitation in September 1992 after the high precipitation in mid-season is still a limiting factor for high NDVI, but to a lesser extent (Figure S10b,c). We observed a similar albeit less acute water limitation in 2000. The years 2009, 2011, 2016 and 2018 show an unseasonal high precipitation peak preceded by too low precipitation in the beginning of the VP (Figures S12g,n, S13g and S15m). The frequency of such occurrences suggests that they may form a new trend in shifting precipitation patterns with climate change, which needs close monitoring in the coming years. However, the existence of such a trend is not certain since 2019 displayed a double peak in precipitation (Figure S17l)—a possible stress factor for the vegetation growth. While there appears to be corresponding fluctuations of NDVI readings within just one month in Figure S17b,e,h,j, we infer that one of the scenes in 2019 may be an outlier (see Section S5 in Supplementary Materials).

In observing these fluctuations, the cross-verification of NDVI with NDWI for the same scene has a growing importance. It can be used to explain spatial differentiation in growth patterns. For example, Figure S8f shows a correlation of growth patterns with NDVI in Figure S8e but also a correlation with the climate model in Figure S8i since the precipitation modeled for early June (Figure S8c) is still visible as evaporating water reserve in the snow on the ridge. Where available, NDWI also demonstrates the spatial distribution of water content in the foliage at a much more fine-grained scale than the climate model's precipitation component (Figure S11d). In combination with NDVI and climate modeling, it can help verify time lags between precipitation and changes in vegetation growth. An example of such verification is the overall correlation of timing between high NDWI and precipitation fluctuations; e.g., the June 2009 simultaneous dip in NDWI and precipitation was followed by increased precipitation and a delayed recovery in July 2009; the opposite correlation of peak precipitation, growth and leaf water content occurred in August 2011—see Figure S12. NDWI is also prone to distortion by clouds (as shown in Figure S13d–f), and to avoid misconstruing the simultaneous dip in NDVI and NDWI as disturbance, there is a need to use improved cloud masks as these become available in newer remote sensing products, observing a longer time series over such spots or using ground measurements from meteorological stations. The usefulness of NDWI grows with the number of high quality satellite images available. For example, in 2018 the interplay of NDVI, NDWI and climate modeling shows the decrease of available water in parallel with vegetation growth in May. While there is almost no precipitation, the available moisture is absorbed by the vegetation and does not evaporate even after the rainfalls in August (Figure S15), indicating resilient ecosystem functioning. Similarly, better understanding the effect of the double dip in precipitation in 2019 is possible once we account for:

(a) the using of water from melting snow in April (higher water content despite lacking precipitation—Figure S16a,c);
(b) a slowdown of growth in August as the June rains are absorbed and no additional precipitation comes while the temperature keeps rising (Figure S17a,c), and
(c) the second surge in growth in the second half of September with the new precipitation—see Figure S17d,f,j,k.

Based on the findings of this and the previous sections, we can find ecologically meaningful links between different datasets and can go beyond the statistical analysis, therefore confirming the first hypothesis on the utility of data fusion on the semantic and methodological levels and utilizing all available data.

We furthermore achieved a co-benefit of testing the hypothesis of data extensibility by developing corresponding candidate indicators that use remote sensing data sources and address a more integrated approach to ecosystem monitoring by widening the data sources used in it. These indicators are adapted for easy remote monitoring with a minimum of ground data, thus contributing towards the ongoing monitoring of changes in ecosystem extent and condition caused by climate change. We settled for proposing for future use two linked indicators describing the ecosystem dynamics related to extent and condition of HME that are sufficiently underpinned by existing data: "Habitat extent increase/reduction attributable to climate change", and "Ecosystem succession attributable to climate change". These two candidate indicators on ecosystem condition may, after verification, become part of the methodologies for terrestrial ecosystems (Table 2) and facilitate automated monitoring of climate change impacts on HME through remote sensing. Both candidate indicators aim at complementing the existing indicators in the methodologies of the respective ecosystems that form the landscape in our study area. Adding them to the Methodological Framework is important with a view to the scientifically established acceleration of climate change.

Table 2. Candidate indicators and their parameters proposed for inclusion in the terrestrial ecosystem methodologies. Data sources: (**a**) Copernicus HRL change product or difference between status products of two consecutive releases (every three years), with spatial resolution of 20 m per pixel or below; (**b**) Difference between Copernicus HRL status products for the two ecosystem types/subtypes.

Candidate Indicators, Parameters and Units of Measurement	Ecosystem Type	Ecosystem Subtype(s)	Score				
			1 (Bad)	2 (Poor)	3 (Moderate)	4 (Good)	5 (Very Good)
(**a**) **Condition indicator**: Habitat extent increase/reduction attributable to climate change **Parameter**: Change in area (%) covered with ecosystem of 10% vegetation cover or above.	Woodland and Forest	Coniferous	>−1.7	−1.7 to −0.75	−0.75 to 1.5	1.5 to 3.5	>3.5
		Deciduous	>−9.6	−9.6 to −5	−5 to 12	12 to 25.9	>25.9
	Grassland	Alpine & Subalpine	>−17.2	−17.2 to−12	−12 to −7	−7 to 0	>0
	Heathland & Shrubs		Only one reference product layer available yet; to be filled in once HRL Small Woody Features 2018 is released				
(b) **Condition indicator**: Ecosystem succession attributable to climate change **Parameter**: Change in area, %, due to climate related succession between two ecosystem types	Woodland and forest: coniferous to deciduous		>−20	−20 to −15	−15 to −10	−10 to −5	>−5
	Grassland to Heathland and Shrubs		Only one reference product layer available yet; to be filled in once HRL Small Woody Features 2018 is released (announced for the end of 2021)				

The proposed indicators require a quite rigorous approach to distinguishing between extent change and succession caused by climate change, and similar changes caused by other factors, notably grazing and land management that are identified as the main factor influencing the speed of changes in the HME grassland–forest ecotone [5,104,105]. Therefore, they are only suitable for forward-looking remote monitoring of the spatial ecosystem extent in our study area and other HMEs with a high degree of naturalness (reusing them in intensively managed HMEs would require additional research and their adaptation to account for human influence on the ecotone and species composition). While our study area has been protected with no large-scale anthropogenic influences since 1933, the data in [77] proves the existence of land management and grazing in the late 1980s. Therefore, we take the expansion of the NATURA 2000 network in Bulgaria (legislation adopted in 2007 and extended in 2008) as the moment in which these influences were legally suspended.

To avoid biases in the indicators, establishing the causes underlying the change in extent is important. At the same time, it would be difficult or even impossible by only looking at ecosystem information since there is a need to consider a time lag after the establishment of the legal protection status of NATURA 2000. Therefore, we verified the lack of disturbances in the ecosystem by other means available, including:

- scientific publications and official documents: The orders determining the management regimes for the Parangalitsa reserve (in force during our entire study period) and NATURA 2000 protected areas for the rest of our study area (in force since 2007, extended in 2008)—both stipulating the seizing of economic activities;
- georeferenced sources: Bondev [77] for the land use and management regimes in the 1980s as well as the subsidy eligibility layers for grassland management (since 2015);
- agricultural data to confirm that the alpine and subalpine grasslands are largely undisturbed after this date and any occurring grazing is extensive (therefore, not significantly influencing the landscape).

In this context, the use of semantically consistent data fusion allows the deduction that the main driver of succession in this landscape since 2008 is climate change.

Based on the above, we can evaluate the parameters for the two candidate indicators only for periods of proven lack of anthropogenic disturbances for at least 8–10 years of undisturbed ecosystem development to ensure that the dynamics of ecosystem processes is exclusively attributable to climate change. This means that, in our study, only the latest part of the study timeframe is useful for deriving reference values for our candidate indicators—a minimum of 5 to 10 years without disturbance, that is from 2015 onwards for the entire study area.

We selected the reference values of the parameters measured to contain easily available but quality controlled remote sensing and expert products. In this manner, changes in data sources will be easily traceable through the metadata of the expert product versions and a reanalysis in the expert products performed by their publisher can be monitored and trigger a reanalysis of our indicators as well.

We estimate the initial values in Table 2 having in mind the observed speed of ecosystem change in Copernicus products produced for two base years, by ecosystem type. However, as more such products will become available in the form of continuous data series, these values will have to be reviewed and verified before including them in the Methodological Framework. The availability of new data or satellite products is very likely to prompt an adjustment of reference percentage values, especially for the proposed succession indicator (red in Table 2). We expect that with establishing the Long-Term Ecosystem Research site Parangalitsa (whose infrastructure development started in 2021), better ground data on biodiversity and abiotic heterogeneity will also become available over time and will be conductive in establishing the proper reference values for these candidate indicators.

Having in mind the observed dynamics in ecosystem condition (Section 3.1.2) and the observed NDVI values in our time series, we propose a new, functional parameter to complement the largely static IP index of the terrestrial ecosystems where a detailed field inventory is not possible (Table 3). A correlation analysis with the IP index calculated using the forest database or a future forest inventory will allow for finding outliers and reducing uncertainty, for a more robust remote monitoring of ecosystem conditions. In this manner, it can also semantically and ontologically connect the ecosystem condition and ecosystem service indicators in the methodologies.

This candidate indicator requires additional verification, including some or all of the following: Cross-verification with NDWI and climate data time series (temperature, precipitation); error correction via NDGI; field verification: Vegetation inventories and IP index if it becomes available for NATURA 2000. In addition, a more precise localization of heathland and shrubs is a prerequisite for determining the missing reference values for this ecosystem type since it has very few polygons in the forest database but according to the Copernicus HRL products is likely to be one of the most dynamically developing ecosystem types.

Table 3. Candidate indicator for remote ecosystem monitoring of the terrestrial ecosystems. Data source: Copernicus and Landsat remote sensing. Due to the very limited extent of heathland and shrubs, we were not able to derive reliable reference values of all reference states and left these reference values empty for further research and verification.

Candidate Indicators, Parameters and Units of Measurement	Ecosystem Type	Ecosystem Subtype(s)	Score				
			1 (Bad)	2 (Poor)	3 (Moderate)	4 (Good)	5 (Very Good)
Condition indicator: Ecosystem functioning **Parameter**: Peak vegetation season NDVI	Woodland and Forest	Coniferous	−1–0	0–0.2	0.2–0.4	0.4–0.6	0.61–0.76
		Deciduous	−1–0	0–0.2	0.2–0.4	0.4–0.6	0.68–0.84
	Grassland	Alpine & Subalpine	−1–0	0–0.2	0.2–0.4	0.4–0.6	0.69–0.87
	Heathland & Shrubs						0.37–0.77

3.2. *Analysis of Climate Parameters and Cross-Validation with Vegetation Indices*

We analyzed the climate parameters 't2m' (temperature 2 m above ground level), 'tp' (total precipitations), 'evpt' (evaporation) and 'v10' (10 m V wind component) (Table S5 in the Supplementary Materials) and plotted graphics of their dynamics for the entire period of available ERA Interim data (40 years) within the vegetation periods (VP). Based on the results by climate parameter, we could establish a trend of increasing values of the t2m parameter during the months May–July and September for the full 40-year period. Only for August, we found a trend of decreasing t2m values (Figure S3 in Supplementary Materials).

In order to test if there is an acceleration in climate change, we divided the 40-year period of ERA Interim data into two shorter periods: (1) 1979–1999 and (2) 1999–2019 (except for September parameters that are not available for 2019 in ERA Interim and therefore end in 2018). In each half period, we calculated the mean values of t2m given the minimal and maximal values for each month within the VP (Table S5 in Supplementary Materials). We observe trends of increasing mean temperatures in May–July and September during the studied period. In the second period, we observed increased mean values of the t2m parameter compared to the first period—for May by 2.2 °C, June by 1.1 °C, July by 0.9 °C and September by 1.2 °C. The only exception is August when we observe a trend towards decrease in the mean values of t2m by 0.6 °C for the study period (Table S6 in Supplementary Materials). We carried out Pearson correlation analysis between the NDVI and the t2m parameter in the study area which revealed sufficiently strong correlation between these variables (correlation coefficient- R^2 = 0.605). The correlation table for this analysis is available in Table S5 in the Supplementary Materials.

The review and analysis of the annual climate plots reveal that, apart from the overall rise in t2m, high temperatures occur about a month earlier at the end of the period as compared to 1977, 1984 and 1985, suggesting a lengthening of the vegetation period. Confirmation by studying even longer-term observations and/or the new Copernicus Phenology HRL product when it becomes available for our study area appears necessary to confirm this conclusion as it may provide insights into the drivers of vegetation change.

Based on regression analyses and temporal dynamics analysis (Figures S1 and S2), we establish trends of tp decline in May (Figure S2a) and June (Figure S2b) within the studied period. In addition, we observe trends of tp increase in July (Figure S2c), August (Figure S2d) and September (Figure S2e) for the same period within the study area. In order to test the existence of a relationship between the NDVI and tp, we conducted Pearson correlation analysis (Table S5) between them in the study area. We found R^2 = 0.0228 and p = 0.907 (i.e., $p > 0.050$). We established negative correlations for both NDVI–evpt and NDVI–v10, with correlation coefficients −0.275 and −0.0184, respectively. Therefore, we did not find definite trends of influence of the evpt and v10 in the NDVI. The findings of Section 3.1.1, however, suggest that such trends may well exist in some parts of the study area where other abiotic factors influence a strong dynamic of the ecosystem development/succession.

In Figures S4 and S5, we present the change dynamics of these parameters. When compared to the changes in the NDVI values throughout the study period, we were not able to establish a connection. Therefore, in the current study we establish that cross-analyzing the influence of the t2m parameter on the NDVI and HME conditions yields the best results as it is the only parameter showing strong correlation to the NDVI ($R^2 = 0.605$).

This semi-qualitative analysis is in line with the complicated interplay of environmental factors that limit vegetation growth in different parts of the vegetation season; climate parameters are only part of these factors. Among climate variables, temperature seems to be the most obvious overall limitation to vegetation growth, while other parameters may influence it in different years without a clear trend.

We performed correlation analysis of NDGI and local parameters from the forestry database using the *corrplot* library in R (Figure S19 of the Supplementary Materials). The overall weak correlations between any two parameters taken across the whole study area is ecologically meaningful since it confirms the complex interrelations between growth and microclimatic environmental parameters, in line with the findings of Section 3.1.1 about the microclimate's importance. It also reveals a scale mismatch for some of the possible correlations as they likely occur only in parts of our study area, as suggested by the data fusion reanalysis in Section 2.3.4. On the current scale of observation, the higher correlations would exist only in areas where favorable microclimatic conditions influence the vegetation dynamics, whereas large areas with relatively homogeneous microclimate or vegetation or weaker changes in vegetation growth (typically alpine grasslands or shrubs) yield weaker signals and therefore weaken the overall correlation. This is visible in Figure S19b, where the strongest observed correlations are to parameters determining the microclimate—elevation and slope. In contrast, strong climate fluctuations lessen the impact of local microclimate—see Figure S19c for year-to-year comparison and Figure S19c for the seasonal comparison. As such, the sensitive NDGI must be used at smaller scale for cross-validation of both NDVI and NDWI.

As to the nature of microclimatic determinants, the spatial analysis of Figure 9 strongly suggests the need for more detailed, field-measured climate data in focal points of particularly dynamic ecosystem change, specifically the southern facing slopes in the higher altitudes and the forest—alpine grassland ecotone, especially in locations with established succession. Such targeted data collection would enable a georeferenced and hopefully much more precise correlation analysis.

4. Discussion

The present study in some of its parts repeats the methodology of earlier work of some of the authors in the same study area [7]. However, it is much wider in terms of data integration and consistent application of the Whole System approach. To our knowledge, this is the first study that makes use of the ontological and semantical power of structured indicator–parameter systems as defined in the Bulgarian Methodological Framework [26] and, in this sense, also contributes to the worldwide research on the Whole System application. A summary of the results follows in this part.

4.1. Methodological Results

With a view to ontological similarity defining one-to-many relations between ecosystems and habitats, forestry data and our field verification (Section 2.3.4), we were able to narrow down the ontological similarity on level 3 (ecosystem subtype to habitat) for the forest ecosystems in the study area by verifying the types of habitats occurring in the forest ecosystems (Figure 7, middle). Due to insufficiently granular data about the other ecosystem types, we were only able to verify the level 2 relation for the other ecosystems present.

4.2. *Hypothesis Verification*

Our study confirms both the working hypotheses we presented in the introduction. In support of our first hypothesis:

- We confirm the semantic consistency of the Methodological Framework by using new data sources in a manner consistent with the Whole System approach. Based on this work, we propose an ecologically meaningful extension by adding a candidate indicator set on climate change impact. These indicators are of particular importance for the sensitive HME. They are balanced on a landscape level and reflect the trade-offs between the extent and condition of different ecosystems in the course of climate change-induced succession since in the limited habitat the expansion of one ecosystem type is at the expense of another. In addition, we explore the scale of observation and confirm its importance in reducing uncertainty when a dataset's scale (in our case, ERA Interim) significantly exceeds the characteristic scale of the object of observation.
- We use an ontologically consistent approach to utilize data collected for different purposes (in this case use habitat data) for verification of a Level 3 crosswalk for forest ecosystems. In addition, through this crosswalk field data on habitats can be useful in conjunction with vegetation indices for a future remote monitoring of forest areas where the dominant vegetation type consists of climate-vulnerable species. This allows for focusing our limited fieldwork resources on problem spots.
- The production pipeline of Copernicus remote sensing products delivered by the European Environment Agency utilizes significant human and financial resources to ensure their methodological consistency within a single information infrastructure; they also undergo rigorous quality checking and control of both data and metadata. Incorporating them in a future remote monitoring within the Bulgarian Methodological Framework would therefore add to the strong methodological and information-technical consistency of the Whole System approach with very low additional costs for exploring data poor biodiversity hotspots like our study area. This seamless incorporation of a new data source as a basis for new candidate indicators further confirms the extensibility clause of our first hypothesis. Data fusion and co-analysis form the basis to confirm the possibility to derive ecologically meaningful information from all available data sources.
- We use data fusion also to cross-check and verify data sources, thus reducing uncertainty and supporting our hypothesis that applying the Whole System approach allows for more reliable data integration. In the course of our research, we confirm the suitability of the selection of vegetation indices and their combination with expert remote sensing products, climate models, publications or other non-georeferenced documents and field data as a set of mutually complementary data sources that allows confirming synergies and identifying outliers.
- Beyond the expected results, we further found that using data integration:
    - allows the additive introduction and use of a higher number of very diverse data sources for a more reliable monitoring of both the ecosystem extent and conditions in data sparse environments;
    - supports finding the appropriate observation scale for different scientific questions; and
    - enables extending the holistic approach of the Methodological Framework to accommodate for future ecologically meaningful linking of ecosystem condition and ecosystem services in the sense of natural capital accounting principles that can replace their current assessment within the Methodological Framework. This, in turn, underpins the use of the Whole System approach in a socio-ecological context.

The analysis of fused data also supports our second hypothesis. We were able to observe the impact of climate change on both ecosystem conditions and functioning, and the landscape level changes in the spatial distribution of different ecosystems and the species composition. Based on the conceptual and instrumental advantages of the Whole System approach, the selected toolbox allows for complex monitoring that goes beyond the observation of species, ecotones or single ecosystems.

In terms of ecosystem extent, our study suggests that climate change may influence the ecosystem succession. A longer time series may confirm an acceleration as suggested by NDGI observations of early succession signals in the 1990s and the more rigorous recent changes found in Copernicus HRL data. In such cases, regular ecosystem extent monitoring may identify the possible approaching of tipping points in extent reduction or species composition for some ecosystems. For the grassland ecosystems in our study area, such speeding decline may also influence the overall ecosystem resilience and reduce the habitat of endemic species. Should the observation of a longer time series confirm such accelerating trends, the observation frequency for some ecosystem types might also need reconsidering in the Monitoring Guide [76] of the Methodological Framework.

On ecosystem condition and functioning, we made a number of observations.

We established the correlation between vegetation indices and the changes of selected climate parameters (temperature, t2m; total precipitation, tp; evaporation, evpt; and 10 m V wind component, v10) throughout the 40-year period—from 1979 to 2019—within the study area. We found an overall increase in the t2m parameter during the vegetation period between the start and end years of observation; with the exception of August the values of the t2m decrease. In addition, based on the regression analysis, we found an overall increase of the tp parameter. These trends suggest that climate change challenges the resilience of HME ecosystems, which are increasingly subject to dryer and hotter weather during the vegetation peak seasons. At the same time, lengthening the vegetation season contributes to increase in biomass production and overall resilience. Seasonal shift in phenology is important to observe closely, as revealed by the NDGI correlation analysis (Section S5 in Supplementary Materials).

With regard to the forests, we could not establish conclusive long-term links between tp, evpt and v10 parameter change and the NDVI changes over the entire study area. The results of detailed analysis in different years suggest that the limiting factor to growth changes depending on the combination of climatic variables and can be precipitation (years 1994, 2000) or temperature (years 1986, 2009). The only direct connection and dependence we were able to confirm is the connection between the t2m parameter and NDVI. Overall, the interplay and combined influence of both the t2m and tp on the NDVI across the study area confirms the findings of previous studies.

At the same time, the spatial analysis of species change patterns in the forest parts of our study area suggest the importance of microclimatic parameters, in particular elevation and slope. This finding points out the need for meteorological ground data and smaller scale data fusion to verify the importance of these factors in the conditions of climate change.

We observed seasonal and annual dynamics arising from these interrelations in the NDVI value distribution in the main ecosystem types and forest ecosystem subtypes, forming the composition of the studied HME. The increasing temperature and total precipitation in the last two decades appear to have led to an increase in the NDVI values, suggesting an improvement in the vegetation functioning and conditions through the last 20 years, as well as changes in species composition accompanied by growth acceleration. The subalpine and alpine grassland and shrub ecosystems may be more susceptible to climate parameter changes compared to the forest ecosystems. Since grassland ecosystem dynamics develop at a higher speed, the collection of time series for these sensitive ecosystems is necessary to facilitate better remote monitoring. On a subtype level, conifer forest ecosystems show resilience to changes in climate parameters and remain functionally stable; there is also a trend of succession towards higher altitudes. The more vigorous vegetation process that we observed in the broadleaf and coppice forests is an indication of future changes in their distribution to higher habitats atypical for them, thus influencing the species composition. We furthermore found indications that the changes in climate parameters led to a longer vegetation season, most clearly seen in 2019. Future phenological observations and analyses are necessary to confirm and specify these trends.

By introducing a rigorous approach to extent and condition observations, our study also creates a basis for further research on the HME's capacity to provide ecosystem services related to the primary production ecosystem function along with the new UN System of Environment Economic Accounting (SEEA) standard. Further analyses based on a combination of field observations of the energy budget and remote sensing are necessary to specify and assess the dynamics of these ecosystem services. A more detailed study of some key services such as pollination will also require additional holistic methods such as pollinator efficiency studies [106–108], observation of environmental stress [109] or eDNA [110,111] to establish a better overview of species composition and within our large study area.

The evaluation of vegetation indices confirms the resilience of the HME ecosystems, whereas the spatial analysis using HRL products demonstrates spatial changes at the landscape level. Since the study area is protected and human intervention is minimal, we expect that climate change will remain the main driver for most of the developments in quickly developing ecosystems (grassland, sparsely vegetated areas). As tree growth is much slower than the retreat of grasslands, we expect these ecosystems' dynamic to benefit from increased observation frequency enabled by the better availability of remote sensing imagery. While both forest and grassland ecosystems show increasing growth dynamics as climate change accelerates (as confirmed by joint analysis of the Copernicus HRL products for the end of our observation period), the relatively late appointment of NATURA 2000 protection for all of our study area means that both succession and change in species composition started before its full protection. Therefore, we cannot fully attribute the succession in the tree line ecotone to climate change for the period 1990–2015. Nonetheless, we can conclude that:

- Observing the gaps between the Copernicus Grassland and Tree Cover Density HRL products is a good way to localize succession areas of particular monitoring interest for both future fieldwork and observation of the newly formulated extent and condition candidate indicators for climate change impact;
- The use of new remote sensing products will allow for a finer grained monitoring. This, in turn, would enable downscaled monitoring, the early identification of potential tipping points and problematic areas resulting from climate change-induced disturbances such as storms, hails or pests. Such upcoming potentially useful new products to add to our remote sensing portfolio are the small woody features for remote localization of heathland and shrub ecosystems and phenology for better tracing the effects of climate change on the ecosystem.
- Changes in forest ecosystems are slower but still observable over longer timescales. Due to the slower processes in forests, both the impact of earlier anthropogenic activity and the results of natural disasters are of stronger local influence. As they require more accurate and frequent observation, the introduction of Synthetic Apreture Radar (SAR) products for monitoring in cloudy days and the upcoming launch of hyperspectral Copernicus missions are prospective important directions for enhancing the remote sensing parameter portfolio of the Methodological Framework.
- Together, these directions will largely enhance the toolbox available for the monitoring of climate change effects on HME. They will be better suited to inform and support:
- future directions for targeted fieldwork,
- scientific products delivered through the European Long-Term Ecosystem Research Network, and
- the implementation of national and regional climate change adaptation policies.

5. Conclusions

The current paper presents, for the first time in a Bulgarian high-mountain ecosystem, a Whole System approach to long-term (42-year period) assessment of the response of the worldwide sensitive HME ecosystems to changes in climate parameters. We demonstrate the usefulness of a toolset of complementary approaches: Selected indices from satellite-derived data (NDVI was verified by the NDWI and the NDGI), HRL products and cross-validation with climate modeling data and ground observations/inventories as an integral part of the Whole System approach.

Studying long time series also presents a number of challenges and uncertainties. In our study area; these consisted in:

- Technical constraints of the sensors used in the earlier years. Since missions of different space agencies overlap, the granularity of available imagery varies with the imaging satellite's sensors, sometimes even within the same year or month. Thus, a source of uncertainty is the need to intercallibrate data series, e.g., between different Landsat sensors and Sentinel—which is only partially alleviated by the use of NDGI.
- Uniform and precise climate modeling is not available for the first years of our study period. The ERA Interim data series started in 1985 and was discontinued in August 2019; the replacing dataset ERA 5 was not available for the entire period at the time of data processing. Using ERA 5 in later studies requires careful cross-checking of available data and identifying (where possible—also assessing) the cross-calibration uncertainties and errors.
- A dearth of suitable satellite images due to the shorter vegetation season and the frequent occurrence of clouds. Therefore, our dataset is imbalanced towards the last years when Landsat was complemented by Sentinel imagery and both the frequency and quality of available image data, as well as the number of vegetation indices retrievable from the new sensors, increased.
- Uncertainty of data in the forestry database. The existing forestry database has no QA information either in the published official data collection guidelines or in the dataset itself. Furthermore, the relatively limited data on species composition suggests limitations in the scope of field surveys over the years.
- We found significant data and research bias towards studying forest ecosystems and, consequently, could not perform the same quality analysis on the grassland, shrubs and sparsely vegetated ecosystems in the study area. To be able to repeat the study uniformly within the entire area of interest, a more detailed mapping, aimed at filling the spatial gaps on an extended study area (yellow colored parts from the outline in Figure 1), would be necessary.
- The great scale disparity of data sources at the appropriate observation scale for our study object limits the current climate data fusion approach to semi-qualitative observations.

All of the above, along with the incorporation of new remote sensing products as specified in the Discussion section above, presents a number of instrumental research directions to improve the methodology of our study as new methods and products become available.

At the same time, this toolbox of complementary methods allows for specifying new ecological research questions that present a wide field for future work:

- Exploring a denser time series of satellite imagery and the corresponding ERA 5 climate parameters along with field data would be conductive to determining the best time slots for reliably identifying ecosystem types and subtypes depending on the best correlations with other environmental factors during the vegetation period.
- Data fusion involving climate models requires additional research in finding the appropriate observation scales of complementary ground and remote sensing data.
- The joint analysis also allows for hypothesizing on a shift in the extent and upper border of the forest ecotone, as evidenced by the surge in the vegetation growth of shrubs (containing coniferous trees) located in the highest parts of the study area. Another key direction, therefore, is the regular observation of phenology shifts and seasonal differences between the reflectance of ecosystem types/subtypes, with broadleaf forests being easiest to spot remotely within the vegetation season.

- Focused dendrochronological studies, in areas where persistent changes in ecosystem extent, conditions or species composition are detected, are necessary to support the automation of monitoring through machine learning and AI.
- Due to the sensitivity of NDGI, a correlation between the NDVI variation and disturbances (e.g., windthorows) in much smaller spots as described by Panayotov [81] in some parts of our study area could be the object of further research.
- With a view to the delay in mapping and assessment of ecosystems within Natura 2000, another important research direction is the field verification and detailed mapping of the location of ecosystems other than forest, by incorporating a wider toolset of field methods such as the ones mentioned in the Discussion section above.
- The assessment of the provisioning capacity for ecosystem services related to the biomass is likely to become easier as more and better quality remote sensing imagery becomes available in conjunction with field data using new standard observation methods.
- The use of climatologic publications may prove useful for cross-checking climate model projections for longer time series before 1985.
- Downscaling of ERA Interim/ERA 5 or obtaining finer grained climate projections—or even better, targeted collection of field data on climate variables—would be beneficial for a geospatially detailed correlation analysis of factors determining changes in the ecosystems. This, in turn, would yield better understanding of the drivers of change over the study area which has significant variation in its relief, slope and elevation. Until such data are available, coping with datasets whose resolution is much coarser than the characteristic scale of HMEs remains an important research direction.
- Scaling up our approach to data integration to cover entire landscapes at a national or regional scale (wall-to-wall mapping).
- Exploring the use of fuzzy logic and fuzzy graphs for automating the data processing.

The greatest research challenge is undoubtedly the smooth and incremental integration of these future work directions in an automated and holistic manner beyond the statistical analysis. The speed with which new open data and research tools become available will make the shift towards machine learning and artificial intelligence an imperative next step for near-real time processing of ensembles containing ground data, climate models and satellite imagery in order to better assess the response of HME to climate change.

Supplementary Materials: The following supporting information can be downloaded at: https://www.mdpi.com/article/10.3390/d14040240/s1: Tables: Table S1: Whole System condition indicator set grouped by subsystem of the ecosystem as defined in Methodological Framework, 2017; Table S2: Cross-reference of indicators between ecosystem types—Biodiversity subsystem; Table S3: Measurement methods for ecosystem parameters according to the Whole System approach as per the Bulgarian Methodological Framework for mapping and assessment of ecosystem condition and services (Methodological Framework, 2017). Preferred single methods are marked in bold. Models (in italic) are only applicable for areas with sufficient and consistent data; Table S4: Satellite data sources. Note that Landsat images before Landsat 7 TM have a coarser grid; Table S5: Data collection of the NDVI and climate parameter values used for Pearson Correlation analyses and correlation coefficients; Table S6: Changes in t2m parameter values during the months May, June, July and August for the period 1979–2019 and 1979–2018 for September. Figures: Figure S1: Long-term trends in tp parameter during the vegetation season: (a) May; (b) June; (c) July; (d) August, and (e) September; Figure S2: Long-term linear regression trends of tp by month: (a) May; (b) June; (c) July; (d) August, and (e) September; Figure S3: Long-term trends in monthly temperatures in the active vegetation months (May to September); Figure S4: Graph of evpt by month: (a) May; (b) June; (c) July; (d) August, and (e) September; Figure S5: Long-term graph of v10 by month: (a) May; (b) June; (c) July; (d) August, and (e) September; Figure S6: 3D graphics of the NDVI values for the period 1977–2019: a,b—in the beginning; c,d—in the middle; e,f—in the end of the period; Figure S7: Data collection for vegetation periods in 1977 and 1984: (a,c,e)—satellite images; (b,d,f)—TM of NDVI: (b) The NDVI values were above 0 all over the study area and the advanced vegetation phase (i.e., leaf biomass production), which suggests overall good functionality of the HME; (g)—diagram of climate parameters dynamics. Due to the lack of t2m data, for 1977 we only show NDVI data; Figure S8: Data collection for vegetation period in 1985 and 1986: (a,d,g)—satellite images; (b,e,h)—TM of NDVI; (f)—TM of NDWI; (c,i)—diagrams of climate parameters dynamics; Figure S9: Data collection for vegetation periods in 1987 and 1990: (a,d)—satellite images; (b,e)—TM of NDVI; (c,f)—diagrams of climate parameters dynamics; Figure S10: Data collection for vegetation periods in 1992 and 1994: (a,d)—satellite images; (b,e)—TM of NDVI; (c,f)—diagrams of climate parameters dynamics; Figure S11: Data collection for vegetation period in 2000: (a)—satellite image; (b)—TM of NDVI; (c)—diagram of climate parameters dynamics; (d)—TM of NDWI; Figure S12: Data collection for vegetation periods in 2009 and 2011: (a,d,h,k)—satellite images; (b,e,i,l)—TM of NDVI; (c,f,j,m)—TM of NDWI; (g,n)—diagrams of climate parameters dynamics; Figure S13: Data collection for vegetation periods in 2012 and 2016: (a,d)—satellite images; (b,e)—TM of NDVI; (f)—TM of NDWI; (c,g)—diagrams of climate parameters dynamics; Figure S14: Data collection for vegetation period in 2017: (a,f,i)—satellite images; (b,d,g,j)—TM of NDVI; (c,e,h)—TM of NDWI; (k)—diagram of climate parameters dynamics; Figure S15: Data collection for vegetation period in 2018: (a,d,g,j)—satellite images; (b,e,h,k)—TM of NDVI; (c,f,i,l)—TM of NDWI; (m)—diagram of climate parameters dynamics; Figure S16: Data collection for vegetation period in 2019-part 1: (a,d,g,j)—satellite images; (b,e,h,k)—TM of NDVI; (c,f,i,l)—TM of NDWI; Figure S17. Data collection for vegetation period in 2019-part 2: (a,d,g,i)—satellite images; (b, e, h, j)—TM of NDVI; (c,f,k)—TM of NDWI and diagram of climate parameters dynamics (l); Figure S18: Relation between NDGI and the respective climate conditions: (a) TM of NDGI between 22/08/1977 and 23/05/1984; (c) TM of NDGI between 03/07/1987 and 11/07/1990; (f) TM of NDGI between 27/06.1985 and 29/06.1994 (i) TM of NDGI between 13/06/2009 and 23/09/2011; (l) TM of NDGI between 12/08/2019 and 13/09/2019; (b,d,e,g,h,j,k,m)—diagrams of climate parameter dynamics for 1984, 1987, 1990, 1985, 1994, 2009, 2011 and 2019, respectively.

Author Contributions: Corresponding author K.K. has substantial contributions to data collection and pre-processing and partially to the results interpretation; the analytical work and interpretation of data and results was carried out by K.G.; S.B.-D. contributed to conceptual design of the current study. All authors have read and agreed to the published version of the manuscript.

Funding: (1) Data collection: National program "Young scientists and Postdoctoral candidates", contract No. 14/01.04.2019 at the Ministry of Education and Science (MES) of Bulgaria. (2) Funding of this publication: "LTER—BG: Upgrading of the distributed scientific infrastructure Bulgarian Long-Term Ecosystem Research Network" under agreement D01-405/18.12.2020 with the Ministry of Education and Science (MES) of Bulgaria. (3) Data analysis and ontology assignment: National Science Program "Environmental Protection and Reduction of Risks of Adverse Events and Natural Disasters", Resolution of the Council of Ministers No. 577/17.08.2018, Agreement No. D01-230/06.12.2018 at the Ministry of Education and Science (MES) of Bulgaria.

Institutional Review Board Statement: Not applicable.

Data Availability Statement: The data produced in this study is contained in the Supplementary material and the underlying datasets can be requested from the lead author. The datasets maintained by Bulgarian authorities that were provided for this research are copyright of the respective agencies and can be obtained subject to permission according to their public data sharing regulations under Bulgarian law.

Acknowledgments: This study's data collection was carried out thanks to National program "Young scientists and Postdoctoral candidates"; the data processing workflow for remote sensing data to produce vegetation indices was continuously supported by the colleagues from "Aerospace Information Department", Space Research and Technology Institute at The Bulgarian Academy of Sciences, in particular their late director, Roumen Nedkov who generously shared their cutting edge research. The second author obtained substantial understanding of the production pipeline and specifics of Copernicus products during her Blue Book internship at the European Environment Agency in March–July 2018 and the excellent tutoring received there. This experience was instrumental for the work on ecosystem extent and links to ecosystem services provision. The analytical work was informed and inspired by the ongoing inventory of diverse data types and classifications. The discussions with colleagues at IBER about the specifics of an IT architecture that could bridge the gaps between them are very important for the methodological and information compatibility concepts developed here. Finally yet importantly, the concept of the Methodological Framework is a further development of the Whole System approach—which was developed within the successful project EnvEurope (https://www.lter-europe.net/news/enveurope-best-award (accessed on 17 March 2022)). Our understanding of the Whole System has evolved immensely since then and through continuous networking in the vivid communities of the International and European Long-Term Ecological Research Networks (ILTER and eLTER) in the course of several past and ongoing projects including the project funding this publication.

Conflicts of Interest: The authors declare no conflict of interest.

References and Note

1. Pettorelli, N.; Schulte to Bühne, H.; Tulloch, A.; Dubois, G.; Macinnis-Ng, C.; Queirós, A.M.; Keith, D.A.; Wegmann, M.; Schrodt, F.; Stellmes, M.; et al. Satellite remote sensing of ecosystem functions: Opportunities, challenges and way forward. *Remote Sens. Ecol. Conserv.* **2017**, *4*, 71–93. [CrossRef]
2. Smith, W.K.; Germino, M.J.; Johnson, D.M.; Reinhardt, K. The Altitude of Alpine Treeline: A Bellwether of Climate Change Effects. *Bot. Rev.* **2009**, *75*, 163–190. [CrossRef]
3. Batllori, E.; Blanco-Moreno, J.M.; Ninot, J.M.; Gutie'rrez, E.; Carrillo, E. Vegetation patterns at the alpine treeline ecotone: The influence of tree cover on abrupt change in species composition of alpine communities. *J. Veg. Sci.* **2009**, *20*, 814–825. [CrossRef]
4. Smith, W.; Germino, M.; Hancock, T.H.; Johnson, D. Another perspective on altitudinal limits of alpine timberlines. *Tree Physiol.* **2003**, *23*, 1101–1112. [CrossRef]
5. Gehrig-Fasel, J.; Guisan, A.; Zimmermann, N. Tree line shifts in the Swiss Alps: Climate change or land abandonment? *J. Veg. Sci.* **2007**, *18*, 571–582. [CrossRef]
6. Holtmeier, F.-K.; Broll, G. Treeline advance—Driving processes and adverse factors. *Landsc. Online* **2007**, *1*, 1–33. [CrossRef]
7. Katrandzhiev, K.; Bratanova-Doncheva, S. Spatial Distribution of High-Mountain Ecosystems—Application of Remote Sensing and GIS: A Case Study in South-Western Rila Mountains (Bulgaria). *Silva Balc.* **2019**, *20*, 57–69. Available online: https://www.researchgate.net/profile/Kostadin-Katrandzhiev/publication/339948096_SPATIAL_DISTRIBUTION_OF_HIGH-MOUNTAIN_ECOSYSTEMS_-_APPLICATION_OF_REMOTE_SENSING_AND_GIS_A_CASE_STUDY_IN_SOUTH-WESTERN_RILA_MOUNTAINS_BULGARIA/links/5e6f51c792851c6ba7066c97/SPATIAL-DISTRIBUTION-OF-HIGH-MOUNTAIN-ECOSYSTEMS-APPLICATION-OF-REMOTE-SENSING-AND-GIS-A-CASE-STUDY-IN-SOUTH-WESTERN-RILA-MOUNTAINS-BULGARIA.pdf (accessed on 17 March 2022).
8. Lindner, M.; Maroschek, M.; Netherer, S.; Kremer, A.; Barbati, A.; Garcia-Gonzalo, J.; Seidl, R.; Delzon, S.; Corona, P.; Kolström, M.; et al. Climate change impacts, adaptive capacity, and vulnerability of European forest ecosystems. *For. Ecol. Manag.* **2009**, *259*, 698–709. [CrossRef]
9. Cudlín, P.; Klopčič, M.; Tognetti, R.; Máliš, F.; Alados, C.L.; Bebi, P.; Grunewald, K.; Zhiyanski, M.; Andonowski, V.; la Porta, N.; et al. Drivers of treeline shift in different European mountains. *Clim. Res.* **2017**, *73*, 135–150. [CrossRef]
10. Körner, C.H. A re-assessment of high elevation treeline positions and their explanation. *Oecologia* **1998**, *115*, 445–459. [CrossRef]
11. Grace, J.; Berninger, F.; Nagy, L. Impacts of Climate Change on the Tree Line. *Ann. Bot.* **2002**, *90*, 537–544. [CrossRef]
12. Körner, C.H.; Paulsen, J. A world-wide study of high altitude treeline temperatures. *J. Biogeogr.* **2004**, *31*, 713–732. [CrossRef]
13. Harsch, M.A.; Hulme, P.H.E.; McGlone, M.S.; Duncan, R.P. Are treelines advancing? A global meta-analysis of treeline response to climate warming. *Ecol. Lett.* **2009**, *12*, 1040–1049. [CrossRef]
14. Leonelli, G.; Pelfini, M.; Morra di Cella, U.; Garavaglia, V. Climate Warming and the Recent Treeline Shift in the European Alps: The Role of Geomorphological Factors in High-Altitude Sites. *AMBIO* **2010**, *40*, 264–273. [CrossRef]
15. Walther, G.R.; Beißner, S.; Pott, R. Climate change and high mountain vegetation shifts. In *Mountain Ecosystems*; Springer: Berlin/Heidelberg, Germany, 2005; pp. 77–96. [CrossRef]
16. Ruiz, D.; Moreno, H.A.; Gutiérrez, M.E.; Zapata, P.A. Changing climate and endangered high mountain ecosystems in Colombia. *Sci. Total Environ.* **2008**, *398*, 122–132. [CrossRef]
17. Körner, C.; Kèorner, C. Alpine Plant Life: Functional Plant Ecology of High Mountain Ecosystems. 1999. Available online: https://link.springer.com/book/10.1007/978-3-642-18970-8 (accessed on 17 March 2022).
18. Grabherr, G.; Gottfried, M.; Gruber, A.; Pauli, H. Patterns and current changes in alpine plant diversity. In *Arctic and Alpine Biodiversity: Patterns, Causes and Ecosystem Consequences*; Springer: Berlin/Heidelberg, Germany, 1995; pp. 167–181. [CrossRef]
19. Makarieva, A.M.; Gorshkov, V.G.; Li, B.L. Precipitation on land versus distance from the ocean: Evidence for a forest pump of atmospheric moisture. *Ecol. Complex.* **2009**, *6*, 302–307. [CrossRef]
20. Makarieva, A.M.; Gorshkov, V.G.; Li, B.L. Revisiting forest impact on atmospheric water vapor transport and precipitation. *Theor. Appl. Climatol.* **2013**, *111*, 79–96. [CrossRef]
21. Ball, L.; Tzanopoulos, J. Interplay between topography, fog and vegetation in the central South Arabian mountains revealed using a novel Landsat fog detection technique. *Remote Sens. Ecol. Conserv.* **2020**, *6*, 498–513. [CrossRef]
22. Su, Q.; Sun, L.; Di, M.; Liu, X.; Yang, Y. A method for the spectral analysis and identification of Fog, Haze and Dust storm using MODIS data. *Atmos. Meas. Tech. Discuss.* **2017**, 1–20. [CrossRef]
23. Skok, G.; Žagar, N.; Honzak, L.; Žabkar, R.; Rakovec, J.; Ceglar, A. Precipitation intercomparison of a set of satellite-and raingauge-derived datasets, ERA Interim reanalysis, and a single WRF regional climate simulation over Europe and the North Atlantic. *Theor. Appl. Climatol.* **2016**, *123*, 217–232. [CrossRef]

24. Solman, S.A.; Sanchez, E.; Samuelsson, P.; da Rocha, R.P.; Li, L.; Marengo, J.; Pessacg, N.L.; Remedio, A.R.C.; Chou, S.C.; Berbery, H.; et al. Evaluation of an ensemble of regional climate model simulations over South America driven by the ERA-Interim reanalysis: Model performance and uncertainties. *Clim. Dyn.* **2013**, *41*, 1139–1157. [CrossRef]
25. Szczypta, C.; Calvet, J.-C.; Albergel, C.; Balsamo, G.; Boussetta, S.; Carrer, D.; Lafont, S.; Meurey, C. Verification of the new ECMWF ERA-Interim reanalysis over France. *Hydrol. Earth Syst. Sci.* **2011**, *15*, 647–666. [CrossRef]
26. Authors. Methodological Framework, 2017. In *Methodological Framework for Assessment and Mapping of Ecosystem Condition and Ecosystem Services in Bulgaria*; Clorind: Sofia, Bulgaria, 2017; Consisting of 12 parts with separate ISBN, as follows:.
    - Part A Bratanova-Doncheva, S.; Chipev, N.; Gocheva, K.; Vergiev, S.; Fikova, R. Methodological framework for assessment and mapping of ecosystem condition and ecosystem services in Bulgaria. In *Conceptual Bases and Principles of Application*; ISBN 978-619-7379-21-1. Available online: http://www.iber.bas.bg/sites/default/files/2018/MAES_2018/A1%20INTRO_ENG%20PRINT.pdf (accessed on 17 March 2022).
    - Parts B1–B9—Methodologies by ecosystem type:
        - B1 Zhiyanski, M.; Nedkov, S.; Mondeshka, M.; Yarlovska, N.; Vassilev, V.; Bratanova-Doncheva, S.; Gocheva, K.; Chipev, N. *Methodology for Assessment and Mapping of Urban Ecosystems Condition and Their Services in Bulgaria*; Clorind: Sofia, Bulgaria, 2017; ISBN 978-619-7379-03-7. Available online: http://www.iber.bas.bg/sites/default/files/2018/MAES_2018/B1%20URBAN_ENG_PRINT.pdf (accessed on 17 March 2022).
        - B2 Yordanov, Y.; Mihalev, D.; Vassiev, V.; Bratanova-Doncheva, S.; Gocheva, K.; Chipev, N. *Methodology for Assessment and Mapping of Cropland Ecosystems Condition and Their Services in Bulgaria*; Clorind: Sofia, Bulgaria, 2017; ISBN 978-619-7379-05-1. Available online: http://www.iber.bas.bg/sites/default/files/2018/MAES_2018/B2%20CROPLAND_ENG_PRINT.pdf (accessed on 17 March 2022).
        - B3 Apostolova, I.; Sopotlieva, D.; Velev, N.; Vasilev, V.; Bratanova-Doncheva, S.; Gocheva, K. *Methodology for Assessment and Mapping of Grassland Ecosystems Condition and Their Services in Bulgaria*; Clorind: Sofia, Bulgaria, 2017; ISBN 978-619-7379-09-9. Available online: http://www.iber.bas.bg/sites/default/files/2018/MAES_2018/B3%20GRASSLAND_ENG%20PRINT.pdf (accessed on 17 March 2022).
        - B4 Kostov, G.; Rafailova, E.; Bratanova-Doncheva, S.; Gocheva, K.; Chipev, N. *Methodology for Assessment and Mapping of Woodland and Forests Ecosystems Condition and Their Services in Bulgaria*; Clorind: Sofia, Bulgaria, 2017; ISBN 978-619-7379-08-2. Available online: http://www.iber.bas.bg/sites/default/files/2018/MAES_2018/B4%20FOREST%20ENG%20PRINT.pdf (accessed on 17 March 2022).
        - B5 Velev, N.; Apostolova, I.; Sopotlieva, D.; Vassilev, V.; Bratanova-Doncheva, S.; Gocheva, K.; Chipev, N. *Methodology for Assessment and Mapping of Heathland and Shrub Ecosystems Condition and Their Services in Bulgaria*; Clorind: Sofia, Bulgaria, 2017; ISBN 978-619-7379-10-5. Available online: http://www.iber.bas.bg/sites/default/files/2018/MAES_2018/B5%20SHRUB_ENG_PRINT.pdf (accessed on 17 March 2022).
        - B6 Sopotlieva, D.; Apostolova, I.; Velev, N.; Bratanova-Doncheva, S.; Gocheva, K.; Chipev, N. *Methodology for Assessment and Mapping of Sparsely Vegetated Land Ecosystems Condition and Their Services in Bulgaria*; Clorind: Sofia, Bulgaria, 2017; ISBN 978-619-7379-13-6. Available online: http://www.iber.bas.bg/sites/default/files/2018/MAES_2018/B6%20SPARS%D0%95LY_ENG_PRINT.pdf (accessed on 17 March 2022).
        - B7 Apostolova, I.; Sopotlieva, D.; Velev, N.; Vassilev, V.; Bratanova-Doncheva, S.; Gocheva, K. *Methodology for Assessment and Mapping of Wetland Ecosystems Condition and Their Services in Bulgaria*; Clorind: Sofia, Bulgaria, 2017; ISBN 978-619-7379-14-3. Available online: http://www.iber.bas.bg/sites/default/files/2018/MAES_2018/B7%20WETLAND%20ENG_PRINT.pdf (accessed on 17 March 2022).
        - B8 Uzunov, Y.; Pehlivanov, L.; Chipev, N.; Vassilev, V.; Nedkov, S.; Bratanova-Doncheva, S. *Methodology for Assessment and Mapping of Freshwater Ecosystems Condition and Their Services in Bulgaria*; Clorind: Sofia, Bulgaria, 2017; ISBN 978-619-7379-17-4. Available online: http://www.iber.bas.bg/sites/default/files/2018/MAES_2018/B8_FRESHWATER%20ENG%20PRINT.pdf (accessed on 17 March 2022).
        - B9 Karamfilov, V.; Berov, D.; Pehlivanov, L.; Nedkov, S.; Vassilev, V.; Bratanova-Doncheva, S.; Chipev, N.; Gocheva, K. *Methodology for Assessment and Mapping of Marine Ecosystems Condition and Their Services in Bulgaria*; Clorind: Sofia, Bulgaria, 2017; ISBN 978-619-7379-18-1. Available online: https://www.iber.bas.bg/sites/default/files/2018/MAES_2018/B9%20MARINE_ENG_PRINT.pdf (accessed on 17 March 2022).
    - Part C Bratanova-Doncheva, S.; Zhiyanski, M.; Mondeshka, M.; Yordanov, Y.; Apostolova, I.; Sopotlieva, D.; Velev, N.; Rafailova, E.; Bobeva, A.; Uzunov, Y.; et al. Methodological framework for assessment and mapping of ecosystem condition and ecosystem services in Bulgaria. In *Guide for in Situ Verification of the Assessment and Mapping of Ecosystems Condition and Services*; Clorind: Sofia, Bulgaria, 2017; ISBN 978-619-7379-23-5. Available online: http://www.iber.bas.bg/sites/default/files/2018/MAES_2018/C_IN%20SITU_ENG%20PRINT.pdf (accessed on 17 March 2022).
    - Part D Chipev, N.; Bratanova-Doncheva, S.; Gocheva, K.; Zhiyanski, M.; Mondeshka, M.; Yordanov, Y.; Apostolova, I.; Sopotlieva, D.; Velev, N.; Rafailova, E.; et al. Methodological framework for assessment and mapping of ecosystem condition and ecosystem services in Bulgaria. In *Guide for Monitoring of Trends in Ecosystem Condition*; ISBN 978-619-7379-25-9. Available online: https://www.iber.bas.bg/sites/default/files/2018/MAES_2018/D_monitor%20book_eng_cmyk.pdf (accessed on 17 March 2022).

27. Bromwich, D.H.; Wilson, A.B.; Bai, L.S.; Moore, G.W.; Bauer, P. A comparison of the regional Arctic System Reanalysis and the global ERA-Interim Reanalysis for the Arctic. *Q. J. R. Meteorol. Soc.* **2016**, *142*, 644–658. [CrossRef]
28. Radeva, K.; Ivanova, I.; Borisova, D. Application of remote sensing for ecosystems monitoring and risk assessment. In Proceedings of the SPIE 2018, Paphos, Cyprus, 6 August 2018. [CrossRef]
29. Pause, M.; Schweitzer, C.; Rosenthal, M.; Keuck, V.; Bumberger, J.; Dietrich, P.; Heurich, M.; Jung, A.; Lausch, A. In Situ/Remote Sensing Integration to Assess Forest Health—A Review. *Remote Sens.* **2016**, *8*, 471. [CrossRef]
30. Kuenzer, C.; Ottinger, M.; Wegmann, M.; Guo, H.; Wang, C.H.; Zhang, J.; Dech, S.; Wikelski, M. Earth observation satellite sensors for biodiversity monitoring: Potentials and bottlenecks. *Int. J. Remote Sens.* **2014**, *35*, 6599–6647. [CrossRef]
31. Capriolo, A.; Boschetto, R.G.; Mascolo, R.A.; Balbi, S.; Villa, F. Biophysical and economic assessment of four ecosystem services for natural capital accounting in Italy. *Ecosyst. Serv.* **2020**, *46*, 101207. [CrossRef]
32. Bagstad, K.J.; Ingram, J.C.; Shapiro, C.D.; La Notte, A.; Maes, J.; Vallecillo, S.; Casey, C.F.; Glynn, P.D.; Heris, M.P.; Johnson, J.A.; et al. Lessons learned from development of natural capital accounts in the United States and European Union. *Ecosyst. Serv.* **2021**, *52*, 101359. [CrossRef]
33. Alcaraz-Segura, D.; Di Bella, C.M.; Straschnoy, J.V. (Eds.) *Earth Observation of Ecosystem Services*; CRC Press: Boca Raton, FL, USA, 2013. [CrossRef]
34. Michener, W.K.; Jones, M.B. Ecoinformatics: Supporting ecology as a data-intensive science. *Trends Ecol. Evol.* **2012**, *27*, 85–93. [CrossRef]
35. Wilkinson, M.D.; Dumontier, M.; Aalbersberg, I.J.J.; Appleton, G.; Axton, M.; Baak, A.; Blomberg, N.; Boiten, J.-W.; Santos, L.B.D.S.; Bourne, P.E.; et al. The FAIR Guiding Principles for scientific data management and stewardship. *Sci. Data* **2016**, *3*, 1–9. [CrossRef]
36. Berners-Lee, T.; Hendler, J.; Lassila, O. The semantic web. *Sci. Am.* **2001**, *284*, 34–43. Available online: http://jmvidal.cse.sc.edu/library/berners-lee01a.pdf (accessed on 17 March 2022). [CrossRef]
37. Needleman, M. The W3C semantic Web activity. *Ser. Rev.* **2003**, *29*, 63–64. [CrossRef]
38. Horrocks, I. Ontologies and the semantic web. *Commun. ACM* **2008**, *51*, 58–67. [CrossRef]
39. Haller, A.; Janowicz, K.; Cox, S.J.; Lefrançois, M.; Taylor, K.; Le Phuoc, D.; Lieberman, J.; García-Castro, R.; Atkinson, R.; Stadler, C. The modular SSN ontology: A joint W3C and OGC standard specifying the semantics of sensors, observations, sampling, and actuation. *Semant. Web* **2019**, *10*, 9–32. [CrossRef]
40. Villa, F.; Balbi, S.; Athanasiadis, I.N.; Caracciolo, C. Semantics for interoperability of distributed data and models: Foundations for better-connected information. *F1000Research* **2017**, *6*, 686. [CrossRef]
41. Chollet, F. On the Measure of Intelligence. *arXiv* **2019**, arXiv:1911.01547. Available online: https://arxiv.org/abs/1911.01547v2 (accessed on 17 March 2022).
42. Kumazawa, T.; Saito, O.; Kozaki, K.; Matsui, T.; Mizoguchi, R. Toward knowledge structuring of sustainability science based on ontology engineering. *Sustain. Sci.* **2009**, *4*, 99–116. [CrossRef]
43. Mellino, S.; Buonocore, E.; Ulgiati, S. The worth of land use: A GIS–emergy evaluation of natural and human-made capital. *Sci. Total Environ.* **2015**, *506*, 137–148. [CrossRef]
44. Pérez-Soba, M.; Elbersen, B.; Braat, L.; Kempen, M.; van der Wijngaart, R.; Staritsky, I.; Rega, C.; Paracchini, M.L. *The Emergy Perspective: Natural and Anthropic Energy Flows in Agricultural Biomass Production*; Publications Office of the European Union: Luxembourg, 2019. [CrossRef]
45. Morin, X.; Fahse, L.; Jactel, H.; Scherer-Lorenzen, M.; García-Valdés, R.; Bugmann, H. Long-term response of forest productivity to climate change is mostly driven by change in tree species composition. *Sci. Rep.* **2018**, *8*, 1–12. [CrossRef]
46. Maass, M.; Equihua, M. Earth stewardship, socioecosystems, the need for a transdisciplinary approach and the role of the International Long Term Ecological Research Network (ILTER). In *Earth Stewardship 2015*; Springer: Cham, Switzerland, 2015; pp. 217–233.
47. Mirtl, M.; Kuhn, I.; Montheith, D.; Bäck, J.; Orenstein, D.; Provenzale, A.; Zacharias, S.; Haase, P.; Shachak, M. Whole System Approach for in-situ research on Life Supporting Systems in the Anthropocene (WAILS). In Proceedings of the Copernicus Meetings 2021, Oline, 19–30 April 2021; No. EGU21-16425.
48. García-Duro, J.; Ciceu, A.; Chivulescu, S.; Badea, O.; Tanase, M.A.; Aponte, C. Shifts in Forest Species Composition and Abundance under Climate Change Scenarios in Southern Carpathian Romanian Temperate Forests. *Forests* **2021**, *12*, 1434. [CrossRef]
49. Ferretti, M. Forest health assessment and monitoring—Issues for consideration. *Environ. Monit. Assess.* **1997**, *48*, 45–72. [CrossRef]
50. Kostov, G.; Rafailova, E.; Vassilev, V.; Bratanova-Doncheva, S.; Gocheva, K.; Chipev, N. *Methodology for Assessment and Mapping of Woodland and Forest Ecosystems Condition and Their Services in Bulgaria*; Part B4 of Methodological Framework 2017; Clorind: Sofia, Bulgaria, 2017; ISBN 978-619-7379-07-5. Available online: http://www.iber.bas.bg/sites/default/files/2018/MAES_2018/B4%20FOREST%20ENG%20PRINT.pdf (accessed on 17 March 2022).
51. Velev, N.; Apostolova, I.; Sopotlieva, D.; Vassilev, V.; Bratanova-Doncheva, S.; Gocheva, K.; Chipev, N. *Methodology for Assessment and Mapping of Shrub Ecosystems Condition and Their Services in Bulgaria*; Part B5 of Methodological Framework 2017; Clorind: Sofia, Bulgaria, 2017; ISBN 978-954-7379-11-2. Available online: http://www.iber.bas.bg/sites/default/files/2018/MAES_2018/B5%20SHRUB_ENG_PRINT.pdf (accessed on 17 March 2022).

52. Apostolova, I.; Sopotlieva, D.; Velev, N.; Vassilev, V.; Bratanova-Doncheva, S.; Gocheva, K.; Chipev, N. *Methodology for Assessment and Mapping of Grassland Ecosystems Condition and Their Services in Bulgaria 2017*; Part B3 of Methodological Framework; Clorind: Sofia, Bulgaria, 2017; ISBN 978-619-7379-06-8. Available online: http://www.iber.bas.bg/sites/default/files/2018/MAES_2018/B3%20GRASSLAND_ENG%20PRINT.pdf (accessed on 17 March 2022).
53. Uzunov, Y.; Pehlivanov, L.; Chipev, N.; Vassilev, V.; Nedkov, S.; Bratanova-Doncheva, S. *Methodology for Assessment and Mapping of Freshwater Ecosystems Condition and Their Services in Bulgaria*; Clorind: Sofia, Bulgaria, 2017; ISBN 978-619-7379-17-4. Available online: http://www.iber.bas.bg/sites/default/files/2018/MAES_2018/B8_FRESHWATER%20ENG%20PRINT.pdf (accessed on 17 March 2022).
54. Sopotlieva, D.; Apostolova, I.; Velev, N.; Bratanova-Doncheva, S.; Gocheva, K.; Chipev, N. *Methodology for Assessment and Mapping of Sparsely Vegetated Land Ecosystems Condition and Their Services in Bulgaria*; Clorind: Sofia, Bulgaria, 2017; ISBN 978-619-7379-13-6. Available online: http://www.iber.bas.bg/sites/default/files/2018/MAES_2018/B6%20SPARS%D0%95LY_ENG_PRINT.pdf (accessed on 17 March 2022).
55. Wu, J.; Li, H. Concepts of scale and scaling. In *Scaling and Uncertainty Analysis in Ecology*; Wu, J., Jones, K.B., Li, H., Loucks, O.L., Eds.; Springer: Dordrecht, The Netherlands, 2006. [CrossRef]
56. Jelinski, D.E.; Wu, J. The modifiable areal unit problem and implications for landscape ecology. *Landsc. Ecol.* **1996**, *11*, 129–140. [CrossRef]
57. Gocheva, K.; Lü, Y.; Li, F.; Bratanova-Doncheva, S.; Chipev, N. Ecosystem restoration in Europe: Can analogies to Traditional Chinese Medicine facilitate the cross-policy harmonization on managing socio-ecological systems? *Sci. Total Environ.* **2019**, *657*, 1553–1567. [CrossRef]
58. Pan, J.Z. A flexible ontology reasoning architecture for the semantic web. *IEEE Trans. Knowl. Data Eng.* **2006**, *19*, 246–260. [CrossRef]
59. Serafini, L.; Borgida, A.; Tamilin, A. Aspects of distributed and modular ontology reasoning. *IJCAI* **2005**, *5*, 570–575. Available online: http://citeseerx.ist.psu.edu/viewdoc/download?doi=10.1.1.76.1677&rep=rep1&type=pdf (accessed on 17 March 2022).
60. Bratanova-Doncheva, S.; Zhiyanski, M.; Mondeshka, M.; Yordanov, Y.; Apostolova, I.; Sopotlieva, D.; Velev, N.; Rafailova, E.; Bobeva, A.; Uzunov, Y.; et al. Methodological framework for assessment and mapping of ecosystem condition and ecosystem services in Bulgaria. In *Guide for In Situ Verification of the Assessment and Mapping of Ecosystems Condition and Services*; Clorind: Sofia, Bulgaria, 2017; ISBN 978-619-7379-23-5. Available online: http://www.iber.bas.bg/sites/default/files/2018/MAES_2018/C_IN%20SITU_ENG%20PRINT.pdf (accessed on 17 March 2022).
61. Haase, P.; Völker, J. Ontology learning and reasoning—Dealing with uncertainty and inconsistency. In *Uncertainty Reasoning for the Semantic Web I*; Springer: Berlin/Heidelberg, Germany, 2008; pp. 366–384. [CrossRef]
62. Zadeh, L.A. Fuzzy logic. *Computer* **1988**, *21*, 83–93. [CrossRef]
63. Rosenfeld, A. Fuzzy graphs. In *Fuzzy Sets and Their Applications to Cognitive and Decision Processes*; Academic Press: Cambridge, MA, USA, 1975; pp. 77–95.
64. Bleiholder, J.; Naumann, F. Data fusion. *ACM Comput. Surv.* **2009**, *41*, 1–41. [CrossRef]
65. Yager, R.R. A framework for multi-source data fusion. *Inf. Sci.* **2004**, *163*, 175–200. [CrossRef]
66. Ren, C.; Ju, H.; Zhang, H.; Huang, J. Forest land type precise classification based on SPOT5 and GF-1 images. In Proceedings of the 2016 IEEE International Geoscience and Remote Sensing Symposium (IGARSS), Beijing, China, 10–15 July 2016; IEEE: Piscataway, NJ, USA, 2016; pp. 894–897. [CrossRef]
67. Lu, M.; Chen, B.; Liao, X.; Yue, T.; Yue, H.; Ren, S.; Li, X.; Nie, Z.; Xu, B. Forest types classification based on multi-source data fusion. *Remote Sens.* **2017**, *9*, 1153. [CrossRef]
68. Maxwell, A.E.; Warner, T.A.; Fang, F. Implementation of machine-learning classification in remote sensing: An applied review. *Int. J. Remote Sens.* **2018**, *39*, 2784–2817. [CrossRef]
69. Van Etten, A.; Lindenbaum, D.; Bacastow, T.M. Spacenet: A Remote Sensing Dataset and Challenge Series. *arXiv* **2018**, arXiv:1807.01232. Available online: https://arxiv.org/abs/1807.01232 (accessed on 17 March 2022).
70. Defries, R.S.; Hansen, M.C.; Townshend, J.R.; Janetos, A.C.; Loveland, T.R. A new global 1-km dataset of percentage tree cover derived from remote sensing. *Glob. Change Biol.* **2000**, *6*, 247–254. [CrossRef]
71. Gislason, P.O.; Benediktsson, J.A.; Sveinsson, J.R. Random forest classification of multisource remote sensing and geographic data. In Proceedings of the IGARSS 2004. 2004 IEEE International Geoscience and Remote Sensing Symposium, Anchorage, AK, USA, 20–24 September 2004; IEEE: Piscataway, NJ, USA, 2004; Volume 2, pp. 1049–1052. [CrossRef]
72. Hooker, J.; Duveiller, G.; Cescatti, A. A global dataset of air temperature derived from satellite remote sensing and weather stations. *Sci. Data* **2018**, *5*, 1–11. [CrossRef]
73. Abdi, A.M. Land cover and land use classification performance of machine learning algorithms in a boreal landscape using Sentinel-2 data. *GISci. Remote Sens.* **2020**, *57*, 1–20. [CrossRef]
74. Vallecillo, S.; La Notte, A.; Polce, C.; Zulian, G.; Alexandris, N.; Ferrini, S.; Maes, J. *Ecosystem Services Accounting: Part I-Outdoor Recreation and Crop Pollination*; Publications Office of the European Union: Luxembourg, 2018. Available online: https://publications.jrc.ec.europa.eu/repository/handle/JRC110321 (accessed on 17 March 2022).

75. Vallecillo, S.; La Notte, A.; Kakoulaki, G.; Roberts, N.; Kamberaj, J.; Dottori, F.; Rega, C.; Maes, J. Ecosystem services accounting. In *Part Ii-Pilot Accounts for Crop and Timber Provision, Global Climate Regulation and Flood Control, 165*; Publications Office of the European Union, 2019. Available online: https://publications.jrc.ec.europa.eu/repository/handle/JRC116334 (accessed on 17 March 2022).
76. Chipev, N.; Bratanova-Doncheva, S.; Gocheva, K.; Zhiyanski, M.; Mondeshka, M.; Yordanov, Y.; Apostolova, I.; Sopotlieva, D.; Velev, N.; Rafailova, E.; et al. Methodological framework for assessment and mapping of ecosystem condition and ecosystem services in Bulgaria. In *Guide for Monitoring of Trends in Ecosystem Condition*; Clorind: Sofia, Bulgaria, 2019; ISBN 978-619-7379-25-9. Available online: https://www.iber.bas.bg/sites/default/files/2018/MAES_2018/D_monitor%20book_eng_cmyk.pdf (accessed on 17 March 2022).
77. Bondev, I. The Vegetation of Bulgaria. In *Map 1:600,000 with Explanatory Text*; St. Kliment Ohridski University Press: Sofia, Bulgaria, 1991.
78. Kuiumdzhieva, T. A soils study in the ecological reserve of Paragalitsa- the Rila Mountains. *Probl. Na Khigienata* **1991**, *16*, 33–43. Available online: https://europepmc.org/article/med/1796107 (accessed on 17 March 2022).
79. Badot, P.M.; Lucot, E.; Sokolovska, M.G. Decline of forest stands in the Massif of Rila (Bulgaria). Ecophysiological Characterization and Research of Potential Causes. Annales Scientifiques de l'Universite de Franche Comte Besancon Biologie Ecologie (France) 1996. Available online: https://agris.fao.org/agris-search/search.do?recordID=FR19970107114 (accessed on 17 March 2022).
80. Panayotov, M.; Dimitrov, D.; Yurukov, S. Extreme climate conditions in Bulgaria—Evidence from Picea abies tree-rings. *Silva Balc.* **2011**, *12.1*, 37–46. Available online: https://www.researchgate.net/profile/Momchil-Panayotov/publication/268363316_Extreme_climate_conditions_in_Bulgaria_-_Evidence_from_Picea_Abies_tree-rings/links/549407140cf240d1cb4d23e4/Extreme-climate-conditions-in-Bulgaria-Evidence-from-Picea-Abies-tree-rings.pdf (accessed on 17 March 2022).
81. Panayotov, M.; Kulakowski, D.; Spiecker, H.; Krumm, F.; Laranjeiro, L.; Bebi, P. Natural disturbance history of the pristine Picea abies forest Parangalitsa. *Forestry* **2011**, *17*, 41. Available online: https://www.academia.edu/download/48853518/Natural_Disturbance_History_of_the_Prist20160915-15172-rumbfd.pdf (accessed on 17 March 2022).
82. Tsvetanov, N.; Nikolova, N.; Panayotov, M. Trees reaction after windthrow recorded in tree rings of pristine Picea abies forest "Parangalitsa". *Tree Rings Archaeol. Climatol. Ecol.* **2011**, *9*, 89–96. Available online: https://www.researchgate.net/profile/Nickolay-Tsvetanov/publication/285145176_Trees_reaction_after_windthrow_recorded_in_tree_rings_of_pristine_Picea_abies_forest_Parangalitsa/links/565c1b4c08ae1ef92981d37e/Trees-reaction-after-windthrow-recorded-in-tree-rings-of-pristine-Picea-abies-forest-Parangalitsa.pdf (accessed on 17 March 2022).
83. Stoyanova, N.; Dimitrov, D.; Georgiev, G.P.; Delkov, A. Biosphere reserves in Bulgaria and their forest genetic resources. *Silva Balc.* **2011**, *12*, 13–24. Available online: https://www.researchgate.net/profile/Dimitar-Dimitrov-6/publication/292797771_Biosphere_reserves_in_Bulgaria_and_their_forest_genetic_resources/links/57ea361b08aed0a291326172/Biosphere-reserves-in-Bulgaria-and-their-forest-genetic-resources.pdf (accessed on 17 March 2022).
84. Bebi, P.; Krumm, F.; Brändli, U.B.; Zingg, A. Dynamik dichter, gleichförmiger Gebirgsfichtenwälder. *Schweiz. Z. Forstwes.* **2013**, *164*, 37–46. [CrossRef]
85. Ivanov, M.A.; Tyufekchiev, K.A. Remote Sensing Based Vegetation Analysis in Parangalitsa Reserved Area. *Ecol. Balk.* **2019**. Available online: http://web.uni-plovdiv.bg/mollov/EB/2019_SE2/187-197_eb.19SE212.pdf (accessed on 17 March 2022).
86. Tuominen, J.; Lipping, T.; Kuosmanen, V.; Haapanen, R. Remote Sensing of Forest Health. In *Geoscience and Remote Sensing*; Pei-Gee, P.H., Ed.; Books on Demand (International): Norderstedt, Germany, 2009. [CrossRef]
87. Gao, B.-C. NDWI—A Normalized Difference Water Index for Remote Sensing of Vegetation Liquid Water from Space. *Remote Sens. Environ.* **1996**, *58*, 257–266. [CrossRef]
88. Nedkov, R. Normalized Differential Greenness Index for Vegetation Dynamics Assessment. *Sci. Cosm.* **2017**, *70*, 1143–1146. Available online: sites.udel.edu/delaware-water-watch/files/2014/06/NDWI-A-Normalized-Difference-Water-Index-for-Remote-Sensing-of-Vegetation-Liquid-Water-From-Space-1ko95nn.pdf (accessed on 17 March 2022).
89. Pettorelli, N.; Vik, J.O.; Mysterud, A.; Gaillard, J.-M.; Tucker, C.J.; Stenseth, N.C. Using the satellite-derived NDVI to assess ecological responses to environmental change. *Trends Ecol. Evol.* **2005**, *20*, 503–510. [CrossRef]
90. Katrandzhiev, K. Application of Remote Sensing for High Mountain Ecosystem Condition Assessment (South West Rila Mountain—Bulgaria). *Ecol. Eng. Environ. Prot.* **2018**, *2*, 35–40. Available online: https://www.researchgate.net/profile/Kostadin-Katrandzhiev/publication/327792948_APPLICATION_OF_REMOTE_SENSING_FOR_HIGH_MOUNTAIN_ECOSYSTEM_CONDITION_ASSESSMENT_SOUTH_WEST_RILA_MOUNTAIN-BULGARIA/links/5ba4c8d892851ca9ed1a829e/APPLICATION-OF-REMOTE-SENSING-FOR-HIGH-MOUNTAIN-ECOSYSTEM-CONDITION-ASSESSMENT-SOUTH-WEST-RILA-MOUNTAIN-BULGARIA.pdf (accessed on 17 March 2022). [CrossRef]
91. Hersbach, H.; Bell, B.; Berrisford, P.; Hirahara, S.; Horanyi, A.; Muñoz-Sabater, J.; Nicolas, J.; Peubey, C.; Radu, R.; Schepers, D.; et al. The ERA5 global reanalysis. *Q. J. R. Meteorol. Soc.* **2020**, *146*, 1999–2049. [CrossRef]
92. Tarek, M.; Brissette, F.P.; Arsenault, R. Evaluation of the ERA5 reanalysis as a potential reference dataset for hydrological modelling over North America. *Hydrol. Earth Syst. Sci.* **2020**, *24*, 2527–2544. [CrossRef]
93. Jiang, Q.; Li, W.; Fan, Z.; He, X.; Sun, W.; Chen, S.; Wen, J.; Gao, J.; Wang, J. Evaluation of the ERA5 reanalysis precipitation dataset over Chinese Mainland. *J. Hydrol.* **2021**, *595*, 125660. [CrossRef]

94. Pavlova, A.; Nedkov, R. Application of the Different Vegetation Indexes Regarding to Forest Physiology and Climatic Seasons. In Proceedings of the Scientific Conference "Space, Ecology, Safety" (S E S) with International Participation, Varna, Bulgaria, 10–13 June 2005; pp. 263–268. Available online: https://www.researchgate.net/publication/240620084 (accessed on 17 March 2022).
95. Tucker, C.J.; Sellers, P.J. Satellite remote sensing of primary production. *Int. J. Remote Sens.* **1986**, *7*, 1395–1416. [CrossRef]
96. Kerr, J.T.; Ostrovsky, M. From space to species: Ecological applications for remote sensing. *Trends Ecol. Evol.* **2003**, *18*, 299–305. [CrossRef]
97. Wang, J.; Rich, P.M.; Price, K.P. Temporal responses of NDVI to precipitation and temperature in the central Great Plains, USA. *Int. J. Remote Sens.* **2003**, *24*, 2345–2364. [CrossRef]
98. Avetisyan, D.; Nedkov, R.; Borisova, D.; Cvetanova, G. Application of spectral indices and spectral transformation methods for assessment of winter wheat state and functioning. In *Remote Sensing for Agriculture, Ecosystems, and Hydrology XXI 2019, Proceedings of the International Society for Optics and Photonics, Strasbourg, France, 21 October 2019*; SPIE: Washington, DC, USA, 2019; Volume 11149, p. 1114929.
99. Avetisyan, D.; Nedkov, R.; Georgiev, N. Monitoring maize (*Zea Mays* L.) phenology response to water deficit using Sentinel-2 multispectral data. In *International Society for Optics and Photonics Proceedings of the Eighth International Conference on Remote Sensing and Geoinformation of the Environment (RSCy2020) 2020, Paphos, Cyprus, 26 August 2020*; SPIE: Washington, DC, USA, 2020; Volume 11524, p. 1152403.
100. Nedkov, R. Quantitative assessment of forest degradation after fire using ortogonalized satellite images from SENTINEL-2. *Comptes Rendus L'academie Bulg. Des Sci.* **2018**, *71*, 83–86.
101. Velizarova, E.; Radeva, K.; Stoyanov, A.; Georgiev, N.; Gigova, I. Post-fire forest disturbance monitoring using remote sensing data and spectral indices. In *International Society for Optics and Photonics 2019, Proceedings of the Seventh International Conference on Remote Sensing and Geoinformation of the Environment (RSCy2019), Paphos, Cyprus, 27 June 2019*; SPIE: Washington, DC, USA, 2020; Volume 11174, p. 111741G.
102. Stoyanov, A.; Borisova, D.; Radeva, K. Application of SAR and optical data from Sentinel satellites for spatial-temporal analysis of the flood in the region of Bregovo-Bulgaria, 11/03/2018. In *Remote Sensing for Agriculture, Ecosystems, and Hydrology XX. Proceedings of the International Society for Optics and Photonics, Berlin, Germany, 10 October 2018*; SPIE: Washington, DC, USA, 2018; Volume 10783, p. 107831K.
103. Radeva, K.; Nedkov, R.; Dancheva, A. Application of remote sensing data for a wetland ecosystem services assessment in the area of Negovan village. In *Remote Sensing for Agriculture, Ecosystems, and Hydrology XX. Proceedings of the International Society for Optics and Photonics, Berlin, Germany, 10 October 2018*; SPIE: Washington, DC, USA, 2018; Volume 10783, p. 107830Y.
104. Palombo, C.; Chirici, G.; Marchetti, M.; Tognetti, R. Is land abandonment affecting forest dynamics at high elevation in Mediterranean mountains more than climate change? *Plant Biosyst. Int. J. Deal. All Asp. Plant Biol.* **2013**, *147*, 1–11. [CrossRef]
105. Peringer, A.; Siehoff, S.; Chételat, J.; Spiegelberger, T.; Buttler, A.; Gillet, F. Past and future landscape dynamics in pasture-woodlands of the Swiss Jura Mountains under climate change. *Ecol. Soc.* **2013**, *18*, 11. [CrossRef]
106. Spears, E.E. A direct measure of pollinator effectiveness. *Oecologia* **1983**, *57*, 196–199. Available online: https://www.jstor.org/stable/4216947 (accessed on 17 March 2022). [CrossRef]
107. Bingham, R.A.; Orthner, A.R. Efficient pollination of alpine plants. *Nature* **1998**, *391*, 238–239. [CrossRef]
108. Richman, S.K.; Levine, J.M.; Stefan, L.; Johnson, C.A. Asynchronous range shifts drive alpine plant–pollinator interactions and reduce plant fitness. *Glob. Change Biol.* **2020**, *26*, 3052–3064. [CrossRef]
109. Yakimov, L.; Tsvetanova, E.; Georgieva, A.; Petrov, L.; Alexandrova, A. Assessment of the Oxidative status of Black Sea Mussels (Mytilus galloprovincialis Lamark, 1819) from Bulgarian coastal areas with introduction of a specific oxidative stress index. *J. Environ. Prot. Ecol.* **2018**, *19*, 1614–1622. Available online: https://www.researchgate.net/profile/Albena_Alexandrova/publication/330601540_ASSESSMENT_OF_THE_OXIDATIVE_STATUS_OF_BLACK_SEA_MUSSELS_Mytilus_galloprovincialis_Lamark_1819_FROM_BULGARIAN_COASTAL_AREAS_WITH_INTRODUCTION_OF_SPECIFIC_OXIDATIVE_STRESS_INDEX/links/5c4a1965a6fdccd6b5c59fc6/ASSESSMENT-OF-THE-OXIDATIVE-STATUS-OF-BLACK-SEA-MUSSELS-Mytilus-galloprovincialis-Lamark-1819-FROM-BULGARIAN-COASTAL-AREAS-WITH-INTRODUCTION-OF-SPECIFIC-OXIDATIVE-STRESS-INDEX.pdf (accessed on 17 March 2022).
110. Rees, H.C.; Maddison, B.C.; Middleditch, D.J.; Patmore, J.R.; Gough, K.C. The detection of aquatic animal species using environmental DNA—A review of eDNA as a survey tool in ecology. *J. Appl. Ecol.* **2014**, *51*, 1450–1459. [CrossRef]
111. Engelstad, M.E. Determining Nature Types in Norway (NiN) by Soil eDNA Metabarcoding. Master's Thesis, Degree-Granting University, Oslo, Norway, 2020. Available online: https://www.duo.uio.no/handle/10852/79675 (accessed on 17 March 2022).

 *diversity*

Article

Differential Effects of Tree Species on Soil Microbiota 45 Years after Afforestation of Former Pastures

Richard Gere, Mikuláš Kočiš, Ján Židó, Dušan Gömöry and Erika Gömöryová *

Faculty of Forestry, Technical University in Zvolen, TG Masaryka 24, SK-96001 Zvolen, Slovakia; r.gere60@gmail.com (R.G.); kocis@is.tuzvo.sk (M.K.); xzido@is.tuzvo.sk (J.Ž.); gomory@tuzvo.sk (D.G.)
* Correspondence: gomoryova@tuzvo.sk

Abstract: Several decades ago, many former pastures in Central Europe were afforested or colonized by trees after being abandoned. Knowledge of the effects of tree species on soil properties is important for planning of the composition of future forests. In this regard, a research location in Vrchdobroč (Central Slovakia), which is former agricultural land used as pasture, enables the exploration of ecosystem processes and properties in stands of different tree species after afforestation. The goal of our study was to find out whether changes in soil properties, including soil microbial activity and diversity among different stands, were already observable 45 years after the afforestation, and how the effects differed among stands of different tree species. The study was conducted in the pure stands of Norway spruce (*Picea abies* L. Karst.), Douglas fir (*Pseudotsuga menziesi* (Mirb.) Franco), European beech (*Fagus sylvatica* L.) and sycamore maple (*Acer pseudoplatanus* L.). Multivariate analyses of physico-chemical soil properties indicated an overlap between the soils under the Douglas fir and the spruce, but a clear separation of beech from sycamore. In general, both microbial activity and diversity were, surprisingly, highest under the Douglas fir, followed by the sycamore, with the beech and the spruce showing mostly lower values.

Keywords: Norway spruce; Douglas fir; beech; maple; soil chemical properties; soil microorganisms

Citation: Gere, R.; Kočiš, M.; Židó, J.; Gömöry, D.; Gömöryová, E. Differential Effects of Tree Species on Soil Microbiota 45 Years after Afforestation of Former Pastures. *Diversity* **2022**, *14*, 515. https://doi.org/10.3390/d14070515

Academic Editors: Michael Wink, Lucian Dinca and Miglena Zhiyanski

Received: 5 June 2022
Accepted: 22 June 2022
Published: 25 June 2022

Publisher's Note: MDPI stays neutral with regard to jurisdictional claims in published maps and institutional affiliations.

1. Introduction

Soil and plants are linked very tightly. Soil properties affect plant growth and, conversely, plants affect soil attributes. The effect of plants on soil properties is very distinct, especially in forest ecosystems because of the long-term influence of forest stand on soil. Trees can affect soil properties directly through the input of organic material (dead organic matter, root exudates), living tissues (roots) and/or indirectly via modification of microclimate, e.g., radiation reaching the ground, evaporation from the soil surface, relative air humidity, air circulation, water input to the soil, etc., [1–5]. The effects differ depending on tree species, because of the different qualities and quantities of organic residues left on the soil surface or going directly into the soil, as well as different crown and root architecture, canopy openness, stemflow rate, etc., [5–7]. Numerous studies have shown that differences in tree cover are reflected especially in the thickness of the surface organic layer, in the soil acidity, in the base saturation, and in the carbon and nitrogen concentrations, and consequently in the responses of the soil microbial biomass, the activity and structure, and the diversity of the microbial communities [8–10].

Human activity has often caused decreases in forest areas—especially in favor of agriculture—in many parts of the world, and deforestation has persisted. In Europe, the minimum of forest cover occurred during the 18th and 19th centuries [1]. Increasing demand for wood production led to the planting and extensive use of conifers in many European countries, often at the expense of native deciduous tree species. Conifers and deciduous trees are traditionally considered to differ in their effect on soil properties, especially because of their different qualities of litter input which, in the case of conifers, can lead to soil acidification, deterioration and podzolization in the long term [11,12]. Nevertheless, there are also contradictory results indicating similar or even better soil properties under certain conifer species in comparison to broadleaves [13,14]. According to a meta-analysis by Augusto, et al. [1], the discrepancy can be partially explained by variation between the soils of some of the study sites; therefore, they emphasized the importance of experimental design. As they stated, there are only a few sufficiently replicated studies with the same stand age and stand management, on the same soil type and with the same land use history.

In Slovakia, the locality of Vrchdobroč is former agricultural land which was used for decades as pasture. In 1960, the area, comprising 26,000 ha, was declared important for water management, as it contained the source of two rivers. During 1960–1985, an area of over 5166 ha was afforested to increase the forest cover from 29 to 49%. Within this area, 283 ha, afforested by 17 tree species, was used for research and demonstration purposes, managed by the Forestry Research Institute of the National Forestry Centre in Zvolen (Slovakia). Some of the plots were damaged by drought, animals, snow, etc., and nowadays conifers and broadleaves represent 85% and 15% of the area, respectively [15].

The research location allows for the exploration of ecosystem processes and properties in stands of different tree species after afforestation. In this study, we used it to evaluate the impact of different tree species on soil, with emphasis on chemical and microbial properties. We focused on pure forest stands of two coniferous and two deciduous species of the same age and management history, growing under almost identical environmental conditions (soil type, altitude, climate, etc.). Our goal was to find out whether changes in soil properties among different stands were already observable 45 years after the afforestation; what had been the respective effects of the different tree species; and whether those effects had also been reflected in the properties of the microbial community (biomass, activity, diversity). We hypothesized that 45 years was sufficiently long enough for the effects of trees on soil properties to be manifested, and that changes would have occurred especially in the surface organic layer, which started to form after afforestation, and probably also in the top 10 cm of the mineral soil horizons. As the quality of the litter of conifers essentially differs from that of broadleaves, we expected differences to be especially notable between these two groups of trees.

2. Materials and Methods

2.1. Study Sites and Soil Sampling

The study area, Vrchdobroč, is located in the Veporské vrchy Mts., Central Slovakia. The mean annual temperature is 5 °C; mean temperature in July 15 °C; and in January −6 °C. The yearly precipitation reaches 920 mm. The dominant soil type is Cambisol with a sandy loam texture formed from porphyric granodiorites and granites. The soil skeleton content is 20–50%.

Pure 45-year-old stands of Norway spruce (*Picea abies* (L.) Karst.), Douglas fir (*Pseudotsuga menziesi* (Mirb.) Franco), European beech (*Fagus sylvatica* L.) and sycamore maple (*Acer pseudoplatanus* L.), randomly distributed across the area, were selected for the study. Three stands of each species were chosen. The stand area of each tree species varied between 0.40 and 1.59 ha. The plots were situated at the altitudes of 815–850 m a. s. l., with S-SE aspect, on a slope of 5–10°. The stands were managed with sylvicultural methods, which are traditional in the Carpathian mountain forests: conifer stands underwent cleaning every 20 years, and thinning from below at around 35 years; broadleaved stands were cleaned twice (at 15 and 25 years), and subsequently thinned from above at the age of 40 years [15].

Five soil samples were collected along linear transects, with 10 m spacing in each stand, in July 2015. Samples from the O-horizon were collected, using a 0.2 m × 0.2 m frame placed on the soil surface, while the humus layer underneath the template was cut by knife from its surroundings. After removing the surface organic layer, mineral soil samples were taken from a depth of 0–0.1 m (the A-horizon), using the knife and shovel with a depth scale indication. We did not use a probe sampler because of high skeleton content in the soil. Approximately 400 g of soil samples were put into plastic bags, i.e., a comparable soil weight from all sampling plots. Visible coarse particles (e.g., roots, fauna, stones) were removed. Samples from deeper layers were not taken because we assumed that the effect of the trees would be notable, especially in the topsoil horizons. In total, 120 soil samples were collected (4 tree species × 3 stands × 2 horizons × 5 replications).

2.2. Laboratory Analyses

After bringing the samples to the laboratory, they were divided into two parts. One part, intended for microbial analyses, was stored in a refrigerator. The second part was air-dried immediately after being brought in, and analyzed for chemical properties. For the determination of soil moisture and dry weight, a gravimetric method was used, based on oven-drying of fresh soil samples from the O-horizon at 60 °C and from the A-horizon at 105 °C until a constant weight. Soil acidity (pH-$CaCl_2$) was determined potentiometrically in suspension prepared from 2.5 g of litter or 10 g of mineral soil and 25 mL of 0.01 M $CaCl_2$. The C, N and S concentrations were measured using a VarioMacro Elemental Analyzer (CNS Version, Elementar Gmbh, Langenselbold, Germany), employing the dry combustion method. Exchangeable cations of Ca^{2+}, Mg^{2+} and K^+ were determined in 1 M extract of NH_4Cl, using atomic absorption spectrometry (GBC Avanta AAS, Dandenong, Victoria, Australia), and evaluated only for samples from the mineral A-horizon.

Determination of basal respiration (BR) was performed according to Isermeyer's method [16]. The amount of CO_2 released from a fresh soil sample in a glass container for 24 h and absorbed in 0.05 M NaOH was measured. The amount of carbonate was determined by titration with 0.05 M HCl after 5 mL of $BaCl_2$ and phenolphthalein were added. For the determination of substrate-induced respiration (SIR), 0.5 g and 0.125 g of glucose was added to samples from the A-horizon and the O-horizon, respectively. The evolved CO_2 was measured, as described above, after 4.5 h [16]. Catalase activity (Cat) was estimated according to the method described by Khaziev [17], based on the measurement of the released O_2, 10 min after 3% H_2O_2 was added to a fresh soil sample. Microbial biomass carbon (Cmic) was determined using the microwave-irradiation procedure [18]. C concentration was quantified titrimetrically, after the oxidation of the extract with $K_2Cr_2O_7/H_2SO_4$. N-mineralization (Nmin) was measured according to the procedure described by Kandeler [19]. Soil samples under anaerobic conditions were incubated at 40 °C for 7 days, and the released NH_4-N was estimated by a colorimetric procedure. Nmin was determined only in the samples from the A-horizon.

The Biolog® method was used to determine the activity of the microbial functional groups [20]. 0.5 g of soil sample was placed into a plastic bank. After the addition of 50 mL 0.85% NaCl solution, the suspension was left for 45 min on automatic shaker, and then filtered. Subsequently, the supernatant was diluted to 1:1000 and 1:10,000 for samples from the O- and A-horizon, respectively. 150 µL of the extract was pipetted into BIOLOG Ecoplates, and incubated at 27 °C for 5 days. Absorbance at 590 nm was measured spectrophotometrically using a Sunrise Microplate reader (Tecan, Salzburg, Austria) every 24 h. Absorbance values were blanked against the control well. The metabolic activity was calculated as the area below the time–absorbance curve, and was used as a measure of the abundance of the respective functional group. The richness of the soil functional groups was assessed as the number of substrates with a non-zero response. The functional diversity of the microbial community was assessed by Hill's diversity index (N_2) Equation (1) [21]:

$$N_2 = 1/\sum \mathrm{p}i^2 \tag{1}$$

where pi is the ratio of the activity on a particular substrate to the sum of activities on all substrates.

2.3. Statistical Analysis

The effect of different tree species on soil and microbial properties was evaluated by one-way ANOVA (tree species being considered a fixed-effect factor), with subsequent Tukey's HSD post-hoc tests conducted separately for the O- and A-horizons, using Statistica 12 software [22].

As microbial richness and diversity do not completely explain how the structure of community changes with stand-forming trees species, a multivariate analysis (canonical correspondence analysis; CCA) was performed using CANOCO 5 software [23]. Analyses were done separately for physico-chemical properties, microbial community parameters and community-level physiological profiles assessed by the Biolog® method.

3. Results

3.1. Soil Chemical Properties

We observed significant differences in the effects of tree species on soil chemical properties; however, the effects differed between soil horizons (Table 1). CCA ordination diagrams (Figure 1), based on physico-chemical properties, clearly show the distinction between sycamore maple, beech and coniferous stands. No significant differences were found in soil chemical characteristics between the spruce and the Douglas fir stands in both horizons; however, litter weight was higher in spruce stands. The differences between the beech and sycamore stands in the O-horizon were less pronounced than in the A-horizon: generally, chemical soil properties (SOC, nutrients, base saturation) were more favorable in sycamore stands. Nevertheless, no general difference in favor of broadleaves compared to conifers was observed.

Table 1. Basic statistics (means ± standard deviations) of physico-chemical soil properties in stands of different tree species.

Physico-Chemical Properties	Horizon	*P*	*Picea*	*Pseudotsuga*	*Fagus*	*Acer*
pH/$CaCl_2$	O	<0.1	4.85 ± 0.40 a	5.08 ± 0.35 a	5.06 ± 0.21 a	5.06 ± 0.35 a
	A	<0.001	3.73 ± 0.09 b	3.94 ± 0.28 b	3.89 ± 0.23 b	4.28 ± 0.26 a
Soil moisture (%)	O	<0.01	29.71 ± 6.90 c	61.62 ± 54.34 a	32.47 ± 12.37 bc	37.63 ± 13.76 b
	A	<0.001	17.10 ± 3.12 b	15.29 ± 5.52 b	23.28 ± 7.22 ab	36.80 ± 10.18 a
C (%)	O	<0.001	32.37 ± 10.05 a	32.53 ± 8.64 a	22.61 ± 7.79 b	21.42 ± 5.01 b
	A	<0.001	4.60 ± 0.85 b	4.15 ± 1.05 b	3.73 ± 0.71 b	7.19 ± 1.07 a
N (%)	O	<0.001	1.65 ± 0.48 a	1.70 ± 0.40 a	1.22 ± 0.35 b	1.48 ± 0.25 ab
	A	<0.001	0.37 ± 0.06 b	0.35 ± 0.08 b	0.34 ± 0.05 b	0.67 ± 0.09 a
S (%)	O	0.92	0.24 ± 0.06 a	0.24 ± 0.05 a	0.23 ± 0.23 a	0.21 ± 0.05 a
	A	<0.001	0.07 ± 0.03 b	0.07 ± 0.02 b	0.06 ± 0.01 b	0.10 ± 0.02 a
C:N ratio	O	<0.001	19.40 ± 1.41 a	18.97 ± 2.07 a	18.19 ± 1.46 a	14.37 ± 1.05 b
	A	<0.001	12.46 ± 0.92 a	11.87 ± 1.02 a	10.81 ± 0.67 b	10.82 ± 0.70 b
Litter weight ($kg{\cdot}m^{-2}$)	O	<0.001	1.46 ± 0.36 a	1.24 ± 0.31 b	0.85 ± 0.27 b	1.45 ± 0.44 a
Ca^{2+} ($mg{\cdot}kg^{-1}$)	A	<0.001	475.1 ± 168.6 b	742.8 ± 413.1 b	692.7 ± 163.5 b	1272.5 ± 379.5 a
Mg^{2+} ($mg{\cdot}kg^{-1}$)	A	<0.001	116.0 ± 33.9 b	131.2 ± 21.9 b	135.6 ± 13.3 b	178.3 ± 29.8 a
K^+ ($mg{\cdot}kg^{-1}$)	A	<0.001	49.13 ± 6.37 b	56.11 ± 36.41 b	67.23 ± 15.61 b	86.99 ± 27.51 a

P—probability associated with ANOVA *F*-test for tree species effect; different letters designate homogeneous groups based on Tukey's HSD post-hoc tests. O—forest floor; A—mineral horizon.

The O-horizon exhibited lower soil acidity, higher C/N ratio and higher concentration of C, N, S than the A-horizon. Generally, higher decline of C and N content from the O- to the A-horizon was observed under conifers in comparison to deciduous stands. Soil pH of the O-horizon, surprisingly, did not differ significantly among tree species. On the other hand, significantly higher C concentration was found in the litter of conifers than the deciduous trees. Beech litter exhibited the lowest N content, and the litter of maple the lowest C/N ratio. However, in the A-horizon the pattern differed. Soils in the maple stands showed the lowest acidity and the highest content of all nutrients, in comparison to other stands. Surprisingly, soil properties under beech stands did not differ significantly from those under the conifers. The C/N ratio was the only parameter that differed between the two groups —conifers and deciduous stands.

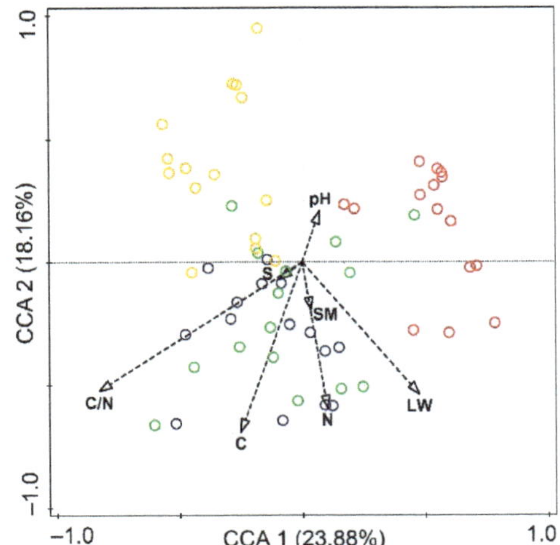

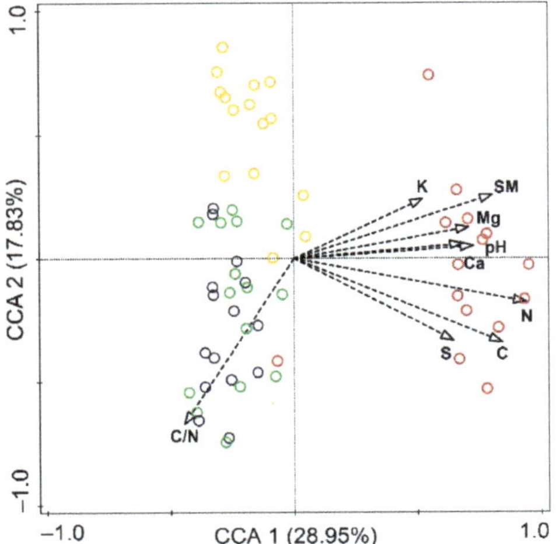

Figure 1. Canonical correspondence analysis for physico-chemical soil properties and different forest stands within the O-horizon (**left**) and the A-horizon (**right**). Arrows represent the direction of the steepest increase of individual physico-chemical properties. Circles represent individual sampling sites, with color distinctions of dominant tree species in the stands (blue—*Pseudotsuga menziesii*; green—*Picea abies*; yellow—*Fagus sylvatica*; red—*Acer pseudoplatanus*).

3.2. Soil Microbial Characteristics

Like chemical properties, soil microbial characteristics also differed significantly between the stands of different tree species (Table 2). However, as shown by the CCA ordination, the pattern of differences did not follow completely that of the soil physico-chemical properties, and especially in the O-horizon, clear differences in microbial characteristics between the conifer stands were obvious (Figure 2). Again, in general, microbial characteristics in deciduous stands did not significantly differ from those of the conifer stands.

Table 2. Basic statistics (means ± standard deviations) of soil microbial characteristics in stands of different tree species.

Microbial Properties	Horizon	P	*Picea*	*Pseudotsuga*	*Fagus*	*Acer*
Basal respiration	O	<0.001	1.37 ± 0.49 b	4.47 ± 2.89 a	1.88 ± 1.51 b	2.61 ± 2.38 ab
($\mu g\ CO_2 \cdot g^{-1} \cdot h^{-1}$)	A	0.11	0.18 ± 0.04 a	0.16 ± 0.09 a	0.20 ± 0.09 a	0.42 ± 0.63 a
Substrate-induced	O	<0.001	5.02 ± 1.97 b	22.00 ± 23.62 a	5.37 ± 2.42 b	9.64 ± 6.67 b
respiration ($\mu g\ CO_2 \cdot g^{-1} \cdot h^{-1}$)	A	<0.001	1.09 ± 0.61 b	0.63 ± 0.29 b	1.11 ± 0.46 b	2.40 ± 0.82 a
Catalase activity	O	<0.05	5.26 ± 0.93 b	6.02 ± 2.26 ab	6.62 ± 1.26 ab	6.76 ± 1.39 a
(ml $O_2 \cdot g^{-1} \cdot min^{-1}$)	A	<0.001	0.65 ± 0.20 a	0.56 ± 0.17 a	0.62 ± 0.21 a	1.06 ± 0.26 a
N mineralization ($\mu g\ NH_4^+ \text{-} N \cdot g^{-1} \cdot d^{-1}$)	A	<0.001	0.74 ± 0.50 b	1.41 ± 1.04 b	4.00 ± 2.17 a	5.38 ± 4.36 a
Microbial biomass carbon	O	<0.001	6440 ± 1250 ab	7869 ± 3339 a	4338 ± 1047 c	4894 ± 1747 bc
($\mu g \cdot g^{-1}$)	A	<0.001	568.2 ± 169.7 b	420.3 ± 99.2 c	361.8 ± 105.3 c	912.4 ± 185.4 a
Richness of functional groups	O	<0.001	26.87 ± 2.13 c	28.93 ± 1.79 a	27.47 ± 1.99 ab	28.40 ± 1.59 ab
	A	<0.001	27.80 ± 1.01 a	26.87 ± 2.13 a	25.07 ± 1.79 b	26.93 ± 1.75 a
Diversity of functional groups	O	<0.001	10.48 ± 2.63 b	17.60 ± 1.97 a	16.55 ± 2.33 a	13.53 ± 3.01 b
	A	0.01	11.45 ± 1.88 c	15.18 ± 2.63 a	12.99 ± 1.37 b	11.18 ± 1.78 c

P—probability associated with ANOVA *F*-test for tree species effect; different letters designate homogeneous groups based on Tukey's HSD post-hoc tests. O—forest floor; A—mineral horizon.

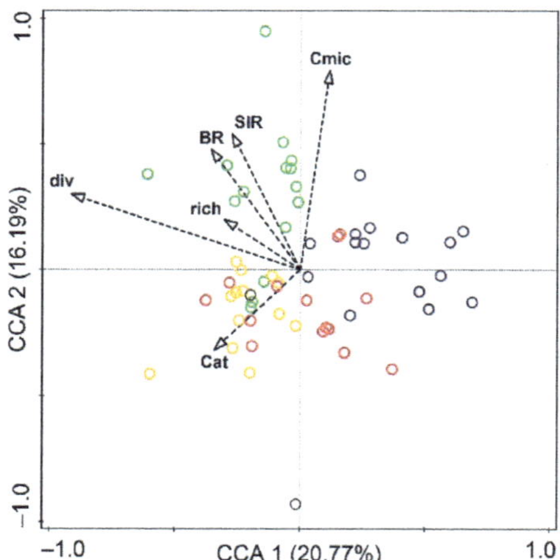

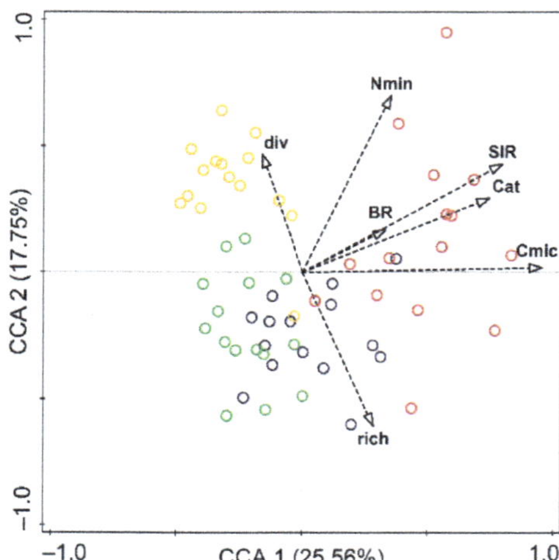

Figure 2. Canonical correspondence analysis for soil microbial community properties in relation to stand tree species composition within the O-horizon (**left**) and A-horizon (**right**). Arrows represent the direction of the steepest increase of individual microbial properties. Circles represent individual sampling sites, with color distinctions of dominant tree species in the stands (blue—*Pseudotsuga menziesii*; green—*Picea abies*; yellow—*Fagus sylvatica*; red—*Acer pseudoplatanus*).

Generally, the Douglas fir litter exhibited higher microbial activity, as well as richness and diversity of functional groups, than the spruce litter. Unlike the conifers, the differences in microbial characteristics (except the diversity index) between the broadleaves were negligible. Interestingly, the litter of deciduous trees exhibited lower microbial biomass than the litter of conifers. The highest microbial biomass activity, as well as richness and diversity of microbial functional groups, were typical for the litter in the Douglas fir stands.

While, in the O-horizon, the Douglas fir litter differed significantly from the others in microbial characteristics, in the A-horizon the situation was quite different and, generally, no differences were found between the soils of the spruce and the Douglas fir stands (except the diversity index). On the other hand, while basal respiration and enzyme activity did not differ between the beech and the maple stands, the microbial biomass, SIR and indices of microbial community structure did. In soils under the maples, higher microbial biomass occurred, and there seemed to be a trend of higher activity as well. Within the A-horizon, N-min was the only microbial property that distinguished deciduous and coniferous forest stands.

The community structure, based on the Biolog® method, showed differences in the utilization of 25 substrates in the O-horizon, and of only 16 in the A-horizon between the evaluated forest stands (Figure 3, Table S1). The Douglas fir litter showed high utilization in the majority of substrates, followed by the litter of the beech and the maple, while the litter of the spruce generally showed the lowest utilization rates. Surprisingly, the most distinct differences in utilization were observed between the litter of two conifers. Glycogen was the only substrate for which utilization was found to be the lowest in the Douglas litter. Xylose showed different utilization between the litter of the conifers and the broadleaves, with higher intensity in coniferous stands. For the sycamore, maple and beech stands, differences in utilization were observed only in five substrates—D-xylose, α-cyclodextrin, glycogen, α-ketobutyric acid and putrescine.

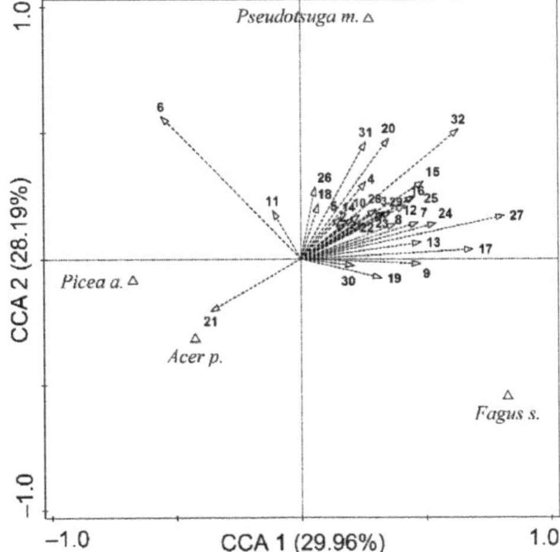

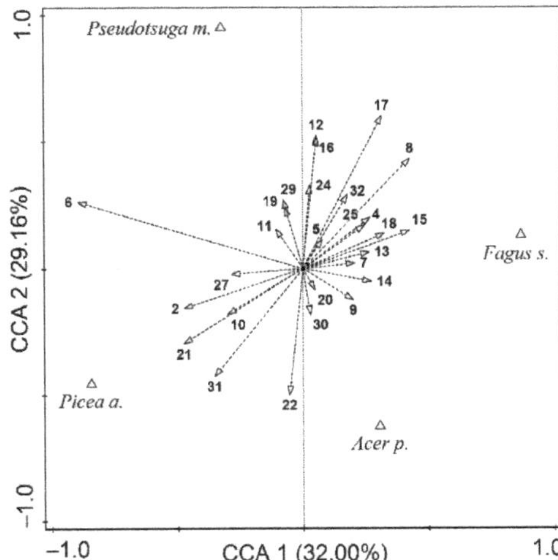

Figure 3. Canonical correspondence analysis based on community-level physiological profiles in different forest stands in O-horizon (**left**) and A-horizon (**right**). Arrows represent the direction of the steepest increase of activity within individual functional groups. Triangles represent individual tree species (2—β-methyl-D-glucoside; 3—D-galactonic acid γ-lactone; 4—L-arginine; 5—pyruvic acid methyl ester; 6—D-xylose; 7—D-galacturonic acid; 8—L-asparagine; 9—Tween 40; 10—i-erythritol; 11—2-hydroxy benzoic acid; 12—L-phenylalanine; 13—Tween 80; 14—D-mannitol; 15—4-hydroxy benzoic acid; 16—L-serine; 17—α-cyclodextrin; 18—N-acetyl-D-glucosamine; 19—γ-hydroxybutyric acid; 20—L-threonine; 21—glycogen; 22—D-glucosaminic acid; 23—itaconic acid; 24—glycyl-L-glutamic acid; 25—D-cellobiose; 26—glucose-1-phosphate; 27—α-ketobutyric acid; 28—phenylethyl-amine; 29—α-D-lactose; 30—D,L-α-glycerol phosphate; 31—D-malic acid; 32—putrescine).

In the A-horizon, the utilization pattern of substrates differed compared to litter. While in the Douglas fir stands, litter exhibited higher utilisation activity than the spruce litter, in the A-horizon four substrates were found (glycogen, D-glucosaminic acid, α-ketobutyric acid, D-malic acid) with higher utilization in soils under the spruce. In the soils under the beech stands, L-asparagine, α-cyclodextrin, glycyl-L-glutamic acid and α-D-lactose were more utilized, while D-xylose was less utilized than in the sycamore maple stands.

4. Discussion

Conversion of agricultural land to forest is known to change soil chemical properties, including carbon stocks serving as the main energy source for soil microbiota [24–26]. Shifts in vegetation cover, whether through afforestation or natural colonization of agricultural areas by trees, cause changes in carbon sequestration [27]. They also affect the storage of soil organic matter by changing the quality and quantity of litter entering the soil [28–31]. This has been the main focus of studies hitherto: the effect of tree species was mostly studied in relation to soil organic matter content, carbon sequestration and soil reaction. Several studies have shown that in forest soils more SOC was stored under coniferous trees in the upper horizons than under broadleaves, while in afforested agricultural soils SOC sequestration did not differ between broadleaf and coniferous trees [32]. Generally, soils under conifers were found to be more acid, with higher thickness of surface organic layer and C:N ratio, and less water-soluble substances leached from litter [8]. This is explained by the chemical composition of conifer litter containing more components recalcitrant to decomposition than broadleaf litter, which can result in litter accumulation on the soil surface, and the formation of acidic compounds [7]. Tree species also exert differential effects on soil fauna (microarthropods, earthworms, nematodes), while litter quality seems to be an important factor [33]. Soil fauna also mediates the effect of trees on soil properties such as layer thickness or carbon accumulation [34]. The dominant tree species have a greater effect on soil biota richness and composition than tree richness *per se* [35]. Changes in soil properties affect the function, structure and activity of the soil microbial community [36,37]. As the quality of litter and root exudates vary considerably among tree species, the effects on soil microbial community composition and microbial activity are expected to differ depending on dominant tree species, stand management and other factors, and need not necessarily be positive in terms of diversity and microbial functions [38].

Our results confirmed that a period of 45 years is long enough for the manifestation of changes in soil properties in the uppermost soil horizons triggered by the tree layer. The succession of all communities associated with trees (including soil microbiota) started in a relatively homogeneous area: pasture grasslands with small variation in the herbaceous vegetation, located within a narrow altitudinal range on a shallow slope with the same aspect, covered by the same soil type. Consequently, soil properties can reasonably be expected to have been quite similar across the area. In spite of this homogeneity, both soil physical and chemical properties and soil microbiota composition diverged during the 45 years, depending on the dominant tree species. However, the differences in successional trajectories cannot be simplified as contrasts between conifers and broadleaves. Multivariate analyses of physico-chemical soil properties indicated an overlap between the soils under the Douglas fir and the spruce, but a clear separation of beech from sycamore. Acidifying effect has frequently been attributed solely to conifers [11,39,40]. However, more recent studies contradict this assumption [14]. Soil pH and base saturation is often lower in stands dominated by beech compared to other temperate broadleaves [41,42]. This was confirmed by our study: no difference was observed in the O-horizon, and beech stands exhibited the same pH in the A-horizon as both conifers, while only the soils under the sycamore stands were less acidic, and richer in base cations. Conifers also exhibited generally higher C and N content in the O-horizon, but in the top 10 cm of mineral soil the C and N content was significantly higher only under the maple stands. Litter quality, especially with regard to the content of recalcitrant substances and decomposition rate, seems to be the main driver of nutrient cycling and soil chemical properties [42–44].

Changes in soil properties also mean changing living conditions for soil microbiota. Plant species are unique in their effects on the belowground system. Providing the matter decomposed by soil microorganisms, trees influence soil microbiota essentially in the same way as other plants, but their effect is potentially stronger because of a greater biomass [45,46]. The effect of afforestation on the composition, biomass and activity of microbial communities after afforestation is thus usually dramatic [31,37,47,48]. As in the case of physico-chemical properties, soil microbial community parameters in our study significantly differed depending on the stand-forming tree species, but again without a clear conifer-broadleaf contrast, and with different patterns in the soil horizons. In general, both microbial activity and diversity were, surprisingly, highest under the Douglas fir, followed by the sycamore, with the beech and the spruce showing mostly lower values. Utilization patterns of Biolog® substrates also differed between tree species, although by no means identically in both soil horizons. This is not surprising, as in the A-horizon the organic fraction represented only a small part of the soil mass compared to the surface organic layer, and had undergone chemical transformation.

5. Conclusions

Land cover in central Europe has undergone dramatic changes during the last few centuries, especially in mountainous areas. Initially, large areas were deforested to gain pastures and partly also arable land. Currently, this trend was reversed, and many former pastures were afforested or colonized by trees after being abandoned. The same applies to tree species composition of forest stands: since the 18th century Norway spruce monocultures gradually replaced natural broadleaved and mixed forests on a large scale because of its relatively fast growth and wood quality, but prolonged drought periods and climate warming during the last decades have led to increasing spruce mortality especially in pure stands. Alternatives to spruce are currently sought also among introduced species; Douglas fir, which seems to be more tolerant to heat and drought, is considered a suitable replacement of Norway spruce in many parts of Europe [49,50]. The knowledge of the effects of tree species on soil processes is thus indispensable for planning of the composition of future forests. Objects such as the Vrchdobroč area, where stands of various species have been planted in a relatively homogeneous area and have undergone various sylvicultural treatments, are invaluable for studying the effects on soil properties and soil microbial community.

Supplementary Materials: The following are available online at https://www.mdpi.com/xxx/s1, Table S1: Means and results of Tukey's HSD tests for differences between tree species of microbial activity of functional groups (utilization of different carbon sources)

Author Contributions: Conceptualization, E.G.; methodology, E.G.; validation, E.G., R.G. and D.G.; formal analysis, M.K., J.Ž. and D.G.; investigation, E.G. and M.K.; data curation, E.G. and J.Ž.; writing—original draft preparation, R.G.; writing—review and editing, E.G. and D.G.; visualization, D.G., R.G. and J.Ž.; supervision, E.G.; project administration, E.G.; funding acquisition, E.G. All authors have read and agreed to the published version of the manuscript.

Funding: This research was funded by the Scientific Grant Agency of Ministry of Education, Science, Research and Sport of the Slovak Republic, project VEGA No. 1/0115/21, and the Slovak Research and Development Agency, project numbers APVV-19-0142 and APVV-19-0183.

Informed Consent Statement: Not applicable.

Data Availability Statement: The data presented in this study are available on request from the corresponding author.

Acknowledgments: We thank Jozef Capuliak and Terézia Dvorská for technical and laboratory assistance.

Conflicts of Interest: The authors declare no conflict of interest.

References

1. Augusto, L.; Ranger, J.; Binkley, D.; Rothe, A. Impact of several common tree species of European temperate forests on soil fertility. *Ann. For. Sci.* **2002**, *59*, 233–253. [CrossRef]
2. Bonan, G.B. Forests and climate change: Forcings, feedbacks, and the climate benefits of forests. *Science* **2008**, *320*, 1444–1449. [CrossRef] [PubMed]
3. Gömöryová, E.; Hrivnák, R.; Janišová, M.; Ujházy, K.; Gömöry, D. Changes of the functional diversity of soil microbial community during the colonization of abandoned grassland by a forest. *Appl. Soil. Ecol.* **2009**, *43*, 191–199. [CrossRef]
4. Augusto, L.; De Schrijver, L.A.; Vesterdal, L.; Smolander, A.; Prescott, C.; Ranger, J. Influences of evergreen gymnosperm and deciduous angiosperm tree species on the functioning of temperate and boreal forests. *Biol. Rev.* **2015**, *90*, 444–466. [CrossRef] [PubMed]
5. Haesen, S.; Lembrects, J.J.; De Frenne, P.; Lenoir, J.; Aalto, J.; Ashcroft, M.B.; Kopecký, M.; Luoto, M.; Maclean, I.; Nijs, I.; et al. Forest Temp–Sub-canopy microclimate temperatures of European forests. *Glob. Chang. Biol.* **2021**, *27*, 6307–6319. [CrossRef] [PubMed]
6. Finzi, A.C.; Canham, C.D. Non-additive effects of litter mixtures on net N mineralization in a southern New England forest. *For. Ecol. Manag.* **1998**, *105*, 129–136. [CrossRef]
7. Berg, B. Litter decomposition and organic matter turnover on northern forest soils. *For. Ecol. Man.* **2000**, *133*, 13–22. [CrossRef]
8. Smolander, A.; Kitunen, V. Soil microbial activities and characteristics of dissolved organic C and N in relation to tree species. *Soil Biol. Biochem.* **2002**, *34*, 651–660. [CrossRef]
9. Saetre, P.; Bååth, E. Spatial variation and patterns of soil microbial community structure in a mixed spruce-birch stand. *Soil Biol. Biochem.* **2000**, *32*, 909–917. [CrossRef]
10. Vesterdal, L.; Schmidt, I.K.; Callesen, I.; Nilsson, L.O.; Gundersen, P. Carbon and nitrogen in forest floor and mineral soil under six common European tree species. *For. Ecol. Manag.* **2008**, *255*, 35–48. [CrossRef]
11. Brand, D.G.; Kehoe, P.; Connors, M. Coniferous afforestation leads to soil acidification in central Ontario. *Can. J. For. Res.* **1986**, *16*, 1389–1391. [CrossRef]
12. Zwanzig, L.; Zwanzig, M.; Sauer, M. Outcomes of a quantitative analysis of 48 soil chronosequence studies in humid mid and high latitudes: Importance of vegetation in driving podsolization. *Catena* **2021**, *196*, 104821. [CrossRef]
13. Mareschal, L.; Bonnaud, P.; Turpault, M.P.; Ranger, J. Impact of common European tree species on the chemical and physicochemical properties of fine earth: An unusual pattern. *Soil Sci.* **2010**, *61*, 14–23. [CrossRef]

14. Achilles, F.; Tischer, A.; Bernhardt-Römermann, M.; Heinze, M.; Reinhardt, F.; Makeschin, F.; Michalzik, B. European beech leads to more bioactive humus forms but stronger mineral soil acidification as Norway spruce and Scots pine—Results of a repeated site assessment after 63 and 82 years of forest conversion in Central Germany. *For. Ecol. Manag.* **2021**, *483*, 118769. [CrossRef]
15. Šály, R.; Capuliak, J.; Pavlenda, P. Vrchdobroč—Sľubný objekt k štúdiu pôdotvorného procesu pod vplyvom zalesnenia. In *Diagnostika, Klasifikácia a Mapovanie Pôd*; Sobocká, J., Ed.; VÚPOP: Bratislava, Slovakia, 2011; pp. 25–31. (In Slovak)
16. Alef, K.; Nannipieri, P. *Methodenhandbuch Bodenmikrobiologie*; Academic Press: London, UK, 1995; 576p.
17. Khazijev, F.C. *Fermentativnaja Aktivnost' Počv*; Metodičeskoje Posobje: Moskva, Russia, 1976; p. 180. (In Russian)
18. Islam, K.R.; Weil, R.R. Microwave irradiation of soil for routine measurement of microbial biomass carbon. *Biol. Fertil. Soils* **1998**, *27*, 408–416. [CrossRef]
19. Kandeler, E. Bestimmung der N-mineralisation im anaeroben Brutversuch. In *Bodenbiologische Arbeitsmethoden*; Schinner, F., Ohlinger, R., Kandeler, E., Margesin, E.R., Eds.; Springer: Heidelberg, Germany, 1993; pp. 160–161. (In German)
20. Insam, H. A new set of substrates proposed for community characterization in environmental samples. In *Microbial Communities: Functional Versus Structural Approaches*; Insam, H., Rangger, A., Eds.; Springer: Berlin, Germany, 1997; pp. 260–261.
21. Hill, M.O. Diversity and evenness: A unifying notation and its consequences. *Ecology* **1973**, *54*, 427–432. [CrossRef]
22. StatSoft, Inc. *STATISTICA Data Analysis Software System, Version 12.0*; StatSoft, Inc.: Tulsa, OH, USA, 2013.
23. Ter Braak, C.J.F.; Smilauer, P. *CANOCO Reference Manual and User's Guide to Canoco for Windows: Software for Canonical Community Ordination (Version 4)*; Centre of Biometry Wageningen: Winnipeg, MA, Canada, 2002; p. 353.
24. Lal, R. Forest soils and carbon sequestration. *For. Ecol. Manag.* **2005**, *220*, 242–258. [CrossRef]
25. Armolaitis, K.; Aleinikoviene, J.; Baniuniene, A.; Lubyte, J.; Zekaite, V. 2007 Carbon sequestration and nitrogen status in Arenosols following afforestation or following abandonment of arable land. *Balt. For.* **2007**, *13*, 169–178.
26. Cerli, C.; Celi, L.; Kaiser, K.; Guggenberger, G.; Johansson, M.B.; Cignetti, A.; Zanini, E. Changes in humic substances along an age sequence of Norway spruce stands planted on former agricultural land. *Org. Geochem.* **2008**, *39*, 1269–1280. [CrossRef]
27. Kuzyakov, Y. Sources of CO2 efflux from soil and review of partitioning methods. *Soil Biol. Biochem.* **2006**, *38*, 425–448. [CrossRef]
28. Hiltbrunner, D.; Zimmermann, S.; Hagedorn, F. Afforestation with Norway spruce on a subalpine pasture alters carbon dynamics but only moderately affects soil carbon storage. *Biogeochemistry* **2013**, *115*, 251–266. [CrossRef]
29. Guidi, C.; Magid, J.; Rodeghiero, M.; Gianelle, D.; Vesterdal, L. Effects of forest expansion on mountain grassland: Changes within soil organic carbon fractions. *Plant Soil* **2014**, *385*, 373–387. [CrossRef]
30. Saiz, G.; Bird, M.; Wurster, C.; Quesada, C.A.; Ascough, P.L.; Domingues, T.; Schrodt, F.; Schwarz, M.; Feldpausch, T.R.; Veenendaal, E.M.; et al. The influence of C_3 and C_4 vegetation on soil organic matter dynamics in contrasting semi-natural tropical ecosystems. *Biogeosciences* **2015**, *12*, 5041–5059. [CrossRef]
31. Ortiz, C.; Fernandez-Alonsó, M.J.; Kitzler, B.; Díaz-Pines, E.; Saiz, G.; Rubio, A.; Benito, M. Variations in soil aggregation, microbial community structure and soil organic matter cycling associated to long-term afforestation and woody encroachment in a Mediterranean alpine ecotone. *Geoderma* **2022**, *405*, 115450. [CrossRef]
32. Hübblová, L.; Frouz, J. Contrasting effect of coniferous and broadleaf trees on soil carbon storage during reforestation of forest soils and afforestation of agricultural and post-mining soils. *J. Environ. Manag.* **2021**, *290*, 112567. [CrossRef] [PubMed]
33. Peng, Y.; Holmstrup, M.; Kappel Schmidt, I.; De Schrijver, A.; Schelfhout, S.; Heděnec, P.; Zheng, H.; Ruggiero Bachega, L.; Yue, K.; Vesterdal, L. Litter quality, mycorrhizal association, and soil properties regulate effects of tree species on the soil fauna community. *Geoderma* **2022**, *407*, 115570. [CrossRef]
34. Frouz, L.; Livečková, M.; Albrechtová, J.; Chroňáková, A.; Cajthaml, T.; Pizl, V.; Háněl, L.; Starý, J.; Baldrian, P.; Lhotáková, Z.; et al. Is the effect of trees on soil properties mediated by soil fauna? A case study from post-mining sites. *For. Ecol. Manag.* **2013**, *309*, 87–95. [CrossRef]
35. Tedersoo, L.; Bahram, M.; Cajthaml, T.; Polme, S.; Hiiesalu, I.; Anslan, S.; Harend, H.; Buegger, F.; Pritsch, K.; Koricheva, J.; et al. Tree diversity and species identity effects on soil fungi, protists and animals are context dependent. *ISME J.* **2016**, *10*, 346–362. [CrossRef]
36. Curiel Yuste, J.; Barba, J.; Fernández-Gonzalez, J.A.; Fernández-López, M.; Mattana, S.; Nolis, P.; Lloret, F. Changes in soil bacterial community triggered by drought-induced gap succession preceded changes in soil C stocks and quality. *Ecol. Evol.* **2012**, *2*, 3016–3031. [CrossRef]
37. Wang, K.; Zhang, Y.W.; Tang, Z.S.; Shangguan, Z.P.; Chang, F.; Jia, F.; Chen, Y.P.; He, X.H.; Shi, W.Y.; Deng, L. Effects of grassland afforestation on structure and function of soil bacterial and fungal communities. *Sci. Total Environ.* **2019**, *676*, 396–406. [CrossRef]
38. Ding, L.L.; Wang, P.C. Afforestation suppresses soil nitrogen availability and soil multifunctionality on a subtropical grassland. *Sci. Total Environ.* **2021**, *761*, 143663. [CrossRef] [PubMed]
39. Hornung, M. Acidification of soils by trees and forests. *Soil Use Manag.* **1985**, *1*, 24–27. [CrossRef]
40. Côté, B.; Fyles, J.W. Nutrient concentration and acid–base status of leaf litter of tree species characteristic of the hardwood forest of southern Quebec. *Can. J. For. Res.* **1994**, *24*, 192–196. [CrossRef]
41. Guckland, A.; Jacob, M.; Flessa, H.; Thomas, F.M.; Leuschner, C. Acidity, nutrient stocks, and organic-matter content in soils of a temperate deciduous forest with different abundance of European beech (*Fagus sylvatica* L.). *J. Plant Nutr. Soil Sci.* **2009**, *172*, 500–511. [CrossRef]
42. Langenbruch, C.; Helfrich, M.; Flessa, H. Effects of beech (*Fagus sylvatica*), ash (*Fraxinus excelsior*) and lime (*Tilia* spec.) on soil chemical properties in a mixed deciduous forest. *Plant Soil* **2012**, *352*, 389–403. [CrossRef]

43. Desie, E.; Vancampenhout, K.; Heyens, K.; Hlava, J.; Verheyen, K.; Muys, B. Forest conversion to conifers induces a regime shift in soil process domain affecting carbon stability. *Soil Biol. Biochem.* **2019**, *136*, 107540. [CrossRef]
44. Haghverdia, K.; Kooch, Y. Effects of diversity of tree species on nutrient cycling and soil-related processes. *Catena* **2019**, *178*, 335–344. [CrossRef]
45. Eisenhauer, N.; Bessler, H.; Engels, C.; Gleixner, G.; Habekost, M.; Milcu, A.; Partsch, S.; Sabais, A.C.W.; Scherber, C.; Steinbeiss, S.; et al. Plant diversity effects on soil microorganisms support the singular hypothesis. *Ecology* **2010**, *91*, 485–496. [CrossRef]
46. Pfeiffer, B.; Fender, A.C.; Lasota, A.; Hertel, D.; Jungkunst, H.F.; Daniel, R. Leaf litter is the main driver for changes in bacterial community structures in the rhizosphere of ash and beech. *Appl. Soil Ecol.* **2013**, *72*, 150–160. [CrossRef]
47. Chen, C.R.; Condron, L.M.; Xu, Z.H. Impacts of grassland afforestation with coniferous trees on soil phosphorus dynamics and associated microbial processes: A review. *For. Ecol. Manag.* **2008**, *255*, 396–409. [CrossRef]
48. Chen, X.D.; Condron, L.M.; Dunfield, K.E.; Wakelin, S.A.; Chen, L.J. Impact of grassland afforestation with contrasting tree species on soil phosphorus fractions and alkaline phosphatase gene communities. *Soil Biol. Biochem.* **2021**, *159*, 108274. [CrossRef]
49. Zeller, B.; Legout, A.; Bienaimé, S.; Gratia, B.; Santenoise, P.; Bonnaud, B.; Ranger, J. Douglas fir stimulates nitrifcation in French forest soils. *Sci. Rep.* **2019**, *9*, 10687. [CrossRef] [PubMed]
50. Schmid, M.; Pautasso, M.; Holdenrieder, O. Ecological consequences of Douglas fir (*Pseudotsuga menziesii*) cultivation in Europe. *Eur. J. Forest. Res.* **2014**, *133*, 13–29. [CrossRef]

 diversity

Article

Which Are the Best Site and Stand Conditions for Silver Fir (*Abies alba* Mill.) Located in the Carpathian Mountains?

Lucian Dinca [1], Mirabela Marin [1], Vlad Radu [2], Gabriel Murariu [3], Romana Drasovean [3], Romica Cretu [3], Lucian Georgescu [3] and Voichiţa Timiş-Gânsac [4,*]

1 "Marin Drăcea" National Institute for Research and Development in Forestry, 13 Closca Street, 500040 Brasov, Romania; dinka.lucian@gmail.com (L.D.); mirabelamarin@yahoo.com (M.M.)
2 "Marin Drăcea" National Institute for Research and Development in Forestry, 73BIS Calea Bucovinei Street, 725100 Campulung Moldovenesc, Romania; vlad.radu2@gmail.com
3 Chemistry, Physics, and Environment Department, Faculty of Sciences and Environment, "Dunărea de Jos" University of Galaţi, No. 111 Street Domnească, 800201 Galaţi, Romania; gmurariu@ugal.ro (G.M.); romana.drasovean@ugal.ro (R.D.); romica.cretu@ugal.ro (R.C.); lucian.georgescu@ugal.ro (L.G.)
4 Department of Forestry and Forest Engineering, University of Oradea, 410048 Oradea, Romania
* Correspondence: timisvoichita@yahoo.com

Abstract: Silver fir (*Abies alba* Mill.) is one of the most valuable and productive tree species across European mountains, that accomplish multiple economic, protective and ecologic functions. Alongside spruce (*Picea abies* (L.) Karst) and beech (*Fagus sylvatica* L.), silver fir is a characteristic species for the Romanian Carpathians. Although silver fir tree is recommended for the diversification of forests in order to increase the resistance to climate change, it is very sensitive to climatic excesses, especially those that proceed rapidly. Therefore, the aim of this study is to investigate both the environmental conditions and stand characteristics of fir from five mountain ranges of the Romanian Carpathians. The study is based on data recorded over a period of 10 years (1990–2000). As such, a total of 77,251 stands that occupy 211,954 hectares have been investigated in regard to silver fir behaviour. MATLAB scripts were used for analysing consistent data volumes as well as the impact of eight factors on the silver fir productivity (altitude, field aspect, field slope, soil type, participation percentage, road distance, structure and consistency). Our analysis has revealed that higher silver fir productivity is found at altitudes of up to 1200 m, on mid and upper slopes, on NW field aspects, on eutric cambisols and dystric cambisols, with a 10–20% participation in stand composition and in relatively-even aged stands with a full consistency. This study offers valuable insights for forest managers that require comprehensive information in adopting effective strategies to enhance forest resilience under climate change.

Keywords: silver fir; Romanian Carpathians; stand productivity; forest sustainability

Citation: Dinca, L.; Marin, M.; Radu, V.; Murariu, G.; Drasovean, R.; Cretu, R.; Georgescu, L.; Timiş-Gânsac, V. Which Are the Best Site and Stand Conditions for Silver Fir (*Abies alba* Mill.) Located in the Carpathian Mountains? *Diversity* **2022**, *14*, 547. https://doi.org/10.3390/d14070547

Academic Editor: Panayiotis Dimitrakopoulos

Received: 20 June 2022
Accepted: 30 June 2022
Published: 7 July 2022

Publisher's Note: MDPI stays neutral with regard to jurisdictional claims in published maps and institutional affiliations.

1. Introduction

Considering the essential role of forests in providing many ecosystem services, the scientific community has been interested in the last years in evaluating the connections between forest diversity and their productivity [1,2]. Climatic changes influence forest productivity both directly (by altering temperature and rainfall patterns) [3,4] as well as indirectly (through the action of different harmful and pathogen agents) [5–7]. Even though forest productivity at European level has followed a growing trend, forecasts regarding the growth of extreme phenomena in intensity and frequency will alter water resources and, implicitly, forest productivity through water-limiting [2,5,8,9].

Silver fir (*Abies alba* Mill.) is considered one of the most valuable and productive species from mountainous areas of Europe, due to its ecologic, economic and soil protection functions [10,11]. Due to its high productivity, silver fir is recommended in forest diversification in order to enhance the resistance towards climatic changes [12] as well as for protecting carbon stocks from European forests during the following centuries [13,14]. European forests, comprised mainly of silver fir, common beech (*Fagus sylvatica* L.) and Norway spruce (*Picea abies* (L.) Karst) and widespread on over 10 million hectares, are very sensitive towards climatic changes, especially those that occur more rapidly and do not allow trees to adapt over time [15,16]. Silver fir is one of the species with high requirements towards stational conditions, having high requests towards soil and air humidity [17–19]. Very sensitive towards climatic excesses, especially during youth, seedlings suffer from drought, insolation, late frosts, excessive frost, cold and dry winds and soil and air dryness. Hence, during the last decades, silver fir faced a European decline due to atmospheric pollution and climatic changes, that led to a decrease up to 80% within the Carpathians [20,21].

Romania owns an important part of Europe's natural resources. Some of the largest compact forests from the temperate area are located in our country, together with the oldest forests from Europe [22,23]. Due to the fact that mountain areas occupy 23% of the total global forest surface, a sustainable management of these ecosystems is emphasized in the 2030 Agenda for Sustainable Development [24,25]. Romania holds over 50% of the Carpathians' surface, considered the largest mountain chain from Central Europe [26]. Silver fir occupies 5% of their afforested surface, being widespread in the Eastern Carpathians, Southern Carpathians, Curvature Carpathians, Apuseni and Banat Mountains [19].

If we consider silver fir's ecologic and economic importance, as well as its sensibility towards climate excesses, numerous international and national studies focused on assessing the influence of stational conditions on the specie's development. As such, some authors [27] pursued to identify the role of soil properties on the growth of silver fir and have shown that soil parameters influence the growth of this species. Furthermore, Pinto [28] mentions that, besides soil properties, climate and stand composition also control the growth and development of silver fir. On the other hand, Bosela's study [29] has shown that the specie's genetic diversity has an important role in the silver fir's growth and resistance towards climatic changes. Mohytych [30] intended to identify the most favourable conditions for the regeneration of this species. In addition, research by Lasch and Sperlich [9,31] have emphasized that silver fir's productivity will decrease in the following years as a consequence of climatic changes.

Studies conducted in Romania have shown that some resinous species, among which *Abies alba* Mill., are affected by *Viscum album* ssp. abietis. Mistletoe attack is among the factors that degrades old silver tree stands [7,17,32]. The study performed by Dincă and his collaborators [33] shows that silver fir stands located on the north slopes of Southern Carpathians evince superior productivity compared to the south slopes. The study realized by Tudoran [34] emphasizes that silver tree record height growths up to the age of 80, reaching a maximum between 50 and 60 years after which, trees accumulate growths in diameter. Dendrochronological growths performed by Kern and Popa [35] have shown that the radial growth of silver tree is positively correlated with both abundant precipitation from spring and summer and mild winters, as well as with temperatures recorded during summer months and early winter [36]. On the other hand, studies conducted by Dinulică shown that a radial growth is influenced by wind. However, an important role in mitigating the impact of climate changes is posed by the genetic variation of local, regional and national silver fir populations, as mentioned by Teodosiu [19,37].

Therefore, the purpose of this study is to identify the key characteristics (stational or stand) that lead to the occurrence of superior productivity classes in silver fir from the five Romanian Carpathians Mountain chains. In order to reach this goal, we have established two specific objectives, namely: (1) evaluating stational conditions, namely altitude, field aspect, slope and soil type in each of the five areas; and (2) evaluating stand conditions, namely participation percentage, distance from the road, stand structure and consistency within the five mountain chains. The research's outcomes are useful for forest managers in establishing the most adequate management measures that can increase the resilience of silver fir in the context of different changes (climatic and environmental).

Climate factors are important for the specific composition of forests, while stational conditions can determine sensible changes at smaller scales. Amongst these, the topographic ones influence both forest structure, species distribution and diversity as well as floristic composition [38].

The studies performed by Sidor and his collaborators in Banat (Romania), have shown that the average surface growth for silver tree presents almost double values, compared with Scots pine and larch. Furthermore, silver fir had a positive response to the precipitations from the beginning of the vegetation season [39]. The positive influence of temperature from the cold season on the radial growth of silver fir was emphasized in other areas from the Romanian Carpathians as well as from Europe [36,40,41].

Studies focused on silver fir behaviour in south-east Carpathians revealed both a negative correlation between radial growth and temperature, as well as a positive correlation between radial growth and precipitation during the growth season.

The aim of this study is to investigate both the environmental conditions and stand characteristics of fir from five mountain ranges of the Romanian Carpathians.

The working hypotheses were:

(a) there is a strong connection and a certain dependence between the production classes from the forest areas of different mountain massifs?

(b) are the productivity classes are influenced by a series of parameters?

As the preliminary investigation the distributions of the studied parameters was a priori performed, it was observed that the obtained results do not always respect the conditions of normal distribution.

2. Materials and Methods

2.1. Study Area

The study area comprehended all silver fir stands from the five Romanian Carpathians Mountain chains, namely: Apuseni Mountains (AM), Banatului Mountains (BM), Eastern Carpathians (EC), Curvature Carpathians (CC) and Southern Carpathians (SC), that comprises silver fir in their composition (Figure 1).

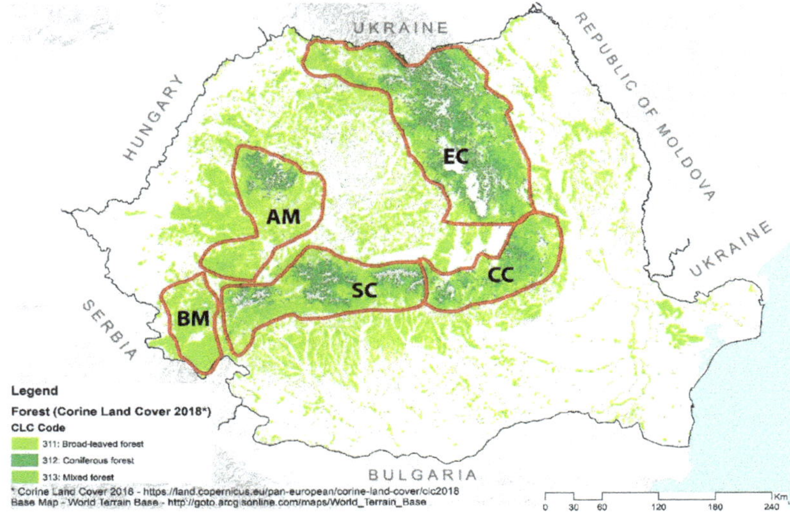

Figure 1. Carpathian Mouintains: Apuseni Mountains (AM), Banatului Mountains (BM), Eastern Carpathians (EC), Curvature Carpathians (CC), Souuthern Carpathians (SC).

The analysed silver fir stands were: in the AM: 4203 stands and 10,143 ha; in the BM: 4185 stands and 11,980 ha; in the EC: 37,385 stands, 103,020 ha; in the CC: 15,304 stands, 41,778 ha; in the SC: 16,174 stands, 45,024 ha. Total Carpathians: 77,251 stands, 211,945 ha.

2.2. Data Collection and Methods

For each stand, we used information from forest management plans for 1990–2000 period. Data regarding productivity classes were extracted in order to analyse the influence of environment factors (altitude, field aspect, field slope and soil type) and stand factors (participation percentage, road distance, structure and consistency) on the productivity of silver fir stands. Data from forest management plans for 1990–2000 period were used [42]. This period was chosen as it precedes the massive transmission of forests towards previous owners, in which case, complete forest management plans are no longer requested.

The altitude was measured on the field with GPS devices, the field aspect was established based on the area's mapping. The fields slope was measured with a hypsometer. The soil type was determined by realizing soil profiles, gathering samples and analysing them in specialty laboratories. The participation percentage in the stand's composition, as well as stand structure and consistency were established visually. Road distance and altitude was analysed from maps with level curves.

In the case of even-aged stands, the site index can be established based on the average height and age. For uneven-aged stands, this value can be determined by the average height that corresponds to the average diameter of 50 cm. The site index reports the height of dominant and co-dominant trees in a stand at a base age. Thus, it is used to measure the site productivity, to determine site management options and to describe the potential of forest trees to grow at a particular site. In our case, the site index was calculated for each stand.

In order to determine the optimal stational and stand conditions for silver fir, we have grouped all values for the first and second productivity classes (CP) and afterwards compared them with the sum of the other values (belonging to classes productivity 3 + 4 + 5).

2.3. Statistical Calculations

The data used in this study was processed with StatSoft Statistica and MATLAB statistical analysis scripts which is specific for a consistent volume of data [43]. In order to verify the distribution type in the case of all parameters we applied the Kolmogorov-Smirnov tests. As it is already known, the Kolmogorov-Smirnov test is used for testing large sets of data [44].

The parametric ANOVA and Non-Parametric Tests, such as Kruskal-Wallace Test, were applied simultaneously to highlight the simultaneously to highlight the consistency of the addiction investigation.

Moreover, the comparison T statistical tests of Fisher two tails type were applied for observing significant statistical differences between the values of the studied parameters. In the case where the p threshold parameter's value was under 0.05, we have considered the differences between averages as significant and added an observation in the presentation's text. If the threshold's value is superior to the imposed value, the observed differences were considered as insignificant.

3. Results

3.1. Altitude

Regarding the distribution of altitude for the silver fir stands for each of the five area (Table 1; Figure 2) we can observe that, in AM, EC, SC and Total, the high productivity silver tree is founded at altitudes between 857–1042. In contrast, in the BM, the upper silver fir productivity class (CP) occurs at high altitudes (over 900 m), while in the case of CC, high attitudes are responsible for the lower-class fir (996 m). For the Total of the five mountain ranges analysed, the upper-class fir is found at low altitudes (800–900 m).

Table 1. Average altitude (m) for silver fir trees located in the Carpathians.

Mountain Area	Altitude (m) for Productivity Classes				
	1	2	3	4	5
Apuseni Mountains = AM	870	1020	1054	1035	1055
Banatului Mountains = BM	939	915	845	864	879
Eastern Carpathians = EC	886	940	962	950	1010
Curvature Carpathians = CC	1011	1003	996	1034	1026
Southern Carpathians = SC	1042	1063	1096	1105	1110
Total Carpathians	857	917	972	986	1033

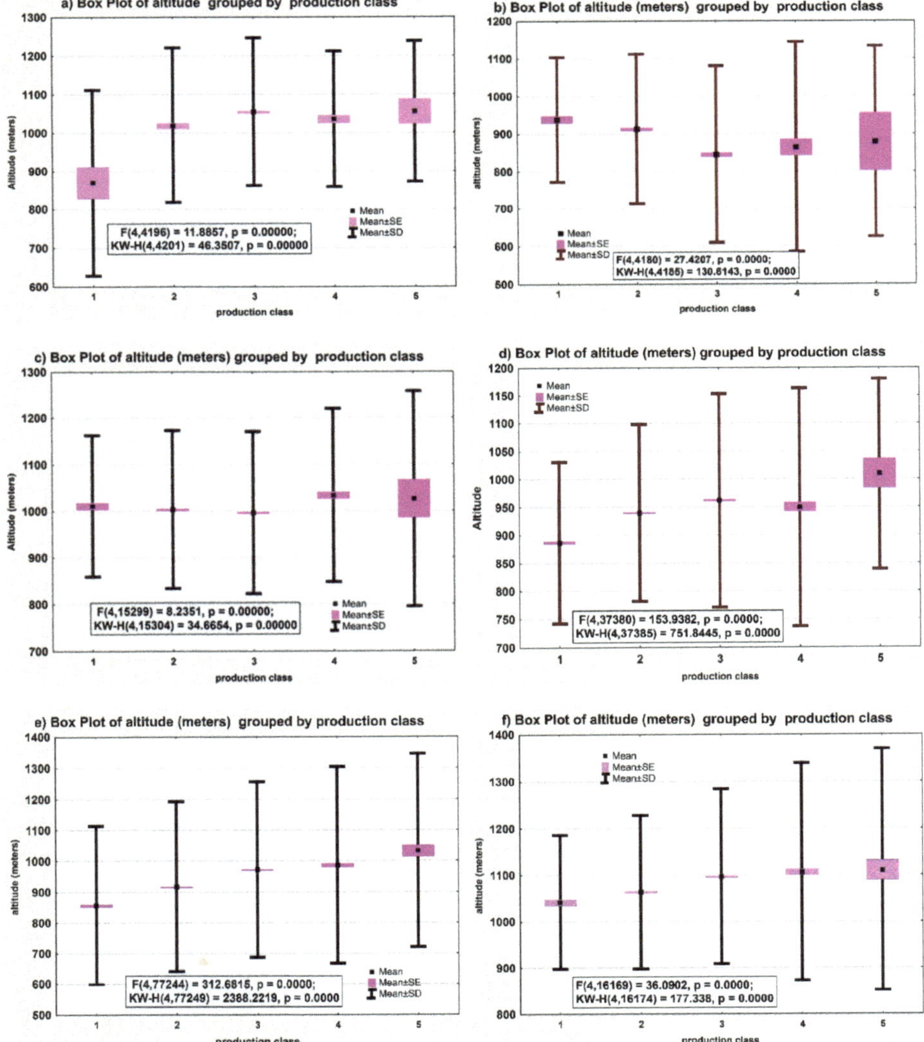

Figure 2. (**a**) Altitude AM. (**b**) Altitude BM. (**c**) Altitude EC. (**d**) Altitude CC. (**e**) Altitude SC. (**f**) Altitude TOTAL.

In this sense, we have applied comparison T statistical tests of Fisher type which shown that the height distribution for the superior class has a significantly lower average than the other distributions ($p = 0.0001$). These tests have proved significant differences for both T one tail tests as well as for two tails.

We investigated the existence of significant differences in the average values obtained for the series of parameters related to each production class. We were able to demonstrate by using Fisher-type tests that there were no significant differences.

In principle—according to the reference [26], the preliminary evaluation of the distribution of the studied parameters was performed. The obtained distributions do not always respect the conditions of normal distribution. Under these conditions, parametric and non-parametric ANOVA analyses were applied.

3.2. Field Aspect

Silver fir has the largest propagation in the Romanian Carpathians on the NW field aspect (Figure 3a). This phenomenon is more significant in superior and average productivity stands. The Curvature Carpathians do not record field aspect differences, probably due to their shape (the curvature comes from their shape, with a transition from N-S to E-W). The same context is present in Carpathians Total (Figure 3b).

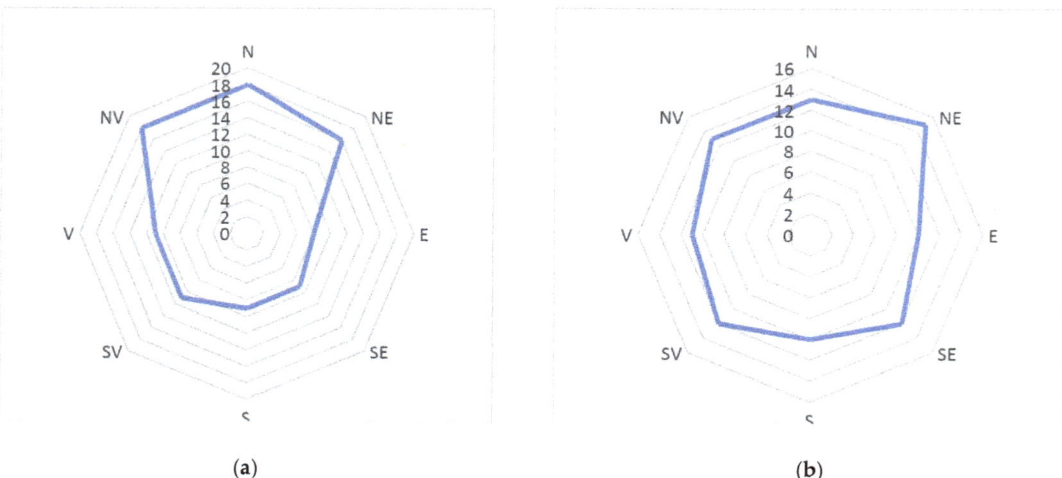

Figure 3. (**a**) AM-field aspect for classes 2 prod. (**b**) Total-Carpathians field aspect.

Analysing the data from Table 2 see can see a preference for shadowed field aspects (for AM, EC and SC). In the case of the BM, on the other hand, the shaded exposures host fir trees of lower productivity classes, while for the CC, there are no differences in exposures, probably due to their shape (the curvature comes from their shape, with a transition from N-S direction to E-V).

3.3. Field Slope

The field slope on productivity classes is given in Figure 4, for each of the five areas. Analysing the figures below, we distinguish the fact that significant differences between the productivity class and the slope of the land appear at the lower productivity classes (CP4 and CP5) in each of the five mountain ranges analysed. The same situation is encountered in the case of Total Carpathians where productivity classes 4 and 5 show the most significant differences.

Table 2. The distribution of silver fir trees from the Carpathians on altitudes (%).

Mountains	Prod Class	N	NE	NV	E	SE	S	SV	V
AM	1	13	13	22	12	16	9	9	6
	2	18	16	18	8	9	9	11	11
	3	16	15	17	9	11	7	13	12
	4	17	14	15	9	11	10	11	13
	5	7	3	23	6	13	23	19	6
	Total	17	15	17	9	10	7	13	12
BM	1	9	17	18	9	16	10	12	9
	2	13	12	15	10	14	13	15	8
	3	14	13	15	9	13	12	13	11
	4	16	12	14	8	13	16	9	12
	5	27	9	-	9	-	28	18	9
	Total	14	13	15	9	14	12	13	10
EC	1	14	16	16	11	13	8	12	10
	2	14	14	17	9	13	10	12	11
	3	14	14	16	10	13	10	12	11
	4	12	12	15	14	11	15	11	10
	5	18	9	12	14	12	12	18	5
	Total	14	14	17	9	13	10	12	11
CC	1	10	14	17	9	16	9	13	12
	2	12	15	17	9	12	10	14	11
	3	15	15	14	9	11	12	12	12
	4	14	12	15	9	11	12	11	16
	5	6	7	13	7	19	19	13	16
	Total	14	15	15	9	11	11	13	12
SC	1	10	17	16	12	12	10	11	12
	2	13	16	19	12	10	8	11	11
	3	12	17	19	10	10	7	12	13
	4	11	13	22	12	10	7	12	13
	5	14	12	26	8	6	9	10	15
	Total	13	16	19	11	10	7	12	12
TOTAL	1	13	16	16	11	13	8	12	11
	2	13	15	17	10	12	10	12	11
	3	14	15	16	9	11	10	13	12
	4	13	12	18	11	11	11	12	12
	5	13	10	21	9	9	13	13	12
	Total	13	15	17	10	12	10	12	11

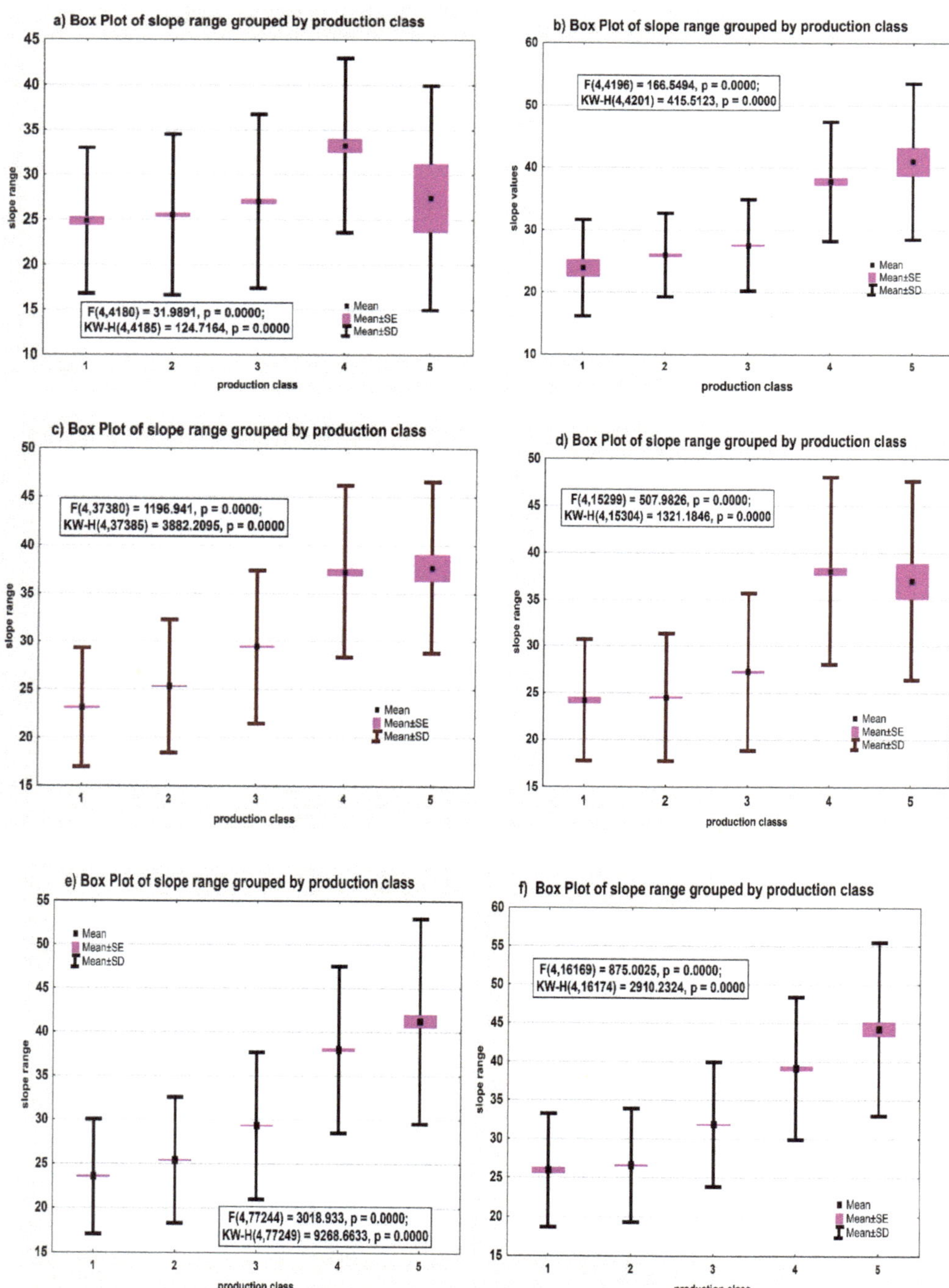

Figure 4. (**a**) Field slope AM. (**b**) Field slope B. (**c**) Field slope EC. (**d**) Field slope CC. (**e**) Field slope SC. (**f**) Field slope Total.

Low productivity silver firs are located in all cases on fields with a very high slope (27–44%) (Table 3). In this regard, we have applied T statistic tests of Fisher type and proved that height distribution on superior classes has a significantly lower average than the other distributions ($p = 0.0001$). These tests showed significant differences for both one tail and two tails *T* tests.

Table 3. Average slope for silver firs located in the Carpathians.

Mountain Area	Slope (%) for Productivity Classes				
	1	2	3	4	5
Apuseni Mountains = AM	23.88	25.97	27.56	37.80	41.00
Banatului Mountains = BM	24.85	25.55	27.05	33.27	27.45
Eastern Carpathians = EC	23.14	25.34	29.46	37.25	37.70
Curvature Carpathians = CC	24.21	24.54	27.26	38.04	37.03
Southern Carpathians = SC	25.94	26.63	31.91	39.17	44.27
Total Carpathians	23.54	25.41	29.33	38.01	41.27

3.4. Soil Type and Subtype

The graphical representation of the main soil types and subtypes is shown in Figure 5. As we can see, there are no significant differences between the 2nd and 3rd productivity classes in any of the studied mountain areas. We can observe that calcic eutric cambisol is largest in BM (9%), while is very few SC (under 1%). Entic podzol is low at BM and more at EC (9%). For Total Carpathians, eutric cambisol and dystric cambisol occupy the same area (38%), with biggest differences between the two soils at AM, where dystric cambisol is larger than eutric cambisol (41% comapred with 28%).

The distribution of silver fir by soil types and subtypes for all five areas is shown in Table 4. Due to the large number of soil subtypes, we have taken into account only the 2nd (CP2) and 3rd productivity classes (CP3) as well as 3 soil types (eutric cambisol, dystric cambisol, entic podzol) + 2 subtypes (calcic eutric cambisol and lythic dystric cambisol), (according to the WRB soil classification).

Table 4. Soil type for silver fir stands located in the Carpathians.

Mountain Area	Soil (%)					
	Eutric Cambisol	Calcic Eutric Cambisol	Dystric Cambisol	Lythic Dystric Cambisol	Entic Podzol	Other Soils
Apuseni Mountains = AM CP2	35	3	43	1	0	18
AM-CP3	26	5	43	4	0	22
AP-Total	28	5	41	5	3	18
Banatului Mountains = BM CP2	37	11	41	0	0	11
BM-CP3	41	8	34	1	0	16
BM-Total	38	9	39	1	0	13
Eastern Carpathians = EC CP2	43	1	44	3	0	9
EC-CP3	30	0	37	12	2	19
EC-Total	38	1	42	5	9	5
Curvature Carpathians = CC CP2	48	0	34	2	1	15
CC-CP3	39	0	44	7	2	8
CC-Total	42	0	38	6	2	12
Southern Carpathians = SC CP2	46	0	25	3	2	24
SC-CP3	32	0	36	7	6	19
SC-Total	35	0	29	7	4	25
Total Carpathians CP2	44	1	41	3	1	10
Total-CP3	33	3	39	8	3	14
Total-Total	38	1	38	5	2	16

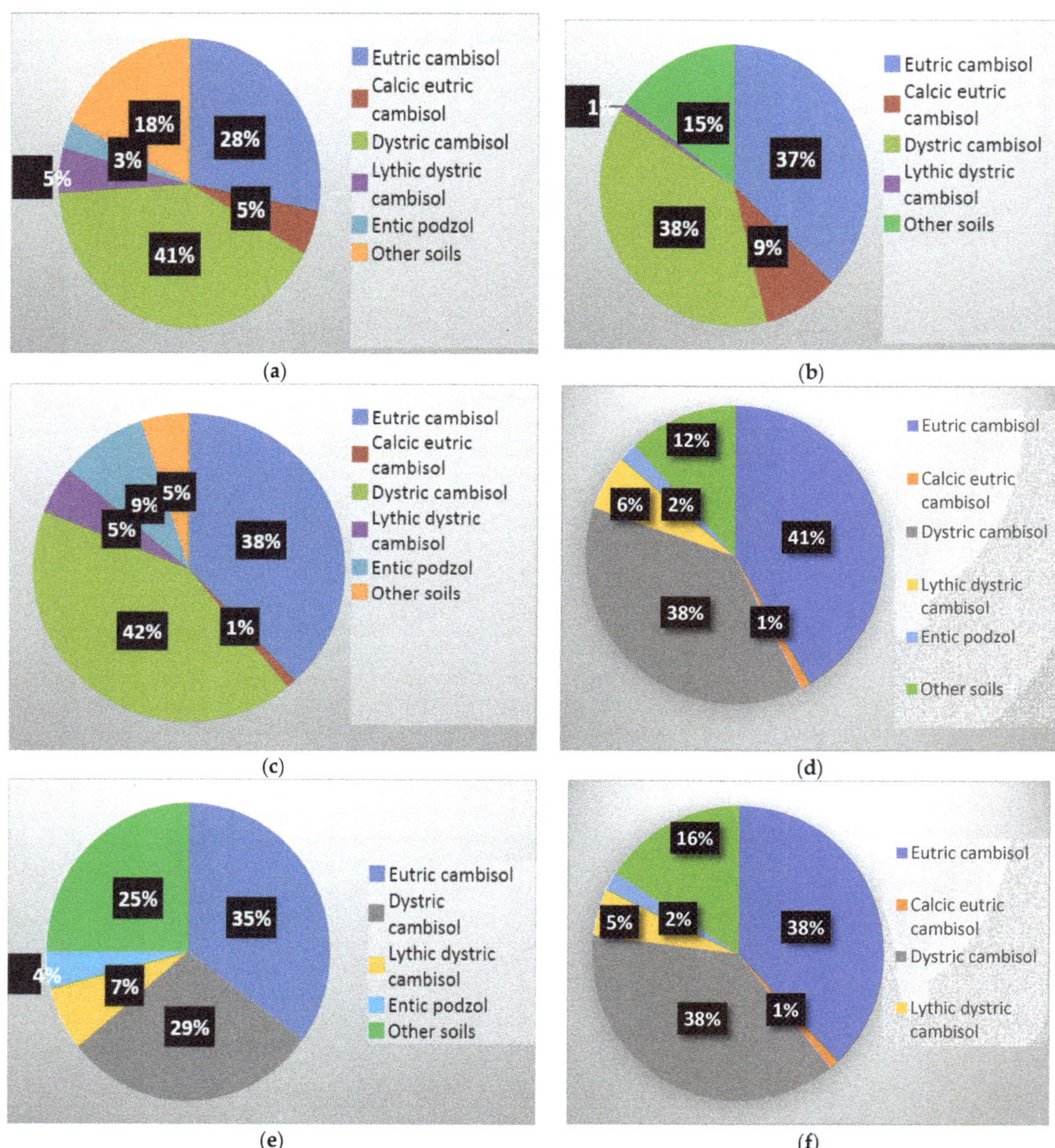

Figure 5. (**a**) Soil AM. (**b**) Soil BM. (**c**) Soil EC. (**d**) Soil CC. (**e**) Soil SC. (**f**) Soil Total.

3.5. The Participation Percentage of Silver Firs in the Stand Composition

The participation percentage of the silver tree in the stand's composition is very similar for all five mountain chains (Table 5). The largest percentages are of 10% and 20%, a fact that confirms the predilection of silver fir for mixed stands. Pure silver fir stands (100% composition) are extremely rare.

Table 5. Participation of silver fir in the stand compositions from the Carpathians (%).

Mountain Area	Participation Percentage (%)									
	10	20	30	40	50	60	70	80	90	100
Apuseni Mountains = AM CP1	44	38	6		3					9
CP2	41	27	15	7	4	2	1	1	1	1
CP3	41	28	15	8	4	2	1			1
CP4	47	29	11	6	3	3	1			
CP5	58	32	3	7						
Total	42	28	14	8	4	2	1			1
Banatului Mountains = BM-CP1	37	24	13	4	3	3	4	4	6	2
CP2	40	26	12	7	5	3	3	2	1	1
CP3	42	26	12	7	5	3	2	1	1	1
CP4	48	30	8	4	3	3	2	1		1
CP5	27	46	9			9			9	
Total	41	26	12	7	4	3	3	2	1	1
Eastern Carpathians = EC	38	30	17	7	3	2	2	1		
CP2	45	31	13	6	3	1	1			
CP3	47	31	12	5	2	2	1			
CP4	48	30	13	5	3	1				
CP5	44	35	12	5	2					2
Total	45	31	13	6	2	2	1			
Curvature Carpathians = CC-CP1	21	25	21	12	9	6	3		1	1
CP2	34	31	15	8	5	3	2	1		1
CP3	40	32	15	6	3	2	1	1		
CP4	44	32	14	6	2	1	1			
CP5	50	34	6	6	4					
Total	38	31	15	7	4	3	1	1		
Southern Carpathians = SC-CP1	25	23	17	10	9	8	4	1	2	1
CP2	35	29	15	8	6	3	2	1	1	
CP3	43	34	14	5	2	1	1			
CP4	42	35	14	5	2	2				
CP5	37	37	14	5	4	2	1			
Total	40	32	14	6	3	2	1	1	1	
Total Carpathians-CP1	35	29	17	7	4	3	2	1	1	1
CP2	41	30	14	7	4	2	1	1		
CP3	44	31	13	6	2	2	1	1		
CP4	45	32	13	5	2	1	1	1		
CP5	42	36	11	5	3	2	1			
Total	42	31	14	6	3	2	1	1		

3.6. Distance from the Road

The obtained data regarding the distance of silver firs from the road are not relevant: superior productivity stands are closer to existent roads in BM, while the situation is opposite in CC and SC where stands are far-off from roads (Table 6).

Table 6. Distance from the road (km) of silver firs located in the Carpathians.

Mountain Area	Distance from the Road (km) for Productivity Classes				
	1	2	3	4	5
Apuseni Mountains = AM	5.63	6.78	7.69	6.84	6.71
Banatului Mountains = BM	5.92	6.28	9.43	10.72	9.10
Eastern Carpathians = EC	7.50	8.13	8.94	8.73	7.51
Curvature Carpathians = CC	7.14	7.12	7.87	10.86	6.94
Southern Carpathians = SC	5.48	5.46	6.42	8.61	8.13
Total Carpathians	7.19	7.34	7.36	8.69	7.28

3.7. Stand Structure

The majority of silver firs from the Carpathians are relatively uneven-aged (Table 7). There are no significant differences between the five analysed areas and between productivity classes.

Table 7. Stand structure of silver firs located in the Carpathians (%).

Mountain Area	Structure			
	1	2	3	4
Apuseni Mountains = AM-CP1	6	60	34	
CP2	5	57	38	
CP3	9	43	48	
CP4	2	33	63	2
CP5		45	55	
Total	7	45	47	1
Banatului Mountains = BM-CP1	5	55	37	3
CP2	8	46	43	3
CP3	19	56	24	1
CP4	7	68	24	1
CP5	9	55	36	
Total	12	51	35	2
Eastern Carpathians = EC-CP1	2	55	42	1
CP2	4	40	55	1
CP3	7	41	51	1
CP4	6	42	51	1
CP5	9	44	47	
Total	5	42	52	1
Curvature Carpathians = CC-CP1	10	27	58	5
CP2	10	24	61	5
CP3	15	25	55	5
CP4	7	27	59	7
CP5	13	34	34	19
Total	12	25	58	5

Table 7. *Cont.*

Mountain Area	Structure			
	1	**2**	**3**	**4**
Southern Carpathians = SC-CP1	9	23	64	4
CP2	8	26	59	7
CP3	25	17	53	5
CP4	19	14	58	9
CP5	29	20	42	9
Total	19	20	55	6
Total Carpathians-CP1	4	50	44	2
CP2	6	36	55	3
CP3	15	31	51	3
CP4	11	28	55	6
CP5	20	30	43	7
Total	10	35	52	3

The meaning of numbers from Table 7: 1= even-aged stand; 2= relatively even-aged stand; 3= relatively uneven-aged stand; 4= uneven-aged stand

3.8. Crown Density

In the following table (Table 8), we have shown the crown density of the fir groves in the Carpathians. We can notice that most of the trees from the five mountain ranges of the Romanian Carpathians are trees of almost full consistency (0.7−0.9). The trees characterized by a degraded crown density (0.3) are located mainly in stands from EC and Total Carpathians.

Over 40% of the fir trees in the BM, EC and CC are almost full-grown trees, while the full consistency is found in up to 3% of the fir trees (Figure 6). Regarding Total Carpathians, about 40% of the fir trees have a consistency of 0.8 (almost full), while the full consistency is found in less than 2% of the fir trees.

In all cases, the predominant consistency is of 0.8, followed by 0.7 and 0.9 (Table 8).

3.9. The Characteristics of Superior Productivity Silver Fir Stands

By grouping productivity classes in three groups (superior, average and inferior), we obtain similar results with those from points 3.1–3.8. As such, it can be observed that the Carpathian silver fir has superior productivity classes at lower altitudes, on shadowed field aspects (North, North-East and North-West), and on fields with lower slopes. Significant altitude and field aspect differences are found at AM, EC and SC.

These are the most representative mountain chains as surface and geographic position.

The other environment (soil type), stand (composition, structure, consistency) or location characteristics (distance from the road) do not differ significantly based on productivity class.

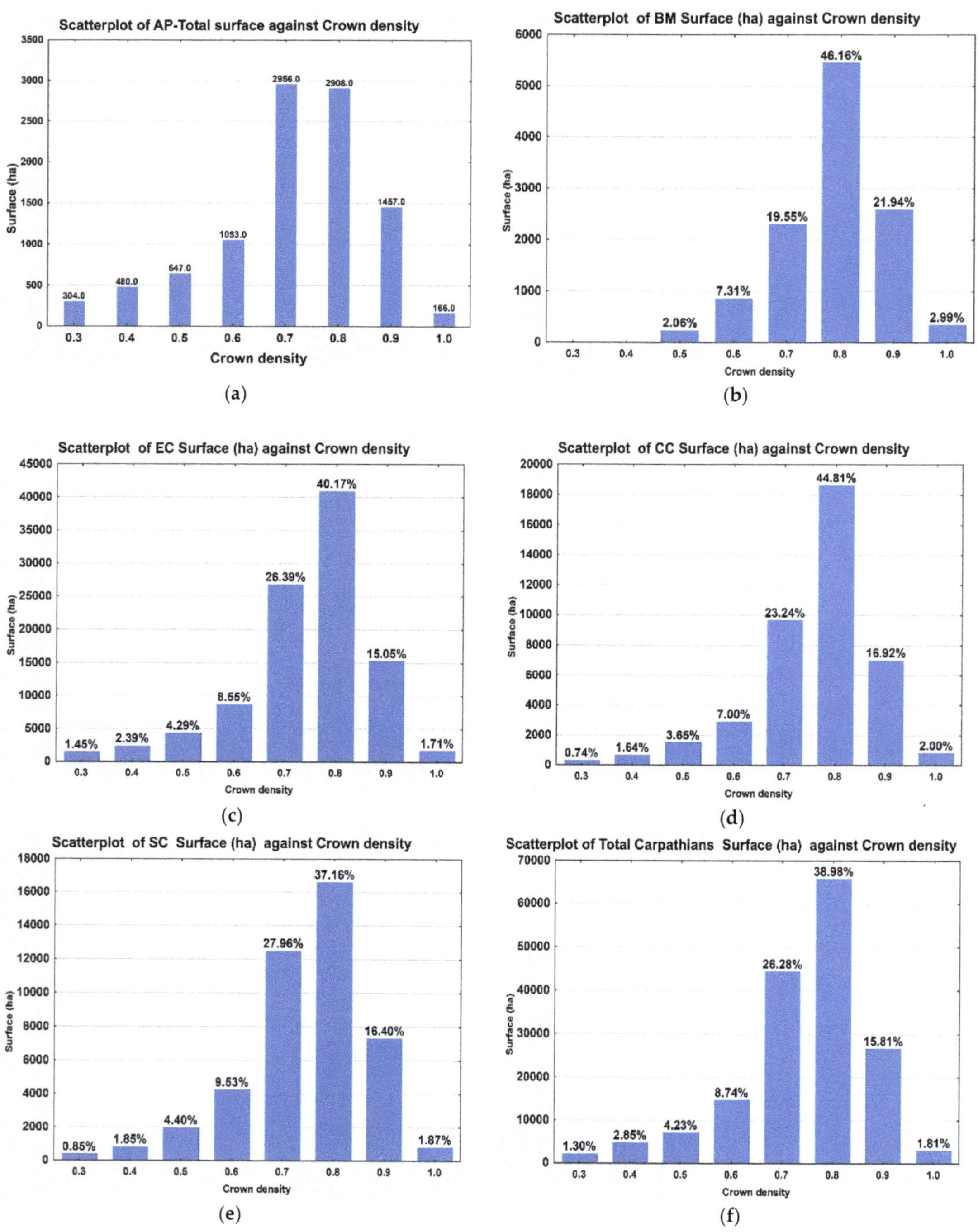

Figure 6. (**a**) AM surface according to crown density. (**b**) BM surface according to crown density. (**c**) EC surface according to crown density. (**d**) CC surface according to crown density. (**e**) Crown density SC. (**f**) Total Carpathians surface according to crown density.

Table 8. The crown density of silver firs located in the Carpathians (ha).

Mountain Area	**Crown Density**							
	0.3	**0.4**	**0.5**	**0.6**	**0.7**	**0.8**	**0.9**	**1.0**
Apuseni Mountains = AM-CP2		85	179	187	459	679	294	
CP3	226	362	374	717	2283	2121	1117	98
CP4			81	132	194	86		
Total	304	480	647	1053	2956	2908	1457	166
Banatului Mountains = BM-Cp1					188	427	227	
CP2			172	576	1301	3299	1135	93
CP3			50	218	723	1610	1139	223
CP4					91	106	90	
Total			243	863	2309	5453	2591	353
Eastern Carpathians = EC-CP1	177	240	575	941	4113	6393	2148	291
CP2	879	1464	2412	5106	15,590	25,267	8601	1091
CP3	403	663	1250	2445	6672	8983	4540	354
CP4			124	211	503	328		
Total	1480	2437	4377	8725	26,913	40,974	15,353	1742
Curvature Carpathians = CC1				147	415	652	185	
CP2	190	263	780	1610	4421	8958	2530	486
CP3		381	636	1057	4345	8474	4189	296
CP4				94	478	531	133	
Total	306	684	1518	2914	9669	18,640	7038	830
Southern Carpathians = SC1			167	161	523	555	86	
CP2	187	268	658	1591	4282	6038	1912	319
CP3	163	497	821	1914	6072	9121	5088	493
CP4			270	488	1399			
Total	379	827	1968	4263	12,511	16,630	7339	837
Total Carpathians-C1	195	277	750	1156	4836	7396	2491	343
CP2	1143	1884	3421	7460	21,632	35,283	11,943	1553
CP3	800	2533	2495	5293	15,750	21,835	11,884	1167
CP4	53	115	486	853	2187	1334	399	
Total	2191	4809	7152	14,762	44,405	65,848	26,717	3063

4. Discussion

Due to the tremendous importance of silver fir in providing the manifolds economic, protective and ecologic functions this study focused on assessing the influence of environmental conditions and stand characteristics of silver fir stands from five mountain ranges of the Romanian Carpathians.

4.1. Altitude

The present study has shown that silver fir of superior productivity class is located at low altitudes (870−1040 m) in most Carpathian chains from Romania. This fact is justified by the fact that stands located at higher altitudes undergo less intense silvicultural interventions so that their productivity is reduced [21]. Furthermore, these stands fulfil different protective functions [33,45]. As a result, stands located at higher altitudes fulfil especially the protection function and are mainly covered by special conservation works that are not aiming the increase of forest productivity. This type of management is similar to the forest's natural structure, whose evolution is controlled by attitudes as key factor of the local climate. Although in the case of silver fir, higher altitudes limit the development of this species because of its shadow requirements, this practice ensures a better resistance towards the disturbing factors [46]. On the other hand, silver fir is present especially in common beech mixtures at altitudes between 700 m and 1200 m [34]. A predominance of silver fir mixture stands at altitudes of up to 1100 m is also reported by Klopčič [21]. Mixtures between evergreen and deciduous species favour a more vigorous silver fir growth due to leaf phenology [47].

Silver fir crowns entrap light all along the year, fact that leads to superior productivity classes as was noticed also in our research [48,49]. The influence of some environmental conditions (especially climatic) over the productivity of silver fir stands is also stated by other authors who emphasize that altitude is strictly correlated with climate conditions [16,35–37]. Hilmers [16] has shown that the productivity of silver fir stands increases as temperature grows at high altitudes while Toromani [50] mentions that the extension of the vegetation season favours the increase of silver fir productivity. These projections are confirmed both for the territory of our country and for the Carpathian area, which may confirm the higher productivity of silver fir in some Carpathian chains such as Eastern and Southern parts [51,52]. Moreover, the silver fir productivity increment in stands located at altitudes of up to 1100 m as we observed in our research, is correlated with the temperatures from the cold season (January-March) as other authors have already shown [40]. This situation can be explained by the temperature increase during winter that reduces the risk of frost so that stand productivity is not affected [50]. On the other hand, the soil water deficit is a limiting factor that can increase the silver fir's vulnerability to drought and a decrease of its productivity if it appears in the warm season [48]. However, other authors have shown that, although silver fir stands are located at lower altitudes, those can record inferior productivity classes [48]. This is the case of stands from BM, that although are located in Western Romania, area which is not exposed to could waves, but in which reduced precipitation influence stands productivity due to the reduced water deficit [53].

4.2. Field Aspect

Our results regarding the preference of superior productivity silver firs for shadowed field aspects are in accordance with the shadow character of this species, mentioned by many authors [33,53–56]. The growing stock of fir in the northern field aspect exceeds that in the southern field aspect [57–59] because the fir is supplied with a more suitable light and water regime on northern field aspect [60,61]. Other authors have shown that, compared with the sunny areas, the shadowed ones favour an increased tree diversity as this is characterized by optimum temperature and humidity conditions [55]. More than that, silver fir is a species resistant to low temperatures, frosts and high precipitation quantities [11,28,61,62]. On the other hand, water availability is a limiting factor in the growth and development of silver fir. South field aspects are characterized by a higher probability for the appearance of water deficit due to a more accentuated solar radiation [46,55,63]. However, when high temperatures do not overlay with a precipitation deficit, the silver fir growth is favoured, thus leading to higher classes of productivity [50].

4.3. Field Slope

Our observation that low productivity silver fir is located, in all cases, on fields with a very high slope is correlated with the fact that those are situated at the highest altitudes. This corresponds in the mountain area with fields near mountain peaks that also have a high slope. As other studies highlighted these fields have a low accessibility degree so that the intervention with silvicultural treatment which sustain silver fir growth is limited [64]. Similar results were also reported by other authors who observed a predisposition of this species for mid to upper and steep gradients, especially when silver fir appears in common beech and Norway spruce mixtures [65]. This situation can be caused by the fact that fields located on steeper slopes are mainly fields with superficial soils characterized by a low soil water content that are not favourable for silver firs, as it was mentioned in other studies [41].

4.4. The Soil

Silver fir prefers acid, deep soils, with a high humidity content [66]. Eutric cambisol and dystric cambisol are the characteristic soils for silver fir. Besides a sufficient depth and humidity, these soils are rich in nutritive substances and are characterized by an intense biologic activity [67–69]. As a consequence, our results show the preponderance of silver fir stands on eutric cambisol and dystric cambisol, a fact that is in line with this specie's ecologic requirements. Other authors have also observed vigorous silver fir growths on acid soils [66]. Tree growth is influenced by soil depth, and thickness of genetic soil horizons [27]. Soils rich in calcium are more favourable for silver fir [15,70,71]. This is the case of calcic eutric cambisol, which has a higher presence in one of the causes of spreading valuable stands in this region [72]. On the other hand, lithic soils are not favourable for silver fir due to its taproot system [34,73,74]. In this contest, silver fir stands from the Southern Carpathians have lower productivities.

4.5. Participation Percentage of Silver Fir in the Stand's Composition

The participation percentage of silver fir in the stand's composition depends not only on the local specific (region, soil, humidity conditions) but also on the applied silvicultural system applied [66]. Our study revealed that the largest proportions of silver fir participation in the stand composition are 10% and 20% respectively. This fact confirms the silver fir's predilection for mixed stands, while pure and even aged silver fir stands (composition: 100%) are extremely rare [66]. Unlike pure stands, mixed stands offer a better resilience against harmful factors (biotic and abiotic) and, implicitly, a better phytosanitary stand state [47,75]. In addition, a silver fir participation percentage of up to 20% in the stand's composition contributes significantly to ensuring the sustainability of forest ecosystems [76].

Similar data with our research were also obtained in other studies: the participation of silver fir does not exceed 0.1–0.4 of stand composition in fir-spruce and lime-fir-spruce forests from Vyatka-Kama biome, Russia [77]; the participation of silver fir in mature stand composition was about 20–30% at altitudes between 900 and 1200 m in Eastern Carpathians. Over this altitude the presence of silver fir in stand composition decreases sharply [17]. Furthermore, a high percentage of silver fir in the stand's composition will lead to the decline of this species especially in mountain areas where the effects of climate changes are more evident and where they diminish the vitality of this species [46,75].

These participation percentages can also be connected with the decline of silver fir from the last decades. Numerous similar situations were identified only in the Carpathian area in Western Carpathians from Slovakia [78,79], in the northern Carpathians (Czech Republic, Slovakia), in Slovenia and Croatia [66], and in the Ukrainian Carpathians [80].

The distance from the road is normal to not be a decisive factor in distributing stands on productivity classes. Other authors [81,82] have also observed that the spatial extent of road effects on plant communities from forests remains unclear. However, closeness to a network of existent roads facilitates the appliance of silvicultural treatments that lead stands towards superior productivity classes, as it was obtained in BM. On the other hand, distance from the road and, implicitly, limiting the intervention of silvicultural or exploitation works leads to a decrease of stand productivity [45]. However, we have obtained an inferior productivity in CC for a minimum distance from the road. We consider that this situation is caused by superior altitudes that limits the appliance of silvicultural interventions and favours the protection function against the productivity one.

Stand structure plays a fundamental role in improving tree resilience, especially in the context of climate changes [83]. Furthermore, it is important to know stand structure for understanding forest management [34,84]. In the case of silver fir stands, their characteristic structure is relatively even-aged and relatively uneven-aged [34,85]. These management practices are economically efficient as they ensure a vigorous tree growth and, consequently superior productivity classes [85]. Many studies have shown that silver fir stands managed in this way are characterized by a better resistance and stability against harmful factors, including SO_2 emissions that, in high percentages, increase silver fir mortality rates [83,86]. Nevertheless, some authors mention that a relatively uneven-aged structure favours silver fir better [21,66].

4.6. Stand Crown Density

The study performed for resinous stands located in the Southern Carpathians, Romania, has shown that structure, consistency, relief, slope and flora type are the most important factors in defining stand structure [26]. Our results show that in the case of the five studied mountain chains, the majority of silver fir stands (approximately 45%) are stands with an almost full consistency. This observation is in accordance with this specie's temperament, the silver fir being a shadow species [32,66]. Alternatively, full consistency stands occupy very small percentages (2–3%). This type of consistency is not recommended in the case of silver fir stands because even though the species has a shadow temperament, a full consistency would endanger the regeneration's success [53,59,83]. Being mainly found in mixtures, silver fir records higher growth rates when compared with other species. This fact can be attributed to a larger amplitude of silver fir crowns that can occupy more than half of the tree's height [21,66]. Crown length and consistency has a decisive role in the growth of tree vitality, as it is also mentioned by other authors [87]. In addition, stands with an almost full consistency have a better resistance against harmful factors such as mistletoe infection [17]. As a consequence, we can consider that stands with an almost full consistency are favourable to the silver fir's sustained growth which, in these conditions realizes superior productivity classes.

The best conditions where silver fir realizes superior productivity classes are:

- ➢ lower altitudes up to approximately 1100−1200 m [26,88];
- ➢ shadowed field aspects, characterized through better temperature and humidity conditions [46,59];
- ➢ fields with a slope up to 30°, which is favourable to a sustained development that favours the superior productivity classes [41,65];
- ➢ a silver fir's tree participation percentage up to 20% in mixtures with other species [76,77];
- ➢ stands characterized by an uneven-aged structure [46] and incomplete canopy closure (0.7–0.9) [17].

5. Conclusions

There is already evidence that silver fir has been declining at European level due to air pollution and the effects of excessive climates. In this context, changes in the forest management are necessary to improve forest resilience under climate change, particularly considering the silver fir's potential to ensure multiple productive and protective functions.

Our analysis revealed that a higher silver fir productivity class is found on sites located at altitudes of up to 1200 m (AM, EC, SC). Regarding the field aspect, we found that the most suitable sites have NW field aspects (AM, EC and SC) because these ensure the most favourable conditions of light and water compared to the southern field aspects. These findings are in line with the specie's shade temperament. Silver fir growth performances are recorded on mid- to upper and steep slopes, which can be explained through the depth and water content of the soil layer. Another important characteristic that controls silver fir productivity is soil, with eutric cambisol and dystric cambisol being the most favourable for silver fir growth performance in all considered sites. Our study suggests that silver fir participation in the stand composition should be between 10−20% in order to ensure an increased silver fir stand resilience on various disturbing factors. Other fundamental characteristics of silver fir are stand structure and consistency. Hence, our findings revealed that the application of relatively even-aged structure and almost full crown density ensures not only a vigorous growth of silver fir and implicitly higher productivity classes but also optimum requirements for species regeneration. However, some characteristics, like positioning (distance from the road) of the silver fir trees did not show important differences in the productivity class.

Overall, we can conclude that the fist hypothesis established at the beginning of this study is confirmed, there is a strong connection and a certain dependence between the production classes from the forest areas of different mountain massifs. Regarding the second hypothesis we can assert that altitude, filed aspect, slope, the percentage of tree participation and stand structure are influencing the silver fir productivity class.

The findings of the present study are consistent with numerous previous studies that concluded that a higher growth performance of silver fir is recorded in certain relief-stand conditions. Under these conditions, mainly mixed silver fir stands would ensure the provision of ecosystem services. Despite its increased adaptation potential to climate change, it is fundamental that forest managers pay more attention to these driving factors when designing future strategies that aim towards forest resilience and sustainability.

Author Contributions: Conceptualization, L.D. and M.M.; methodology, L.D. and V.R.; software, G.M.; validation, R.D. and R.C.; formal analysis, G.M. and V.R.; investigation, R.D. and L.G.; resources, R.C.; data curation, L.D.; writing—original draft preparation, L.D., M.M. and V.T.-G.; writing—review and editing, L.D., M.M., V.R. and V.T.-G.; visualization, V.T.-G.; supervision, L.D.; project administration, L.D.; funding acquisition, L.D. All authors have read and agreed to the published version of the manuscript.

Funding: This paper was carried out through the project "Increasing the institutional capacity and performance of INCDS "Marin Drăcea" in the activity of RDI—CresPerfInst" (Contract no. 34PFE./30.12.2021) financed by the Ministry of Research, Innovation and Digitization through Program 1—Development of the national research—development system, Subprogramme 1.2—Institutional performance—Projects to finance excellence in RDI.

Institutional Review Board Statement: Not applicable.

Informed Consent Statement: Not applicable.

Data Availability Statement: The data presented in this study are available on request from the corresponding author. The data are not publicly available due to the project through which the research was funded is not yet completed.

Acknowledgments: The work of Gabriel Murariu was supported by the "An Integrated System for the Complex Environmental Research and Monitoring in the Danube River Area", REXDAN, SMIS code 127065, co-financed by the European Regional Development Fund through the Competitiveness Operational Programme 2014-2020; contract no. 309/10.07.2020.

Conflicts of Interest: The authors declare no conflict of interest.

References

1. Barredo, J.I.; Brailescu, C.; Teller, A.; Sabatini, F.M.; Mauri, A. *Mapping and Assessment of Primary and Old-Growth Forests in Europe*; Publications Office of the European Union: Luxemburg, 2021.
2. Morin, X.; Fahse, L.; Jactel, H.; Scherer-Lorenzen, M.; García-Valdés, R.; Bugmann, H. Long-term response of forest productivity to climate change is mostly driven by change in tree species composition. *Sci. Rep.* **2018**, *8*, 5627. [CrossRef]
3. Ducci, F.; De Rogatis, A.; Proietti, R.; Curtu, L.A.; Marchi, M.; Belletti, P. Establishing a baseline to monitor future climate-change-effects on peripheral populations of *Abies alba* in central Apennines. *Ann. For. Res.* **2021**, *64*, 33–66. [CrossRef]
4. Kutnar, L.; Kermavnar, K.; Pintar, A.M. Climate change and disturbances will shape future temperate forests in the transitionzone between Central and SE Europe. *Ann. For. Res.* **2021**, *64*, 67–86.
5. Medlyn, B.E.; Duursma, R.A.; Zeppel, M.J.B. Forest productivity under climate change: A checklist for evaluating model studies. *WIREs Clim. Chang.* **2011**, *2*, 332–355.
6. Trumbore, S.; Brando, P.; Hartmann, H. Forest health and global change. *Science* **2015**, *349*, 814–818. [CrossRef]
7. Barbu, I.; Barbu, C. *Silver Fir (Abies alba Mill.) in Romania*; Editura Tehnică Silvică: Bucharest, Romania, 2005; p. 220.
8. Pretzsch, H.; Biber, P.; Schütze, G.; Uhl, E.; Rötzer, T. Forest stand growth dynamics in Central Europe have accelerated since 1870. *Nat. Commun.* **2014**, *5*, 4967. [CrossRef]

9. Sperlich, D.; Nadal-Sala, D.; Gracia, C.; Kreuzwieser, J.; Hanewinkel, M.; Yousefpour, R. Gains or losses in forest productivity under climate change? The uncertainty of CO_2 fertilization and climate effects. *Climate* **2020**, *8*, 141. [CrossRef]
10. Mauri, A.; de Rigo, D.; Caudullo, G. *Abies alba* in Europe: Distribution, habitat, usage and threats. In *European Atlas of Forest Tree Species*; San-Miguel-Ayanz, J., de Rigo, D., Caudullo, G., Houston Durrant, T., Mauri, A., Eds.; Publications Office of the European Union: Luxemburg, 2016; p. e01493b+.
11. Rolland, C.; Michalet, R.; Desplanque, C.; Petetin, A.; Aimé, S. Ecological requirements of *Abies alba* in the French Alps derived from dendro-ecological analysis. *J. Veg. Sci.* **1999**, *10*, 297–306. [CrossRef]
12. Kerr, G.; Stokes, V.; Peace, A.; Jinks, R. Effects of provenance on the survival, growth and stem form of European silver fir (*Abies alba* Mill.) in Britain. *Eur. J. For. Res.* **2015**, *134*, 349–363. [CrossRef]
13. Kempf, M.; Zarek, M.; Paluch, J. The pattern of genetic variation, survival and growth in the *Abies Alba* Mill. Population within the introgression zone of two refugial lineages in the carpathians. *Forests* **2020**, *11*, 849. [CrossRef]
14. Ruosch, M.; Spahni, R.; Joos, F.; Henne, P.D.; Van der Knaap, W.O.; Tinner, W. Past and future evolution of *A. alba* forests in Europe–comparison of a dynamic vegetation model with palaeo data and observations. *Glob. Chang. Biol.* **2016**, *22*, 727–740. [CrossRef]
15. Forzieri, G.; Girardello, M.; Ceccherini, G.; Spinoni, J.; Feyen, L.; Hartmann, H.; Beck, P.S.A.; Camps-Valls, G.; Chirici, G.; Mauri, A.; et al. Emergent vulnerability to climate-driven disturbances in European forests. *Nat. Commun.* **2021**, *12*, 1081. [CrossRef]
16. Hilmers, T.; Avdagi, A.; Bartkowicz, L.; Bielak, K.; Binder, F.; Bonina, A.; Dobor, L.; Forrester, D.I.; Hobi, M.L.; Ibrahimspahi, A.; et al. The productivity of mixed mountain forests comprised of *Fagus sylvatica, Picea abies,* and *Abies alba* across Europe. *For. Int. J. For. Res.* **2019**, *92*, 512–522. [CrossRef]
17. Barbu, C.O. Impact of white mistletoe (*Viscum album* ssp. Abietis) infection on needles and crown morphology of silver fir (*Abies alba* Mill.). *Not. Bot. Horti Agrobot.* **2012**, *40*, 152–158. [CrossRef]
18. Clinovschi, F. *Dendrologie*; Editura Universităţii Suceava: Suceava, Romania, 2005; p. 299.
19. Teodosiu, M.; Mihai, G.; Fussi, B.; Ciocîrlan, E. Genetic diversity and structure of silver fir (*Abies alba* Mill.) at the south-eastern limit of its distribution range. *Ann. For. Res.* **2019**, *62*, 139–156. [CrossRef]
20. Bošela, M.; Lukac, M.; Castagneri, D.; Sedmák, R.; Biber, P.; Carrer, M.; Konôpka, B.; Nola, P.; Nagel, A.T.; Popa, I.; et al. Contrasting effects of environmental change on the radial growth of co-occurring beech and fir trees across Europe. *Sci. Total Environ.* **2018**, *615*, 1460–1469.
21. Klopčič, M.; Simoncic, T.; Boncina, A. Comparison of regeneration and recruitment of shade-tolerant and light-demanding tree species in mixed uneven-aged forests: Experiences from the Dinaric region. *For. Int. J. For. Res.* **2015**, *88*, 552–563.
22. Dumitras, M.; Kucsicsa, G.; Dumitrică, C.; Popovici, E.-A.; Vrînceanu, A.; Mitrică, B.; Mocanu, I.; Serban, P.R. Estimation of future changes in aboveground forest carbon stock in Romania. A prediction based on forest-cover pattern Scenario. *Forests* **2020**, *11*, 914.
23. Fedorca, A.; Fedorca, M.; Ionescu, O.; Jurj, R.; Ionescu, G.; Popa, M. Sustainable landscape planning to mitigate wildlife-vehicle collisions. *Land* **2021**, *10*, 737.
24. Mountain Partnership. Implementing the 2030 Agenda for Mountains [WWW Document]. 2017. Available online: https://www.fao.org/mountain-partnership/our-work/advocacy/2030-agenda-for-sustainable-development/en/ (accessed on 26 January 2022).
25. Mountain Partnership. *Mountains and the Sustainable Development Goals*; FAO: Rome, Italy, 2014.
26. Murariu, G.; Dincă, L.; Tudose, N.; Crisan, V.; Georgescu, L.; Munteanu, D.; Dragu, M.D.; Rosu, B.; Mocanu, G.D. Structural characteristics of the main resinous stands from Southern Carpathians, Romania. *Forests* **2021**, *12*, 1029. [CrossRef]
27. Kobal, M.; Grčman, H.; Zupan, M.; Levanič, T.; Simončič, P.; Kadunc, A.; Hladnik, D. Influence of soil properties on silver fir (*Abies alba* Mill.) growth in the Dinaric Mountains. *For. Ecol. Manag.* **2015**, *337*, 77–87. [CrossRef]
28. Pinto, P.E.; Gégout, J.C.; Hervé, J.C.; Dhôte, J.F. Respective importance of ecological conditions and stand composition on *Abies alba* Mill. dominant height growth. *For. Ecol. Manag.* **2008**, *255*, 619–629. [CrossRef]
29. Bosela, M.; Popa, I.; Gömöry, D.; Longauer, R.; Tobin, B.; Kyncl, J.; Kyncl, T.; Nechita, C.; Petráš, R.; Sidor, C.G.; et al. Effects of post-glacial phylogeny and genetic diversity on the growth variability and climate sensitivity of European silver fir. *J. Ecol.* **2016**, *104*, 716–724. [CrossRef]
30. Mohytych, V.; Sułkowska, M.; Klisz, M. Reproduction of silver fir (*Abies alba* Mill) forests in the Ukrainian Carpathians. *Folia For. Pol.* **2019**, *61*, 156–158. [CrossRef]
31. Lasch, P.; Lindner, M.; Erhard, M.; Lasch, P.; Lindner, M.; Erhard, M.; Suckow, F.; Wenzel, A. Regional impact assessment on forest structure and functions under climate change-The Brandenburg case study. *Artic. For. Ecol. Manag.* **2002**, 17. [CrossRef]
32. Dincă, L.; Murariu, G.; Enescu, C.M.; Achim, F.; Georgescu, L.; Murariu, A.; Timis-Gânsac, V.; Holonec, L. Productivity differences between southern and northern slopes of Southern Carpathians (Romania) for Norway spruce, silver fir, birch and black alder. *Not. Bot. Horti Agrobot.* **2020**, *48*, 1070–1084. [CrossRef]
33. Dincă, L.; Achim, F. The management of forests situated on fields susceptible to landslides and erosion from the Southern Carpathians. *Sci. Pap. Ser. Manag. Econ. Eng. Agric. Rural. Dev.* **2019**, *19*, 183–188.
34. Tudoran, G.M.; Cicsa, A.; Boroeanu, M.; Dobre, A.C.; Pascu, I.S. Forest Dynamics after Five Decades of Management in the Romanian Carpathians. *Forests* **2021**, *12*, 783. [CrossRef]

35. Kern, Z.; Popa, I. Climate–growth relationship of tree species from a mixed stand of Apuseni Mts., Romania. *Dendrochronologia* **2007**, *24*, 109–115. [CrossRef]
36. Popa, I.; Cheval, S. Early winter temperature reconstruction of Sinaia area (Romania) derived from tree-rings of Silver Fir (*Abies alba* Mill.). *Rom. J. Meteorol.* **2007**, *9*, 47–54.
37. Dinulică, F.; Marcu, V.; Borz, S.A.; Vasilescu, M.-M.; Petritan, I.C. Wind contribution to yearly silver fir (*Abies alba* Mill.) compression wood development in the Romanian Carpathians. *iForest* **2016**, *9*, 927–936. [CrossRef]
38. Sellan, G.; Thompson, J.; Majalap, N.; Brearley, F. Soil characteristics influence species composition and forest structure differentially among tree size classes in a Bornean heath forest. *Plant Soil* **2019**, *483*, 173–185. [CrossRef]
39. Sidor, G.C.; Popa, I.; Vlad, R.; Cherubini, P. Different tree-ring responses of Norway spruce to air temperature across an altitudinal gradient in the Eastern Carpathians (Romania). *Trees* **2015**, *29*, 985–997. [CrossRef]
40. Maaten-Theunissen, M.; Kahle, H.P.; van der Maaten, E. Drought sensitivity of Norway spruce is higher than that of silver fir along an altitudinal gradient in southwestern Germany. *Ann. For. Sci.* **2013**, *70*, 185–193. [CrossRef]
41. Bouriaud, O.; Popa, I. Site and species influence on tree growth response to climate in Vrancea Mountains. *Proc. Rom. Acad.* **2007**, *9*, 63–72.
42. Forest management plans of the forest districts: Campina (2020), Azuga (1999), Sinaia (2002), Pietrosita (2005), Rucar (1996), Câmpulung (2006), Aninoasa (2005), Domnesti (2004), Musatesti (1994), Vidraru (2005), Cornet (1993), Suici (2008), Brezoi (1991), Voineasa (2003), Latorița (1994), Bumbești (2002), Polovragi (2001), Lupeni (2000), Petrosani (2001), Runcu (2000), Novaci (2002), Râșnov (1993), Zarnesti (1993), Șercaia (1986), Fagaras (1985), Voila (1985), Arpas (1986), Avrig (2005), Talmaciu (1980), Valea Sadului (1982), Valea Cibinului (1982), Cugir (1993), Bistra (1999), Orastie (1993), Gradiste (2004), Petrila (2000), Baru (1996), Pui (2005), Retezat (1996), Dobra (1997), Hunedoara (1993), Ana Lugojana (1999), Oțelul Roșu (2000), Rusca Montană (1991), Caransebeș (1995), Teregova (1992), Mehadia (1993), Băile Herculane (2001), Bocșa Română (2004), Păltiniș (1998), Reșița (2001), Văliug (1992), Oravița (2008), Anina (2007), Nera (2004), Deva (1991), Ilia (1998), Cosava (2002), Berzeasca (1995), Bocsa Montana (2003), Bozovici (2004), Moldova Noua (2006), Sasca Montana (2007), Remeți (2005), Huedin (2003), Sudrigiu (1998), Vașcău (2004), Beliș (2002), Someșul Rece (2005), Turda (2007), Gilau (2007), Gârda (2001), Valea Arieșului (1990), Baia de Arieș (1998), Câmpeni (1987), Baia de Cris (2007), Brad (1997), Beius (2002), Codru Moma (2002), Dobresti (1997), Barzava (1996), Gurahont (1994), Halmagiu (1994), Sebis-Moneasa (1994), Valea Mare (1997), Breaza (2003), Brodina (2004), Putna (2000), Tomnatec (2002), Frasin (1988), Moldovita (1994), Pojorata (2003), Vama (1990), Crucea (1999), Mălini (1998), Stulpicani (2000), Broșteni (1999), Borca (1999), Galu (1999), Ceahlau (2000), Tarcău (2001), Brates (2001), Bicaz (2003), Agas (1985), Comanesti (1994), Darmanesti (2005), Targu Ocna (2005), Oituz (2005), Manastirea Casin (2006), Sânmartin (2004), Miercurea Ciuc (1981), Izvorul Mureș (2002), Zetea (2007), Gheorgheni (2002), Borsec (2002), Tulghes (2002), Sovata (2000), Fâncel (2000), Lunca Bradului (1999), Răstolița (1999), Gurghiu (2000), Toplita (1991), Coșna (1999), Iacobeni (1998), Cârlibaba (1998), Rodna (1996), Sângiorz Bai (1992), Năsăud (1993), Ilva Mica (1995), Prundu Bargaului (1990), Sălăuța (1993), Strâmbu Băiuți (2007), Dragomirești (1996), Borsa (1999), Viseu (2008), Mara (2005), Sighet (2005), Poieni (2006), Ruscova (2005), Baia Mare (2002), Baia Sprie (1991), Tauti Magheraus (2001), Negresti-Oas (2005), Brasov (1993), Sacele (1993), Teliu (1993), Maneciu (1999), Nehoiasu (1999), Nehoiu (1999), Gura Teghii (2001), Intorsura Buzaului (1982), Comandau (1991), Covasna (1999), Bretcu (1995), Naruja (1991), Nereju (2004), Tulnici (1981), Lepșa (1980), Soveja (2001).
43. Iticescu, C.; Murariu, G.; Georgescu, L.P.; Burada, A.; Țopa, M.C. Seasonal variation of the physico-chemical parameters and water quality index (WQI) of Danube Water in the transborder Lower Danube area. *Rev. Chim.* **2016**, *67*, 1843–1849.
44. Iticescu, C.; Georgescu, L.P.; Murariu, G.; Topa, C.; Timofti, M.; Pintilie, V. Lower Danube Water Quality Quantified through WQI and Multivariate Analysis. *Water* **2019**, *11*, 1305. [CrossRef]
45. Bebi, P.; Kienast, F.; Schönenberger, W. Assessing structures in mountain forests as a basis for investigating the forests' dynamics and protective function. *For. Ecol. Manag.* **2001**, *145*, 3–14. [CrossRef]
46. Čater, M.; Levanič, T. Beech and silver fir's response along the Balkan's latitudinal gradient. *Sci. Reports.* **2019**, *91*, 1–14. [CrossRef]
47. Versace, S.; Garfì, V.; Dalponte, M.; Febbraro, D.; Frizzera, L.; Damiano, G.; Tognetti, R. Species interactions in pure and mixed-species stands of silver fir and European beech in Mediterranean mountains. *iForest Biogeosciences For.* **2021**, *14*, 1–11. [CrossRef]
48. Schwarz, J.A.; Bauhus, J. Benefits of Mixtures on Growth Performance of Silver Fir (*Abies alba*) and European Beech (*Fagus sylvatica*) Increase with tree size without reducing drought tolerance. *Front. For. Glob. Chang.* **2019**, *2*, 79. [CrossRef]
49. Vallet, P.; Pérot, T. Silver fir stand productivity is enhanced when mixed with Norway spruce: Evidence based on large-scale inventory data and a generic modelling approach. *J. Veg. Sci.* **2011**, *22*, 932–942. [CrossRef]
50. Toromani, E.; Sanxhaku, M.; Pasho, E. 2011. Growth responses to climate and drought in silver fir (*Abies alba*) along an altitudinal gradient in Southern Kosovo. *Can. J. For. Res.* **2011**, *41*, 1795–1807. [CrossRef]
51. Micu, D.M.; Amihaesei, V.A.; Milian, N.; Cheval, S. Recent changes in temperature and precipitation indices in the Southern Carpathians, Romania (1961–2018). *Theor. Appl. Climatol.* **2021**, *144*, 691–710. [CrossRef]
52. Cheval, S.; Bulai, A.; Croitoru, A.-E.; Dorondel Ștefan Micu, D.; Mihăilă, D.; Sfîcă, L.; Tișcovschi, A. Climate change perception in Romania. *Theor. Appl. Climatol.* **2022**, *149*, 253–272. [CrossRef] [PubMed]
53. Busuioc, A.; Baciu, M.; Breza, T.; Dumitrescu, A.; Stoica, C.; Baghina, N. Changes in intensity of high temporal resolution precipitation extremes in Romania: Implications for Clausius-Clapeyron scaling. *Clim. Res.* **2017**, *7*, 239–249. [CrossRef]

54. Oliva, J.; Colinas, C. Decline of silver fir (*Abies alba* Mill.) stands in the Spanish Pyrenees: Role of management, historic dynamics and pathogens. *For. Ecol. Manag.* **2007**, *252*, 84–97. [CrossRef]
55. Hu, S.; Ma, J.; Shugart, H.H.; Yan, X. Evaluating the impacts of slope aspect on forest dynamic succession in Northwest China based on FAREAST model. *Environ. Res. Lett.* **2008**, *13*, 034027. [CrossRef]
56. Fontaine, M.; Aerts, R.; Özkan, K.; Mert, A.; Gülsoy, S.; Süel, H.; Waelkens, M.; Muys, B. Elevation and field aspect rather than soil types determine communities and site suitability in Mediterranean mountain forests of southern Anatolia, Turkey. *For. Ecol. Manag.* **2007**, *247*, 18–25. [CrossRef]
57. Diaci, J.; Rozenbergar, D.; Anic, I.; Mikac, S.; Saniga, M.; Kucbel, S.; Visnjic, C.; Ballian, D. Structural dynamics and synchronous silver fir decline in mixed old-growth mountain forests in Eastern and Southeastern Europe. *Forestry* **2011**, *84*, 479–491. [CrossRef]
58. Pernar, N.; Matic, S.; Baksic, D.; Klimo, E. The accumulation and properties of surface humus layer in mixed selection forests of fir on different substrates. *Ecology* **2008**, *27*, 41.
59. Kara, F.; Topaçoğlu, O. Effects of canopy structure on growth and belowground/aboveground biomass of seedlings in uneven-aged trojan fir stands. *Cerne* **2018**, *24*, 312–322. [CrossRef]
60. Salehiarjmand, H.; Ebrahimi, S.N.; Hadian, J.; Ghorbanpour, M. Essential oils main constituents and antibacterial activity of seeds from Iranian local landraces of dill (*Anethum graveolens* L.). *J. Hortic. For. Biotechnol.* **2014**, *18*, 1–9.
61. Lebourgeois, F. Climatic signal in annual growth variation of silver fir (*Abies alba* Mill.) and spruce (*Picea abies* Karst.) from the French Permanent Plot Network (RENECOFOR). *Ann. For. Sci.* **2007**, *64*, 333–343. [CrossRef]
62. Bałazy, R.; Kamińska, A.; Ciesielski, M.; Socha, J.; Pierzchalski, M. Modeling the Effect of Environmental and Topographic Variables Affecting the Height Increment of Norway Spruce Stands in Mountainous Conditions with the Use of LiDAR Data. *Remote Sens.* **2019**, *11*, 2407. [CrossRef]
63. Vandekerckhove, L.; Poesen, J.; Wijdenes, D.O.; Gyssels, G. Short-term bank gully retreat rates in Mediterranean environments. *Catena.* **2001**, *44*, 133–161. [CrossRef]
64. Filipiak, M. Distribution of silver-fir (*Abies alba* Mill.) in the Sudeten Mts. *Biodivers. Res. Conserv.* **2006**, *3–4*, 294–299.
65. Abrudan, I.V.; Mather, R.A. The influence of site factors on the composition and structure of semi-natural mixed-species stands of beech (*Fagus sylvatica*), silver fir (*Abies alba*) and Norway spruce (*Picea abies*) in the upper draganul watershed of North-West Romania. *Forestry* **1999**, *72*, 87–93. [CrossRef]
66. Dobrowolska, D.; Bončina, A.; Klumpp, R. Ecology and silviculture of silver fir (*Abies alba* Mill.): A review. *J. For. Res.* **2017**, *22*, 326–335. [CrossRef]
67. Li, J.; Cao, X.; Wang, Y.; Yan, W.; Peng, Y.; Chen, X. Effects of thinning on soil nutrients in a chronosequence of Chinese fir in subtropical Chinaforests. *Ann. For. Res.* **2021**, *64*, 147–158.
68. Onet, A.; Dincă, L.C.; Grenni, P.; Laslo, V.; Teusdea, A.C.; Vasile, D.L.; Enescu, R.E.; Crisan, V.E. Biological indicators for evaluating soil quality improvement in a soil degraded by erosion processes. *J. Soils Sediments* **2019**, *19*, 2393–2404. [CrossRef]
69. Spârchez, G.; Dincă, L.; Marin, G.; Dincă, M.; Enescu, R.E. Variation of eutric cambisols' chemical properties based on altitudinal and geomorphologic zoning. *Environ. Eng. Manag. J.* **2017**, *16*, 2911–2918.
70. Elling, W.; Dittmar, C.; Pfaffelmoser, K.; Rötzer, T. Dendroecological assessment of the complex causes of decline and recovery of the growth of silver fir (*Abies alba* Mill.) in Southern Germany. *For. Ecol. Manag.* **2009**, *257*, 1175–1187. [CrossRef]
71. Potočić, N.; Ćosić, T.; Pilaš, I. The influence of climate and soil properties on calcium nutrition and vitality of silver fir (*Abies alba* Mill.). *Environ. Pollut.* **2005**, *137*, 596–602. [CrossRef]
72. Budeanu, M.; ApostoL, E.N.; Dincă, L.; Pleșca, I.M. In situ conservation of narrow crowned Norway spruce ideotype (*Picea abies* Pendula Form and Columnaris Variety) in Romania. *Int. J. Conserv. Sci.* **2021**, *12*, 1139–1152.
73. Bert, G.D. Impact of ecological factors, climatic stresses, and pollution on growth and health of silver fir (*Abies alba* Mill.) in the Jura mountains: An ecological and dendrochronological study. *Acta Oecol.* **1993**, *14*, 229–246.
74. Zielonka, A.; Drewnik, M.; Musielok, Ł.; Dyderski, M.K.; Struzik, D.; Smułek, G.; Ostapowicz, K. Biotic and abiotic determinants of soil organic matter stock and fine root biomass in mountain area temperate forests—Examples from cambisols under European Beech, Norway Spruce and Silver Fir (Carpathians, Central Europe). *Forests* **2021**, *12*, 823. [CrossRef]
75. Ugarković, D.; Jazbec, A.; Seletković, I.; Tikvić, I.; Paulić, V.; Ognjenović, M.; Marušić, M.; Potočić, N. Silver fir decline in pure and mixed stands at western edge of spread in croatian dinarides depends on some stand structure and climate factors. *Sustainability* **2021**, *13*, 6060. [CrossRef]
76. Šach, F.; Švihla, V.; Černohous, V.; Kantor, P. Management of mountain forests in the hydrology of a landscape, the Czech Republic-Review. *J. For. Sci.* **2014**, *60*, 42–50. [CrossRef]
77. Kadetov, N.G. Fir-spruce and lime-fir-spruce forests of Vyatka-Kama biome. In *BIO Web of Conferences*; EDP Sciences: Les Ulis, France, 2018; Volume 11, p. 00020.
78. Kucbel, S.; Jaloviar, P.; Saniga, M.; Vencurik, J.; Klimaš, V. Canopy gaps in an old-growth fir-beech forest remnant of Western Carpathians. *Eur. J. For. Res.* **2010**, *129*, 249–259. [CrossRef]
79. Paule, L. Biodiversity of the Western Carpathians' Forest ecosystems. In *Conservation of Forests in Central Europe*; Paulenka, J., Paule, L., Eds.; Arbora Publishers: Zvolen, Slovakia, 1994; pp. 31–38.
80. Shparyk, Y.S.; Berkela, Y.Y.; Viter, R.M.; Losiuk, V.P. Main types of forest stands dynamics in the Ukrainian Carpathians. *Nat. Carpathians Annu. Sci. J. CBR Inst. Ecol. Carpathians NAS Ukr.* **2018**, *1*, 50–57.

81. Avon, C.; Bergès, L.; Dumas, Y.; Dupouey, J.L. Does the effect of forest roads extend a few meters or more into the adjacent forest? A study on understory plant diversity in managed oak stands. *For. Ecol. Manag.* **2010**, *259*, 1546–1555. [CrossRef]
82. Bunnell, F.L.; Goward, T.; Houde, I.; Björk, C. Larch seed trees sustain arboreal lichens and encourage recolonization of regenerating stands. *West. J. Appl. For.* **2007**, *22*, 94–98. [CrossRef]
83. Podlaski, R. Can forest structural diversity be a response to anthropogenic stress? A case study in old-growth fir *Abies alba* Mill. stands. *Ann. For. Sci.* **2018**, *75*, 75–99. [CrossRef]
84. Keren, S.; Diaci, J.; Motta, R.; Govedar, Z. Stand structural complexity of mixed old- growth and adjacent selection forests in the Dinaric Mountains of Bosnia and Herzegovina. *For. Ecol. Manag.* **2017**, *400*, 531–541. [CrossRef]
85. Szmyt, J.; Barzdajn, W.; Kowalkowski, W.; Korzeniewicz, R. Moderate diversity in forest structure and its low dynamics are favored by uneven-aged silviculture—The lesson from medium-term experiment. *Forests* **2020**, *11*, 57. [CrossRef]
86. Uhl, E.; Hilmers, T.; Pretzsch, H. From acid rain to low precipitation: The role reversal of norway spruce, silver fir, and european beech in a selection mountain forest and its implications for forest management. *Forests* **2021**, *12*, 894. [CrossRef]
87. Spathelf, P. Reconstruction of crown length of Norway spruce (*Picea abies* (L.) Karst.) and Silver fir (*Abies alba* Mill.)-technique, establishment of sample methods and application in forest growth analysis. *Ann. For. Sci.* **2003**, *60*, 833–842. [CrossRef]
88. Vitali, V.; Buntgen, U.; Bauhus, J. Silver fir and Douglas fir are more tolerant to extreme droughts than Norway spruce in south-western Germany. *Glob. Chang. Biol.* **2017**, *23*, 5108–5119. [CrossRef]

Article

Understory Vegetation Dynamics in Non-Native Douglas Fir Forests after Management Abandonment—A Case Study in Two Strict Forest Reserves in Southwest Germany

Steffi Heinrichs [1,2,*], Michaela Dölle [1,3], Torsten Vor [1,2], Patricia Balcar [4] and Wolfgang Schmidt [1]

[1] Silviculture and Forest Ecology of the Temperate Zones, University of Göttingen, Büsgenweg 1, 37077 Göttingen, Germany
[2] Faculty of Resource Management, HAWK University of Applied Sciences and Arts, Büsgenweg 1a, 37077 Göttingen, Germany
[3] Biodiversity, Macroecology and Biogeography, University of Göttingen, Büsgenweg 1, 37077 Göttingen, Germany
[4] Ecological Forest Development, Research Institute of Forest Ecology and Forestry, Rhineland-Palatinate, Hauptstraße 16, 67705 Trippstadt, Germany
* Correspondence: sheinri@gwdg.de

Citation: Heinrichs, S.; Dölle, M.; Vor, T.; Balcar, P.; Schmidt, W. Understory Vegetation Dynamics in Non-Native Douglas Fir Forests after Management Abandonment—A Case Study in Two Strict Forest Reserves in Southwest Germany. *Diversity* **2022**, *14*, 795. https://doi.org/10.3390/d14100795

Academic Editors: Lucian Dinca and Miglena Zhiyanski

Received: 30 August 2022
Accepted: 22 September 2022
Published: 24 September 2022

Publisher's Note: MDPI stays neutral with regard to jurisdictional claims in published maps and institutional affiliations.

Abstract: The non-native Douglas fir (*Pseudotsuga menziesii*) is widely distributed in Europe and promoted by forestry due to its assumed resistance against climate change. An increasing cultivation area is, however, viewed critically by nature conservation as negative effects on native biodiversity and naturalness are expected. We investigated plant species diversity and composition in two strict forest reserves (SFR) dominated by Douglas fir in southwest Germany. These reserves were established in the years 2001/2002 to study the development of Douglas fir forests after management abandonment. Vegetation surveys were conducted in 2005 and repeated in 2017. We used re-survey data from a nearby SFR dominated by native tree species as a reference. The understory vegetation showed consistent development after management abandonment, irrespective of tree species identity and origin. It became less diverse and more shade-tolerant over time due to missing soil disturbance and decreasing light availability. In contrast to a native canopy, though, Douglas fir promoted the share of generalist species. Regeneration of Douglas fir largely decreased in the SFRs underlining its competitive weakness against native tree species, mainly against European beech (*Fagus sylvatica*). Thereby, regeneration patterns of Douglas fir in the SFR were similar to those observed in the native range.

Keywords: competitive strength; forest management; nature conservation; naturalness; European beech; species diversity; invasiveness; homogenization

1. Introduction

The potentials and risks of integrating non-native tree species into native forests of Central Europe are highly debated [1]. Stand-replacing disturbances caused by drought and windthrow in recent decades [2], as well as consequent reforestation plans that aim to complement natural succession, intensified this debate [3–5]. Among non-native species cultivated in Central Europe, the North American Douglas fir (*Pseudotsuga menziesii*) is one of the most widely distributed species [6] that is currently further promoted by forestry. The species is characterized by a higher productivity and resistance against drought compared to native coniferous tree species, mainly compared to Norway spruce (*Picea abies*), and is therefore considered as an important replacement to mitigate the effects of climate change and to secure main ecosystem services [7–12].

At the same time, Douglas fir is regarded critically among nature conservationists who see a potential for invasion with negative effects for native biodiversity in the future [13,14], further discussed in [15]. With a larger cultivation area, an increased reproduction and spread of the early reproducing Douglas fir into conservation-relevant native ecosystems is expected, as well as a shift in floristic and faunistic species composition towards conditions far from the potential natural vegetation. In native mountain forests of Spain, Broncano et al. [16] recorded a Douglas fir invasion that started already 30 years after planting. Invasion was also detected in heathland ecosystems of Europe [17]. Bindewald et al. [18] assessed national forest inventory data of Germany and detected an increase in the total stand area of Douglas fir between the years 2002 and 2012 and an increasing number of inventory plots with Douglas fir regeneration. Among conservation-relevant forest ecosystems, Douglas fir regeneration had the highest share (18% on total habitat area) in open and acidic oak forests (*Quercus* spec.) supporting results by Knoerzer [19]. For most of the inventory plots considered by Bindewald et al. [18], though, regeneration was mainly restricted to a Douglas fir canopy showing the impact of propagule pressure and dispersal limitation [20] as well as the importance of competition by other tree species in restricting the establishment of Douglas fir in Central Europe [21]. Though the potential invasiveness is evaluated differently among forestry and nature conservation [15], there is unity about the need for management and cultivation concepts that prevent a further spread into conservation-relevant ecosystems as well as potential negative effects on native biodiversity [22–24]. These concepts include the establishment of buffer zones around susceptible ecosystems and the establishment of Douglas fir only in mixture with native tree species up to a proportion of 30% [25].

The fear of invasion of Douglas fir is based on the generally strong invasiveness of species of Pinaceae (including Douglas fir) around the world that affected native flora and fauna and led to the establishment of "novel ecosystems" especially in the southern hemisphere [26–28]. The invasive success of Pinaceae species can be explained by its low seed mass associated with wind distribution, its short juvenile period, and the short intervals between mast years [27]. For Douglas fir in Central Europe, though, the ecological impact has been less severe up until now [29–31]. It was for example found that species diversity and composition often resemble native Norway spruce forests [32–34]. Nevertheless, when focusing on distinct species groups, research on the effect of Douglas fir on species diversity revealed mixed results. A reduced species diversity compared to native tree species was found for fungi [29,35], for spiders [36], for arthropods and dependent birds in the Douglas fir canopy [37], and for early successional saproxylic beetles [38]. Ground beetles, on the other hand, were more abundant and diverse in pure Douglas fir compared to European beech (*Fagus sylvatica*, hereafter beech) and Norway spruce forests and mixtures of these tree species [39]. However, the authors also state that the non-native Douglas fir seems to provide no habitat for specialized beech-associated species [39]. The understory vegetation was found to mostly benefit in terms of abundance and species richness from a Douglas fir canopy compared to native beech forests [34,40,41]; but see [42]. If this finding results from a (different) forest management of Douglas fir and beech stands or from individual traits of both tree species is, however, not completely clear yet.

Despite growing data on the effects of Douglas fir on different species groups, investigations on long-term development of Douglas fir stands and its associated diversity are largely missing in Germany and Central Europe until now. According to Eberhard and Hasenauer [43], Douglas fir regeneration requires silvicultural management to survive indicating that a natural forest development represents a natural barrier for a further spread of Douglas fir. Here, monitoring in strict forest reserves (SFR) can deliver valuable insights and might help to disentangle effects of forest management and tree species identity. SFR are formerly managed forests, where management was abandoned to conserve natural forest ecosystems and to monitor and investigate the natural forest development without human influence under changing abiotic conditions. Another important aim is to deduce nature-based management and conservation concepts for managed forests [44,45]. In SFR of Bavaria, southern Germany, regeneration of Douglas fir was recorded in 27 of 160 reserves with small abundances only, supporting a low regeneration potential in unmanaged forests and a competitive strength of native tree species [46]. There, however, Douglas fir only comprised on average 2% of the basal area. A dominance of Douglas fir in the overstorey may result in a different regeneration pattern.

Two strict forest reserves (the SFR Grünberg and SFR Eselskopf), established in the years 2001 and 2002 in the federal state of Rhineland-Palatinate in southwest Germany with areas dominated by Douglas fir can provide evidence on long-term development of Douglas fir forests without forest management in the introduced range. Next to forest inventories, vegetation surveys are part of the regular monitoring program in many SFR. The understory vegetation is the key component of plant diversity in temperate forests and contributes to element cycling and functioning of above and belowground food webs [47–49]. Due to the specific environmental requirements of most plant species, the understory vegetation is also an important indicator of abiotic conditions and its changes. Based on compositional differences in the understory vegetation found between non-native Douglas fir and native beech forests [34], contrasting vegetation dynamics after management abandonment can be expected indicating effects of tree species identity on native biodiversity.

To investigate the diversity dynamics of non-native Douglas fir forests after forest management abandonment also in comparison to native forests, we conducted vegetation surveys at two points in time in the SFRs Grünberg and Eselskopf dominated by Douglas fir and in a nearby SFR dominated by native tree species (the SFR Adelsberg-Lutzelhardt) in southwest Germany. In detail, we wanted to know (1) how vegetation structure, composition, and species richness have changed within a time span of 12 years in Douglas fir dominated unmanaged forests up to 35 years after forest management abandonment and (2) if this development differs from vegetation dynamics in a SFR dominated by native tree species. Furthermore, we focused (3) on the development of Douglas fir in different vegetation layers (tree, shrub and herb layer) compared to native tree species to draw conclusions on its natural development and regeneration over time, its competitiveness against native tree species and its potential invasiveness. Our study will contribute to the increasing knowledge on the effects of Douglas fir on native forest ecosystems and can give important indications on necessary forest management activities for mitigating the potential impact of Douglas fir on native biodiversity.

2. Materials and Methods

2.1. Study Areas

We studied vegetation dynamics in two strict forest reserves (SFR) partly dominated by Douglas fir in southwestern Germany in the federal state of Rhineland-Palatinate, the SFRs Grünberg (GB-DF) and Eselskopf (EK-DF, Table 1). With 6.4% on total forest area, the federal state of Rhineland-Palatinate has the highest share of Douglas fir in Germany (2% for whole Germany; [50]). We contrasted the vegetation development to another SFR of the region, the SFR Adelsberg-Lutzelhardt, that is dominated by native tree species (abbreviation AB-NAT; see Table 1).

The SFRs GB-DF and EK-DF are geologically formed on acidic bedrock that developed to podzolic brown soils in GB-DF, while the Lower Devonian argillite with insertions of fine sand in EK-DF weathered to mesotrophic brown soils (Table 1). The landscape around both SFR would be naturally dominated by acidic beech forests admixed with sessile oak (*Quercus petraea*, [51]). Due to the mesotrophic site character, the EK-DF area is partly in transition to slightly more nutrient rich beech forest communities ([52]; Table 1).

In both SFR, we investigated a fenced representation area of 1 ha in size (= a core area with a forest fence of 2 m height preventing entry for ungulates and European brown hares (*Lepus europaeus*)) that was established within Douglas fir dominated stands. These stands were established as pure stands ca. 120 years ago. Today, naturally regenerating beech accompanies Douglas fir in the core area of GB-DF. In the core area of EK-DF, Douglas fir was still largely dominant in the tree layer in the year 2005 with minor contributions of other tree species, mainly Norway spruce, that established after several small windthrow events in the early 1990s [52]. Windthrow and ice break also affected the core area in GB-DF in 1972 and in 1990. Regular forest management operations in the core areas had already ceased ca. 20 years before the official management abandonment in the years 2001 and 2002 (Table 1).

Table 1. Characteristics of the investigated strict forest reserves and its investigated core areas (following Gauer and Aldinger [51], BLE [53]). DF and NAT indicate the tree species dominance in the core area being Douglas fir or native tree species.

Strict Forest Reserve (SFR)	Grünberg (GB-DF)	Eselskopf (EK-DF)	Adelsberg (AB-NAT)
Geographic location	49°20′8″ N 7°58′12″ E	50°2′51″ N 6°37′11″ E	49°3′32″ N 7°30′24″ E
Forest ecoregion (and subregion)	Pfälzerwald (Middle Pfälzerwald)	Nordwesteifel (Islek and Oesling)	Pfälzerwald (Southern Pfälzerwald, Wasgau)
Reserve total size	64 ha	30 ha	192 ha
Year of establishment	2001	2002	1976
Geology	Bunter (Trifels subdivision)	Lower Devonian argillite	Bunter (Rehberg subdivision)
Soil type	Dystric Cambisol	Spodic Cambisol	Dystric Cambisol
Nutrient status	Oligotrophic	Mesotrophic	Oligotrophic
Elevation	210–420 m a. s. l.	310–440 m a. s. l.	245–399 m a. s. l.
Potential natural vegetation	Acidic beech forests (Luzulo-Fagetum)	Acidic to mesotrophic beech forests (Luzulo-Fagetum to Galio-Fagetum [52])	Acidic beech forests (Luzulo-Fagetum)
Mean annual air temperature (subregion)	8.4 °C	7.6 °C	8.8 °C
Mean annual precipitation (subregion)	933 mm	928 mm	926 mm
Tree species composition of core area (>7 cm diameter at breast height (DBH))	~73% *Pseudotsuga menziesii*, 17% *Fagus sylvatica*, 7% *Pinus sylvestris*, 3% *Picea abies* (Year 2004)	~95% *Pseudotsuga menziesii*, 4% *Picea abies*, others (Year 2006)	54% *Quercus petraea*, 30% *Tilia cordata*, 14% *Fagus sylvatica*, 2% *Carpinus betulus* (Year 2006)
Years of vegetation surveys (core area)	2005 and 2017	2005 and 2017	2000 and 2016
Number of permanent subplots (400 m^2) surveyed in the core area	13	14	29

Beyond the Douglas fir dominated core area, the SFR GB-DF was characterized by mixed forests of beech with Scots pine (*Pinus sylvestris*) or Norway spruce. According to the last forest inventory in 1995, tree species shares were 37.6% Scots Pine, 28.3% Douglas fir, 21.0% beech, 11.5% Norway spruce and others. In the SFR EK-DF, mixed forests of sessile oak with beech and hornbeam (*Carpinus betulus*) existed next to stands composed of Norway spruce and different non-native tree species (e.g., *Larix kaempferi*, *Thuja* spec., *Tsuga* spec.) that had been planted ca. 50 to 70 years ago. The last forest inventory in 1997 gave the following tree species composition in the SFR EK-DF: 49% Douglas fir, 30% Norway spruce, 5% Japanese larch, 4% hornbeam, 3% sessile oak, 3% beech and others including the mentioned not-natives.

We used a 1.5 ha fenced core area of the third SFR AB-NAT as a natural reference (Table 1). AB-NAT lies in the same forest ecoregion and is characterized by similar site conditions as the SFR GB-DF, but the reserve was already established in 1976. The tree species composition represents the potential natural tree species composition for the mountain ridges of this region with sessile oak, small-leaved lime (*Tilia cordata*) and beech [54]. The share of sessile oak was, however, largely promoted by forest management in former times [51,54]. This also accounts for the whole reserve where sessile oak was the dominant tree species (48%) in the year 2006 followed by Scots pine (25%) and beech (11%). Douglas fir made up 3% of the tree species share in the SFR AB-NAT.

2.2. Data Sampling and Analysis

The three core areas were separated into 13 (GB-DF), 14 (EK-DF) and 29 (AB-NAT) subplots of 400 m^2 (20 × 20 m) that represented the sampling unit for original vegetation surveys conducted in the years 2000 (AB-NAT) and 2005 (GB-DF, EK-DF), respectively, and for resurveys conducted 12 to 16 years later (Table 1). For the old and recent surveys, plant species were recorded separately for the tree layer (woody species >5 m), the shrub layer (woody species > 0.5 m ≤ 5 m), the herb layer (woody species ≤0.5 m and non-woody vascular plants) and the moss layer considering all soil dwelling bryophyte species. For the vegetation layers and all species within vegetation layers, the cover value was visually estimated directly in percent by horizontally projecting the area covered by a certain species or vegetation layer on the 400 m^2 subplots.

We contrasted the vegetation structure using vegetation layer cover values and the subplot-based species richness of vegetation layers between the first and the second survey. We also calculated the total species richness across plots for vascular plants and bryophytes. To characterize potential changes in environmental conditions over time, we used species indicator values taken from Ellenberg et al. [55]. The so called Ellenberg Indicator Values (EIV) of plant species are widely used in applied ecology in Central Europe [56]. Depending on their realized niche along ecological gradients that was defined based on field experience (=expert opinion), concurrent recordings of species and environmental variables and experimental tests, plant species of Central Europe have been given indicator values representing ordinal numbers between one and nine for moisture (M; ordinal number goes up to 12), nutrients/nitrogen (N), soil reaction (R), light (L), continentality (C) and temperature (T; [55]). By calculating averages of species indicator values at a plot-scale, the mean EIVs can represent rough surrogates for environmental conditions that are often used when direct measurements are not available [56]. For the different plant species, a value of 1 is representing species indicating dry, nutrient poor or acidic conditions as well as deep shade. Continentality and temperature values relate to the geographic ranges of species with continentality indicating the distance to the sea. C = 1 characterizes extreme oceanic species and C = 9 extreme continental species that are nearly absent from Central Europe. C values were found to correlate with frost resistance of plant species [57]. The temperature values relate to the distribution of species along elevational and latitudinal gradients with T = 1 comprising cold-adapted species of high mountains or boreal-arctic regions and T = 9 very warm-adapted species spreading from the Mediterranean into warmer places of Central Europe. An increase in the mean temperature value may indicate a thermophilization effect in the forest understory caused by global warming [58]. For vascular plant and bryophyte species of Central Europe, EIVs can be taken from Ellenberg et al. [55]. N-values for bryophytes are, however not provided. Therefore, we used the values given by Simmel et al. [59].

For analyzing dynamics in species composition, we grouped species of the field layer (=shrub, herb and moss layer) according to their forest affinity [60] and their association to broad plant communities according to Ellenberg et al. [55], Oberdorfer [61] and Nebel and Philippi [62]. In their list of forest-associated species, Schmidt et al. [60] have characterized typical forest species (category 1) into those that mainly occur in closed forests (sub-category 1.1) and those of edges and clearings (sub-category 1.2). In category 2, they have grouped species with a wider habitat preference into those species that occur both in forests and in open habitats (sub-category 2.1) and into those species that occasionally occur in forests but predominantly in open habitats (sub-category 2.2). Such species may be regarded as disturbance indicators within forests (e.g., *Cirsium vulgare*, *Stellaria media*, [63]). Species not listed in Schmidt et al. [60] were considered as open habitat species (O). As broad plant communities, we defined oak and beech (-mixed) forests (=class Querco-Fagetea), coniferous forests including the class Vaccinio-Piceetea, shrubby vegetation mainly comprising *Rubus fruticosus* agg., edges and clearings represented by the class Epilobietea angustifolii, herbaceous vegetation of disturbed sites mainly of the class Artemisietea and grass- and heathlands ranging from the class Nardo-Callunetea to mesophilous grassland of the class Molinio-Arrhenatheretea. Species that occur in a wide range of communities were categorized as indifferent (e.g., *Taraxacum officinale* agg., *Rubus idaeus*; see Table S1 in Supplementary Materials). For non-native tree species, a community preference is not defined. For each species group, we quantified their relative contribution to the species richness per subplot and contrasted numbers between surveys.

To quantify a general change in species composition within the understories of the three SFRs, we calculated a pairwise presence/absence based Jaccard dissimilarity index by contrasting each subplot to all other subplots sampled within each SFR and survey year. Changes across survey periods may indicate a homogenization or a differentiation within the understory based on local colonization or extinction of species. With diversity partitioning, we quantified the species replacement component showing real species turnover among subplots (indicating that one species is directly replaced by another species) and the nestedness component showing that a dissimilarity among subplots is caused by species richness differences [64]. A dissimilarity mainly caused by the nestedness component indicates that species in species poor subplots represent a subset of the species occurring in species richer subplots. To also integrate the abundance of species and to visualize the direction of species compositional changes between surveys, we conducted a non-metric multidimensional scaling (NMDS) on two dimensions based on Bray–Curtis dissimilarity. We conducted two different NMDS ordinations, one with data of the two Douglas fir SFRs (GB-DF and EK-DF) only and another with data from all three SFRs. We correlated the NMDS axes values with EIVs to identify the main environmental drivers of compositional changes. We additionally correlated the axes values with species richness values and the cover values of beech and Douglas fir in the shrub and herb layer to visualize changes in biodiversity and tree species regeneration.

All analyses were based on the field layer combining the shrub, herb and moss layer. Differences between surveys were tested using the Wilcoxon signed rank test or the paired t-test depending on normality distribution and variance homogeneity. For calculating the Jaccard dissimilarity and its components, we used the betapart package of the R-software [65] and the function beta.pair. The NMDS was conducted using the function metaMDS of the R Package vegan [66] and the function envfit for axes correlations. If not stated otherwise, statistical significance was assumed for $p < 0.05$. The species nomenclature follows Oberdorfer [61] for vascular plants and Nebel and Philippi [62] for bryophytes.

3. Results

3.1. Changes in Vegetation Structure, Species Richness and Environmental Conditions

In the two strict forest reserves (SFRs) dominated by Douglas fir, Grünberg (GB-DF) and Eselskopf (EK-DF), the cover of the herb layer decreased, while the moss layer increased between surveys. This pattern was contrary in the SFR Adelsberg (AB-NAT) that is dominated by native tree species. The canopy cover increased in all three SFRs but significantly only in AB-NAT (Table 2).

Consistent across all three SFRs was the decrease in mean species richness of the herb and field layer, as well as the total vascular plant species richness across plots. Total richness of vascular plants was 11 species lower in AB-NAT at the second compared to the first survey, 14 species lower in EK-DF and 18 species lower in GB-DF (Table 2).

There was no consistent pattern in the dynamics of the Ellenberg Indicator Values (EIVs) and temporal changes were generally small (<0.5 units; Table 2). In all SFRs, the light value decreased (not significantly in GB-DF). For GB-DF and AB-NAT, results show a significant increase in the moisture value, for AB-NAT also the nutrient value increased between surveys as was the temperature value, while the continentality decreased (Table 2). The three SFRs differed in temperature and continentality mainly because of differences in tree species composition also in the field layer, e.g., with Norway spruce (C = 6, T = 3) being abundant in GB-DF, while EK-DF and AB-NAT showed a higher frequency of hornbeam (C = 4, T = 6) or small-leaved lime (C = 4, T = 5).

Table 2. Vegetation layer characteristics for the first and second survey. Given are mean values with standard error in parentheses. Significantly higher values ($p < 0.05$) comparing both survey years are written in bold.

Strict Forest Reserve (SFR)	Grünberg (GB-DF)		Eselskopf (EK-DF)		Adelsberg (AB-NAT)	
Survey year	2005	2017	2005	2017	2000	2016
N	13	13	14	14	29	29
Cover value [%]						
Tree layer	84.4 (2.6)	88.8 (1.5)	75.2 (4.2)	76.8 (3.7)	71.9 (2.9)	**82.3 (1.9)**
Shrub layer	25.7 (3.4)	20.5 (3.1)	**35.9 (5.4)**	15.4 (6.2)	10.3 (1.6)	13.9 (3.1)
Herb layer	**6.0 (1.3)**	1.9 (0.5)	**14.9 (2.8)**	5.8 (1.3)	7.6 (2.1)	**11.1 (2.0)**
Moss layer	1.9 (0.4)	**2.6 (0.4)**	17.7 (4.5)	**25.3 (6.3)**	**5.2 (1.0)**	0.9 (0.2)
Species richness/400 m^2						
Tree layer	2.9 (0.1)	2.9 (0.2)	3.2 (0.3)	3.5 (0.4)	3.0 (0.1)	**3.4 (0.3)**
Shrub layer	3.9 (0.3)	3.4 (0.2)	**5.6 (0.5)**	4.6 (0.5)	1.8 (0.2)	**2.6 (1.9)**
Herb layer	**9.5 (1.5)**	2.2 (0.4)	**14.9 (1.7)**	8.4 (0.7)	**8.5 (0.7)**	7.5 (0.5)
Moss layer	3.6 (0.6)	**4.8 (0.5)**	8.3 (0.9)	9.1 (0.8)	**4.2 (0.3)**	2.5 (0.3)
Field layer (shrub + herb + moss)	**14.7 (2.0)**	8.8 (0.8)	**27.6 (2.2)**	21.7 (1.4)	**13.2 (0.7)**	10.5 (0.7)
Total species richness						
Vascular plants	27	9	48	34	39	28
Bryophytes	11	13	16	18	24	17
EIVs [1] (Field layer)						
Temperature	3.8 (0.1)	3.8 (0.1)	4.4 (0.1)	4.3 (0.1)	4.8 (0.1)	**5.0 (0.1)**
Continentality	4.1 (0.1)	4.3 (0.1)	3.7 (0.1)	3.7 (0.1)	**3.6 (0.1)**	3.3 (0.1)
Light	4.7 (0.1)	4.5 (0.0)	**5.0 (0.1)**	4.8 (0.1)	**4.8 (0.1)**	4.7 (0.1)
Moisture	5.2 (0.1)	**5.4 (0.1)**	5.2 (0.0)	5.1 (0.0)	4.9 (0.0)	**5.1 (0.0)**
Acidity	3.7 (0.1)	3.8 (0.2)	4.2 (0.1)	4.1 (0.1)	4.1 (0.1)	4.3 (0.2)
Nutrients (Nitrogen)	4.3 (0.2)	4.4 (0.1)	5.0 (0.1)	4.8 (0.1)	4.5 (0.1)	**4.9 (0.1)**

[1] EIV = Ellenberg Indicator Value.

3.2. Changes in Field Layer Species Composition

The share of species representing different forest affinity groups developed similarly in all three reserves, even though no significant changes could be detected in AB-NAT (Figure 1). In GB-DF and EK-DF, the share of (predominantly) open habitat species significantly decreased, while in EK-DF the share of closed forest species significantly increased. Though not significant, such an increase was also found in the two other reserves.

In terms of plant community association, the two SFRs dominated by Douglas fir differed from the SFR AB-NAT dominated by native tree species (Figure 2). While the share of species characteristic for deciduous forests of the class Querco-Fagetea significantly increased in the native forest of AB-NAT, the contribution of these species decreased in GB-DF (significantly) and EK-DF (non-significantly). Indifferent species showed the opposite pattern (significant increase in GB-DF and EK-DF and decrease in AB-NAT). In all three reserves, though, there was a reduction in the accumulated share of open habitat species comprising species of disturbed sites, of edges and clearings and of heath- and grasslands (GB-DF: $p = 0.023$ for paired t-test; EK-DF: $p < 0.001$ for paired t-test; AB-NAT: $p = 0.005$ for Wilcoxon signed rank test). The latter two showed a significant decrease in EK-DF (Figure 2). In AB-NAT, an increasing frequency of *Rubus fruticosus* agg. increased the share of shrubby vegetation. Due to an overall decrease in the species richness of the field layer, the contribution of non-native tree species (Douglas fir and Weymouth pine (*Pinus strobus*) detected in two subplots of the SFR GB-DF) with no defined community association increased significantly in GB-DF (Figure 2).

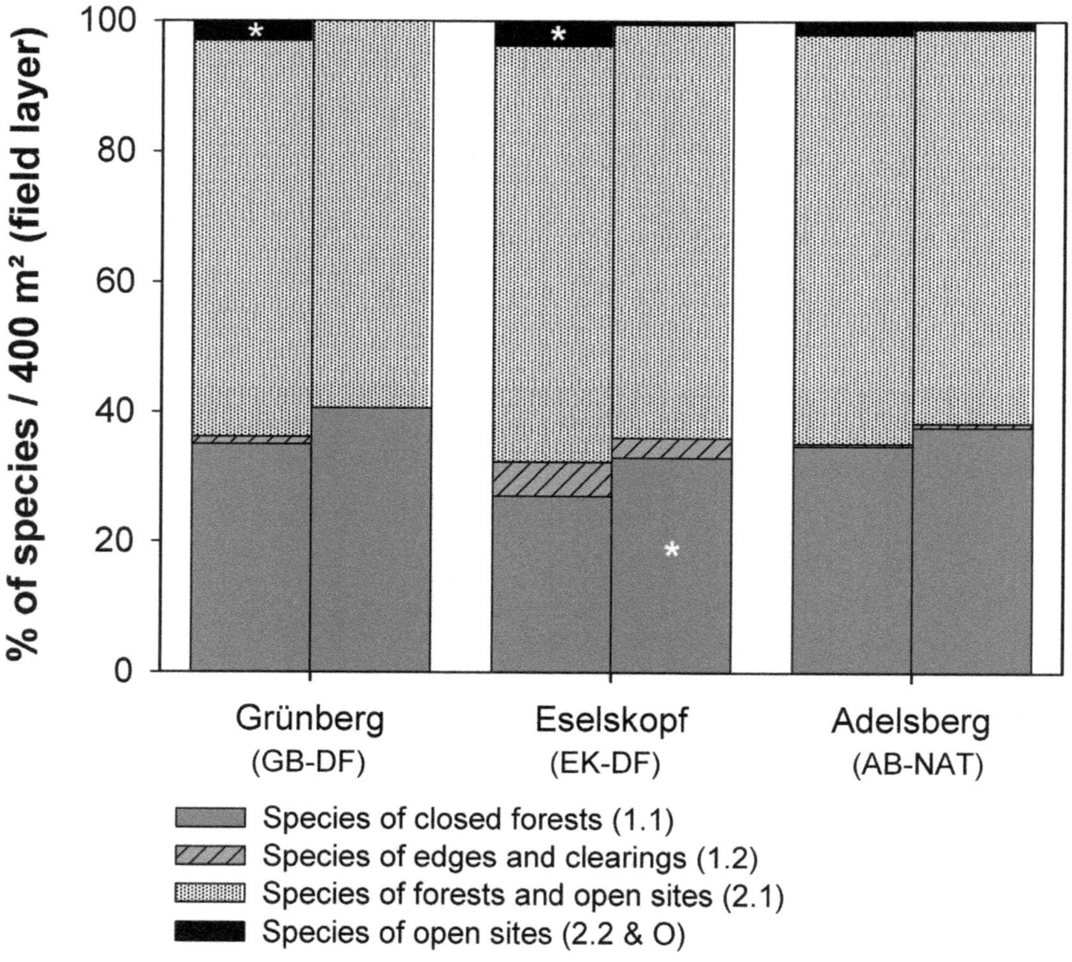

Figure 1. Percentage share of species of the field layer grouped according to their forest affinity in the three strict forest reserves. Left column = old survey, right column = recent survey. * marks significant differences between surveys ($p < 0.05$).

All three SFRs showed a significant change in species dissimilarity over time (Figure 3). While GB-DF and AB-NAT showed a homogenization in the understory with a significant reduction in the Jaccard dissimilarity index, EK-DF was characterized by a differentiation among subplots over time. The largest decrease in dissimilarity was detected in GB-DF that also showed the largest reduction in species richness (Table 2). AB-NAT additionally showed a significant increase in the nestedness component indicating that species were lost from some but not from all subplots (Figure 3).

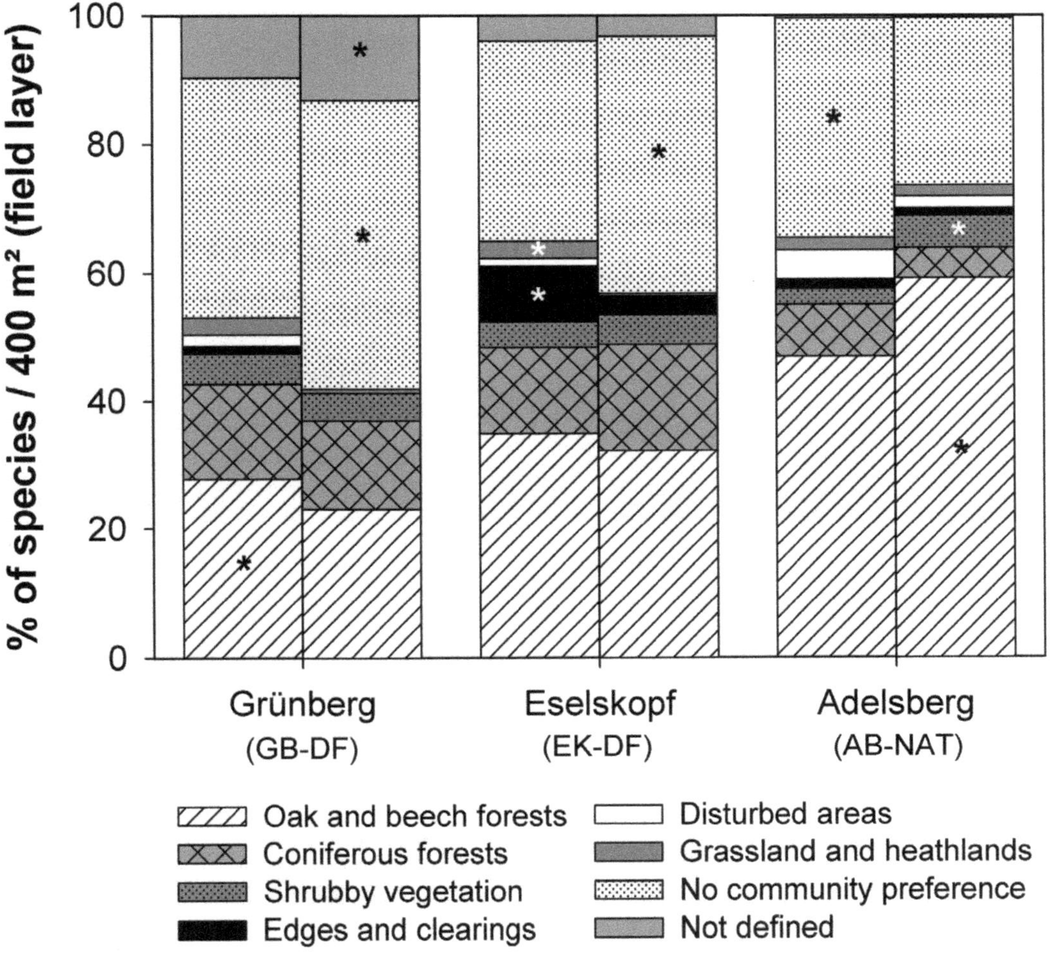

Figure 2. Percentage share of species of the field layer grouped according to their forest community preference. Left column = old survey, right column = recent survey. * marks significant differences between surveys ($p < 0.05$). See text for details.

The differing dissimilarity patterns of GB-DF and EK-DF (homogenization vs. differentiation) were confirmed by the NMDS ordination (Figure 4a). However, the ordination showed a similar direction of species compositional change with time (Figure 4a). For both Douglas fir SFRs, the axes values show a significant shift along the second axis (Table 3). GB-DF and EK-DF mainly differentiated along the first axis explained by contrasting continentality and temperature values due to differences in tree species composition in the regeneration that remained over time. Regeneration of Douglas fir (in the shrub and herb layer) and beech (in the herb layer) was associated with the first survey. The same is true for species richness and the light indicator value that was highest in EK-DF at the first survey.

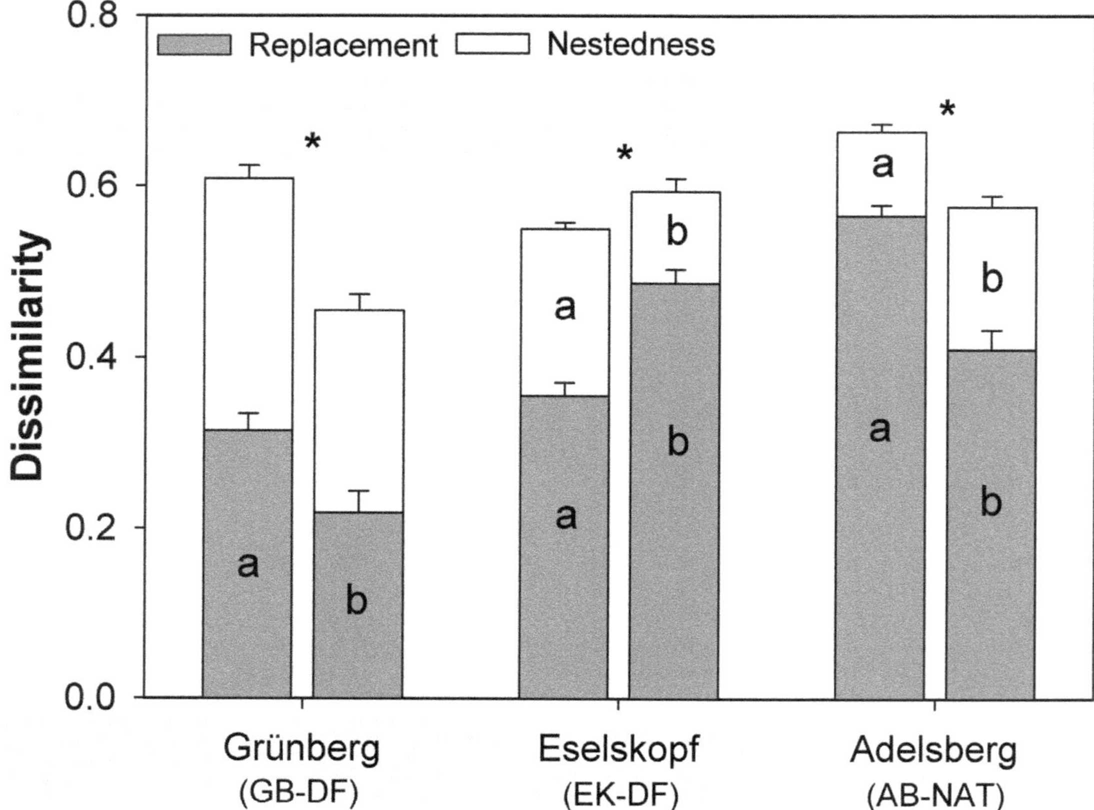

Figure 3. Mean pairwise Jaccard dissimilarity for the old (left bar) and more recent (right bar) survey in the three SFR. The dissimilarity was partitioned into the replacement and nestedness component according to Baselga [64]. * Marks a significant difference ($p < 0.05$) between the surveys for the overall Jaccard-dissimilarity. Different letters show significant differences ($p < 0.05$) for replacement and nestedness component between the old and recent survey.

The NMDS diagram of the three SFR shows a general shift of all three reserves towards a lower species richness, less light and an increasing cover of beech in the shrub layer, while Douglas fir regeneration in the shrub layer was associated with the first survey (Figure 4b). When combining both NMDS axes, we found a shift in the same direction for all three reserves, though a significant shift was only confirmed for GB-DF and AB-NAT (Table 3). The compositional separation between the two SFRs dominated by Douglas fir (GB-DF and EK-DF) and the SFR characterized by native tree species (AB-NAT) remained along the first axis over time.

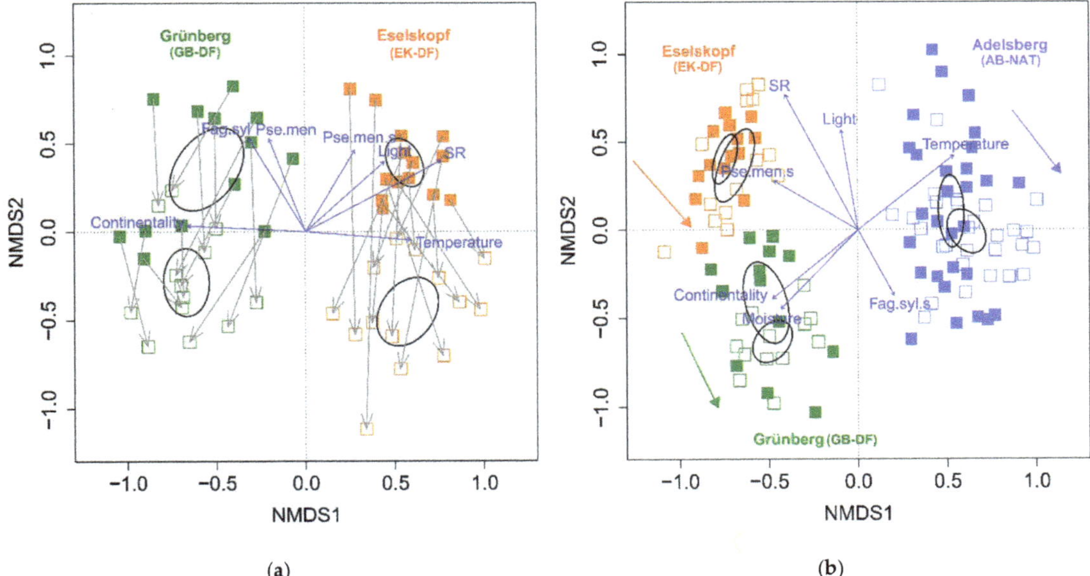

Figure 4. Two-dimensional NMDS ordination diagram of species abundance data of the field layer (shrub, herb and moss layer) for (**a**) the two SFR dominated by Douglas fir (stress = 0.2013) and (**b**) for all three investigated SFR (stress = 0.1899) for the two survey times. Grey arrows in (**a**) indicate the shift of individual subplots between the survey times. The colored arrows in (**b**) indicate the general shift in species composition for the respective SFRs. The ellipses give the standard error around the centroids for each observation and SFR. A bi-plot was created by correlating axes values with mean EIVs, the species richness (SR) and the cover of *F. sylvatica* (Fag.syl) and *P. menziesii* (Pse.men) in the shrub (_s) and herb layer. Only significant correlations with $p < 0.001$ are displayed.

Table 3. Mean axes scores (±SE) of the NMDS ordination for the first and second survey displayed in Figure 4. Significantly higher scores between the first and second survey according to paired t-tests are written in bold. For the ordination of the three SFR (Figure 4b), both axes were additionally combined into one axis score (NMDS1-NMDS2) to identify significant shifts across axes.

	First Survey	Second Survey	*p*-Value
NMDS of Figure 4a			
GB-DF NMDS 1	−0.552 ± 0.085	−0.668 ± 0.052	0.095
GB-DF NMDS 2	**0.355 ± 0.096**	−0.287 ± 0.078	**<0.001**
EK-DF NMDS 1	0.559 ± 0.044	0.573 ± 0.071	0.848
EK-DF NMDS 2	**0.391 ± 0.056**	−0.454 ± 0.079	**<0.001**
NMDS of Figure 4b			
GB-DF NMDS1	−0.514 ± 0.053	−0.480 ± 0.045	0.593
GB-DF NMDS2	**−0.412 ± 0.094**	−0.631 ± 0.048	**0.005**
GB-DF NMDS1-NMDS2	−0.101 ± 0.124	**0.151 ± 0.054**	**0.034**
EK-DF NMDS1	−0.741 ± 0.029	−0.687 ± 0.046	0.152
EK-DF NMDS2	0.404 ± 0.057	0.367 ± 0.085	0.613
EK-DF NMDS1-NMDS2	−1.145 ± 0.049	−1.055 ± 0.068	0.235
AB-NAT NMDS1	0.524 ± 0.030	0.611 ± 0.045	0.071
AB-NAT NMDS2	0.105 ± 0.084	−0.009 ± 0.051	0.095
AB-NAT NMDS1-NMDS2	0.419 ± 0.092	**0.620 ± 0.078**	**0.002**

3.3. Tree Species Dynamics in the Three SFRs

The cover of Douglas fir regeneration was mainly associated with the first survey period (Figure 4). Douglas fir decreased in frequency and/or cover in both SFRs dominated by Douglas fir in almost all vegetation layers, except for the tree layer in EK-DF (Table 4). In EK-DF, the shrub layer cover of Douglas fir largely decreased. In GB-DF, this was the case for both the tree and the shrub layer. Regeneration of Douglas fir in the herb layer was generally absent in all three SFRs at time of the second survey.

A different trend was found for beech. This tree species increased in the shrub layer of EK-DF and in the tree layer of GB-DF (Table 4). While beech and Douglas fir were similar in tree layer coverage at the first survey in GB-DF, beech dominated the tree layer in 2017, though underneath Douglas fir. In EK-DF, beech established in every subplot in the shrub layer from the first to the second survey. The increasing trend of beech in the tree and shrub layer is in line with the development in the SFR AB-NAT. Here, also hornbeam and sessile oak expanded in the tree layer contributing to an increase in tree layer cover (Table 2).

In contrast to Douglas fir, Norway spruce expanded in the shrub layer of GB-DF and remained constant in frequency in EK-DF.

Most tree species in the tree and shrub layer were recorded in the SFR EK-DF for both surveys. There, most species remained relatively constant or showed a shift from the shrub to the tree layer (e.g., hornbeam, hazelnut). For both SFRs dominated by Douglas fir, most woody species in the herb layer showed a significant reduction. In AB-NAT, the abundance of seedlings of sessile oak and small-leaved lime increased in the herb layer (Table 4).

Table 4. Frequency of occurrence (F) and mean cover values in % with standard error (mCV (SE)) for all woody species surveyed in the 400 m^2 subplots in the core areas of the three strict forest reserves (SFR) Grünberg, Eselskopf (both characterized by Douglas fir (DF) and Adelsberg (dominated by native tree species (NAT)) for the two surveys. Species were grouped into the tree layer, shrub layer and herb layer and were ordered according to their temporal dynamics. Significant higher cover values between surveys according to the Wilcoxon signed rank test and higher frequency values of at least 20% comparing both surveys are written in bold, +: <0.05%.

Strict Forest Reserve (SFR)	Grünberg (GB-DF)				Eselskopf (EK-DF)				Adelsberg (AB-NAT)			
N	13				14				29			
Year	2005		2017		2005		2017		2001		2016	
	F	mCV (SE)	F	mCv (SE)	F	mCv (SE)	F	mCv (SE)	F	mCv (SE)	F	mCv (SE)
Tree layer (>5 m height)												
Increasing												
Fagus sylvatica	100	50.1 (3.6)	100	**75.0 (5.6)**	50	1.3 (0.4)	64	3.2 (1.1)	72	17.0 (3.2)	86	**31.0 (4.1)**
Carpinus betulus					43	1.3 (0.5)	57	2.4 (0.8)	45	5.0 (1.6)	**66**	**9.4 (2.0)**
Quercus petraea					21	0.4 (0.2)	7	0.1 (0.1)	86	31.5 (3.6)	97	**40.5 (4.7)**
Corylus avellana					7	1.0 (1.0)	**50**	1.5 (0.5)				
Decreasing												
Pseudotsuga menziesii	100	**48.6 (4.5)**	100	35.9 (4.0)	100	67.6 (3.3)	100	69.1 (3.3)				
No significant change												
Pinus sylvestris	69	4.3 (1.5)	54	4.8 (1.8)								
Picea abies	23	0.7 (0.4)	38	0.9 (0.4)	29	1.8 (1.1)	36	1.8 (1.4)				
Betula pendula					21	0.4 (0.2)	14	0.3 (0.2)				
Salix caprea					21	0.4 (0.2)	7	0.1 (0.1)				
Sorbus aucuparia					21	0.4 (0.3)	7	0.2 (0.2)				
Tilia cordata									93	28.0 (3.6)	93	29.0 (3.3)
Shrub layer (≤5 m height)												
Increasing												
Fagus sylvatica	100	5.6 (0.9)	100	4.3 (1.0)	79	2.0 ± 0.5	**100**	2.6 ± 0.5	72	6.8 ± 1.5	**93**	6.6 ± 1.2
Picea abies	100	7.7 (1.6)	100	**14.8 (3.2)**	64	2.3 ± 1.0	64	0.8 ± 0.2	3	+	7	0.1 ± 0.1
Rubus fruticosus agg.			**23**	0.2 (0.1)			7	+			**41**	**6.8 ± 2.8**
Tilia cordata									66	1.3 ± 0.3	**86**	1.6 ± 0.4

Table 4. *Cont.*

Strict Forest Reserve (SFR)	**Grünberg (GB-DF)**				**Eselskopf (EK-DF)**				**Adelsberg (AB-NAT)**			
N	**13**				**14**				**29**			
Year	**2005**		**2017**		**2005**		**2017**		**2001**		**2016**	
	F	**mCV (SE)**	**F**	**mCv (SE)**	**F**	**mCv (SE)**	**F**	**mCv (SE)**	**F**	**mCv (SE)**	**F**	**mCv (SE)**
Decreasing												
Pseudotsuga menziesii	100	**11.4 (1.9)**	92	1.2 (0.2)	**100**	**20.7 ± 3.4**	71	4.6 ± 2.8				
Pinus sylvestris	**54**	0.7 (0.3)										
Carpinus betulus					**86**	7.0 ± 2.9	50	5.0 ± 3.1	34	**2.2 ± 0.7**	21	0.6 ± 0.3
Quercus robur					**21**	0.2 ± 0.1						
No significant change												
Corylus avellana					86	2.8 ± 0.6	79	1.5 ± 0.3				
Quercus petraea					43	0.6 ± 0.3	43	0.7 ± 0.3			**7**	+
Acer pseudoplatanus					21	0.1 ± 0.1	14	0.1 ± 0.0				
Sorbus aria					21	0.2 ± 0.1	7	+				
Herb layer (≤0.5 m height)												
Increasing												
Quercus petraea					14	0.1 (0.0)	14	+	93	0.4 (0.0)	100	**1.9 (0.6)**
Tilia cordata									66	0.2 (0.0)	**93**	**0.4 (0.0)**
Decreasing												
Rubus idaeus	**77**	0.8 ± 0.2			**36**	0.2 ± 0.1						
Fagus sylvatica	**69**	0.4 ± 0.1	31	0.1 ± 0.1	**21**	+			**86**	0.6 ± 0.1	66	0.6 ± 0.1
Pseudotsuga menziesii	**69**	0.4 ± 0.1			**29**	+						
Cytisus scoparius	**31**	0.1 ± 0.1			**21**	+					3	+
Sambucus racemosa					**71**	0.2 ± 0.1						
Contrary development												
Rubus fruticosus agg.	**69**	1.5 ± 0.8	46	1.1 ± 0.5	100	3.2 ± 0.5	93	1.6 ± 0.5	38	1.7 ± 1.4	48	**3.6 ± 1.2**
No significant change												
Picea abies	85	0.9 ± 0.3	100	0.6 ± 0.1			14	+			7	+
Carpinus betulus					29	0.2 ± 0.1	14	+	38	0.1 ± 0.0	52	0.2 ± 0.0

4. Discussion

4.1. Vascular Plant Species Richness Declines in All Strict Forest Reserves

We detected a general decrease in the species richness of vascular plants across all three strict forest reserves (SFRs). This is in line with other studies showing a decreasing plant species richness following management abandonment [67–69]. We could further show that forest reserves either dominated by non-native coniferous species or by native deciduous tree species respond similarly to management abandonment. Reasons for a plant species richness decline are a lack of disturbance in the soil and canopy and an increasing canopy cover with a consequent reduction in light availability. A slight increase in tree layer cover, including an increase in beech in all SFRs (not significantly in EK-DF), and a decrease in light indicator values across reserves confirm an impact of reduced light availability. In addition, the share of shade-tolerant species typical for closed forests increased, while the share of light-demanding, open site species decreased between surveys.

We observed the largest reduction in species richness in the SFR Grünberg (GB-DF), where beech was the dominant tree species in the tree layer at time of the second survey (Figure S1a,c). Beech is highly competitive [70] and its spread and expansion after management abandonment can accelerate a reduction in plant species numbers [67,71]. The very low species richness observed in the SFR GB-DF is thereby typical for beech-dominated forests on acidic soils after management abandonment [67]. Low light levels below a dense beech canopy, a lack of soil disturbance, and a high leaf layer thickness due to low decomposition rates and high root competition by beech are mainly responsible for the low herb layer diversity. In mixture with other tree species, beech can even benefit from a lower intraspecific competition and can for example respond with an increased horizontal crown expansion and light absorption [72]. In the SFR GB-DF, Douglas fir was established as a pure stand and beech regenerated naturally from surrounding beech forests. The fast growth of Douglas fir and the subsequent colonization by beech led to a vertical separation of both tree species allowing sufficient light for the rather shade-intolerant Douglas fir in the upper canopy and for the shade-tolerant beech in the lower canopy. According to Thurm and Pretzsch [73], this optimized use of canopy space due to vertical niche differentiation of both tree species may result in maximum light interception. While this can increase productivity (mainly for Douglas fir [73]) and maintain a moist microclimate potentially mitigating effects of global warming [74], maximum light interception by the tree layer reduces the light availability for the herb layer and leads to a decline in its species richness and abundance.

Similar to aboveground, there is also growing evidence for belowground complementarity in mixed forests. Mixtures of complementary tree species such as early and late-successional or coniferous and deciduous species can enhance fine-root productivity by a more complete filling of the environment, including a higher horizontal volume filling [75]. Even though Lwila et al. [76] could not detect large differences in fine root biomass between pure and mixed forests of beech and Douglas fir, they found a general high belowground plasticity of beech in nutrient poor sites and a shift of beech fine roots to deeper soil layers across site conditions in mixed stands. These results confirmed research by Hendriks and Bianchi [77] that found a higher root density in deeper soil strata in mixed Douglas fir/beech stands compared to pure stands. While this belowground niche differentiation between tree species can reduce the interspecific competition and increase the exploitation of available resources for tree growth, it may increase root competition for the understory particularly on acidic sites with a low nutrient availability.

The reduction in species richness was lower in EK-DF and AB-NAT compared to GB-DF, where tree species richness was slightly higher and the share of beech much lower. This confirms results by Mölder et al. [78] who showed an increasing reduction in herb layer species richness with increasing beech share mainly due to a higher litter layer thickness.

A reduction in field and herb layer diversity and abundance was, however, also observed in the SFR EK-DF with the lowest abundance of beech in the canopy among the three SFR and a dominance of Douglas fir in the tree layer. Both in the non-native [79] and native range of Douglas fir [80], unthinned stands were species poorer in the understory compared to thinned stands. In the SFR EK-DF, Douglas fir was dominant in the shrub layer at the first survey. At the second survey, Douglas fir regeneration had largely decreased due to self-thinning processes in the shrub layer [81]. In the native range, this dense stem exclusion stage (Figure S1b,d) is characterized by a decrease in understory species richness and abundance [82].

In contrast to the herb layer, the moss layer increased in abundance and species richness (only in GB-DF) in both SFR with Douglas fir. This also supports results of the meta-analysis by Paillet et al. [68] with bryophytes benefiting from management abandonment. A stable or increasing canopy cover can create a moist microclimate including a stable soil moisture that is beneficial for the species richness of bryophytes [83,84]. An increasing EIV for moisture in GB-DF underlines an effect of management abandonment on microclimate. In addition, soil bryophytes can benefit from a coniferous canopy and a thin deciduous leaf litter layer [85–87]. This explains the highest cover values of the moss layer in the SFR EK-DF, where Douglas fir needles dominate the litter layer. The strict forest reserve AB-NAT, dominated by native tree species, showed a contrasting pattern in the moss layer with a reduction in species richness and abundance. Here, the larger amount of deciduous leaf litter because of a significant increase in canopy cover can reduce the moss layer [85]. However, also observer effects have to be taken into account as the separation between substrates (soil, deadwood and rocks) during vegetation sampling was sometimes difficult in the SFR AB-NAT. Long-term vegetation sampling beyond the fenced core area in the SFR AB-NAT showed, however, that also in this reserve the moss layer increased in species richness in plots that had been dominated by conifers between the years 2006 and 2016 [86]. On the other hand, plots dominated by deciduous tree species (oak and beech) showed no change.

4.2. *Contrasting Patterns in Community Composition and Homogenization among the Three Reserves*

Besides a consistent decrease in field layer species richness and in the share of light-demanding species across the three investigated SFRs, there were also distinct differences in compositional dynamics between the Douglas fir forests and the native reference. Between surveys, species typical of deciduous forests (Querco-Fagetea) showed a relative reduction under Douglas fir in GB-DF and EK-DF, but an increase in the SFR AB-NAT with native tree species. On the other hand, the share of generalist species with no community preference increased in the Douglas fir reserves. This is in line with Leitl [88], who detected less species of natural forest communities but more ruderal species within Douglas fir stands compared to native deciduous forests. Similar patterns were also observed for arthropods with Douglas fir mainly supporting generalist species [30,39]. Compositional differences in understory vegetation between forest types with and without Douglas fir have also been confirmed by other studies [29,34,40]. In forests in the Czech Republic, especially nitrophilous and ruderal species benefited from Douglas fir compared to beech, oak or Norway spruce, due to a better litter quality of Douglas fir and higher nitrogen contents in the humus and topsoil [40]. In our study, we did not find an increasing nutrient indicator value in the Douglas fir reserves, but in the SFR AB-NAT, dominated by native tree species. In the SFR AB-NAT, *Rubus fruticosus* agg. expanded between surveys and presumably benefited from fencing and the heterogeneous light conditions due to a higher species diversity in the tree layer. The lower and decreasing share of Querco-Fagetea species in the SFRs with Douglas fir compared to the SFR AB-NAT, on the other hand, may be the result of the management history. Most SFRs in Germany and Central Europe were established in ancient native forests with no tree species change in the past [45]. Thus, they have provided a continuous habitat for species of natural forest communities. Forest conversion from deciduous to coniferous forests, however, is known to cause a shift in plant species composition [89]. Further monitoring within the SFRs will show if the already observed expansion of native tree species in the shrub and tree layer of the Douglas fir reserves will cause a shift to a more natural species composition in the future. Until now, the NMDS ordination showed no convergence of field layer species composition between SFRs with and without Douglas fir, even though all three SFRs showed a similar direction of change over time towards a lower light availability and lower species richness.

We found, however, an effect of beech on understory homogenization over time. The two SFRs with a high share of beech in the tree layer, GB-DF and AB-NAT, showed a clear homogenization among subplots, while the SFR EK-DF displayed a differentiation. As all three SFRs were characterized by a decrease in species richness of the field layer, this suggests different mechanisms of local extinction and colonization as well as tree identity effects. The loss of species occurred directional in GB-DF and AB-NAT with different infrequent species disappearing from the subplots leading to a decrease in the species turnover component. Few new species colonized due to the decreasing light availability and lack of soil disturbance [90]. The SFR EK-DF was characterized by both processes. While the same species disappeared from the subplots of EK-DF (e.g., *Sambucus racemosa*, *Senecio ovatus*; Table S1), they were partly replaced by new and different species in some subplots, among them many bryophyte species that can benefit from the needle litter [34]. In total, 28 species disappeared from the field layer in EK-DF, while 16 new species colonized. Only 7 and 9 species newly occurred in the subplots of GB-DF and AB-NAT, respectively compared to 23 and 28 species that disappeared (Table S1). Besides missing disturbance, this underlines the negative effect of an increasing beech share on understory species richness [78] and supports evidence that increasing shade can lead to homogenization effects [91] also at a local scale. Natural disturbances that may affect the SFR in higher frequency and intensity in the future will presumably change this pattern.

4.3. Douglas Fir Regeneration Is Decreasing

Regeneration of Douglas fir decreased both in the shrub and in the herb layer in both Douglas fir SFRs, while the main native tree species beech and Norway spruce showed increasing or stable trends in the shrub layer. Different reasons are responsible for the decreasing regeneration of Douglas fir. In the SFR GB-DF the decreasing light availability due to an expansion of beech in the overstory seems to be the most decisive factor. According to Montigny and Smith [92], minimum gap sizes of roughly 0.3 ha are needed for a successful regeneration of Douglas fir. Mailly and Kimmins [93] mention a minimum relative light availability of 40% for a survival and growth of Douglas fir regeneration. As no larger gaps occurred between survey years, Douglas fir was clearly inferior in competition with the shade-tolerant beech. In addition, Petriţan et al. [94] showed that Douglas fir regeneration is more sensitive to root competition than beech.

Even though, the SFR EK-DF was not colonized by shade-tolerant tree species to the same degree as the SFR GB-DF, self-thinning also led to a large reduction in Douglas fir in the shrub layer (Figure S1d). According to He and Duncan [81], mortality is mainly observed in dense patches of Douglas fir regeneration leading to a regular spatial distribution of trees. Surviving Douglas fir trees partly reached the lower tree layer in the SFR EK-DF, as did some native tree species such as hornbeam, hazelnut or beech. Due to the lower share of beech in the whole area of the SFR EK-DF compared to GB-DF (according to the forest inventory from 1997 and 1995, 3% of beech in EK-DF vs. 21% of beech in GB-DF), the colonization of shade-tolerant species presumably shows a time lag and may accelerate in upcoming years at the further expense of Douglas fir.

In general, the dynamics of Douglas fir observed in the investigated SFRs reflect the dynamics in mature and old growth stands of the native range. As a pioneer tree species, Douglas fir stands often originate from catastrophic wildfires or other stand replacing disturbances [95]. In course of succession, these pioneer forests are largely shaped by self-thinning processes and are invaded by shade-tolerant tree species such as western hemlock (*Tsuga heterophylla*). The colonization by shade-tolerant tree species was found to be independent of openings that occur due to the stem exclusion of Douglas fir [81]. The increasing competition by shade tolerant tree species reaching the overstorey then hampers the successful further recruitment of Douglas fir [81]. Even in small canopy gaps within mature (100–150 years) and old-growth (>200 years) Douglas fir forests in the native range, no successful regeneration of Douglas fir was observed anymore, while the gaps were colonized by the shade-tolerant hemlock [96]. Thus, the decreasing regeneration success within the SFRs reported here is in line with the dynamics in natural Douglas fir forests in the native range and underlines the pioneer character of this tree species. As also no seedlings were found during the second survey in the SFRs, a successful germination is presumably also hampered by a dense leaf or needle litter and a dense moss layer preventing the seedlings from reaching the mineral soil [10,19,21,97]. However, most tree species showed a reduction in frequency and cover in the herb layer also suffering from decreasing light availability and an increasing litter layer thickness after management abandonment.

A competitive inferiority of Douglas fir compared to native tree species was also confirmed by Frei et al. [21] for a study in Switzerland for which 39 sites with Douglas fir planted mostly in mixture with native tree species were investigated for the occurrence and abundance of Douglas fir and native tree species regeneration. Except for some dry, nutrient-deficient sites with an open canopy where Douglas fir was able to dominate the seedling layer (<130 cm heigh) and comprised up to 23% of saplings ($\geq$130 cm), saplings of Douglas fir were on all sites outnumbered by regeneration of beech, Norway spruce, silver fir (*Abies alba*) or other broadleaved tree species. In the SFR GB-DF, Norway spruce turned out to be more shade-tolerant in terms of survival than Douglas fir as the cover value of Norway spruce in the shrub layer significantly increased between surveys. Already in 2005, Vor [98] recorded a higher percentage (26%) of dead Douglas fir trees in the regeneration (dbh < 7 cm) compared to Norway spruce (4%). Douglas fir was, however, characterized by a greater height growth. The reduction in Douglas fir in the shrub layer between surveys confirms the finding that the light requirement of Douglas fir increases with age [10]. Despite survival, the Norway spruce regeneration showed a low vitality under the low light conditions and remained mainly below 2 m in height at time of the second survey. Mortality of Douglas fir regeneration in the SFR EK-DF showed a similar percentage as observed in GB-DF (27%) in 2005 but was much lower compared to Norway spruce (76%, [98]). Here, the faster growth rate of Douglas fir under a more open canopy seems to have outcompeted Norway spruce. However, subsequent self-thinning within the dense regeneration led to a decrease in Douglas fir in the shrub layer up to the second survey (see above).

The impact of deer is also mentioned as a limiting factor for the successful establishment of Douglas fir regeneration [10], as the tree species is often affected by browsing on spring-flush growth, by fraying from roe bucks (*Capreolus capreolus*) and by bark stripping [99]. Frei et al. [21] found browsing intensity on Douglas fir to be lower compared to ash, maple, oak or silver fir, but higher compared to beech and Norway spruce. While Douglas fir made up 10% of the number of seedlings recorded by Frei et al. [21], the species only comprised 3% of the number of saplings. A similar pattern was found for the browsing sensitive silver fir (dropping from 17 to 10%), while beech and Norway spruce showed the opposite (beech: from 29 to a 48% share, spruce: from 6 to 9%). This supports an impact of deer on the regeneration establishment of Douglas fir. According to Spellmann et al. [10], deer browsing and bark fraying or stripping can cause a complete loss of Douglas fir natural regeneration. However, as the core areas of the SFRs investigated here were fenced, our results indicate that deer seems not to be a major factor explaining failing Douglas fir regeneration. Rather light availability, litter layer thickness and the successional position of Douglas fir in relation to other tree species are likely causes. Outside fenced areas, though, the impact of deer seems to further decrease the competitive strength of Douglas fir compared to native tree species [21]. Outside the fenced areas of the SFRs, few individuals of Douglas fir (>0.5 m) were detected, most were affected by browsing (personal observation).

While most tree species decreased in abundance in the herb layer in the Douglas fir SFRs, sessile oak and small-leaved lime increased in the herb layer in the SFR AB-NAT. The increasing frequency of masting years observed in recent years is one reason for the increased abundance of seedlings [100]. The increasing frequency of both tree species in the regeneration is also responsible for an increasing temperature value in the SFR AB-NAT (T-values of 5 to 6). Further monitoring in this and in other SFRs will show if climate change with increasing temperatures and changes in the precipitation regime will promote a further growth of oak and lime at the expense of beech.

5. Conclusions

We observed similar vegetation dynamics in Douglas fir-dominated forests and forests dominated by native tree species after management abandonment irrespective of the tree species identity and origin, as diminished soil disturbance and a decreasing light availability were more important. The understory vegetation became less species-rich and more shade-tolerant over time.

Tree species identity, however, affected species compositional changes across survey periods and had an effect on understory homogenization. While a Douglas fir canopy promoted the relative share of generalist species, a native canopy promoted typical species of the class Querco-Fagetea. In addition, the expansion of beech led to a homogenization of the understory mainly due to species losses, while a canopy dominated by Douglas fir led to a differentiation mainly due to the colonization of bryophytes. This suggests that certain tree species characteristics, such as the shade-casting ability of beech and its dense leaf litter vs. a well decomposable needle litter of Douglas fir, shape the understory vegetation characteristics irrespective of forest management. However, all strict forest reserves (SFR) showed an increasing share of typical closed forest species.

Dynamics of the tree species Douglas fir in the unmanaged SFRs reflect its successional position in the native range. As a pioneer species, it does not show a successful regeneration in mature stands of the native range and similarly, regeneration also decreased between observation periods in the investigated SFRs. This supports the finding that a successful regeneration of Douglas fir requires silvicultural management or severe disturbances [43] and that the competitive strength against native, particularly shade-tolerant tree species, is low. However, in the native range, Douglas fir can build self-perpetuating forests at very dry sites, where shade-tolerant species such as hemlock fail [81]. This underlines the invasive potential of Douglas fir at dry sites in Europe [18,19,21] and stresses the need for further monitoring of Douglas fir stands under changing abiotic conditions also without forest management. Under current site conditions though, our study shows that, even in pure stands, the regeneration potential of Douglas fir seems to be low when left for natural forest development. As the seed production of the old Douglas fir trees may not have reached its maximum yet [101], uncertainties concerning its future regeneration potential remain. Regarding the long lifespan of Douglas fir in the native range [96], it can be expected that Douglas fir will be dominant in the upper canopy of the SFR in the future with native, shade-tolerant tree species colonizing below and shaping the environment for the understory. Regular monitoring in the SFRs with and without Douglas fir will provide important knowledge on this assumption in the future and on the competitive strength of Douglas fir under changing abiotic conditions.

Supplementary Materials: The following supporting information can be downloaded at: https://www.mdpi.com/article/10.3390/d14100795/s1, Table S1. Frequency of occurrence (F) in % and mean cover values and standard error in % (mCV±SE) for all species surveyed on the 400 m^2 subplots in the core areas of the three strict forest reserves Grünberg, Eselskopf and Adelsberg for two observation times; Figure S1. Impressions from the core areas of the strict forest reserves (SFRs) dominated by Douglas fir.

Author Contributions: Conceptualization, S.H.; methodology, S.H., M.D., T.V., P.B. and W.S.; formal analysis, S.H.; data collection, S.H., M.D. and T.V.; data curation, S.H.; writing—original draft preparation, S.H.; writing—review and editing, S.H., M.D., T.V., P.B. and W.S.; project administration, P.B. and W.S.; funding acquisition, S.H., M.D. and W.S. All authors have read and agreed to the published version of the manuscript.

Funding: This research was partly funded by the Federal State of Rhineland-Palatinate, Germany.

Institutional Review Board Statement: Not applicable.

Informed Consent Statement: Not applicable.

Data Availability Statement: The data presented in this study are available on request from the corresponding author.

Acknowledgments: Our thanks go to Helmut Adam for the provision of map and data material as well as background knowledge about the forests studied. We acknowledge support by the Open Access Publication Funds of the Göttingen University.

Conflicts of Interest: The authors declare no conflict of interest. The funders had no role in the design of the study; in the collection, analyses, or interpretation of data; in the writing of the manuscript, or in the decision to publish the results.

References

1. Pötzelsberger, E.; Spiecker, H.; Neophytou, C.; Mohren, F.; Gazda, A.; Hasenauer, H. Growing non-native trees in European forests brings benefits and opportunities but also has its risks and limits. *Curr. For. Rep.* **2020**, *6*, 339–353. [CrossRef]
2. Senf, C.; Seidl, R. Mapping the forest disturbance regimes of Europe. *Nat. Sustain.* **2021**, *4*, 63–70. [CrossRef]
3. Jandl, R.; Spathelf, P.; Bolte, A.; Prescott, C.E. Forest adaptation to climate change—Is non-management an option? *Ann. For. Sci.* **2019**, *76*, 48. [CrossRef]
4. Kamp, J.; Trappe, J.; Dübbers, L.; Funke, S. Impacts of windstorm-induced forest loss and variable reforestation on bird communities. *For. Ecol. Manag.* **2020**, *478*, 118504. [CrossRef]
5. Hazarika, R.; Bolte, A.; Bednarova, D.; Gaviria, J.; Kanzian, M.; Kowalczyk, J.; Lackner, M.; Lstibůrek, M.; Longauer, R.; Nagy, L.; et al. Multi-actor perspectives on afforestation and reforestation strategies in Central Europe under climate change. *Ann. For. Sci.* **2021**, *78*, 60. [CrossRef]
6. Brus, R.; Pötzelsberger, E.; Lapin, K.; Brundu, G.; Orazio, C.; Straigyte, L.; Hasenauer, H. Extent, distribution and origin of non-native forest tree species in Europe. *Scand. J. For. Res.* **2019**, *34*, 533–544. [CrossRef]
7. Spellmann, H. Ertragskundliche Aspekte des Fremdländeranbaus. *Allg. Forst Jagdztg.* **1994**, *165*, 27–34.
8. Hermann, R.K.; Lavender, D.P. Douglas-fir planted forests. *New For.* **1999**, *17*, 53–70. [CrossRef]
9. Lévesque, M.; Rigling, A.; Bugmann, H.; Weber, P.; Brang, P. Growth response of five co-occurring conifers to drought across a wide climatic gradient in Central Europe. *Agric. For. Meteorol.* **2014**, *197*, 1–12. [CrossRef]
10. Vor, T.; Spellmann, H.; Bolte, A.; Ammer, C. Potentiale und Risiken eingeführter Baumarten—Baumartenportraits mit naturschutzfachlicher Bewertung. *Göttinger Forstwiss.* **2015**, *7*, 187–217.
11. Vitali, V.; Büntgen, U.; Bauhus, J. Silver fir and Douglas fir are more tolerant to extreme droughts than Norway spruce in south-western Germany. *Glob. Chang. Biol.* **2017**, *23*, 5108–5119. [CrossRef] [PubMed]
12. Schueler, S.; George, J.-P.; Karanitsch-Ackerl, S.; Mayer, K.; Klumpp, R.T.; Grabner, M. Evolvability of drought response in four native and non-native conifers: Opportunities for forest and genetic resource management in Europe. *Front. Plant Sci.* **2021**, *12*, 648312. [CrossRef] [PubMed]
13. Höltermann, A.; Nehring, S.; Herberg, A.; Krug, A. Die Douglasie aus Sicht des Bundesamtes für Naturschutz. *AFZ-Der Wald* **2016**, *71*, 34–37.
14. Krüger, U. Gefährdet die Douglasie die Biodiversität wirklich nicht?—Fragen zu den Folgen forstlichen Handelns. *Nat. Landsch.* **2017**, *49*, 110–111.
15. Vor, T.; Nehring, S.; Bolte, A.; Höltermann, A. Assessment of invasive tree species in nature conservation and forestry—Contradictions and coherence. In *Introduced Tree Species in European Forests: Opportunities and Challenges*; Krumm, F., Vitkova, L., Eds.; European Forest Institute: Freiburg, Germany, 2016; pp. 148–157.
16. Broncano, M.J.; Vilà, M.; Boada, M. Evidence of *Pseudotsuga menziesii* naturalization in montane Mediterranean forests. *For. Ecol. Manag.* **2005**, *211*, 257–263. [CrossRef]
17. Fagúndez, J. Heathlands confronting global change: Drivers of biodiversity loss from past to future scenarios. *Ann. Bot.* **2013**, *111*, 151–172. [CrossRef]
18. Bindewald, A.; Miocic, S.; Wedler, A.; Bauhus, J. Forest inventory-based assessments of the invasion risk of *Pseudotsuga menziesii* (Mirb.) Franco and *Quercus rubra* L. in Germany. *Eur. J. For. Res.* **2021**, *140*, 883–899. [CrossRef]

19. Knoerzer, D. *Zur Naturverjüngung der Douglasie im Schwarzwald. Inventur und Analyse von Umwelt- und Konkurrenzfaktoren Sowie eine Naturschutzfachliche Bewertung*; Dissertationes Botanicae 306; Gebrüder Borntraeger Verlagsbuchhandlung: Stuttgart, Germany, 1999; pp. 1–283.
20. Lange, F.; Ammer, C.; Leitinger, G.; Seliger, A.; Zerbe, S. Is Douglas fir (*Pseudotsuga menziesii* [Mirbel] Franco) invasive in Central Europe? A case study from south-west Germany. *Front. For. Glob. Chang.* **2022**, *5*, 844580. [CrossRef]
21. Frei, E.R.; Moser, B.; Wohlgemuth, T. Competitive ability of natural Douglas fir regeneration in central European close-to-nature forests. *For. Ecol. Manag.* **2022**, *503*, 119767. [CrossRef]
22. Ammer, C.; Bolte, A.; Herberg, A.; Höltermann, A.; Krüß, A.; Krug, A.; Nehring, S.; Schmidt, O.; Spellmann, H.; Vor, T. Empfehlungen für den Anbau eingeführter Waldbaumarten—Gemeinsames Papier von Forstwissenschaft und Naturschutz. *Nat. Landsch.* **2016**, *48*, 168–172.
23. Brundu, G.; Richardson, D.M. Planted forests and invasive alien trees in Europe—A code for managing existing and future plantings to mitigate the risk of negative impacts from invasions. *NeoBiota* **2016**, *30*, 5–47. [CrossRef]
24. Brundu, G.; Pauchard, A.; Pyšek, P.; Pergl, J.; Bindewald, A.M.; Brunori, A.; Canavan, S.; Campagnaro, T.; Celesti-Grapow, L.; de Sá Dechoum, M.; et al. Global guidelines for the sustainable use of non-native trees to prevent tree invasions and mitigate their negative impacts. *NeoBiota* **2020**, *61*, 65–116. [CrossRef]
25. Thomas, F.M.; Rzepecki, A.; Werner, W. Non-native Douglas fir (*Pseudotsuga menziesii*) in Central Europe: Ecology, performance and nature conservation. *For. Ecol. Manag.* **2022**, *506*, 119956. [CrossRef]
26. Frank, D.; Finckh, M. Impact of Douglas-fir plantations on vegetation and soil in south-central Chile. *Rev. Chil. Hist. Nat.* **1997**, *70*, 191–211. (In Spanish)
27. Richardson, D.M.; Rejmánek, M. Conifers as invasive aliens: A global survey and predictive framework. *Divers. Distrib.* **2004**, *10*, 321–331. [CrossRef]
28. Orellana, I.A.; Raffaele, E. The spread of the exotic conifer *Pseudotsuga menziesii* in *Austrocedrus chilensis* forests and shrublands in northwestern Patagonia, Argentina. *N. Z. J. For. Sci.* **2010**, *40*, 199–209.
29. Schmid, M.; Pautasso, M.; Holdenrieder, O. Ecological consequences of Douglas fir (*Pseudotsuga menziesii*) cultivation in Europe. *Eur. J. For. Res.* **2014**, *133*, 13–29. [CrossRef]
30. Tschopp, T.; Holderegger, R.; Bollmann, K. Auswirkungen der Douglasie auf die Waldbiodiversität. *Schweiz. Z. Forstwes.* **2015**, *166*, 9–15. [CrossRef]
31. Wohlgemuth, T.; Moser, B.; Pötzelsberger, E.; Rigling, A.; Gossner, M.M. Über die Invasivität der Douglasie und ihre Auswirkungen auf Boden und Biodiversität. *Schweiz. Z. Forstwes.* **2021**, *172*, 118–127. [CrossRef]
32. Gossner, M. Insektenwelten—Die Douglasie im Vergleich mit der Fichte. *LWF Wissen* **2008**, *59*, 70–73.
33. Ziesche, T.M.; Roth, M. Influence of environmental parameters on small-scale distribution of soil-dwelling spiders in forests: What makes the difference, tree species or microhabitat? *For. Ecol. Manag.* **2008**, *255*, 738–752. [CrossRef]
34. Budde, S.; Schmidt, W.; Weckesser, M. Impact of the admixture of European beech (*Fagus sylvatica* L.) on plant species diversity and naturalness of conifer stands in Lower Saxony. *Waldökol. Landsch. Nat.* **2011**, *11*, 49–61.
35. Buée, M.; Maurice, J.-P.; Zeller, B.; Andrianarisoa, S.; Ranger, J.; Courtecuisse, R.; Marcais, B.; Le Tacon, F. Influence of tree species on richness and diversity of epigeous fungal communities in a French temperate forest stand. *Fungal Ecol.* **2011**, *4*, 22–31. [CrossRef]
36. Schuldt, A.; Scherer-Lorenzen, M. Non-native tree species (*Pseudotsuga menziesii*) strongly decreases predator biomass and abundance in mixed-species plantations of tree diversity experiment. *For. Ecol. Manag.* **2014**, *327*, 10–17. [CrossRef]
37. Gossner, M.; Utschick, H. Douglas fir stands deprive overwintering bird species of food resource. *NeoBiota* **2004**, *3*, 105–121.
38. Gossner, M.M.; Wende, B.; Levick, S.; Schall, P.; Floren, A.; Linsenmair, K.E.; Steffan-Dewenter, I.; Schulze, E.-D.; Weisser, W.W. Deadwood enrichment in European forests—Which tree species should be used to promote saproxylic beetle diversity? *Biol. Conserv.* **2016**, *201*, 92–102. [CrossRef]
39. Kriegel, P.; Matevski, D.; Schuldt, A. Monoculture and mixture-planting of non-native Douglas fir alters species composition, but promotes the diversity of ground beetles in a temperate forest system. *Biodivers. Conserv.* **2021**, *30*, 1479–1499. [CrossRef]
40. Podrázský, V.; Martiník, A.; Matejka, K.; Viewegh, J. Effects of Douglas-fir (*Pseudotsuga menziesii* [Mirb.] Franco) on understorey layer species diversity in managed forests. *J. For. Sci.* **2014**, *60*, 263–271. [CrossRef]
41. Heinrichs, S.; Ammer, C.; Mund, M.; Boch, S.; Budde, S.; Fischer, M.; Müller, J.; Schöning, I.; Schulze, E.-D.; Schmidt, W.; et al. Landscape-scale mixtures of tree species are more effective than stand-scale mixtures for biodiversity of vascular plants, bryophytes and lichens. *Forests* **2019**, *10*, 73. [CrossRef]
42. Kostić, O.; Jarić, S.; Gajić, G.; Pavlović, D.; Marković, M.; Mitrović, M.; Pavlović, P. The effects of Douglas fir monoculture on stand characteristics in a zone of montane beech forest. *Arch. Biol. Sci.* **2016**, *68*, 753–766. [CrossRef]
43. Eberhard, B.; Hasenauer, H. Modeling regeneration of Douglas fir forests in Central Europe. *Austrian J. For. Sci.* **2018**, *135*, 33–51.
44. Wolf, G.; Bohn, U. Naturwaldreservate in der Bundesrepublik Deutschland und Vorschläge zu einer bundesweiten Grunddatenerfassung. *Schriftenr. Veg.* **1991**, *21*, 9–19.
45. Parviainen, J.; Bücking, W.; Vandekerkhove, K.; Schuck, A.; Päivinen, R. Strict forest reserves in Europe: Efforts to enhance biodiversity and research on forests left for free development in Europe (EU-COST-Action E4). *Forestry* **2000**, *73*, 107–118. [CrossRef]

46. Endres, U.; Förster, B. Die Douglasie in Naturwaldreservaten—Passt das zusammen? Vorkommen der Douglasie in bayerischen Naturwaldreservaten. *LWF Aktuell* **2013**, *93*, 37–139.
47. Gilliam, F.S. The ecological significance of the herbaceous layer in temperate forest ecosystems. *BioScience* **2007**, *57*, 845–858. [CrossRef]
48. Handa, I.T.; Aerts, R.; Berendse, F.; Berg, M.P.; Bruder, A.; Butenschoen, O.; Chauvet, E.; Gessner, M.O.; Jabiol, J.; Makkonen, M.; et al. Consequences of biodiversity loss for litter decomposition across biomes. *Nature* **2014**, *509*, 218–221. [CrossRef]
49. Neff, F.; Brändle, M.; Ambarli, D.; Ammer, C.; Bauhus, J.; Boch, S.; Hölzel, N.; Klaus, V.H.; Kleinebecker, T.; Prati, D.; et al. Changes in plant-herbivore network structure and robustness along land-use intensity gradients in grasslands and forests. *Sci. Adv.* **2021**, *7*, eabf3985. [CrossRef]
50. BMEL (Bundesministerium für Ernährung und Landwirtschaft) Bundeswaldinventur. 2022. Available online: https://www.bundeswaldinventur.de/ (accessed on 4 March 2022).
51. Gauer, J.; Aldinger, E. Waldökologische Naturräume Deutschlands—Forstliche Wuchsgebiete und Wuchsbezirke. *Mitt. Ver. Forstl. Standortskde. Forstpflanzenz.* **2005**, *43*, 1–324.
52. Vor, T.; Schmidt, W. Auswirkungen des Douglasienanbaus auf die Vegetation der Naturwaldreservate "Eselskopf" (Nordwesteifel) und "Grünberg" (Pfälzer Wald). *Forstarchiv* **2006**, *77*, 169–178.
53. BLE (Bundesanstalt für Landwirtschaft und Ernährung). Naturwaldreservate—Urwälder von Morgen (Datenbank zu Naturwaldreservaten in Deutschland). 2022. Available online: https://fgrdeu.genres.de/naturwaldreservate/ (accessed on 4 March 2022).
54. Balcar, P. Waldstrukturen im grenzüberschreitenden Naturwaldreservat Adelsberg-Lutzelhardt. *Ann. Sci. Rés. Bios. Trans. Vosges Nord-Pfälzerwald* **2008**, *14*, 27–45.
55. Ellenberg, H.; Weber, H.E.; Düll, R.; Wirth, V.; Werner, W. *Zeigerwerte von Pflanzen in Mitteleuropa*, 3rd ed.; Scripta Geobot. 18; Verlag Erich Goltze GmbH & Co KG: Göttingen, Germany, 2001; pp. 1–262.
56. Diekmann, M. Species indicator values as an important tool in applied plant ecology—A review. *Basic Appl. Ecol.* **2003**, *4*, 493–506. [CrossRef]
57. Bartelheimer, M.; Poschlod, P. Functional characterizations of Ellenberg indicator values—A review on ecophysiological determinants. *Funct. Ecol.* **2016**, *30*, 506–516. [CrossRef]
58. Van der Veken, S.; Bossuyt, B.; Hermy, M. Climate gradients explain changes in plant community composition of the forest understorey. *Belg. J. Bot.* **2004**, *137*, 55–69. [CrossRef]
59. Simmel, J.; Ahrens, M.; Poschlod, P. Ellenberg N values of bryophytes in Central Europe. *J. Veg. Sci.* **2021**, *32*, e12957. [CrossRef]
60. Schmidt, M.; Kriebitzsch, W.-U.; Ewald, J. Waldartenliste der Farn- und Blütenpflanzen, Moose und Flechten Deutschlands. *BfN-Skr.* **2011**, *299*, 1–111.
61. Oberdorfer, E. *Pflanzensoziologische Exkursionsflora*, 8th ed.; Ulmer: Stuttgart, Germany, 2001.
62. Nebel, M.; Philippi, G. (Eds.) *Die Moose Baden-Württembergs, Band 1–3*; Ulmer: Stuttgart, Germany, 2000; Volume 2001, p. 2005.
63. Schmidt, M.; Bedarff, U.; Meyer, P. Einfluss von Störungen auf die Vegetation von Buchenwäldern. *AFZ-Der Wald* **2018**, *73*, 20–22.
64. Baselga, A. The relationship between species replacement, dissimilarity derived from nestedness, and nestedness. *Glob. Ecol. Biogeogr.* **2012**, *21*, 1223–1232. [CrossRef]
65. Baselga, A.; Orme, D.; Villeger, S.; De Bortoli, J.; Leprieur, F.; Logez, M. Betapart: Partitioning Beta Diversity into Turnover and Nestedness Components. R Package Version 1.5.4. 2021. Available online: https://CRAN.R-project.org/package=betapart (accessed on 17 September 2022).
66. Oksanen, J.; Blanchet, F.G.; Friendly, M.; Kindt, R.; Legendre, P.; McGlinn, D.; Minchin, R.R.; O'Hara, R.B.; Simpson, G.L.; Solymos, P.; et al. Vegan: Community Ecology Package. R Package Version 2.5-7. 2020. Available online: https://CRAN.R-project.org/package=vegan (accessed on 17 September 2022).
67. Fischer, C.; Parth, A.; Schmidt, W. Vegetationsdynamik in Buchen-Naturwäldern: Ein Vergleich aus Süd-Niedersachsen. *Hercynia-Okol. Umw. Mitteleur.* **2009**, *42*, 45–68.
68. Paillet, Y.; Bergès, L.; Hjältén, J.; Ódor, P.; Avon, C.; Bernhardt-Römermann, M.; Bijlsma, R.-J.; De Bruyn, L.; Fuhr, M.; Grandin, U.; et al. Does biodiversity differ between managed and unmanaged forests? A meta-analysis on species richness in Europe. *Conserv. Biol.* **2010**, *24*, 101–112. [CrossRef]
69. Heinrichs, S.; Schulte, U.; Schmidt, W. Veränderung der Buchenwaldvegetation durch Klimawandel? Ergebnisse aus Naturwaldzellen in Nordrhein-Westfalen. *Forstarchiv* **2011**, *82*, 48–61.
70. Leuschner, C. Mechanismen der Konkurrenzüberlegenheit der Buche. *Ber. D. Reinh.-Tüxen-Ges.* **1998**, *10*, 5–18.
71. Mölder, A.; Streit, M.; Schmidt, W. When beech strikes back: How strict nature conservation reduces herb-layer diversity and productivity in Central European deciduous forests. *For. Ecol. Manag.* **2014**, *319*, 51–61. [CrossRef]
72. Forrester, D.; Ammer, C.; Annighöfer, P.; Barbeito, I.; Bielak, K.; Bravo-Oviedo, A.; Coll, L.; del Río, M.; Drössler, L.; Heym, M.; et al. Effects of crown architecture and stand structure on light absorption in mixed and monospecific *Fagus sylvatica* and *Pinus sylvestris* forests along a productivity and climate gradient through Europe. *J. Ecol.* **2018**, *106*, 746–760. [CrossRef]
73. Thurm, E.A.; Pretzsch, H. Improved productivity and modified tree morphology of mixed versus pure stands of European beech (*Fagus sylvatica*) and Douglas-fir (*Pseudotsuga menziesii*) with increasing precipitation and age. *Ann. For. Sci.* **2016**, *73*, 1047–1061. [CrossRef]

74. Jucker, T.; Bouriaud, O.; Coomes, D.A. Crown plasticity enables trees to optimize canopy packing in mixed-species forests. *Funct. Ecol.* **2015**, *29*, 1078–1086. [CrossRef]
75. Brassard, B.W.; Chen, H.Y.H.; Cavard, X.; Langanière, J.; Reich, P.B.; Bergeron, Y.; Paré, D.; Yuan, Z. Tree species diversity increases fine root productivity through increased soil volume filling. *J. Ecol.* **2013**, *101*, 210–219. [CrossRef]
76. Lwila, A.; Mund, M.; Ammer, C.; Glatthorn, J. Site conditions more than species identity drive fine root biomass, morphology and spatial distribution in temperate pure and mixed forests. *For. Ecol. Manag.* **2021**, *499*, 119581. [CrossRef]
77. Hendriks, C.M.A.; Bianchi, F.J.J.A. Root density and root biomass in pure and mixed forest stands of Douglas-fir and beech. *Neth. J. Agric.* **1995**, *11*, 321–331. [CrossRef]
78. Mölder, A.; Bernhardt-Römermann, M.; Schmidt, W. Herb-layer diversity in deciduous forests: Raised by tree richness or beaten by beech? *For. Ecol. Manag.* **2008**, *256*, 272–281. [CrossRef]
79. Augusto, L.; Dupouey, J.-L.; Ranger, J. Effects of tree species on understory vegetation and environmental conditions in temperate forests. *Ann. For. Sci.* **2003**, *60*, 823–831. [CrossRef]
80. Bailey, J.D.; Mayrsohn, C.; Doescher, P.S.; Pierre, E.S.; Tappeiner, J.C. Understory vegetation in old and young Douglas-fir forests of western Oregon. *For. Ecol. Manag.* **1998**, *112*, 289–302. [CrossRef]
81. He, F.; Duncan, R.P. Density-dependent effects on tree survival in an old-growth Douglas fir forest. *J. Ecol.* **2000**, *88*, 676–688. [CrossRef]
82. Oliver, C.D. Forest development in North America following major disturbances. *For. Ecol. Manag.* **1980**, *3*, 153–168. [CrossRef]
83. Friedel, A.; Oheimb, G.v.; Dengler, J.; Härdtle, W. Species diversity and species composition of epiphytic bryophytes and lichens—A comparison of managed and unmanaged beech forests in NE Germany. *Feddes Repert.* **2006**, *117*, 172–185. [CrossRef]
84. Raabe, S.; Müller, J.; Manthey, M.; Dürhammer, O.; Teuber, U.; Göttlein, A.; Förster, B.; Brandl, R.; Bässler, C. Drivers of bryophyte diversity allow implications for forest management with a focus on climate change. *For. Ecol. Manag.* **2010**, *260*, 1956–1964. [CrossRef]
85. Márialigeti, S.; Németh, B.; Tinya, F.; Ódor, P. The effects of stand structure on ground-floor bryophyte assemblages in temperate mixed forests. *Biodivers. Conserv.* **2009**, *18*, 2223–2241. [CrossRef]
86. Heinrichs, S.; Dölle, M.; Balcar, P.; Schmidt, W. NWR Adelsberg-Lutzelhardt: Keine Chance für die Eiche. *AFZ-Der Wald* **2018**, *73*, 29–32.
87. Müller, J.; Boch, S.; Prati, D.; Socher, S.A.; Pommer, U.; Hessenmöller, D.; Schall, P.; Schulze, E.-D.; Fischer, M. Effects of forest management on bryophyte species richness in Central European forests. *For. Ecol. Manag.* **2019**, *432*, 850–859. [CrossRef]
88. Leitl, R. Artenvielfalt und Bestandesform am Beispiel der Bodenvegetation. *Ber. Bayer Landesanst. Wald Forstwirtsch.* **2001**, *33*, 9–13.
89. Verstraeten, G.; Baeten, L.; De Frenne, P.; Vanhellemont, M.; Thomaes, A.; Boonen, W.; Muys, B.; Verheyen, K. Understorey vegetation shifts following the conversion of temperate deciduous forest to spruce plantation. *For. Ecol. Manag.* **2013**, *289*, 363–370. [CrossRef]
90. Härdtle, W.; Oheimb, G.v.; WestphaL, C. The effects of light and soil conditions on the species richness of the ground vegetation of deciduous forests in northern Germany (Schleswig-Holstein). *For. Ecol. Manag.* **2003**, *182*, 327–338. [CrossRef]
91. Keith, S.A.; Newton, A.C.; Morecroft, M.D.; Bealey, C.E.; Bullock, J.M. Taxonomic homogenization of woodland plant communities over 70 years. *Proc. Biol. Sci.* **2009**, *276*, 3539–3544. [CrossRef] [PubMed]
92. Montigny, L.E.D.; Smith, N.J. The effect of gap size in a group selection silvicultural system on the growth response of young, planted Douglas-fir. A sector plot analysis. *Forestry* **2017**, *90*, 426–435. [CrossRef]
93. Mailly, D.; Kimmins, J.P. Growth of *Pseudotsuga menziesii* and *Tsuga heterophylla* seedlings along a light gradient: Resource allocation and morphological acclimation. *Can. J. Bot.* **1997**, *75*, 1424–1435. [CrossRef]
94. Petriţan, I.C.; Lüpke, B.V.; Petriţan, A.M. Effects of root trenching of overstorey Norway spruce (*Picea abies*) on growth and biomass of underplanted beech (*Fagus sylvatica*) and Douglas fir (*Pseudotsuga menziesii*) sapling. *Eur. J. For. Res.* **2010**, *130*, 813–828. [CrossRef]
95. Spies, T.A.; Franklin, J.F. The structure of natural young, mature, and old-growth Douglas-fir forests. In *Wildlife and Vegetation of Unmanaged Douglas-fir Forests*; Ruggiero, L.F., Aubry, K.B., Carey, A.B., Huff, M.H., Eds.; USDA For. Serv. Gen. Tech. Rep. PNW-GTR-285; US Department of Agriculture, Forest Service: Portland, OR, USA, 1991; pp. 91–110.
96. Spies, T.A.; Franklin, J.F.; Klopsch, M. Canopy gaps in Douglas-fir forests of the Cascade Mountains. *Can. J. For. Res.* **1990**, *20*, 649–658. [CrossRef]
97. Caccia, F.D.; Ballaré, C.L. Effects of tree cover, understory vegetation, and litter on regeneration of Douglas-fir (*Pseudotsuga menziesii*) in southwestern Argentina. *Can. J. For. Res.* **1998**, *28*, 683–692. [CrossRef]
98. Vor, T. Bodenvegetation und Naturverjüngung in Douglasien-Altbeständen. *Forstarchiv* **2011**, *82*, 159–160. [CrossRef]
99. Gill, R. A review of damage by mammals in north temperate forests: 1. Deer. *Forestry* **1992**, *65*, 145–169. [CrossRef]
100. Nussbaumer, A.; Waldner, P.; Etzold, S.; Gessler, A.; Benham, S.; Thomsen, I.M.; Jørgensen, B.B.; Timmermann, V.; Verstraeten, A.; Sioen, G.; et al. Patterns of mast fruiting of common beech, sessile and common oak, Norway spruce and Scots pine in Central and Northern Europe. *Forest Ecol. Manag.* **2016**, *363*, 237–251. [CrossRef]
101. Lavender, D.P.; Hermann, R.K. *Douglas-fir—The Genus Pseudotsuga*; Oregon State University Press: Corvallis, OR, USA, 2014; pp. 1–352.

Article

Selection of Elms Tolerant to Dutch Elm Disease in South-West Romania

Dănuț Chira [1], Florian G. Borlea [2,*], Florentina Chira [1], Costel Ș. Mantale [1], Mihnea I. C. Ciocîrlan [1,3], Daniel O. Turcu [4], Nicolae Cadar [4], Vincenzo Trotta [5], Ippolito Camele [5], Carmine Marcone [6] and Ștefania M. Mang [5,*]

1 Station of Brașov, "Marin Drăcea" National Research and Development Institute in Forestry, 13 Cloşca Str., 500040 Brașov, Romania
2 Faculty of Agriculture, Banat's University of Agricultural Sciences and Veterinary Medicine, "King Michael I of Romania" of Timișoara, 119 Calea Aradului, 300645 Timișoara, Romania
3 Faculty of Silviculture and Forest Engineering, "Transilvania" University of Brașov, Șirul Beethoven 1, 500123 Brașov, Romania
4 Station of Timișoara, "Marin Drăcea" National Research and Development Institute in Forestry, 8 Pădurea Verde Str., 300310 Timișoara, Romania
5 School of Agricultural, Forestry, Food and Environmental Sciences (SAFE), Università degli Studi della Basilicata, Viale dell'Ateneo Lucano 10, 85100 Potenza, Italy
6 Department of Pharmacy, University of Salerno, Via Giovanni Paolo II 132, 84084 Salerno, Italy
* Correspondence: fborlea@yahoo.com (F.G.B.); stefania.mang@unibas.it (Ș.M.M.)

Citation: Chira, D.; Borlea, F.G.; Chira, F.; Mantale, C.Ș.; Ciocîrlan, M.I.C.; Turcu, D.O.; Cadar, N.; Trotta, V.; Camele, I.; Marcone, C.; et al. Selection of Elms Tolerant to Dutch Elm Disease in South-West Romania. *Diversity* **2022**, *14*, 980. https://doi.org/10.3390/d14110980

Academic Editors: Miglena Zhiyanski and Michael Wink

Received: 10 October 2022
Accepted: 11 November 2022
Published: 15 November 2022

Abstract: *Ophiostoma novo-ulmi* continues to be one of the most dangerous invasive fungi, destroying many autochthonous elm forests and cultures throughout the world. Searching for natural genotypes tolerant to Dutch elm disease (DED) is one of the main objectives of silviculturists all over the northern hemisphere in order to save the susceptible elms and to restore their ecosystem biodiversity. In this regard, the first trial was established between 1991 and 1994, in south-west Romania (*Pădurea Verde*, Timișoara), using three elm species (*Ulmus minor*, *U. glabra*, and *U. laevis*) with 38 provenances. A local strain of *Ophiostoma novo-ulmi* was used to artificially inoculate all elm variants and the DED evolution was observed. Furthermore, in 2018–2021 the trial was inventoried to understand the local genotype reaction to DED in the local environmental conditions after almost 30 years. The outcomes of the present study proved the continuous presence of the infections in the comparative culture and its proximity, but the identified pathogen had a new hybrid form (found for the first time in Romania) between *O. novo-ulmi* ssp. *americana* x *O. novo-ulmi* ssp. *novo-ulmi*. Wych elm (*U. glabra*) was extremely sensitive to DED: only 12 trees (out of 69 found in 2018) survived in 2021, and only one tree could be selected according to the adopted health criteria (resistance and vigour). The field elm (*U. minor*) was sensitive to the pathogen, but there were still individuals that showed good health status and growth. In contrast, the European white elm (*U. laevis*) proved constant tolerance to DED: only 15% had been found dead or presented severe symptoms of dieback. Overall, the results of this study report the diverse reactions of the Romanian regional elm genotypes to DED over the last three decades, providing promising perspectives for improving the presence of elms in the forest ecosystems of the Carpathian basin.

Keywords: *Ophiostoma novo-ulmi*; subspecies; *Ulmus minor*; *U. glabra*; *U. laevis*; provenances; tolerance to DED; climate

1. Introduction

Several major risks are threatening the forest ecosystems, of which climate change [1–4] and invasive biotic agents (pathogens, insects, etc.) [5–10] produce the most severe perturbation. For example, the bark beetle insect-wood pathogen relationship is involved in many vascular diseases which can create large outbreaks that are difficult to manage, especially if they are accompanied by climate extremes [11–14].

One of the most lethal forest diseases of the last century is the Dutch elm disease (DED), produced firstly by *Ophiostoma ulmi* (Buisman), Melinand Nannf. (1934), and subsequently by *O. novo-ulmi* Brasier (1991). The pathogens have diversified in time and space and *O. ulmi* and *O. novo-ulmi* are able to hybridise [15]. *O. novo-ulmi* has two subspecies: ssp. *americana* (SSAM), originally spread in North America (more recently, throughout Europe and beyond), and ssp. *novo-ulmi* (SSNU), originally extending in Europe (now identified in different continents) [16]. Within SSAM, two distinct genetic lineages occur [17,18] and the two subspecies hybridise when they come into contact [19]. Additionally, *Ophiostoma himal-ulmi* occurs in Asia [20].

The two pandemics of DED had a large impact both in North America and Europe, destroying large quantities of high-quality wood and severely diminishing elm's role in forest ecosystems [21,22].

In Romania, elm dieback was recorded in the beginning of the last century [23] and SSNU in the 1940s [24]. *O. ulmi* gradually infected the elm forests, but the most significant elm dieback was recorded in the 1960–1970s, following the mass infections of SSNU. Step by step, autochthonous elms have significantly disappeared from the forest composition, in terms of their economic and ecological role, being recorded mainly as elements of biodiversity (young trees) in all plains, hills, and mountainous regions [25–27]. Elm forest recoveries have been observed [28], but the infections reoccurred in forests when the new natural regeneration became favourable to the DED vector *Scolytus* spp. and to *Ophiostoma* ssp. [29–31]. Due to the high impact of DED, many national and international research programs have been developed to better understand the host-disease relationship and to breed disease-resistant elms [32–36].

The variable reaction of elms to *O. novo-ulmi* infections is influenced by a long series of factors, e.g., species/individual genetic resistance/tolerance, the virulence of different pathogen genotypes, environmental features (light, temperature, drought, etc.), elm morphological wood characteristics, host chemical reaction, the bark and wood microbiome, the momentary attractiveness for bark beetles, individual phenotypic plasticity, etc. [37–40]. The effect of these factors can vary with time, therefore long-term tests are needed to understand the tolerance/resistance to DED of the European elm genotypes.

Destroyed elm habitats need to be restored using resistant plants and all the additional knowledge of the host-pathogen-environment relationship [41,42]. However, information on the long-term evolution of the native elm genotype is hardly available [36]. The high level of naturalness in the forests of Romania [43] could be a good opportunity to select the survived natural elm provenances/individual trees with potential resistance/tolerance to DED [26,44].

The main objective of this study was to evaluate the tolerance of the elm species and provenances to DED in the local conditions of south-west Romania. More specifically, the present work aimed to: (a) assess the presence and development of *O. novo-ulmi* after almost 30 years from the inoculation time and identify the fungal pathogen responsible for DED both morphologically and molecularly; (b) perform an inventory of the existent trees, checking their health status; and (c) select the most performant trees for each elm species investigated (*U. glabra*, *U. minor*, and *U. laevis*)-based simultaneously on their health status and growth vigour—to be used in further breeding programs for DED control.

2. Materials and Methods

2.1. Location and Experimental Trial Set up and Biological Material

The first (nursery) test, to identify elm genotype tolerance to DED, was established in *Pădurea Verde* (Forest Research Station of Timișoara), in the eastern part of the Pannonian Field, between the Timiș and Bega River meadows, in south-western Romania starting at the beginning of May 1991 [26]. Four elm species (*Ulmus minor*, *U. glabra*, *U. laevis*, and *U. pumila*) from 38 provenances were sown in the above-described test. Subsequently, the seedlings were inoculated repeatedly in the first three years (1992, 1993, and 1994) with a local strain of *O. novo-ulmi* in accordance with the methodology proposed by Heybroek [26].

A phenomenon of dieback was observed, which progressed over the years and affected the majority of the seedlings and trees [26]. Using the healthy individuals that survived the first inoculations, a second test, in the same location (containing 25 provenances for the four elm species) was carried out in which the trees were planted more distantly (5 × 2 m) in order to ensure proper development over a longer period of time. This allowed the long-term assessment of the regional elm species/provenances for resistance/tolerance to DED, as well as their development and behaviour under the influence of the local conditions. Some of the dead trees were replaced with younger seedlings in the trial, but they are not included in the present assessment.

In 2018–2021 the surviving 453 trees were inventoried as follows: 171 (299 with substitutes) trees of *U. minor*, provenances (name/location)13 (Gurahont 1), 14 (Băneasa 1), 15 (Timișoara 1), 29 (Timișoara 2), and 30 (Băneasa 2); 69 (71) of *U. glabra*: 1 (Retezat CA1), 2 (Retezat CA2), 3 (Gurahonț), 4 (Văliug), 5 (Bozovici), 6 (Anina), 7 (Retezat Rotunda), 8 (Herculane); in the case of the provenances 9 (Sebiș), 10 (Retezat Gura Zlata), and 28 (Săvârșin), all trees died before 2018; 82 of *U. laevis*: 35 Timișoara and 36 Pecica (Ceala) and 1 tree of *U. pumila*: 37 Timișoara (Figure 1).

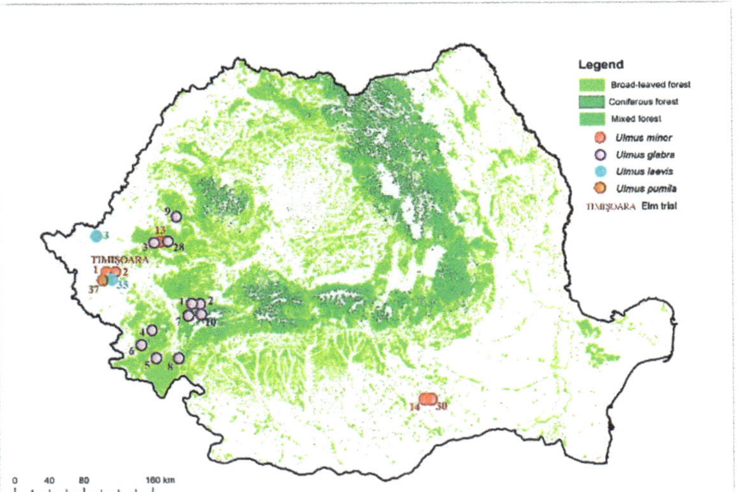

Figure 1. Elm provenances tested in the *Pădurea Verde* trial, Timișoara (based on CORINE Land Cover 2018 [45]).

2.2. Methods

2.2.1. Sampling

In order to identify the pathogen, both symptomatic and asymptomatic elm trees were collected from the investigation site. Samples (10 twigs/species) were randomly collected from different parts of the tree crown with or without visible symptoms, brought to the laboratory, stored at 4 °C, and processed within 7 days.

2.2.2. Fungal Isolation

Ten twigs from each elm species were randomly collected from symptomatic and asymptomatic investigated trees, cut into smaller pieces, and further processed for fungal isolation as described by Mang et al. [46]. Briefly, the twigs were processed under laminar flow sterile conditions where they were surface sterilised by soaking in 70% ethanol, then in 1% NaOCl and sterile water, and finally dried on sterile paper. Surface sterilised twigs were cut using a sterile scalpel into small pieces (0.5–1 cm) and placed on Petri (9 cm ø) dishes containing Potato Dextrose Agar (PDA) (Oxoid Ltd., Hants, UK) as media amended with streptomycin sulphate (40 mg L^{-1}, MerckKGaA, Darmstadt, Germany) and incubated at 23 °C, in the dark, until the mycelial growth became visible. The obtained colonies were subsequently transferred to PDA to obtain pure fungal cultures (PFCs) which were maintained by subcultures on PDA.

2.2.3. Fungal Growth Rate and Colony Morphology

The fungal growth rate at 23 °C and colony morphology were determined according to a modified protocol of Brasier, consisting of the use of PDA media instead of Malt Extract Agar (MEA) [47]. All analyses were performed in triplicate for each isolate. To check the growth rate of the fungus, 9 cm ø Petri plates containing PDA were centrally inoculated with a small (2 mm ø) mycelial plug taken from the edge of a PFC and incubated at 23 °C in darkness. After 2 and 7 days, the diameter of each culture was measured at the right angles of the Petri dish. To examine colony morphology, the plates were exposed to diffuse daylight for another 10 days and subsequently observed.

2.2.4. DNA Isolation, PCR, and Sequencing

Genomic DNAs (gDNAs) were extracted with the NucleoSpin Plant II (Macherey-Nagel, Düren, Germany) genetic material purification kit from the freeze-dried mycelium of PFCs (7–10 days old). For each isolate, 100 mg of mycelium was extracted using a protocol described by Mang et al. [48]. Genomic DNA purity and concentration were checked using a Nanodrop ND-1000 spectrophotometer (Thermo Scientific Inc., Willmington, DE, USA), and the material was stored at −20 °C in 1.5 mL Eppendorf tubes until used. In order to determine the fungal species, four different genes/regions {internal transcribed spacers (ITS1 and ITS2) of ribosomal RNA (ITS), β-tubulin (tub-2), cerato-ulmin hydrophobin (cu), and colony 1 (col1)} were amplified by PCR. The primers used for PCR amplifications were chosen based on their utility in the identification of *O. novo-ulmi* already reported by other authors: ITS5/ITS4 [49], Bt2a/Bt2b [50], CU1-Fwd/CU2-Rev [51], and Col1-F/Col1-R [19], respectively (Table 1).

Table 1. Details of the primer pairs used for PCR in this study.

Locus *	Primer Name	Primer Sequence	Reference
ITS	ITS5	5′-GGAAGTAAAAGTCGTAACAAGG-3′	[49]
	ITS4	5′-TCCTCCGCTTATTGATATGC-3′	
TUB-2	Bt2a	5′-GGTAACCAAATCGGTGCTGCTTTC-3′	[50]
	Bt2b	5′-ACCCTCAGTGTAGTGACCCTTGGC-3′	
Cu	CU1-Fwd	5-GGGCAGCTTACCAGAGTGAAC-3′	[51]
	CU2-Rev	5-GCGTTATGATGTAGCGGTGGC-3′	
col1	Col1-F	5-GCAGTTGTTGACATGTA G-3′	[19]
	Col1-R	5-TGCTTGACGTAGATCTCG-3′	

* Loci: *ITS* (internal transcribed spacer regions and intervening 5.8S rRNA gene); *TUB-2* (partial beta tubulin gene); *cu* (cerato-ulmin hydrophobin); and *col1* (colony 1).

PCR amplifications were performed as explained in Mentana et al. [52] for *ITS*, while for *TUB-2* were performed following a protocol described in Frissulo et al. [53]. For the other two genes (*cu* and *col1*), the PCR reactions were performed with Phire Direct PCR Master mix (Thermo Scientific Inc., USA), following the manufacturer's instructions with some modifications. The PCR mixtures were composed of 10 μL of 2X Phire Plant PCR Buffer (including 1.5 mM MgCl2 and 20 μM of dNTPs); primers, 0.5 μM each; 0.4 μL of Phire Hot Start II DNA polymerase enzyme; 5 μL of template DNA (20 ng/μL); and double distilled water up to 20 μL. The PCR cycling protocols for the *ITS* and *TUB-2*genes are described in Mang et al. [46,48]. Instead, the PCR cycling conditions for the *cu* and *col1* genes consisted of initial denaturation at 98 °C for 5 min for 1 cycle; then 40 cycles of denaturation at 98 °C for 5 s; annealing at 62 °C for 5 s; and extension at 72 °C for 20 s, followed by a final extension at 72 °C for 1 min for 1 cycle. All PCR products were separated in 1.5% agarose gels in Tris-Acetic acid-EDTA (TAE) buffer and visualised under UV after staining with SYBR Safe DNA Gel Stain (ThermoFisher Scientific™, Carlsbad, CA, USA). A 100-bp GeneRuler Express DNA Ladder (Thermo Fisher Scientific™ Baltics UAB, Vilnius, Lithuania) was used as a molecular weight marker. Direct sequencing, in both directions, of all PCR products was performed by BMR Genomics (Padua, Italy), using a 3130xl automatic sequencer and the same primers as for the PCR. Annotations were based on BLAST searches with a minimum of 99–100% identity over at least 80% of the length of the nucleotide sequence, which are the commonly used thresholds for reliable sequence annotation [54]. Nucleotide sequence primary identification was carried out using the BLASTn search tool program [54,55] of the NCBI by comparing all sequences obtained in this study with those already present in the database (accessed on 12 August 2022).

2.2.5. RFLP Analysis of Cu and Col1 Gene

To differentiate between the two subspecies (SSNU and SSNA) of *O. novo-ulmi*, the PCR products of cerato-ulmin hydrophobin (*cu*) and *colony 1(col1)* gene were digested with two restriction enzymes, Hph1 (Thermo Scientific Inc., USA) and BfaI (Thermo Scientific Inc., USA), following the manufacturer instructions and a protocol described by Konrad et al. [19]. Restriction Fragment Length Polymorphism (RFLP) reactions were performed in a 30 µL volume and the digestion mixture consisted of 10 µL (~0.2 µg) of PCR product, 1 µL of enzyme, 2 µL of 10x buffer, 17 µL of nuclease-free sterile water for digestion with the BfaI (FspBI) enzyme and 10 µL (~0.2 µg) of PCR product, 10 U 1 µL of enzyme, 2 µL of 10x buffer and 18 µL of nuclease-free sterile water for digestion with the Hph1 enzyme. Incubation was done at 37 °C for 5 min for the BfaI (FspBI) enzyme and 1 h at the same temperature in the case of digestion with the HphI enzyme. A total of 5 µL of the RFLP samples were separated on 2% agarose gel pre-stained with SYBR Safe DNA stain (Thermo Scientific Inc., USA) in 1x TAE buffer at 5 V/cm for 90 min alongside an O'Gene Ruler Express DNA Ladder (Thermo Scientific Inc., USA) as DNA ladder. Images of the gel were visualised on a UV transilluminator (EuroClone S.p.A, Milan, Italy) and data were registered in a table. Based on the specific pattern obtained for each subspecies and, as Konrad et al. [19] depicted, when the pattern was mixed between those of the above-mentioned subspecies, the isolate was considered a subspecies hybrid.

2.2.6. Virulence Test

The virulence of all *O. novo-ulmi* isolates from this study was tested using the apple assay described by Plourde and Bernier [56]. For this purpose, chemically untreated apples of the cv. "Golden Delicious", obtained from a local store, were first kept for 2–3 days at room temperature (24–26 °C) and visually inspected daily to eliminate any damaged fruits. Prior to the inoculation, they were washed with soap and the surface was disinfected with 70% ethanol. A hole (10 mm deep) was created on each fruit using a sterilised metal corkborer (ø 9 mm). The holes were then filled with PDA plugs bearing fungal mycelium, placing the mycelium inwards. Apple fruits with holes filled only with PDA plugs served as negative controls. The holes of all the inoculated fruits were covered with transparent tape in order to prevent desiccation and contamination from the environment. After inoculation, the apples were kept in the dark at room temperature. The diameters of the lesions were measured on the apples at 14- and 28-days post-inoculation (dpi). The whole apple assay was repeated three times with five replications per isolate.

2.2.7. Inventory

The experimental plot was inventoried, annually, from 2018 to 2021. The tree's health status was evaluated yearly based on crown defoliation [57], leaf wilting/browning, shoot/twig/stem/tree dieback, and tree vigour based on breast height diameter (measured in 2019 and 2021).

2.2.8. Selection of Individual Trees

The most performant trees were selected based on the following simultaneous two criteria:

(i) Health status (which reflects the tree tolerance to DED): only healthy trees showing 0–25% defoliation and without any symptoms of infections were selected;

(ii) Growth vigour: only trees with a larger diameter (D) in the three levels of selection, I (first class of selection: D > Dm + 2 s), II (Dm + 2 s > D > Dm + s), and III (Dm + s > D > Dm), were selected.

2.2.9. Statistical Analysis

For each of the three *Ulmus* species considered, the data on the diameter and the defoliation (after the arcsine transformation) were independently analysed by linear mixed-effects models (LMMs) fitted with REML (restricted maximum likelihood).

The *p*-values for the differences between treatments, sampling dates, and their interactions were obtained through ANOVAs (Type II Wald chi-square tests) and the following model was applied:

$$Y = \mu + Provenance * Year + Repetition \{Provenance \{Year\}\} + \varepsilon$$

where *Y* is the measured variable, Provenance (7 levels for *U. glabra*, 5 levels for *U. minor*, and 2 levels for *U. laevis*), and Year (2 levels for diameter measures and 4 levels for defoliation measures) are the main fixed factors, and the Repetition is the random effect consisting of the three experimental repetitions nested in the Provenance and Year. This model accounts for the non-independence of the data (pseudo replication of measures) due to the different experimental repetitions (the random effect) that are part of the present design.

To test for differences among the *Ulmus* species, LMMs were performed on the diameter values and the defoliation, pooling the data of the three species and considering the Provenance as the random factor. The following model was applied:

$$Y = \mu + Specie * Year + Provenance + \varepsilon$$

This model accounts for the variation in the data due to the Provenance.

For all the analyses described so far, the model distributions were also chosen as the best fitting, based on AIC criteria [58], and the full models were presented. All statistical analyses were performed in R version 4.1.2 "Bird Hippie" [59], with lme4 [60], lmerTest [61] packages.

3. Results

3.1. Fungal Isolation from Different Elm Species

From all symptomatic elm twigs, twelve pure fungal cultures (PFCs) isolates, named Oph_TM1 to Oph_TM12, were obtained and further investigated in this study (Table 2).

3.2. Colony Morphology and Fungal Growth Rate

The colony morphology and growth rate (ranging between 3.2–4.7 mm/day) showed an aerial mycelium with a fibrous striate appearance and a petaloid pattern and presented a strong diurnal zonation on PDA at 22 °C, for all isolates. Their morphological features closely matched those of *O. novo-ulmi* [62,63] (Figure 2).

Table 2. Fungal isolates of *O. novo-ulmi* obtained in this study from four *Ulmus* species located in the experimental field of Pădurea Verde-Timisoara, Romania.

		Genes/GenBank Accession Number			
Isolate Name	**Host**	***ITS***	***TUB-2***	***Cu***	***Col1***
Oph_TM1	*Ulmus glabra*	OP748295	OP817142	OP787366	OP794345
Oph_TM2	*Ulmus glabra*	OP748296	OP817143	OP787367	OP794346
Oph_TM3	*Ulmus glabra*	OP748297	OP817144	OP787368	OP794347
Oph_TM4	*Ulmus glabra*	OP748298	OP817145	OP787369	OP794348
Oph_TM5	*Ulmus minor*	OP748299	OP817146	OP787370	OP794349
Oph_TM6	*Ulmus minor*	OP748300	OP817147	OP787371	OP794350
Oph_TM7	*Ulmus minor*	OP748301	OP817148	OP787372	OP794351
Oph_TM8	*Ulmus minor*	OP748302	OP817149	OP787373	OP794352
Oph_TM9	*Ulmus minor*	OP748303	OP817150	OP787374	OP794353
Oph_TM10	*Ulmus laevis*	OP748304	OP817151	OP787375	OP794354
Oph_TM11	*Ulmus laevis*	OP748305	OP817152	OP787376	OP794355
Oph_TM12	*Ulmus laevis*	OP748306	OP817153	OP787377	OP794356

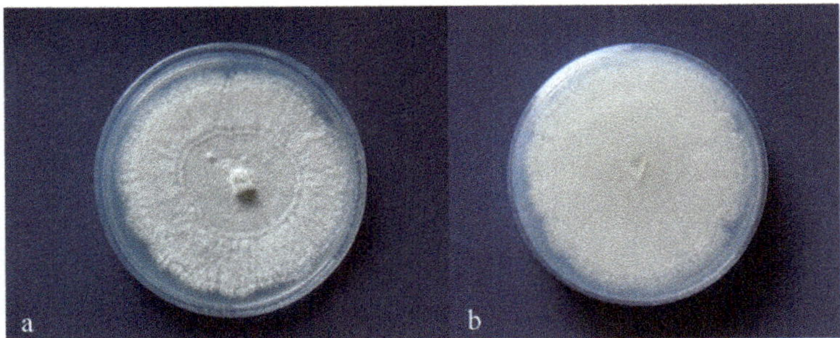

Figure 2. Colony of *O. novo-ulmi* grown on PDA at 22 °C two weeks after inoculation: (**a**) Upper surface of the colony; (**b**) reverse surface of the colony.

3.3. DNA Isolation, PCR, and Sequencing of Pure Fungal Cultures of O. novo-ulmi

High-quality total genomic DNA was obtained from all twelve PFCs of *O. novo-ulmi* using the Macherey Nagel GmbH & Co. (Düren, Germany) extraction kit. The results of the PCR reaction performed with the rDNA ITS5/ITS4 primers yielded, for all investigated fungal isolates, a unique amplicon of ~700 base pairs which, after direct sequencing and taxonomical annotation (based on >99% sequence similarity with other similar species from the NCBI database), was identified as belonging to *O. novo-ulmi* (Acc. nos.: MK990095; MK990096; KF854005; KF854006; KF854008, andKF854009). The PCR results performed with primers Bt2a/Bt2b, of the β-tubulin gene, also successfully produced amplicons of the expected size (~500 bp) for all isolates. The nucleotide sequences of the amplicons were identified as *O. novo-ulmi* (Acc. nos.: MH283243; MH283246; EU977486; MH055739; and MH055740) after their direct sequencing and sequence taxonomically annotations, using the same parameters described above. Details about the fungal isolates obtained in the present study are shown in Table 2.

Although the symptoms of DED are often seen in Romania, including in the south-west part, the causal agent has not been recently or precisely identified at the subspecies level; the only information to date is the mention of SSNU presence in the 1940s by Brasier et al. (1993, 2010) [24,63]. CU1-Fwd and CU2-Rev primers amplified a unique product of

934 base pairs (bp) while the PCR using oligonucleotides of the colony 1 gene (col1), Col1-F, and Col1-R produced a single amplicon of 482 bp, which are both the expected sizes for *O. novo-ulmi* isolates reported in previous studies by Konrad et al. [19] and Katanić et al. [64]. The amplicons described above were obtained for all isolates investigated in this study. Results of the PCR for one representative isolate from this study is shown in Table 2 and Figure 3.

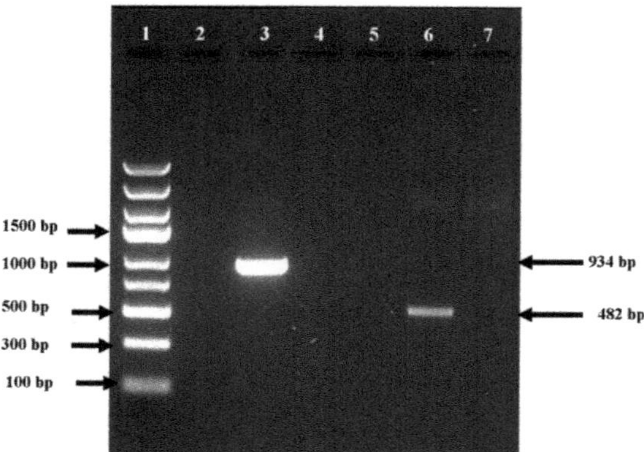

Figure 3. Gel image showing the PCR results of *O. novo-ulmi* genomic DNA amplification with primers CU1-Fwd and CU2-Rev for the cerato-ulmin gene (934 bp) and col1F and col1R for the colony 1 gene (482 bp). Lane 1 = O' Gene Ruler DNA Ladder (ThermoFisher Scientific Inc., Waltham, MA, USA); Lane 3 = amplicon obtained with the CU1-Fwd and CU2-Rev primers; Lane 6 = amplicon obtained with the col1F and col1R primers pairs; Lane 7: Negative control (without DNA template).

3.4. PCR-RFLP Analysis of Cu and Col1 Gene

The PCR-RFLP results of all *O. novo-ulmi* isolates obtained in this study indicate the presence of one single hybrid between *O. novo-ulmi* ssp. *novo-ulmi* x *O. novo-ulmi* ssp. *americana* in the experimental field of Pădurea Verde-Timisoara, Romania which is reported in Romania and at the investigated site for the first time in our study. After the digestion of the PCR products of the cerato-ulmin gene with the HphI enzyme, an RFLP pattern typical of the SSNU containing three specific bands (101 bp, 161 bp, and 632 bp) was observed. The digestion of the colony 1 PCR products with the FspBI enzyme produced an RFLP pattern typical of the SSNA, also displaying three specific bands (100 bp, 156 bp, and 236 bp) (Table 3). The PCR-RFLP results were confirmed by sequencing the *cu* and *col1* genes from all isolates obtained in the present study (Table 2).

3.5. Virulence Test

All *O. novo-ulmi* hybrids isolated from *Ulmus* species were investigated and inoculated on cv. "Golden Delicious" apples, with induced lesions ranging between 29–32 mm (data not shown) at 28 days post inoculation (dpi) and closely matching those reported in the literature by Plourde and Bernier [56] (Figure 4). Outcomes of the virulence test on apples demonstrated that all isolates of *O. novo-ulmi* investigated in this study showed similarly high virulence.

Table 3. Results of the PCR-RFLP analysis of the *cu* and *col1* genes of *O. novo-ulmi* isolates obtained from the experimental field of Pădurea Verde—Timisoara, Romania.

Isolate Name	Digestion Enzyme	Size (bp) and Presence/Absence * of the Bands after Digestion					
		100	101	156	161	236	632
Oph_TM1	HphI	-	+	-	+	-	+
Oph_TM2	HphI	-	+	-	+	-	+
Oph_TM3	HphI	-	+	-	+	-	+
Oph_TM4	HphI	-	+	-	+	-	+
Oph_TM5	HphI	-	+	-	+	-	+
Oph_TM6	HphI	-	+	-	+	-	+
Oph_TM7	HphI	-	+	-	+	-	+
Oph_TM8	HphI	-	+	-	+	-	+
Oph_TM9	HphI	-	+	-	+	-	+
Oph_TM10	HphI	-	+	-	+	-	+
Oph_TM11	HphI	-	+	-	+	-	+
Oph_TM12	HphI	-	+	-	+	-	+
Oph_TM1	FspBI	+	-	+	-	+	-
Oph_TM2	FspBI	+	-	+	-	+	-
Oph_TM3	FspBI	+	-	+	-	+	-
Oph_TM4	FspBI	+	-	+	-	+	-
Oph_TM5	FspBI	+	-	+	-	+	-
Oph_TM6	FspBI	+	-	+	-	+	-
Oph_TM7	FspBI	+	-	+	-	+	-
Oph_TM8	FspBI	+	-	+	-	+	-
Oph_TM9	FspBI	+	-	+	-	+	-
Oph_TM10	FspBI	+	-	+	-	+	-
Oph_TM11	FspBI	+	-	+	-	+	-
Oph_TM12	FspBI	+	-	+	-	+	-

* Note:(+) = band present; (-) = band absent.

Figure 4. Apple fruits of cv. "Golden Delicious" inoculated with the virulent hybrid of *O. novo-ulmi* from Pădurea Verde-Timișoara showed the typically golden-brown circular lesion surrounding the inoculation point (**left**), whereas the control apples (**right**), inoculated with only sterile PDA plugs, did not display any fungal growth.

3.6. Health Status Recent Evolution

3.6.1. *Ulmus minor*

Observations performed in the final period of the study demonstrated that tree debilitation was a continuous process, occurring at a relatively constant rate, i.e., dead or dying trees increased from 41.5% to 53.2% between 2018 and 2021 (Figure 5a).

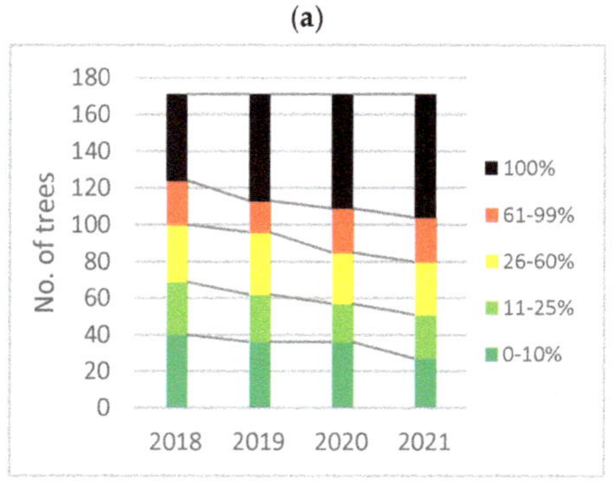

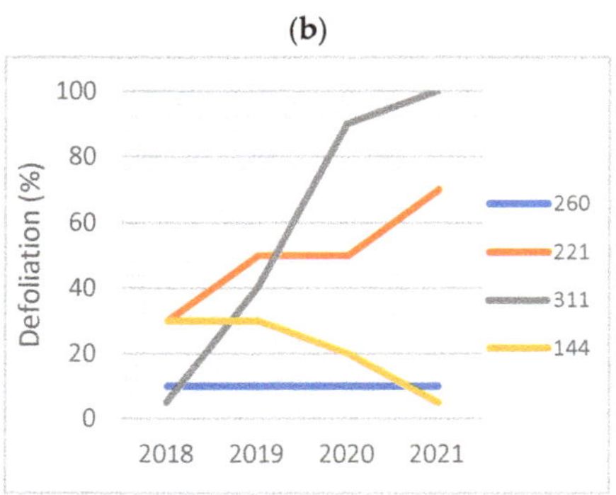

Figure 5. (**a**) Recent defoliation evolution of *Ulmus minor* (all provenances); (**b**) Different types of individual evolution of *U. minor* defoliation (tree numbers: 260, 221, 311, and 144).

All five provenances of field elm were affected, even though significant variation among them was found (Table 4).

Table 4. Defoliation and diameter of *Ulmus minor* provenances.

Provenance	No	Def				D	
		2018	2019	2020	2021	2019	2021
13 (Gurahont)	48	57.3	59.9	66.8	71.9	13.1	14.0
14 (Băneasa 1)	46	41.7	46.8	51.1	53.9	12.4	13.9
15 (Timișoara 1)	13	38.1	42.3	45.0	47.7	12.0	13.5
29 (Timișoara2)	37	58.9	59.3	63.1	62.4	12.2	13.5
30 (Băneasa 2)	27	54.1	61.1	62.8	64.8	14.1	15.4
Average		51.5	55.1	59.5	62.1	12.8	14.0

No: number of trees; Def: defoliation (%); D: diameter (cm); Average of all trees.

Individual evolutions varied a lot. Generally, we observed a relatively constant status (defoliation) of a large number of trees, especially in the healthy/relatively healthy ones. The relatively slow degradation of the elm trees was registered in many cases; the rapid degradation and the relatively slow improvement of health status were rarely noticed (Figure 5b).

The LMMs show that the diameter was not affected by the provenance, although the probability value was close to being statistically significant ($\chi^2 = 8.7$, df = 4, $p = 0.07$), but it is affected by the year ($\chi^2 = 23.2$, df = 1, $p < 0.01$). In contrast, the defoliation was affected by the provenance ($\chi^2 = 16.7$, df = 4, $p < 0.01$) but not by the year ($\chi^2 = 0.495$, df = 3, $p = 0.17$). The interaction "Provenance x Year" was not significant for the diameter or the defoliation ($\chi^2 = 0.54$, df = 4, $p = 0.97$ and $\chi^2 = 0.8$, df = 12, $p = 0.99$, respectively).

3.6.2. *Ulmus glabra*

The *U. glabra* (wych elm or Scots elm) was extremely sensitive to DED in the local conditions of the *Pădurea Verde* trial. All trees from five provenances were already dead before 2018. In the following years, the dieback process was dramatic: only 12 trees of four provenances (Retezat CA1, Retezat CA2, Anina, and Gurahonț) survived, and only 8 of them presented relatively good health conditions (Table 5 and Figure 6).

Table 5. Defoliation and diameter of *Ulmus glabra* provenances.

Provenance	No	Def				D	
		2018	2019	2020	2021	2019	2021
1 (Retezat CA1)	15	60.3	68.0	70.0	74.0	9.3	9.7
2 (Retezat CA2)	15	60.7	66.3	74.3	77.7	7.9	8.5
3 (Gurahonț)	18	81.4	91.1	94.2	94.7	6.5	6.58
4 (Văliug)	5	96.0	96.0	100	100	8.5	8.5
5 (Bozovici)	5	75.0	86.0	96.0	100	5.6	5.6
6 (Anina)	7	85.7	85.7	86.4	86.4	7.7	8.0
7 (Retezat Rotunda)	1	100	100	100	100	15.2	15.2
8 (Herculane)	3	100	100	100	100	17.0	17.0
Average		74.4	80.7	84.9	86.7	8.2	8.5

No: number of trees; Def: defoliation (%); D: diameter (cm); Average of all trees.

(a)

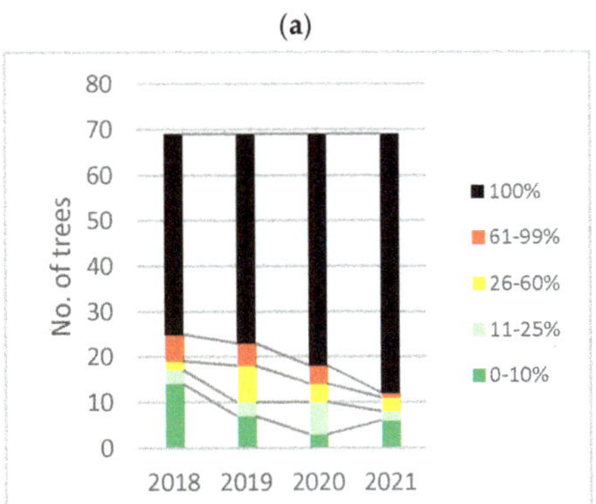

(b)

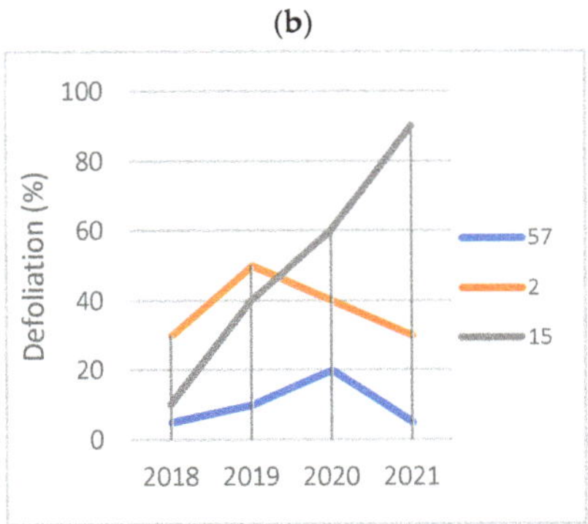

Figure 6. (**a**) Recent defoliation evolution of *Ulmus glabra* (all provenances); (**b**) Different types of individual evolution of *U. glabra* defoliation (tree numbers: 57, 2, and 15).

The LMMs shows that the #diameter# and the #defoliation# were affected by the provenance (χ^2 = 17.6, df = 6, $p < 0.01$ and χ^2 = 26.1, df = 6, $p < 0.001$ respectively) but not by the year (χ^2 = 0.45, df = 1, p = 0.49 and χ^2 = 4.9, df = 3, p = 0.17, respectively), nor by the interaction "Provenance x Year" (χ^2 = 0.9, df = 3, p = 0.99 and χ^2 = 1.5, df = 18, p = 0.99, respectively).

3.6.3. *Ulmus laevis*

Both provenances (Timișoara and Pecica-Ceala) of European white elm proved to be tolerant to *O. novo-ulmi*. Only 12 trees (out of 82) were dead or dying in 2021 (Table 6 and Figure 7). The degradation of crown defoliation and dying process have been recorded in the last two years, but this process does not change the big difference between *U. laevis* and the other two elm species investigated in this study.

Table 6. Defoliation and diameter evolution of *Ulmus laevis* provenances.

Provenience	No	Def				D	
		2018	2019	2020	2021	2019	2021
35 (Timișoara)	41	18.8	22.6	25.6	27.7	10.3	12.2
36 (Pecica-Ceala)	41	16.7	19.3	28.4	36.6	8.4	10.5
Average		17.7	20.9	27.0	32.1	9.2	11.2

No: number of trees; Def: defoliation (%); D: diameter (cm); Average of all trees.

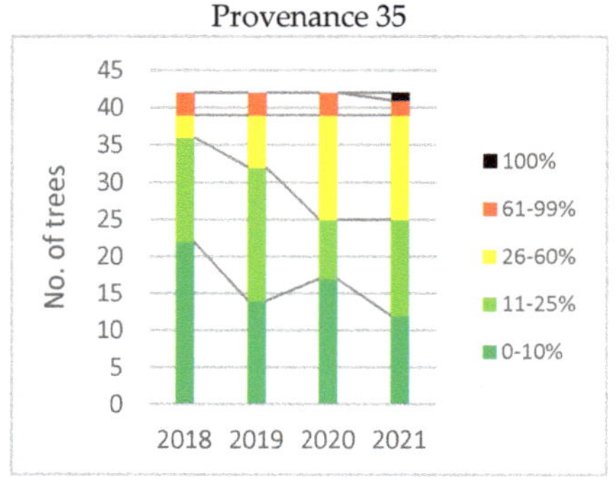

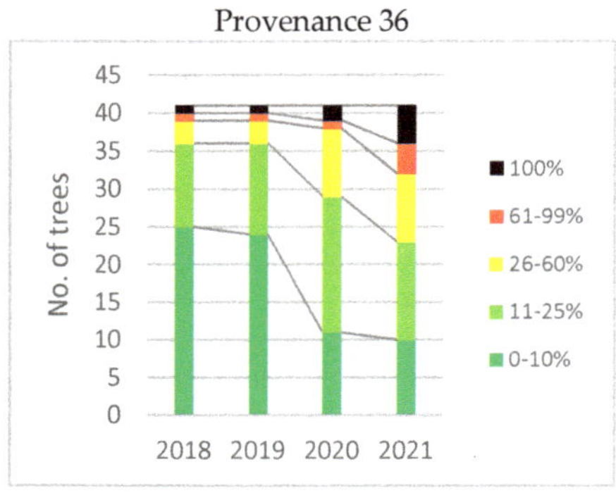

Figure 7. Recent defoliation evolution of the provenances 35 and 36 of *Ulmus laevis*.

The LMMs show that the diameter was affected by the provenance ($\chi^2 = 5.7$, df = 1, $p < 0.05$) and by the year ($\chi^2 = 4.7$, df = 1, $p < 0.05$) and the defoliation was affected by the year ($\chi^2 = 16.8$, df = 3, $p < 0.001$), but not by the provenance ($\chi^2 = 1.8$, df = 1, $p = 0.18$). The interaction "Provenance x Year" was not found to be significant for the diameter or the defoliation ($\chi^2 = 0.03$, df = 1, $p = 0.87$ and $\chi^2 = 3.6$, df = 3, $p = 0.3$, respectively).

3.6.4. Differences among Elm Species

For the data on the diameter and defoliation, the LMMs show that the differences among the three species were highly significant ($\chi^2 = 26.1$, df = 2, $p < 0.001$ and $\chi^2 = 64.9$, df = 2, $p < 0.001$), as well as the differences between years ($\chi^2 = 36$, df = 3, $p < 0.001$ and $\chi^2 = 25.9$, df = 3, $p < 0.001$). The interaction "Species x Year" was not found to be significant for the diameter or the defoliation.

3.7. Selection of DED-Tolerant Plants

Even though the *U. minor* was sensitive to DED, several trees were both tolerant to *O. novo-ulmi* (low crown defoliation and lack of infection symptoms) and had vigorous growth. Out of the 171 *U. minor* trees investigated, only 5 trees could be included in the first class of selection, 7 trees in the second class, and 14 trees in the third class (Figure 8).

In the case of *U. glabra*, very intense dieback was observed (78% of the investigated trees were dead), therefore it was difficult to select among the remaining trees. Despite this, three trees were selected, as they represent precious material for further breeding operations (Figure 9).

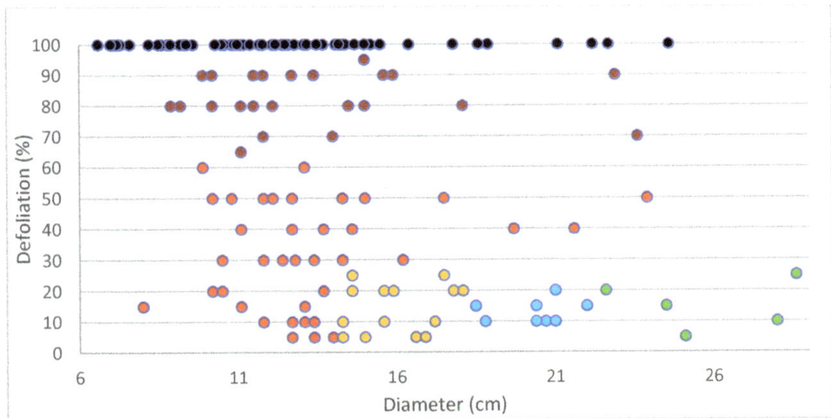

Figure 8. Selection of *U. minor* trees that are tolerant to *O. novo-ulmi* and the local environmental conditions (forest field) {I-green dots: D > Dm + 2 s, II-blue dots: D > Dm + s, III-orange dots: D > Dm, black dots: dead, brown dots: dying individuals, and red dots: not selected trees}.

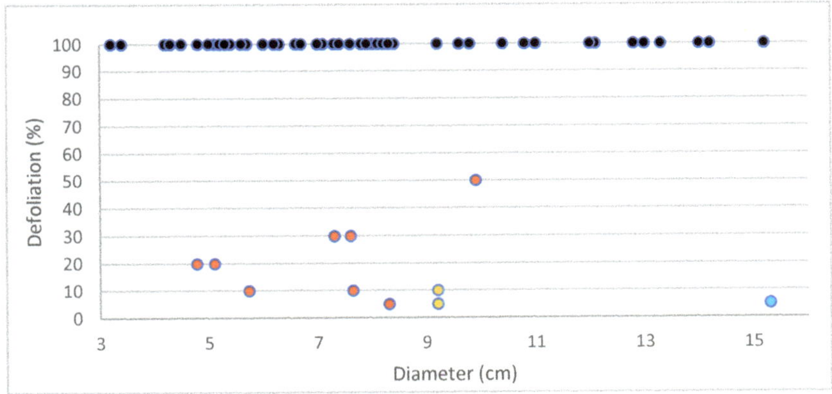

Figure 9. Selection of *U. glabra* trees that are tolerant to *O. novo-ulmi* and the local environmental conditions (forest field) {II-blue: D > Dm + s, III-orange: D > Dm, black dots: dead, and red dots: not selected trees}.

U. laevis was found to be tolerant to DED under the local climate of the *Pădurea Verde* trial. In fact, after 30 years of testing, the majority of trees presented good health conditions, so the selection focused especially on the most vigorous individuals. In particular, 11 healthy trees of *U. laevis* could be included in the second class of selection and another 13 trees of the same species (just thicker than the species average) were included in the third class of selection (Figure 10).

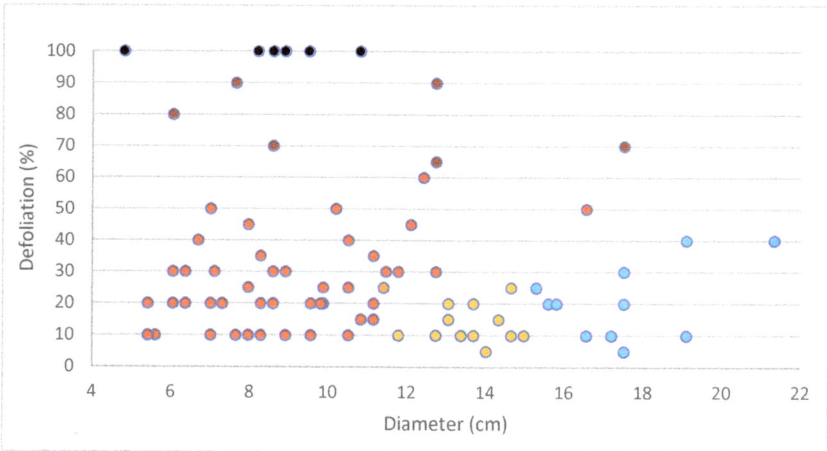

Figure 10. Selection of the trees of *U. laevis* trees that are tolerant to *O. novo-ulmi* and the local environmental conditions (forest field) {II-blue: D > Dm + s, III-orange: D > Dm, black dots: dead, brown dots: dying individuals, and red dots: not selected trees}.

4. Discussion

The evolution of the two subspecies of *O. novo-ulmi* in Europe (ssp. *americana*, introduced in western Europe and migrated eastward; ssp. *novo-ulmi*, spread from Romania/Moldova/Ukraine towards the west and east) was described by Brasier and Kirk [24]. After almost half a century of presence, *O. novo-ulmi* replaced the entire population of *O. ulmi* by 1990 in the Carpathian basin [24,63]. Field inventory in the second pandemic of the 1950–1960s counted more than 0.9 million trees killed by DED in Romania and revealed *U. minor* and *U.* × *hollandica* as the most sensitive to DED, followed by *U. laevis*. *U. glabra* was relatively tolerant, while *U. pumila* was almost not affected by the disease [25].

In the *Pădurea Verde* trial, in the early 1990s, the inoculation was made with a local strain of *O. novo-ulmi*, which most probably belonged to ssp. *novo-ulmi*, since that was the only one widely present at that time between the Black Sea region and Central Europe [24,63]. The presence of *O. novo-ulmi* ssp. *americana* × *O. novo-ulmi* ssp. *novo-ulmi* in *Pădurea Verde* reported in this study is the first identification of the hybrid in Romania. Its occurrence was predicted by previous studies which reported that the two subspecies tend to intensely hybridise when they meet, and thus, the hybrids can eventually replace the subspecies [15,17,19,24,62,64–66].

All species/provenances tested in the "Pădurea Verde" trial were susceptible to *O. novo-ulmi*. Those that appeared relatively healthy had developed various symptoms of DED (wood necrosis, dead twigs, etc.), therefore we used the term "tolerance" to define their reaction (susceptible to infections but without negative consequences) [41]. However, we kept the term "resistance" when we cited the previous international results/opinions.

In the juvenile stage, *U. glabra* was relatively tolerant to DED, while *U. minor* was very sensitive to SSNU in *Pădurea Verde* infections [26]. After three decades, the elms' phytosanitary status very much changed. In fact, almost all *U. glabra* died, while *U. minor* showed a high sensitivity to the *O. novo-ulmi* hybrid with some apparently healthy individuals in the comparative trial. Only *U. laevis* showed constant significant tolerance to *O. novo-ulmi*.

Previous studies showed variable DED resistance of the elm species/clones. In some cases, *U. glabra* was found to be the most sensitive [34,65,67–69]. Other studies consider this species to be less affected, probably due to the fact that the main vector *Scolytus scolytus* (Fabricius) is less attracted to *U. glabra* [70–72] or the natural host microbiome reduces the DED rate of transmission [73–75]. However, *Scolytus multistriatus* (Marsham), which attacks the younger trees/thinner axes, feeds equally on all European elm species, at least if the beetles cannot preferentially choose between the elm hosts [76].

U. minor showed a higher susceptibility to the infections of both subspecies of *O. novo-ulmi*, being considered the most harmed elm species by DED pandemics, but at the same time, many field elm genotypes showed a promising tolerance to DED, which may be useful for breeding programs and elm habitat reconstruction [34,36,41,64]. The selection of moderately tolerant native phenotypes of *U. minor* proved to be useful for the breeding of DED resistance and the reconstruction of elm habitats [77,78].

U. laevis showed high susceptibility to DED in some early trials [79], especially after artificial inoculations [80], sometimes showing similar sensitivity to *U. minor* [34,65]. More recently, however, it was generally (including in the present study) considered relatively resistant/tolerant to DED, being less attractive for elm bark beetles, and having a stronger activation of physiological defence mechanisms [70,80–84].

The local testing conditions of the present study were characteristic of the south-eastern part of the Pannonian Basin, with a transition to the Mediterranean-continental temperate climate (C.f.a.x./Koppen) (12.4 °C/717 mm), in an open field (90 m altitude), on Preluvo soil [85]. The natural forest type was a mixture of oaks and various broadleaves including natural *U. minor* and *U. laevis* (habitat 91YO *Dacian oak-hornbeam forests*). Furthermore, the local conditions are favourable for *U. minor* and partially suitable for *U. laevis* and not so much for *U. glabra*, which demands a colder and more humid climate.

The dynamic of the pathogen genotype (occurrence of ssp. *americana*, then substitution with the hybrid form) may have influenced the elm genotype evolution in *Pădurea Verde* [41,62].

There are some hopes that, in time, *O. novo-ulmi* will become less virulent, being gradually infected by mycoviruses, such as the chestnut blight [86]. Relationships among host genotype, local climate, microbiome, and vectors/pathogens could also diminish the disease virulence as it happened in ash dieback [87–89].

The hybridisation of sensitive European elms with Asian species produced resistant cultivars, which are available for urban green areas but they are restricted for forest use, due to new environmental politics [35,90]. None of the native European genotypes proved to be fully resistant to DED [33,91]. Therefore, the only accepted method of ecosystem restoration is to use the local partial resistant/tolerant genotypes [42,78,92].

Most cultivars have relatively constant resistance/sensitivity to DED, but some genotypes have important plasticity in their response to DED, depending on environmental factors, location, inoculation year, etc. [34,38]. Hence, the selected elm trees of the *Pădurea Verde* trial may be used for local restoration initiatives, but at the same time, they need to be further tested in several different locations in Romania, under new climate-changing conditions, to evaluate their suitability [36,93–96].

5. Conclusions

After more than 100 years of presence, and two pandemics, DED continues to produce intense dieback on elms all over Europe. Fortunately, elm species still survive in the forest and urban areas, at least as young regenerations, due to their ability to resprout/produce early fructification, their partial host tolerance, and the DED organism (vector and pathogen) requirements.

The Romanian local natural elm provenances, inoculated with *O. novo-ulmi*, showed some interesting behaviour in our long-term experimental trials. More specifically, a relatively good tolerance was observed in the juvenile stage for all species and a very high sensitivity was detected in *U. glabra* trees older than 10 years. In the case of *U. minor*, good long-term tolerance for a few genotypes and generally high sensitivity was seen. In addition, a very good tolerance of *U. laevis* was observed.

The initial inoculation was performed in 1992–1994, with a local strain of *O. novo-ulmi* (thought to belong to ssp. *novo-ulmi*). However, the results from the present study showed that the pathogen has changed (by natural spreading) in the last three decades and only one hybrid between SSNA and SSNU was found in the trial.

In the new context of environmental politics in Europe, the use of some local European partial resistant/tolerant genotypes or those with high plasticity in response to DED could play an important role in future elm species re-extension/ecological reconstruction or for use in urban green areas, in close relation with the local environmental condition demands.

Author Contributions: Conceptualisation, D.C., F.G.B., Ş.M.M. and I.C.; methodology, D.C., F.G.B. and Ş.M.M.; software, V.T.; validation, V.T., Ş.M.M. and C.M.; formal analysis, V.T., F.C., N.C. and Ş.M.M.; investigation, C.Ş.M., M.I.C.C., D.O.T., F.G.B., F.C., F.G.B., N.C., D.C., Ş.M.M., C.M. and I.C.; resources, D.C.; data curation, D.C. and Ş.M.M.; writing—original draft preparation, D.C., F.G.B. and Ş.M.M.; writing—review and editing, D.C., F.G.B., D.O.T., C.Ş.M., M.I.C.C., C.M.,V.T., I.C. and Ş.M.M.; visualisation, D.C, F.G.B., F.C., I.C., M.I.C.C., C.Ş.M., N.C. and Ş.M.M.; supervision, D.C. and Ş.M.M.; project administration, D.C.; funding acquisition, D.C. All authors have read and agreed to the published version of the manuscript.

Funding: This research was funded by the Ministry of Research, Innovation and Digitisation (MCID), Romania, through the: (i) "Nucleu" Program(project no. 19070206) and (ii) Program 1-Development of the National R & D System (project *CresPerfInst;* Contract no. 34PFE/30.12.2021).

Institutional Review Board Statement: Not applicable.

Data Availability Statement: The data presented in this study are available on request from the lead author. The data are not publicly available because the project through which the research was funded is not yet completed.

Acknowledgments: We thank all our colleagues at Timişoara Research Station, especially I. Chisăliţă, O. Merce, I. Albu, C. Ciontu, I. Cântar, and F. Morar, among others, for maintaining the trial and helping us with the activities in the field.

Conflicts of Interest: The authors declare no conflict of interest.

References

1. Kutnar, L.; Kermavnar, K.; Pintar, A.M. Climate change and disturbances will shape future temperate forests in the transition zone between Central and SE Europe. *Ann. For. Res.* **2021**, *64*, 67–86. [CrossRef]
2. Anderson-Teixeira, K.J.; Herrmann, V.; Rollinson, C.R.; Gonzalez, B.; Gonzalez-Akre, E.B.; Pederson, N.; Zuidema, P.A. Joint effects of climate, tree size, and year on annual tree growth derived from tree-ring records of ten globally distributed forests. *Glob. Chang. Biol.* **2022**, *28*, 245. [CrossRef] [PubMed]
3. Sajjad, H.; Kumar, P.; Masroor, M.; Rahaman, M.H.; Rehman, S.; Ahmed, R.; Sahana, M. Forest Vulnerability to Climate Change: A Review for Future Research Framework. *Forests* **2022**, *13*, 917. [CrossRef]
4. Zhang, L.; Sun, P.; Huettmann, F.; Liu, S. Where should China practice forestry in a warming world? *Glob. Chang. Biol.* **2022**, *28*, 2461–2475. [CrossRef] [PubMed]
5. Clark, K.L.; Skowronski, N.; Hom, J. Invasive insects impact forest carbon dynamics. *Glob. Chang. Biol.* **2010**, *16*, 88–101. [CrossRef]
6. Santini, A.; Ghelardini, L.; De Pace, C.; Desprez-Loustau, M.L.; Capretti, P.; Chandelier, A.; Stenlid, J. Biogeographical patterns and determinants of invasion by forest pathogens in Europe. *New Phytol.* **2013**, *197*, 238–250. [CrossRef]
7. Garnas, J.R.; Auger-Rozenberg, M.A.; Roques, A.; Bertelsmeier, C.; Wingfield, M.J.; Saccaggi, D.L.; Slippers, B. Complex patterns of global spread in invasive insects: Eco-evolutionary and management consequences. *Biol. Invasions* **2016**, *18*, 935–952. [CrossRef]
8. Chira, D.; Bolea, V.; Chira, F.; Mantale, C.; Tăut, I.; Şimonca, V.; Diamandis, S. Biological control of *Cryphonectria parasitica* in Romanian protected sweet chestnut forests. *Not. Bot. Horti Agrobot.* **2017**, *45*, 632–638. [CrossRef]
9. Seebens, H.; Bacher, S.; Blackburn, T.M.; Capinha, C.; Dawson, W.; Dullinger, S.; Essl, F. Projecting the continental accumulation of alien species through to 2050. *Global Chang. Biol.* **2021**, *27*, 970–982. [CrossRef]
10. Anderson-Teixeira, K.J.; Herrmann, V.; Cass, W.B.; Williams, A.B.; Paull, S.J.; Gonzalez-Akre, E.B.; McShea, W.J. Long-Term Impacts of Invasive Insects and Pathogens on Composition, Biomass, and Diversity of Forests in Virginia's Blue Ridge Mountains. *Ecosystems* **2021**, *24*, 89–105. [CrossRef]
11. Schlarbaum, S.E.; Hebard, F.; Spaine, P.C.; Kamalay, J.C. Three American Tragedies: Chestnut Blight, Butternut Canker, and Dutch Elm Disease. In *Exotic Pests of Eastern Forests Conference Proceedings*; Britton, K.O., Ed.; US Forest Service and Tennessee Exotic Pest Plant Council: Nashville, TN, USA, 1998; pp. 45–54.
12. Chira, D.; Dănescu, F.; Roşu, C.; Chira, F.; Mihalciuc, V.; Surdu, A.; Nicolescu, N.V. Some recent issues regarding the European beech decline in Romania. *Ann. ICAS* **2003**, *46*, 167–176.
13. Fernández-Fernández, M.; Naves, P.; Musolin, D.L.; Selikhovkin, A.V.; Cleary, M.; Diez, J.J. Pine Pitch Canker and Insects: Regional Risks, Environmental Regulation, and Practical Management Options. *Forests* **2019**, *10*, 649. [CrossRef]

14. Hyblerová, S.; Medo, J.; Barta, M. Diversity and prevalence of entomopathogenic fungi (Ascomycota, Hypocreales) in epidemic populations of bark beetles (Coleoptera, Scolytinae) in spruce forests of the Tatra National Park in Slovakia. *Ann. For. Res.* **2021**, *64*, 129–145. [CrossRef]
15. Brasier, C.M.; Kirk, S.A.; Pipe, N.D.; Buck, K.W. Rare interspecific hybrids in natural populations of the Dutch elm disease pathogens *Ophiostoma ulmi* and *O. novo-ulmi*. *Mycol. Res.* **1998**, *102*, 45–57. [CrossRef]
16. Brasier, C.M. Designation of the EAN and NAN races of *Ophiostoma novo-ulmi* as subspecies. *Mycol. Res.* **2001**, *105*, 547–554. [CrossRef]
17. Solla, A.; Dacasa, M.C.; Nasmith, C.; Hubbes, M.; Gil, L. Analysis of Spanish populations of *Ophiostoma ulmi* and *O. novo-ulmi* using phenotypic characters and RAPD markers. *Plant Pathol.* **2008**, *57*, 33–44. [CrossRef]
18. Hessenauer, P.; Fijarczyk, A.; Martin, H.; Prunier, J.; Charron, G.; Chapuis, J.; Landry, C.R. Hybridization and introgression drive genome evolution of Dutch elm disease pathogens. *Nat. Ecol. Evol.* **2020**, *4*, 626–638. [CrossRef]
19. Konrad, H.; Kirisits, T.; Riegler, M.; Halmschlager, E.; Stauffer, C. Genetic evidence for natural hybridization between the Dutch elm disease pathogens *Ophiostoma novo-ulmi* ssp. *novo-ulmi* and *O. novo-ulmi* ssp. *americana*. *Plant Pathol.* **2002**, *51*, 78–84. [CrossRef]
20. Brasier, C.M.; Mehrotra, M.D. *Ophiostoma himal-ulmi* sp. nov., a new species of Dutch elm disease fungus endemic to the Himalayas. *Mycol. Res.* **1995**, *99*, 205–215. [CrossRef]
21. Mackenthun, G. Native Elms of Saxony, Germany. In *The Elms: Breeding, Conservation, and Disease Management*; Dunn, C., Ed.; Kluwer Academic Publisher: Boston, MA, USA, 2000; pp. 305–314.
22. Hauer, R.J.; Hanou, I.S.; Sivyer, D. Planning for active management of future invasive pests affecting urban forests: The ecological and economic effects of varying Dutch elm disease management practices for street trees in Milwaukee, WI, USA. *Urban Ecosyst.* **2020**, *23*, 1005–1022. [CrossRef]
23. Alexandri, A. Elm dieback in Romania. *Bul. Sc. Stud. Şt. Naţ. Rom.* **1933**, *III*, 1–16.
24. Brasier, C.M.; Kirk, S.A. Rapid emergence of hybrids between the two subspecies of *Ophiostoma novo-ulmi* with a high level of pathogenic fitness. *Plant Pathol.* **2010**, *59*, 186–199. [CrossRef]
25. Petrescu, M.; Diţu, I.; Poleac, E.; Cucuianu, E. Dutch elm disease in Romania. *St. Cerc. ICF* **1967**, *XXIII B*, 135–151.
26. Borlea, G.F. Research on Breeding Resistance of Autochthonous elms to *Ophiostoma novo-ulmi*. Ph.D. Thesis, University Transilvania of Braşov, Braşov, Romania, 1996.
27. Simionescu, A.; Mihalache, G. *Forest Protection*; Muşatinii: Suceava, Romania, 2000.
28. Dinca, L. Field elm (*Ulmus minor* Mill.) stands the Moldavian plain. *Land Reclam. Earth Obs. Surv. Environ. Eng.* **2022**, *11*, 115–119.
29. Simionescu, A.; Mihalciuc, V.; Lupu, D.; Vlăduleasa, A.; Badea, O.; Fulicea, T. *Health Status of Romanian Forest in 1986–2000*; Muşatinii: Suceava, Romania, 2001.
30. Simionescu, A.; Chira, D.; Mihalciuc, V.; Ciornei, C.; Tulbure, C. *Health Status of Romanian Forest in 2001–2010*; Muşatinii Group: Suceava, Romania, 2012; 600p.
31. Constandache, C.; Tudor, C.; Popovici, L.; Dincă, L. Ecological reconstruction of the stands affected by drought from meadows of inland rivers. *Land Reclam. Earth Obs. Surv. Environ. Eng.* **2022**, *11*, 76–84.
32. Heybroek, H.M. The Dutch elm breeding program. In *Dutch Elm Disease Research: Cellular and Molecular Approaches*; Sticklen, M.B., Sherald, J., Eds.; Springer: New York, NY, USA, 1993; pp. 16–25.
33. Smalley, E.B.; Guries, R.P. Breeding elms for resistance to Dutch elm disease. *Annu. Rev. Phytopatol.* **1993**, *31*, 325–354. [CrossRef]
34. Solla, A.; Bohnens, J.; Collin, E.; Diamandis, S.; Franke, A.; Gil, L.; Broeck, A.V. Screening European elms for resistance to *Ophiostoma novo-ulmi*. *For. Sci.* **2005**, *51*, 134–141.
35. Santini, A.; Pecori, F.; Ghelardini, L. The Italian elm breeding program for Dutch elm disease resistance. In Proceedings of the Fourth International Workshop on the Genetics of Host-Parasite Interactions in Forestry: Disease and Insect Resistance in Forest Trees, Eugene, OR, USA, 31 July–5 August 2011; Sniezko, R.A., Yanchuk, A.D., Kliejunas, J.T., Palmieri, K.M., Alexander, J.M., Frankel, S.J., Eds.; Gen. Tech. Rep. PSW-GTR-240. US Department of Agriculture: Albany, CA, USA, 2012; Volume 240, pp. 326–335.
36. Collin, E.; Pozzi, T.; Joyeau, C.; Matz, S.; Rondouin, M.; Joly, C. Monitoring of the incidence of Dutch Elm Disease and mortality in experimental plantations of French *Ulmus minor* clones. *iForest* **2022**, *15*, 289. [CrossRef]
37. Venturas, M.; López, R.; Martín, J.A.; Gascó, A.; Gil, L. Heritability of *Ulmus minor* resistance to Dutch elm disease and its relationship to vessel size, but not to xylem vulnerability to drought. *Plant Pathol.* **2014**, *63*, 500–509. [CrossRef]
38. Martín, J.A.; Domínguez, J.; Solla, A.; Brasier, C.M.; Webber, J.F.; Santini, A.; Gil, L. Complexities underlying the breeding and deployment of Dutch elm disease resistant elms. *New For.* **2021**, 1–36. [CrossRef]
39. Moravčík, M.; Mamoňová, M.; Račko, V.; Kováč, J.; Dvořák, M.; Krajňáková, J.; Ďurkovič, J. Different Responses in Vascular Traits between Dutch Elm Hybrids with a Contrasting Tolerance to Dutch Elm Disease. *J. Fungi* **2022**, *8*, 215. [CrossRef] [PubMed]
40. Islam, M.T.; Coutin, J.F.; Shukla, M.; Dhaliwal, A.K.; Nigg, M.; Bernier, L.; Saxena, P.K. Deciphering the Genome-Wide Transcriptomic Changes during Interactions of Resistant and Susceptible Genotypes of American Elm with *Ophiostoma novo-ulmi*. *J. Fungi* **2022**, *8*, 120. [CrossRef] [PubMed]
41. Martín, J.A.; Sobrino-Plata, J.; Rodríguez-Calcerrada, J.; Collada, C.; Gil, L. Breeding and scientific advances in the fight against Dutch elm disease: Will they allow the use of elms in forest restoration? *New For.* **2019**, *50*, 183–215. [CrossRef]

42. Brasier, C.M.; Webber, J.F. Is there evidence for post-epidemic attenuation in the Dutch elm disease pathogen *Ophiostoma novo-ulmi*? *Plant Pathol.* **2019**, *68*, 921–929. [CrossRef]
43. Giurgiu, V. Saving of natural forests. In *Protection and Sustainable Development of Romanian Forests*; Giurgiu, V., Ed.; Arta Grafică: București, Romania, 1995; pp. 104–109.
44. Borlea, G.F. Ecology of elms in Romania. *For. Syst.* **2004**, *13*, 29–35.
45. Copernicus Europe's Eye on Earth. Land Monitoring Service. Available online: https://land.copernicus.eu/pan-european/corine-land-cover/clc2018 (accessed on 7 November 2022).
46. Mang, S.M.; Racioppi, R.; Camele, I.; Rana, G.L. Use of volatile metabolite profile to distinguish three *Monilia* species. *J. Plant Pathol.* **2015**, *97*, 55–59. [CrossRef]
47. Brasier, C.M. Laboratory investigation of *Ceratocystis ulmi*. In *Compendium of Elm Diseases*; Stipes, R.J., Campana, R.J., Eds.; American Phytopathological Society: St Paul, MN, USA, 1981; pp. 76–79.
48. Mang, S.M.; Figluolo, G. Species delimitation in *Pleurotus eryngii* species-complex inferred from ITS and EF1-α gene sequences. *Mycology* **2010**, *1*, 269–280. [CrossRef]
49. White, T.J.; Bruns, T.; Lee, S.; Taylor, J. Amplified and direct sequencing of fungal ribosomal RNA genes for phylogenetics. In *PCR Protocols: A Guide to Methods and Applications*; Innis, M.A., Gelfand, D.H., Sninsky, J.J., White, T.J., Eds.; Academic: San Diego, CA, USA, 1990; pp. 315–322.
50. Glass, N.L.; Donaldson, G.C. Development of primer sets designed for use with the PCR to amplify conserved genes from filamentous ascomycetes. *Appl. Environ. Microbiol.* **1995**, *61*, 1323–1330. [CrossRef]
51. Pipe, N.D.; Buck, K.W.; Brasier, C.M. Comparison of the cerato-ulmin (*cu*) gene sequences of the Himalayan Dutch elm disease fungus *Ophiostoma himal-ulmi* with those of *O. ulmi* and *O. novo-ulmi* suggests that the *cu* gene of *O. novo-ulmi* is unlikely to have been acquired recently from *O. himal-ulmi*. *Mycol. Res.* **1997**, *101*, 415–421. [CrossRef]
52. Mentana, A.; Camele, I.; Mang, S.M.; De Benedetto, G.E.; Frisullo, S.; Centonze, D. Volatolomics approach by HS-SPME-GC-MS and multivariate analysis to discriminate olive tree varieties infected by *Xylella fastidiosa*. *Phytochem. Anal.* **2019**, *30*, 623–634. [CrossRef]
53. Frisullo, S.; Elshafie, H.S.; Mang, S.M. First report of two *Phomopsis* species on olive trees in Italy. *J. Plant Pathol.* **2015**, *97*, 401. [CrossRef]
54. Altschul, S.F.; Gish, W.; Miller, W.; Myers, E.W.; Lipman, J.D. Basic Local Alignment Search Tool. *J. Mol. Biol.* **1990**, *215*, 403–410. [CrossRef]
55. Johnson, M.; Zaretskaya, I.; Raytselis, Y.; Merezhuk, Y.; Mcginnis, S.; Madden, T.L. NCBI BLAST: A better web interface. *Nucleic Acids Res.* **2008**, *36* (Suppl. 2), W5–W9. [CrossRef] [PubMed]
56. Plourde, K.V.; Bernier, L. A rapid virulence assay for the Dutch elm disease fungus *Ophiostoma novo-ulmi* by inoculation of apple (*Malus* × *domestica* 'Golden Delicious') fruits. *Plant Pathol.* **2014**, *63*, 1078–1085. [CrossRef]
57. ICP Forests. Visual Assessment of Crown Condition and Damaging Agents. 2021. Available online: http://icp-forests.net/page/icp-forests-manual (accessed on 25 July 2021).
58. Burnham, K.P.; Anderson, D.R. Multimodel Inference: Understanding AIC and BIC in Model Selection. *Sociol Methods Res.* **2004**, *33*, 261–304. [CrossRef]
59. R Core Team. *R: A Language and Environment for Statistical Computing*; R Foundation for Statistical Computing: Vienna, Austria, 2021; Available online: https://www.R-project.org (accessed on 15 December 2021).
60. Bates, D.; Mächler, M.; Bolker, B.M.; Walker, S.C. Fitting linear mixed-effects models using lme4. *J. Stat. Softw.* **2015**, *67*. [CrossRef]
61. Kuznetsova, A.; Brockhoff, P.B.; Christensen, R.H.B. lmerTest Package: Tests in Linear Mixed Effects Models. *J. Stat. Softw.* **2017**, *82*, 1–26. [CrossRef]
62. Brasier, C.; Franceschini, S.; Forster, J.; Kirk, S. Enhanced outcrossing, directional selection and transgressive segregation drive evolution of novel phenotypes in hybrid swarms of the Dutch elm disease pathogen *Ophiostoma novo-ulmi*. *J. Fungi* **2021**, *7*, 452. [CrossRef]
63. Brasier, C.M.; Bates, M.R.; Charter, N.W.; Buck, K.W. DNA Polymorphism, Perithecial Size and Molecular Aspects of D Factors in *Ophiostoma-ulmi* and *O. novo-ulmi*. In *Dutch Elm Disease Research*; Springer: New York, NY, USA, 1993; pp. 308–321.
64. Katanić, Z.; Krstin, L.; Ježić, M.; Ćaleta, B.; Stančin, P.; Zebec, M.; Ćurković-Perica, M. Identification and characterization of the causal agent of Dutch elm disease in Croatia. *Eur. J. For. Res.* **2020**, *139*, 805–815. [CrossRef]
65. Santini, A.; Montaghi, A.; Vendramin, G.G.; Capretti, P. Analysis of the Italian Dutch elm disease population. *J. Phytopathol.* **2005**, *153*, 73–79. [CrossRef]
66. Dvořák, M.; Tomšovský, M.; Jankovský, L.; Novotný, D. Contribution to identify the causal agents of Dutch elm disease in the Czech Republic. *Plant Prot. Sci.* **2007**, *43*, 142–145. [CrossRef]
67. Townsend, A.M. Relative resistance of diploid *Ulmus* species to *Ceratocystis ulmi*. *Plant Dis. Rep.* **1971**, *55*, 980–982.
68. Brasier, C.M. Inheritance of pathogenicity and cultural characters in *Ceratocystis ulmi*; hybridization of protoperithecial and non-aggressive strains. *Trans. Br. Mycol. Soc.* **1977**, *68*, 45–52. [CrossRef]
69. Jürisoo, L.; Selikhovkin, A.V.; Padari, A.; Shevchenko, S.V.; Shcherbakova, L.N.; Popovichev, B.G.; Drenkhan, R. The extensive damage to elms by Dutch elm disease agents and their hybrids in northwestern Russia. *Urban For. Urban Green.* **2021**, *63*, 127214. [CrossRef]

70. Sacchetti, P.; Tiberi, R.; Mittempergher, L. Preference of *Scolytus multistriatus* (Marsham) during the gonad maturation phase between two species of elm. *Redia* **1990**, *73*, 347–354.
71. Webber, J. Insect vector behaviour and the evolution of Dutch elm disease. In *The Elms: Breeding, Conservation and Disease Management*; Dunn, C.P., Ed.; Kluwer: Boston, MA, USA, 2000; pp. 47–60.
72. Pajares, J.A.; García, S.; Díez, J.J.; Martín, D.; García-Vallejo, M.C. Feeding responses by *Scolytus scolytus* to twig bark extracts from elms. *For. Syst.* **2004**, *13*, 217–225.
73. Dvořák, M.; Palovčíková, D.; Jankovský, L. The occurrence of endophytic fungus *Phomopsis oblonga* on elms in the area of southern Bohemia. *J. For. Sci.* **2012**, *52*, 531–535. [CrossRef]
74. Macaya-Sanz, D.; Witzell, J.; Collada, C.; Gil, L.; Martín, J.A. Structure of core fungal endobiome in *Ulmus minor*: Patterns within the tree and across genotypes differing in tolerance to Dutch elm disease. *bioRxiv* **2020**. [CrossRef]
75. Martínez-Arias, C.; Sobrino-Plata, J.; Ormeño-Moncalvillo, S.; Gil, L.; Rodríguez-Calcerrada, J.; Martín, J.A. Endophyte inoculation enhances *Ulmus minor* resistance to Dutch elm disease. *Fungal Ecol.* **2021**, *50*, 101024. [CrossRef]
76. Anderbrant, O.; Yuvaraj, J.K.; Martin, J.A.; Gil, J.L.; Witzell, J. Feeding by *Scolytus* bark beetles to test for differently susceptible elm varieties. *J. Appl. Entomol.* **2017**, *141*, 417–420. [CrossRef]
77. Collin, E.; Bozzano, M. Implementing the dynamic conservation of elm genetic resources in Europe: Case studies and perspectives. *iForest* **2015**, *8*, 143–148. [CrossRef]
78. Domínguez, J.; Macaya-Sanz, D.; Gil, L.; Martín, J.A. Excelling the progenitors: Breeding for resistance to Dutch elm disease from moderately resistant and susceptible native stock. *For. Ecol. Manag.* **2022**, *511*, 120113. [CrossRef]
79. Smucker, S.J. Comparison of susceptibility of American elm and several exotic elms to *Ceratostomella ulmi*. *Phytopathology* **1941**, *31*, 758–759.
80. Webber, J.F. Experimental studies on factors influencing the transmission of Dutch elm disease. *For. Syst.* **2004**, *13*, 197–205.
81. Mittempergher, L.; La Porta, N. *Ophiostoma ulmi*/Olmo. In *Miglioramento Genetico Delle Piante per Resistenza a Patogeni e Parassiti*; Crino, P., Ed.; Edagricole: Bologne, Italy, 1993; pp. 412–432.
82. Collin, E. Strategies and guidelines for the conservation of the genetic resources of *Ulmus* spp. In *Noble Hardwoods Network*; International Plant Genetic Resources Institute: Rome, Italy, 2002; pp. 50–67.
83. Torre, S.; Sebastiani, F.; Burbui, G.; Pecori, F.; Pepori, A.L.; Passeri, I.; Santini, A. Novel Insights into Refugia at the Southern Margin of the Distribution Range of the Endangered Species *Ulmus laevis*. *Front. Plant Sci.* **2022**, *13*, 826158. [CrossRef]
84. Kännaste, A.; Jürisoo, L.; Runno-Paurson, E.; Talts, E.; Drenkhan, R.; Niinemets, Ü. Impacts of Dutch elm disease-causing fungi on foliage photosynthetic characteristics and volatiles in *Ulmus* species with different pathogen resistance. *Tree Physiol.* **2022**. *accepted for publication*. [CrossRef]
85. Climate Timișoara (Romania). 2021. Available online: https://en.climate-data.org/europe/romania/timis/timisoara-997/ (accessed on 12 August 2022).
86. Rigling, D.; Prospero, S. *Cryphonectria parasitica*, the causal agent of chestnut blight: Invasion history, population biology and disease control. *Mol. Plant Pathol.* **2018**, *19*, 7–20. [CrossRef]
87. Landolt, J.; Gross, A.; Holdenrieder, O.; Pautasso, M. Ash dieback due to *Hymenoscyphus fraxineus*: What can be learnt from evolutionary ecology? *Plant Pathol.* **2016**, *65*, 1056–1070. [CrossRef]
88. Chira, D.; Chira, F.; Tăut, I.; Popovici, O.; Blada, I.; Doniţă, N.V.; Dinu, C. Evolution of ash dieback in Romania. In *Dieback of European Ash (Fraxinus spp.)-Consequences and Guidelines for Sustainable Management*; Vasaitis, R., Enderle, R., Eds.; Swedish University of Agricultural Sciences: Uppsala, Sweden, 2017; pp. 185–194.
89. Griffiths, S.M.; Galambao, M.; Rowntree, J.; Goodhead, I.; Hall, J.; O'Brien, D.; Antwis, R.E. Complex associations between cross-kingdom microbial endophytes and host genotype in ash dieback disease dynamics. *J. Ecol.* **2020**, *108*, 291–309. [CrossRef]
90. Buiteveld, J.; Van Der Werf, B.; Hiemstra, J.A. Comparison of commercial elm cultivars and promising unreleased Dutch clones for resistance to *Ophiostoma novo-ulmi*. *iForest* **2015**, *8*, 158. [CrossRef]
91. Townsend, A.M.; Bentz, S.E.; Douglass, L.W. Evaluation of 19 American elm clones for tolerance to Dutch elm disease. *J. Environ. Hortic.* **2005**, *23*, 21–24. [CrossRef]
92. Collin, E.; Rondouin, M.; Joyeau, C.; Matz, S.; Raimbault, P.; Harvengt, L.; Bilger, I.; Guibert, M. Conservation and use of elm genetic resources in France: Results and perspectives. *iForest* **2020**, *13*, 41–47. [CrossRef]
93. Budeanu, M.; Apostol, E.N.; Radu, R.G.; Ioniță, L. Genetic variability and juvenile–adult correlations of Norway spruce (*Picea abies*) provenances, tested in multisite comparative trials. *Ann. For. Res.* **2021**, *64*, 105–122. [CrossRef]
94. Ducci, F.; De Rogatis, A.; Proietti, R.; Curtu, A.L.; Marchi, M.; Belletti, P. Establishing a baseline to monitor future climate-change-effects on peripheral populations of *Abies alba* in central Apennines. *Ann. For. Res.* **2021**, *64*, 33–66. [CrossRef]
95. Mihai, G.; Alexandru, A.M.; Nita, I.A.; Birsan, M.V. Climate Change in the Provenance Regions of Romania over the Last 70 Years: Implications for Forest Management. *Forests* **2022**, *13*, 1203. [CrossRef]
96. Prăvălie, R.; Sîrodoev, I.; Nita, I.A.; Patriche, C.; Dumitraşcu, M.; Roşca, B.; Birsan, M.V. NDVI-based ecological dynamics of forest vegetation and its relationship to climate change in Romania during 1987–2018. *Ecol. Indic.* **2022**, *136*, 108629. [CrossRef]

Article

Plant Diversity along an Urbanization Gradient of a Tropical City

Balqis Aqila Alue [1], Noraine Salleh Hudin [1,2,*], Fatimah Mohamed [1,2], Zahid Mat Said [1,2] and Kamarul Ismail [3]

1 Department of Biology, Faculty of Science and Mathematics, Universiti Pendidikan Sultan Idris, Tanjong Malim 35900, PRK, Malaysia
2 Centre of Biodiversity and Conservation, Universiti Pendidikan Sultan Idris, Tanjong Malim 35900, PRK, Malaysia
3 Department of Geography and Environment, Faculty of Human Science, Universiti Pendidikan Sultan Idris, Tanjong Malim 35900, PRK, Malaysia
* Correspondence: noraine@fsmt.upsi.edu.my; Tel.: +60-015-4879-7438

Abstract: This study aimed to investigate the plant diversity, plant traits, and environmental variables along the tropical urbanization gradient in Ipoh, Perak, Malaysia. The study areas comprised 12 sampling plots sized 1 km^2 that represented different urbanization intensities. Urbanization intensity was quantified as the percentage of the built-up area within a 1 km^2 area. A total of 96 woody plant species belonging to 71 genera and 42 families were found in the study areas. In general, species diversity, richness, and evenness declined significantly as urbanization intensity increased. The number of native species reduced by 67.6% when urbanization intensity increased from wildland to suburban while the non-native species remained stable along the urbanization gradient. Regarding the plant traits, tree height decreased with increasing urbanization intensity, while no significant result was found for specific leaf areas. All environmental factors were significantly associated with urbanization where air temperature and light intensity showed a positive relationship with increasing urbanization intensity while the opposite trend was found for air humidity. This study emphasizes the importance of built-up areas as the predictor of native species in the tropics. The findings of this study may help town planners and policymakers to create more sustainable urban development in the future.

Keywords: built-up area; geographical information systems (GIS); native species; plant diversity; urbanization gradient

Citation: Alue, B.A.; Salleh Hudin, N.; Mohamed, F.; Mat Said, Z.; Ismail, K. Plant Diversity along an Urbanization Gradient of a Tropical City. *Diversity* **2022**, *14*, 1024. https://doi.org/10.3390/d14121024

Academic Editors: Lucian Dinca and Mario A. Pagnotta

Received: 16 October 2022
Accepted: 21 November 2022
Published: 24 November 2022

1. Introduction

The rate of world urbanization is increasing, and by 2050 urbanization will account for 65% of the global population [1]. While urbanization frequently improves social life and living standards [2], it also has severe repercussions on humans and the natural environment [3]. Rapid urbanization can result in changes in environmental conditions [4] and degradation of the natural environment [5]. As a result of anthropogenic activities, there are drastic shifts in land use types [6,7] and this may put ecosystem services in jeopardy [8]. Ecosystem services and ecological processes depend primarily on plants, so plants must be adequately maintained and conserved [9]. Due to the importance of plants, it is also advocated that plant vegetation components should be used as a more effective metric of the ecological condition [10]. Such can be completed through vegetation monitoring which relies heavily on the analysis of plant species composition and structure [9].

Many biological processes depend on plants, and people would benefit greatly from them. For instance, plants may assist in lowering the temperature [11] controlling air pollution [12], improving the mental health of city dwellers (e.g., through horticulture and nature therapy) [13], encouraging social interactions (e.g., in community gardens) [14], and encouraging the preservation of biodiversity [15]. Given that people benefit from ecosystems, these advantages are known as ecosystem services. Therefore, it is important to maintain the plants, especially in metropolitan regions. Therefore, even though urbanization is continuously growing, we need to ensure that the ecological cycle is maintained.

Prior research on the impact of urbanization on plant diversity yielded inconsistent and contradictory results. Even though urbanization is expected to impact plant diversity [16] plant diversity patterns along an urbanization gradient vary greatly from study to study, rendering that plant diversity investigations and their relationship with urbanization gradient remained unclear [17]. For example, some studies discovered that plant diversity increases in tandem with urbanization, e.g., [17–19] which could be attributable to the unintended introduction of non-native plant species into urban areas [20]. Meanwhile, other research revealed that plants tend to peak, particularly around intermediate levels of urbanization, i.e., Godefroid and Koedam [21]; Ranta and Viljanen [22]; Zerbe [23]. This research validated the intermediate perturbation theory, suggesting that moderate perturbation would enable more species to survive than high or low perturbation [24]. On the contrary, other studies found that plant diversity increases with increasing distance from the metropolitan area [25], and this could be related to the decline of habitable spaces in urban areas, which generally leads to a decrease in plant diversity [26]. Because of the severe environmental changes brought about by urbanization, various plant diversity patterns emerge as urbanization increases.

Previous studies on plant diversity patterns along an urbanization gradient have mainly been conducted in temperate regions, e.g., Belgium [21], Finland [22], Zerbe [23], Kazakhstan [25], North America [26], Germany [27], Switzerland [28], subtropical region-e.g., from China by Wang et al. [17], Tian et al. [29], Yang et al. [30], and only a few were performed in the tropical region, e.g., Nigeria [10], India [19] and Malaysia [31]. Those in Malaysia investigated plant diversity only in wildlands, such as studies by Suratman [32], Onrizal et al. [33], and Ghollasimood et al. [34], while others were limited to green spaces in city centers only, such as studies by Nabilla et al. [35], Kanniah [36], Kanniah & Ho [37], and Rostam et al. [38]. Moreover, an earlier study in Malaysia by Rahmad & Akomolafe [31] that compared plant diversity between different urbanization intensity categorized the urbanization levels only based on visual observation and subjective perception, thus causing difficulty to compare the findings. This lack of knowledge becomes a concern because the tropical region does not only serve as habitats for a high number of plant species, but some tropical countries, especially those in the Southeast Asia, are experiencing urbanization at an accelerating rate [37]. Since previous studies showed mixed results on how urbanization affects plant diversity, there is a need to confirm how plant communities respond to urbanization in tropical cities.

Fundamentally, plant traits are correlated to plant strategies [38], which are responsible for the adaptations of plants to the environmental changes [3,17]. Palma et al. [39] have stated that plant traits determine whether the plant species will survive in the long run especially in a harsh condition. Such harsh conditions could be induced by changes in environmental conditions when natural landscapes undergo disturbance caused by urban development. Eventually, this scenario will affect the abundance and diversity of plants in the cities. In response to the ecosystem change, the functional traits of organisms that stimulate their development and growth are essential in determining which species thrive and cease to exist [40].

Hence, the aims of this study were (i) to explore plant diversity patterns along a tropical urbanization gradient; (ii) to examine the relationship between urbanization intensity with plant traits (tree height and specific leaf area (SLA)); and (iii) to investigate the variations in environmental conditions (air temperature, air humidity, and light intensity) along the urbanization gradient. To address these aims, data on plant diversity, plant traits, and environmental variables were obtained from 12 sampling plots that sized 1 km^2. These plots represented different urbanization intensities according to the percentage of the built-up area of the plots. Regression analyses were later performed, which involved the data on plant diversity, plant traits, and environmental variables against the percentage of the built-up area of each plot.

2. Materials and Methods

2.1. Study Area

We conducted fieldwork in the Ipoh city council in the Perak State of Peninsular Malaysia. Perak State was chosen because it is able to serve as a "regional state", with Ipoh serving as the core urban center [41]. Additionally, anthropogenic activities have resulted in massive landscape modifications and abrupt changes in land use in Perak [42]. For example, Perak state has lost about 16 percent (189,423 ha) of its forest cover in the last 29 years [42]. Moreover, Ipoh is known as a phytogeography ally unique area with three elements, including limestone flora [43], the Perak sub-province that influences the Sumatran flora [44], and the Seasonal Asiatic Intrusion that is enclave invasion by Burmese-Thai floristic elements [43]. Thus, Ipoh is an ideal place to study the impacts of urbanization on plant diversity.

This research was performed in four distinct urbanization settings, i.e., wildland, rural, suburban, and urban areas in Ipoh City Council in the Perak state of Peninsular Malaysia (101°3′57.118747″–101°3′57.118747″ E, 4°28′26.148383″–4°28′26.148383″ N, Figure 1). Ipoh is located between the two major cities in Malaysia, which are Kuala Lumpur and Penang, and thus serves as a significant hub for road transportation across west Malaysia. Most economic ventures concentrate in Ipoh due to the large population and high volume of road traffic [45]. The Ipoh city is bordered by limestone hills which can also be found throughout the northeast, east, and southeast suburban areas. Ipoh has the typical climate of a rainforest. The monthly relative humidity of the city shows slight differences while the temperature varies between 20.7 °C to 30.6 °C throughout the year. With an average of 200 mm (7.9 in) of rain per month and 2427.9 mm (95.59 in) of rain annually, Ipoh experiences high rainfall over the year. October is the most humid month for Ipoh city, with rain falling at an average of 297.2 mm (11.70 in). In addition to that, the driest month is January, with an average of 132.3 mm (5.21 in) of rainfall.

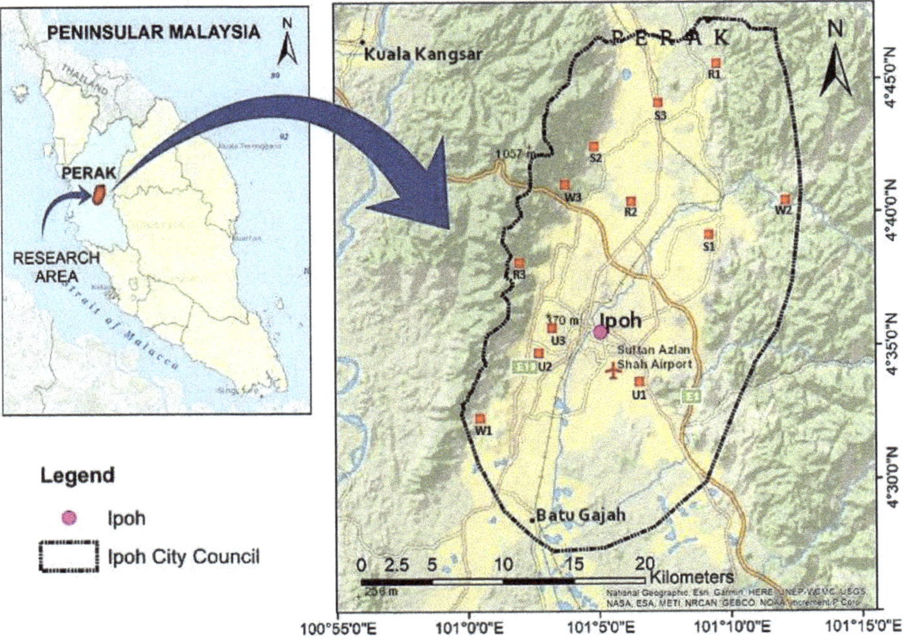

Figure 1. Study area in Ipoh, Perak in Peninsular Malaysia where 12 sampling plots represented an urbanization gradient. W: wildland, R: rural, S: suburban, U: urban.

2.2. Urbanization Intensity Quantification

The urbanization intensity of sampling sites was determined based on the percentage of the built-up area within a 1 km^2 area. As built-up percentage can be quantified in any urban environment, regardless of geographical, cultural, and historical variation, the built-up percentage is an adequate proxy of urbanization intensity. The urbanization intensity quantification was performed in ArcGIS version 10.8 software using the Ipoh land use map issued by Malaysia's Federal Department of Town and Country Planning. The map is composed of 12 distinct attributes of land use. These characteristics were classified into two categories, which were the green areas and built-up areas. The built-up areas consisted of seven features which were manufacturing, service and infrastructure, commercial, institutional, mixed construction, transport and mobility, and residential or housing areas. On the other hand, forests, parks, recreation centers, undeveloped land, and farming areas made up the green areas. The water body was exempted from all categories. To achieve unbiased sampling, tessellation of the map consisting of 1 km $\times$ 1 km grid squares was completed where these squares corresponded to the size of the study plot. Then, the percentage of built-up area of each square was calculated. Following the urbanization categorization using the built-up percentage of the 1 km^2 plots by [46], four categories of urbanization intensity were identified, which were the wildland (0–2% built-up percentage), rural (5–20%), suburban (30–50%) and urban areas (>50%). The study plots were selected randomly, and each category was replicated twice, which created a total of 12 sampling plots.

2.3. Field Sampling

Twelve sampling plots sized 1 km^2 were randomly established in Ipoh, Perak, to represent an urbanization gradient (Figure 1). Previous study suggested that an area that sized 500–1000 m^2 with a minimum aspect ratio of 1:20 of a rectangular plot is sufficient for studies on tropical plant diversity [47]. To increase coverage of area sampled within a plot, this study used 40 subplots sized 1 m $\times$ 25 m that were randomly positioned within the plot, which produced a total sampling area of 1000 m^2 per plot. Any plants with a diameter at breast height (DBH, 1.3 above the ground) of 5 cm and above in the subplot were included as samples. The number of individuals per species was recorded and each species was identified by a plant taxonomist from the Forest Research Institute Malaysia (FRIM). We determined the nativity status of plant species based on the Royal Botanic Garden database.

2.4. Plant Traits

Plant height and SLA (mm^2 mg^{-1}) represented the functional traits of the plant species. The maximum value of tree height and SLA were quantified using a standardized method [48]. To calculate SLA, one healthy leaf was collected from each of the five mature individuals [49]. The leaf samples should be relatively young (presumably more photosynthetically active) and carefully selected from fully expanded and hardened leaves of adult plants growing in light or directly proportional optimal conditions [48]. When feasible, leaves with visible signs of a disease or a thick cover of epiphyllous were avoided. Following that, carefully selected leaves were wrapped in moist paper, placed in sealed plastic bags to minimize water loss during transpiration, and kept in a cool box before being transferred to the laboratory to measure leaf area [48]. The selected leaves were digitally scanned, and the leaf area was estimated using the ImageJ software. The leaves were then oven-dried for 72 h at 60 °C before being weighed [49], and the SLA was calculated as area divided by dry mass.

2.5. Environmental Conditions

Air temperature, relative humidity, and light intensity were measured at the study sites to represent the environmental variables. At the height of 1.3 m above the ground [50], the readings were taken at noon [51] on the day with no clouds visible and clear skylight [52].

The 1 km^2 plot was divided into five equal sections, each with a measurement station spaced 200 m apart. The environmental variables were measured three times at each location.

2.6. Plant Species Richness, Diversity and Evenness along the Urbanization Gradient

Three indices were used to define the plant species diversity in each plot: the Shannon–Wiener species diversity index (H′), plant species richness, evenness index.

2.6.1. Species Richness

The species richness of plant was determined by Margalef's index of richness, Dmg [53]. Given as:

$$\mathrm{Dmg} = \frac{(\mathrm{S}-1)}{\ln \mathrm{N}} \tag{1}$$

where S = number of species in total, and N = total number of individual species in a sampling plot.

2.6.2. Species Diversity

Since species diversity is regularly used as an ecological indicator, various diversity indices can be used to examine various features of group structure; nonetheless, the Shannon–Weiner index is the most frequently used as a diversity indicator [54]. This study measured species diversity between urbanization intensity using the Shannon–Weiner index since it is widely used. This index takes both species abundance and species richness into account.

Shannon–Weiner diversity index (H′) was calculated using the following equation [54].

$$\mathrm{H}' = -\sum_{i=1}^{\mathrm{S}} \mathrm{Pi} \ln \mathrm{Pi} \tag{2}$$

where H′ = the species diversity, and Pi = the proportion of species i relative to the total number of species. The resulting product was summed across species and multiplied by −1.

2.6.3. Species Evenness

Species evenness index (J) was used to calculate how evenly the species were distributed within the study area [55]. Species evenness was calculated as:

$$\mathrm{J} = \frac{\mathrm{H}'}{\mathrm{H}'\,\mathrm{max}} \tag{3}$$

where J = Pielou's evenness, H′ = Shannon diversity index, H′ max = In S (number of species).

2.7. Statistical Analysis

Linear regression was used to analyze the influence of urbanization intensity on plant diversity, richness, and evenness using IBM SPSS Statistics version 24. To understand how urbanization affects the number of native and non-native plant species along the urbanization gradient, two separate analyses of linear regression were performed with the percentage of built-up area and number of native and non-native species as the variables. Since the regression analysis involving the native species obtained a significant result, a one-way ANOVA was performed to compare the number of native species between different urbanization settings (urban, suburban, rural, and wildland). Then, we performed Tukey HSD test to determine which urbanization level significantly reduce the number of native species. Regarding plant functional traits (tree height and SLA) and environmental factors (air temperature, relative humidity, and light intensity), these variables were correlated with the percentage of built-up area using Pearson's correlation.

3. Results

Overall, in all 12 sampling plots which covered an area of 0.1 hectares, a total of 96 woody plant species families with a DBH of 5.0 cm or greater belonging to 71 genera, and 42 were found from the urban area, suburban, rural, and wildland (Table 1). Based on the total number of families obtained from this study, it encompassed 16.94% of the entire families of flora recorded by Turner [56] in Peninsular Malaysia while the 71 genera represented 4.24% of the total of 1674 genera reported. The number of species found in this study comprised 1.16% of the total 8290 species documented in Peninsular Malaysia. However, it should be noted that Turner [56] had included all types of plant species in the checklist, whereas the present study only collected the species with a DBH of 5 cm or greater.

Table 1. Plant species and their abundance in urban, suburban, rural, and wildland areas in Ipoh, Perak, Malaysia.

No.	Family	Plant Species	Abundance				Nativity Status
			U	S	R	W	
1	Anacardiaceae	*Buchanania arborescens* (Blume) Blume				31	N
2	Anacardiaceae	*Mangifera indica* L.	55	62	71	102	A
3	Anacardiaceae	*Swintonia floribunda* Griff.				26	N
4	Annonaceae	*Monocarpia marginalis* (Scheff.) J. Sinclair				17	N
5	Annonaceae	*Polyalthia cauliflora* Hook.f. & Thomson				35	N
6	Apocynaceae	*Alstonia angustiloba* Miq.		7			N
7	Apocynaceae	*Alstonia spatulata* Blume				76	N
8	Apocynaceae	*Plumeria alba* L.	7				A
9	Araliaceae	*Arthrophyllum diversifolium* Blume				47	N
10	Asteraceae	*Chromolaena odorata* (L.) R. M. King & H. Rob.		17			A
11	Bignoniaceae	*Tabebuia rosea* (Bertol.) DC.	10			21	A
12	Calophyllaceae	*Mesua ferrea* L.				28	N
13	Cannabaceae	*Trema tomentosa* (Roxb.) Hara				28	N
14	Caricaceae	*Carica papaya* L.	11	12			A
15	Celastraceae	*Salacia maingayi* M. A. Lawson				41	N
16	Combretaceae	*Terminalia mantaly* H. Perrier	8	11			A
17	Dipterocarpaceae	*Dipterocarpus oblongifolius* Blume				28	N
18	Dipterocarpaceae	*Shorea bracteolata* Dyer				17	N
19	Dipterocarpaceae	*Shorea hopeifolia* (F.Heim) Symington				39	N
20	Dipterocarpaceae	*Shorea multiflora* (Burck) Symington			17		N
21	Euphorbiaceae	*Hura crepitans* L.	22		27		A
22	Euphorbiaceae	*Hevea brasiliensis* (Willd. ex A.Juss.) Müll.Arg.			58	98	A
23	Euphorbiaceae	*Macaranga denticulata* (Blume) Müll.Arg.		6			N
24	Euphorbiaceae	*Macaranga tanarius* (L.) Müll. Arg.		13	101	67	A
25	Euphorbiaceae	*Mallotus muticus* (Müll. Arg.) Airy Shaw		13	28		N
26	Euphorbiaceae	*Microdesmis caseariifolia* Planch. ex Hook.				29	N
27	Fabaceae	*Adenanthera pavonina* L.	33	14			N
28	Fabaceae	*Acacia auriculiformis* A. Cunn. ex Benth.		35	38	48	A
29	Fabaceae	*Acacia mangium* Willd.		41		35	A
30	Fabaceae	*Aganope thyrsiflora* (Benth.) Polhill				28	N
31	Fabaceae	*Bauhinia purpurea* L.			18		A
32	Fabaceae	*Caesalpinia sappan* L.	12				A
33	Fabaceae	*Millettia pinnata* (L.) Panigrahi				33	N
34	Fabaceae	*Parkia speciosa* Hassk.		10			N
35	Fabaceae	*Samanea saman* (Jacq.) Merr.	15				A
36	Guttiferae	*Garcinia mangostana* L.			24	33	N
37	Hypericaceae	*Cratoxylum formosum* (Jack) Benth. & Hook.f. ex Dyer				27	N
38	Hypericaceae	*Cratoxylum maingayi* Dyer				24	N
39	Ixonanthaceae	*Ixonanthes icosandra* Jack				30	N
40	Ixonanthaceae	*Ixonanthes reticulata* Jack			16		N
41	Lamiaceae	*Vitex pinnata* L.				26	N
42	Lauraceae	*Beilschmiedia perakensis* Gamble				47	N

Table 1. *Cont.*

No.	Family	Plant Species	Abundance				Nativity Status
			U	S	R	W	
43	Lauraceae	*Cinnamomum javanicum* Blume				20	N
44	Lauraceae	*Cinnamomum iners* Reinw. ex Blume			42	53	N
45	Lauraceae	*Cinnamomum verum* J.Presl			31		A
46	Lecythidaceae	*Barringtonia racemosa* (L.) Spreng.				29	N
47	Malvaceae	*Durio zibethinus* Murray		13	70	43	N
48	Malvaceae	*Hibiscus rosa-sinensis* L.		8			A
49	Malvaceae	*Microcos tomentosa* Sm.		15	22	39	A
50	Melastomataceae	*Pternandra coerulescens* Jack				37	N
51	Meliaceae	*Azadirachta indica* A.Juss.	8	11			A
52	Meliaceae	*Sandoricum koetjape* Merr.				23	N
53	Meliaceae	*Swietenia macrophylla* G.King				40	A
54	Moraceae	*Artocarpus elasticus* Reinw. ex Blume		12			N
55	Moraceae	*Artocarpus heterophyllus* Lam.		5	19	27	A
56	Moraceae	*Artocarpus integer* (Thunb.) Merr.			46	98	N
57	Moraceae	*Ficus aurata* (Miq.) Miq.		35		47	N
58	Moraceae	*Ficus benjamina* L.		6		22	N
59	Moraceae	*Ficus elastica* Roxb. ex Hornem.			16		N
60	Moraceae	*Ficus hispida* L.fil.			41	29	N
61	Moraceae	*Ficus magnoliifolia* Blume				21	N
62	Moraceae	*Ficus racemosa* L.				23	N
63	Moraceae	*Ficus religiosa* L.			16	22	N
64	Moraceae	*Ficus sinuata* Thunb.				28	N
65	Moraceae	*Streblus elongatus* (Miq.) Corner				92	N
66	Moringaceae	*Moringa oleifera* Lam.	9				A
67	Muntingiaceae	*Muntingia calabura* L.		33	39		A
68	Myrtaceae	*Syzygium aqueum* (Burm.fil.) Alston	7			32	N
69	Myrtaceae	*Syzygium grande* (Wight) Walp.			31	39	N
70	Myrtaceae	*Syzygium myrtifolium* Walp.				29	N
71	Myrtaceae	*Syzygium valdevenosum* (Duthie) Merr. & Perry				25	N
72	Myrtaceae	*Syzygium zeylanicum* (L.) DC.			25		N
73	Olacaceae	*Ochanostachys amentacea* Mast.				61	N
74	Opiliaceae	*Champereia manillana* (Blume) Merr.				24	N
75	Ochnaceae	*Ochna kirkii* Oliv.		8			A
76	Oxalidaceae	*Sarcotheca griffithii* (Planch. ex Hook.fil.) Hallier fil.				19	N
77	Passifloraceae	*Paropsia vareciformis* (Griff.) Mast.				25	N
78	Pentaphylacaceae	*Eurya acuminata* DC.				39	N
79	Phyllanthaceae	*Antidesma cuspidatum* Müll.Arg.				22	N
80	Phyllanthaceae	*Aporosa penangensis* (Ridl.) Airy Shaw				19	N
81	Phyllanthaceae	*Aporosa symplocoides* (Hook.f.) Gage				20	N
82	Phyllanthaceae	*Baccaurea parviflora* (Müll.Arg.) Müll.Arg.				34	N
83	Phyllanthaceae	*Commersonia bartramia* (L.) Merr.			13		A
84	Polygalaceae	*Xanthophyllum affine* Korth. ex Miq.				30	N
85	Rhamnaceae	*Ziziphus mauritiana* Lam.			21		N
86	Rhizophoraceae	*Pellacalyx saccardianus* Scort.				33	N
87	Rubiaceae	*Aidia densiflora* (Wall.) Masam.	7				N
88	Rubiaceae	*Morinda citrifolia* L.	15		22	25	N
89	Rubiaceae	*Morinda elliptica* (Hook.f.) Ridl.				28	N
90	Rubiaceae	*Pertusadina eurhyncha* (Miq.) Ridsdale				19	N
91	Sapindaceae	*Nephelium lappaceum* L.		25	50		N
92	Sapindaceae	*Pometia pinnata* J.R.Forst. & G.Forst.				29	A
93	Sapotaceae	*Mimusops elengi* L.				35	N
94	Sapotaceae	*Palaquium gutta* (Hook.) Baill.				60	N
95	Symplocaceae	*Symplocos cochinchinensis* (Lour.) Moore				33	N
96	Ulmaceae	*Gironniera nervosa* Planch.				38	N

Note: U = urban, S = suburban, R = rural and W = wildland; N = native species, A = non-native species.

3.1. Relationship between Urbanization and Plant Diversity, Richness, and Evenness

Wildland had the greatest number of species, and the highest species richness, evenness, and diversity (d = 8.58, J′= 0.97, H′ = 4.11). Rural was the second highest in species richness, evenness, and diversity (d = 3.68, J′ = 0.95, H′ = 3.10), followed by suburban (d = 3.49, J′ = 0.93, H′ = 2.95) and urban (d = 2.41, J′ = 0.90, H′ = 2.38) (Table 2). This study found that plant diversity ($r = -0.781$; $p = 0.003$), richness ($r = -0.0842$; $p = 0.001$), and evenness ($r = -0.901$; $p < 0.001$) significantly decreased when urbanization intensity increased (Figure 2).

Table 2. Diversity indices in urban, suburban, rural, and wildland areas in Ipoh, Perak, Malaysia. See Table S1 for details on the native and non-native species.

	Urban	Suburban	Rural	Wildland
Number of species	19	28	36	79
Number of native species	6	15	22	68
Number of non-native species	13	13	14	11
Number of individuals	219	412	902	2472
Number of families	11	13	16	35
Shannon, H′	2.38	2.95	3.10	4.11
Evenness, J	0.90	0.93	0.95	0.97
Margalef′s, Dmg	2.41	3.49	3.68	8.58

In addition, native species was the highest in the wildland while the non-native species were dominant in the urban area. A regression analysis on the number of native plant species against the built-up percentage yielded a significant negative relationship ($r = -0.730$; $p = 0.007$) whereby the built-up percentage explained 53.3% of the variation in the number of native species. To find out at what urbanization level the number of native species significantly reduce, a one-way ANOVA analysis was performed. Overall comparison between the number of native species in urban, suburban, rural, and wildland found a significant difference (F (3,8) = 38.146, $p < 0.001$). Further analysis using the Tukey HSD test indicated that the number of native species significantly reduced in the rural area (M = 7.333, SD = 1.528) in comparison to the wildland (M = 22.667, SD = 4.509) (Table 3) by 67.6%. No further decline was found since the numbers of native species between rural, suburban (M = 5.000, SD = 1.732), and urban (M = 2.000, SD = 1.000) areas were statistically similar (Table 3).

Table 3. Multiple comparisons on the number of native species in urban, suburban, rural, and wildland areas in Ipoh, Perak, Malaysia.

Urbanization Setting		Mean Difference (I-J)	Std. Error	Sig.
I	J			
Wildland	Rural	15.33333	2.10819	0.000 *
	Suburban	17.66667	2.10819	0.000 *
	Urban	20.66667	2.10819	0.000 *
Rural	Suburban	2.33333	2.10819	0.696
	Urban	5.33333	2.10819	0.129
Suburban	Urban	3.00000	2.10819	0.521

* The mean difference is significant at the 0.05 level.

On the other hand, no significant relationship between the number of non-native species and the built-up percentage was found (F (3,8) = 0.083, $p = 0.967$). The non-native species which were found in great numbers in urban areas of Ipoh were *Terminalia mantaly*, *Mangifera indica*, *Caesalpinia sappan*, *Azadirachta indica*, *Moringa oleifera*, *Tabebuia rosea*, *Samanea saman*, *Hura crepitans*, *Carica papaya*, and *Plumeria alba*. Meanwhile, *Morinda citrifolia*, *Syzygium aqueum*, *Adenanthera pavonina*, *Aidia densiflora*, and *Morinda citrifolia* were the native species that successfully survived in the urban environment.

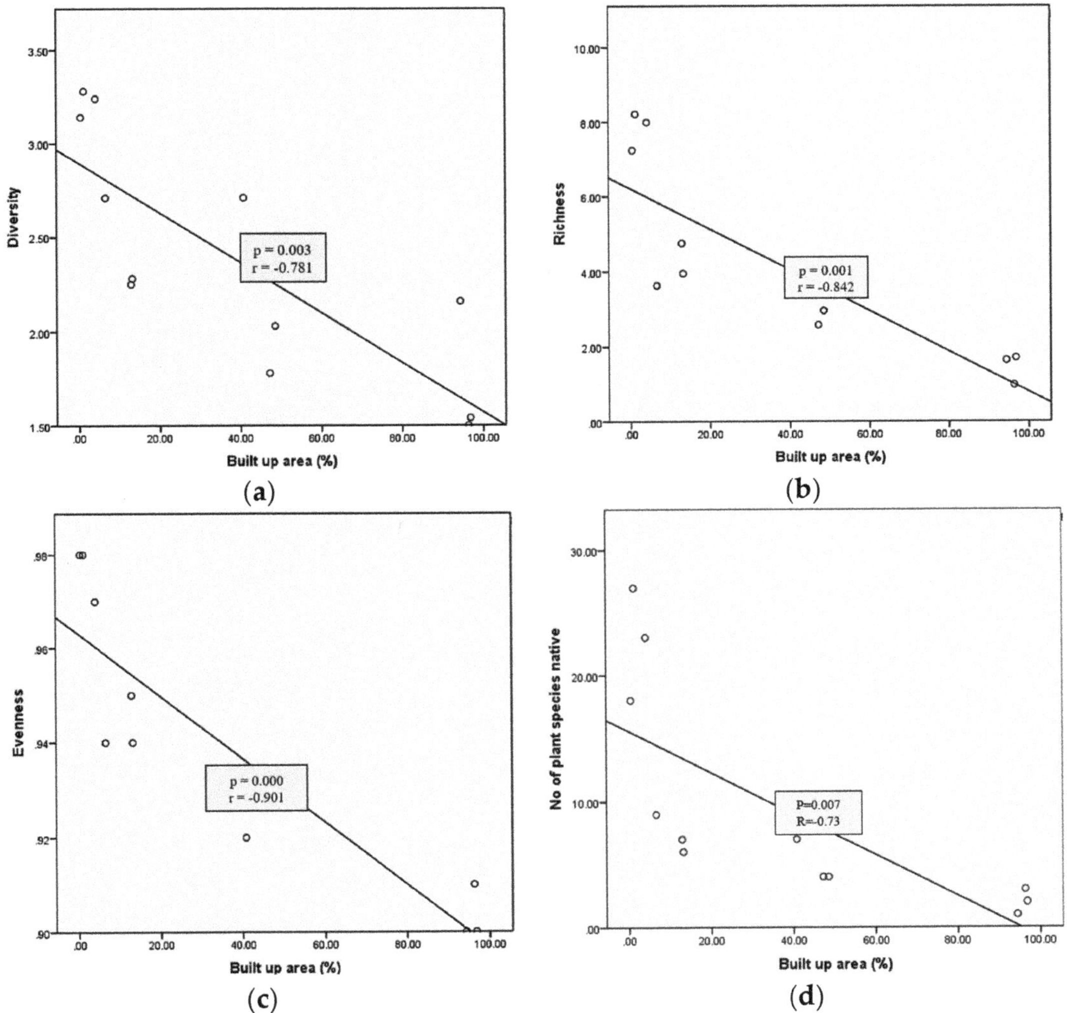

Figure 2. Relationships between the percentage of built-up area and (**a**) plant diversity; (**b**) plant richness; (**c**) plant evenness; and (**d**) plant native (species).

3.2. Relationship between Urbanization with Plant Traits and Environmental Conditions

Regarding plant trait, we found a significant relationship between the percentage of built-up area and plant height ($r = -0.656$, $p = 0.021$; Figure 3a) but not SLA ($r = 0.354$, $p = 0.260$; Figure 3b). All environmental conditions (temperature: $r = 0.809$, $p = 0.001$, Figure 3c; air humidity: $r = -0.859$, $p < 0.01$, Figure 3d; and light intensity: $r = 0.769$, $p = 0.003$, Figure 3e) were significantly correlated with the percentage of the built-up area.

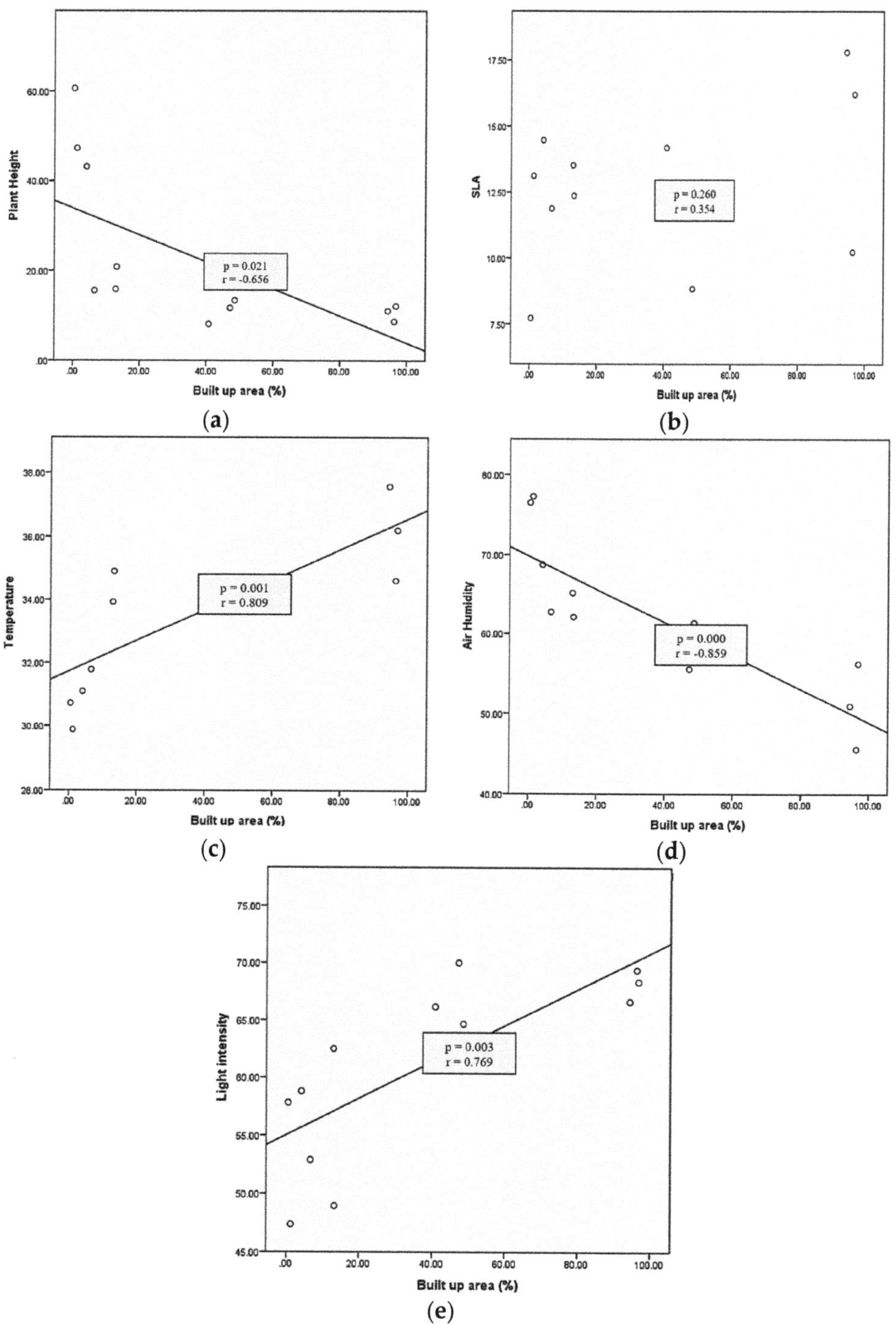

Figure 3. Relationship between percentage of built-up area and (**a**) plant height (m); (**b**) SLA; (**c**) temperature (°C); (**d**) air humidity (%); and (**e**) light intensity (Klux).

4. Discussion

4.1. Effect of Urbanization on Plant Diversity

The study revealed the pattern of plant diversity as a function of the urbanization gradient, and the findings showed that as urbanization increases, plant diversity decreases. Most notably, the findings of this study contradict the prior reports on diversity gradient in cities with rising plant species richness in urban areas and most of the studies are coming from temperate regions such as China and Germany as Wang [17], Čeplová [18], and McKinney [57]. The increase in plant diversity in urban areas could be attributed to various factors. For instance, the urban landscape is highly heterogeneous, providing diverse habitats for various plant species [58]. Urban areas typically include a variety of land use and land cover types. Hence, these land attributes and their spatial compositions are likely to promote plant diversity [59]. Following that, urban socioeconomic conditions, cultural diversity, and land use management are all essential elements determining the plant diversity pattern along the urbanization gradient [60]. Furthermore, the unintended introduction of many non-native species into urban regions has been identified as one of the most fundamental causes contributing to high plant variety in urban environments [61]. Hence, all these factors may work in tandem to establish plant diversity [62].

According to earlier research, cities appear to enhance plant diversity [19]. However, this is not the case in Ipoh, Perak. From 1850 to 1930, the tin industry expanded and resulted in the massive migration of Chinese workers to tin mining industrial towns in Perak. Thus, Ipoh, Perak was recognized as one of the greatest cities in the country from 1911 to 1931 and was primarily involved with tin mining. Tin production was favorably associated with urban population development; hence, as Malaysia develops, more people will relocate to and dwell in places like Ipoh city [63]. In accordance with the mission of becoming a developed country, these countries' industrial, transportation infrastructure, and major construction activities have resulted in massive land clearing. As a result of the high population, the percentage of built-up areas in urban areas increased; hence, Perak has experienced massive landscape changes and drastic changes in land use/cover due to anthropogenic activities [41]. Meanwhile, Hosni et al. [64] stated that during the last 20 years, a significant quantity of forested land had been lost by 183.12 hectares, while construction and development land has increased significantly by 157.12 hectares in Ipoh Perak. A common outcome of urbanization is the substantial loss of plants, leading to further negative impacts on the natural ecology [6].

Based on the regression analysis between plant diversity and the percentage of built-up area, it appears to be negatively related, like what was found by Vakhlamova et al. [25], Aronson et al. [61], and McKinney [65]. This demonstrates that as the percentage of built-up area increases, plant diversity decreases in Ipoh, Perak. Similarly, the number of native species was significantly lower in areas with higher urbanization intensity. This result is aligned with the earlier findings by Aronson et al. [16], Blouin [20], Ranta [22], and Vakhlamova [25] that also highlighted that native species were more prevalent in locations with less development, such as rural area or wildland. Previous studies have demonstrated that the intensively "constructed" landscape of urban cores has the poorest species diversity along the urbanization gradient [65]. This could be caused by a significant reduction inhabitable land for plants in highly populated places, high coverage of impermeable surfaces in urban areas, which diminish and fragment plant-able space, and the frequent stomping of vegetated areas [25]. Moreover, changes in landscape patterns due to population growth and urbanization significantly impact the dispersion of plant diversity [66]. Most crucially, urban fragments reduce the remaining space for plant species [24]; thus, plant species requiring much space have difficulty surviving [67]. Furthermore, habitat fragmentation results in smaller patches, lowering habitat quality and cutting down the number of plants in the patches, leading to loss of plant diversity [68].

Only 25 plant species are listed as non-native out of 96 plant species found in our study areas, but these species were distributed evenly along the urbanization gradient. Human introductions and plants' ability to utilize new resources in urban settings could probably be the reasons for the occurrence of non-native species. The non-native species that thrived in the rural, suburban, and urban areas of Ipoh were ornamental plants; hence, human introduction plays a vital role in the establishment of these plant species in urban areas. Those found in wildlands could also be the result of human activities because the three wildland areas were recreation forests. Thus, the unintentional introduction by humans such as through transportation may explain this finding.

Non-native species are considered an emerging risk of harm to biodiversity at spatial and temporal scales [69]. However, it is unlikely that the decline of native species with the increasing urbanization intensity was caused by the competition with the non-native species since the number of non-native species was similar along the urbanization gradient. Hence, this emphasizes the importance of built-up areas as the predictor of plant diversity along the urbanization gradient.

4.2. Changes in Plant Functional Traits along Urbanization Gradient

Urbanization could act as a filter for plant species by altering the physical environment since it may affect the plant structure and functional traits which will result in the alteration of ecosystem services [3]; thus, an analysis of plant traits would be beneficial to understand how plants respond towards urbanization. In different ecosystems, functional traits are increasingly considered good indicators of the effect of biodiversity on natural ecosystems and ecosystem services [70]. Furthermore, studying the relationship between plant traits at community levels is also pivotal for connecting the processes that determine composition and function [71].

Plant height was negatively correlated with urbanization intensity in this study, resulting in only shorter plants being discovered in urban areas, whereas taller plant species were more common in the wildland. Temperatures are expected to be higher in cities compared to non-urban areas as mentioned in past studies [72,73]. Referring to the previous studies by Lüttge & Buckeridge [74], in these low-altitude urban plains, top overgrowth is getting more apparent and height restrictions are in place. This is explained by water stress, which is more intense at the tops of trees due to the influence of gravity. Therefore, the solution to temperature-related water stress is a reduction in tree top heights from the ground. The tops of trees at different heights have similar leaf water relations. They are controlled by tree top height, under the stress of drought brought on by temperature.

Other than that, as temperature increases in urban areas, plant growth and development will be adversely affected, resulting in smaller and shorter plants, slower reproductive production, lower yield capability [75], and short-lived species [4]. However, several studies, such as those by Song et al. [3], Cochard et al. [49], and Williams et al. [76] are not parallel with this finding. According to earlier studies, urbanization favors taller species in response to urban pressure, which have a substantial competitive advantage, whereas short species are more prone to extinction in urban environments [76]. The variations in these results could be caused by other factors such as local landscape management. For instance, plants in urban areas are normally pruned for ecstatic and safety purposes, therefore tall trees as less common in this environment. In this study, we found that urbanization was significantly positively associated with air temperature. Gong and Gao [77] showed that temperature significantly influenced SLA, therefore we would predict that SLA would relate to urbanization intensity. However, this prediction was not supported by our result as SLA was not statistically associated with the percentage of built-up area. Even though this study found a rise in temperature with the increasing urbanization intensity, it is possible that the increase is rather moderate and not significant enough to have an impact on the SLA. Moreover, analysis of SLA from various climatic regions indicated a significant impact of temperature on SLA occurred in higher latitudes but less in the tropics [77]. Hence, the use of SLA in studies related to urbanization and plant traits in the topics should be carefully considered.

This study also found that SLA was not statistically associated with the percentage of built-up area. In opposition to our finding, a past study by Song et al. [3] from China stated that SLA was statistically significant with urbanization gradient and that they claimed urban areas had increased plant diversity, since the study area contain high nitrogen which could be attributable to an increase in nitrogen affinity as urbanization increases [78]. Furthermore, earlier research has found significant links between SLA and nutrient accessibility such as nitrogen and phosphorus [79]. However, a study by Cochard et al. [49] shows the same result as the present study and this was due to the uneven distribution of soil nutrients such as nitrogen and phosphorus along the urbanization gradient. In this study, Ipoh city appeared to be a limestone area [43]; thus, soils in Ipoh that originated from limestone parent material are primarily rich in calcium and high in pH [80]. Additionally, calcium and pH are two elements that significantly impact the vegetation pattern in the limestone region [81]. Our findings suggest that the concentration of calcium and pH are probably similar in all our study plots or that calcium and pH are not necessarily important factors that influence SLA.

However, the scarcity of apparent plant trait trends in Malaysia should also be emphasized, and more in-depth investigations are essential to grasp better the relevance of these traits in urbanization gradient contexts.

4.3. Environmental Impacts on Filtering Diversity Species

Environmental factors (air temperature, air humidity, and light intensity) were significantly associated with the percentage of built-up area in this study. In Ipoh, Perak, the most built-up area had the highest temperature, the least moisture, and the greatest light intensity. Conversely, the temperature in the wildland area was the lowest, with the highest air humidity and the lowest light intensity.

Findings from previous studies also showed that temperature increases with increasing urbanization intensity. For instance, Morris et al. [72] found that the temperature increased towards the urban center of Klang Valley, Malaysia, where the highest temperature was recorded in the commercial area. Similarly, a study by Saha et al. [73] which focused on one tropical and two subtropical urban agglomerations in India also discovered that areas with more built-up coverage had higher temperatures. Moreover, the projection of the urban climate in the tropical city of Ho Chi Minh City, Vietnam in the 2050s revealed that temperature will increase in both urban and rural areas, but the more urbanized areas will experience an additional 0.5 °C increment than the rural areas [82]. Such changes in environmental conditions can occur when vegetated areas are converted into impervious surfaces, particularly in urbanized areas [83,84].

Climate change has an impact on plants' growth [85]; hence, plants must deal with a variety of stressful situations during their lives. Due to urbanization, only a few plant species can survive in urban environments, which are known as alcalinophilous (light-loving plants) [86], thermophilous (warm-loving plants) [76], and drought-tolerant species [4]. Warm-loving, drought-tolerant, and light-loving plants are only found in big cities or otherwise; they will become extinct due to their inability to cope with harsh environmental conditions [75].

The results of this study showed that when urbanization intensity increased, air temperature also increased, while plant diversity reduced. Studies that investigated the direct effects of urbanization-related temperature on plant species diversity are scarce but a study by Lososová et al. [87] found a similar result that showed a reduction in the species diversity of native and non-native plant species due to the warmer climate of cities. The contradicting results of the non-native species between this study and Lososová et al. [87] are likely to be caused by the considerable influence of humans in introducing non-native species in Ipoh. Regarding the native species, a similar outcome to this study was obtained although Lososová et al. [87] focused on plant diversity in European cities. This suggests that climatic region is a less important factor in determining the effects of urban-related climate conditions on plant diversity. The increased temperature in either tropical or temperate cities induces stressful conditions for plants which eventually limits their survival.

On the contrary, a study by Zhang et al. [88] which involved 672 worldwide cities, including 99 from tropical countries found that urbanization in the tropical region barely affects plant growth due to temperature. The difference in this result compared to ours could be because the climatic data of the tropical region in Zhang et al. [88] were taken from three countries which were Malaysia, Brazil, and India while this study focused on local climate in the study plot. In addition to that, the vegetation data in this study were collected in the field but those in Zhang et al. [88] were the satellite-sensed data where there are possibilities that some of the satellite images may be of low confidence due to conditions such as snow and cloud. Therefore, the specificity of data (regional vs. local) should be taken into account when analyzing results from different studies. In this study, one of the plant species found in wildland, rural and urban areas is *Morinda citrofolia* which was discovered to be capable of surviving extreme environmental circumstances and prolonged periods of drought conditions [89]. Moreover, *Morinda citrofolia* can thrive in both acidic and alkaline soils, conditions with climates that range from extremely dry to excessively moist and shaded conditions (>80% shade) [90]. *Morinda citrofolia* also has a deep and robust taproot and extensive root system, making it exceedingly difficult to eliminate once established [91]. This could explain how *Morinda citrofolia* can adapt to urban, rural, and wildland areas.

Referring to Table 1, *Mangifera indica* is the only species discovered in all levels of urbanization in this study. A human has deliberately moved mango sp. for ages. At the same time, *Mango* sp. is a drought-tolerant species that can resist the seasonal dry season for up to 8 months. Its deep tap with sinker roots and long-lived and rigid leaves with a thick cuticle nutrient uptake are all drought-tolerant properties of *Mango* sp. [92]. *Mangifera indica* plants can withstand a wide range of climatic conditions as they may continue living in swampy areas as well as hot and humid climates [93]. Bally [92] also mentioned that mango trees are found well in regions with a well-defined, generally cold dry season with a high-temperature accumulation of full sunlight during the flowering and fruit development phase.

According to Table 1, *Caesalpinia sappan* was only be found in the urban areas of Ipoh; however, based on Hung et al. [94], *Caesalpinia sappan* is a species that requires high intensities of light and known as drought tolerant. Therefore, more drought tolerance species prefer with warmer and dry conditions [95] especially in urban areas. Many plant species that cannot survive or adapt to harsh environments face greater extinction risk since urbanization act as a filter of plant species. This could be mainly attributable to excessive light with warmer temperatures affecting plant productivity, growth, and development [75]. However, according to the Royal Botanic Garden, *Caesalpinia sappan* is a non-native species to peninsular Malaysia, and most non-native species are environmentally sustainable and resistant to drought [25].

Plant species that could not sustain in urban areas can be found in rural and wildland areas since rural and wildland have lower temperatures, with higher air humidity and lower light intensity compared to urban areas. For instance, in this study, referring to Table 1, *Shorea* sp. was found in rural and wildland areas. Since *Shorea* sp. grows at different rates; seedlings are sensitive to light intensity and must be grown in the shade for a while before being exposed to sunlight which makes it grow faster [96] and this would make sense why *Shorea* sp. could not survive in urban areas and this type of forest species is presumed to be adapted to shady conditions [4].

Based on the case of *Morinda citrofolia, Mangifera indica, Caesalpinia sappan,* and *Shorea* sp., we implied that urbanization favors plants with a wide range of environmental tolerance where it allows the plants to withstand the warmer, drier, and sunnier conditions of the urban environment. Such characteristic is common in non-native species [21] and it may explain their ability to survive in cities after being introduced by humans.

5. Conclusions

Plant diversity is essential to ecosystem consistency and functionality. Acknowledging the plant diversity pattern across urbanization gradient and the correlation between the built-up area with plant diversity, plant nativity, plant traits, and environmental factors are very important for urban planners towards new green infrastructure and how to preserve plant diversity in urban areas. Our study found that rapid urbanization and the high number of built-up areas in Ipoh, Perak reduced plant diversity and filters a few species of plants by adjusting physical surroundings, resulting in species decreasing along the urbanization gradient. The noticeable drastic changes in the environment caused by urbanization imply a tendency to produce different types of patterns for plant diversity as urbanization increases. In addition, non-native plants can be found at all levels of urbanization intensity, thus highlighting the role of human in dispersing species.

As the temperature rises, the intensity of the light and humidity in the environment decreases, creating stressful conditions for plant species. Only some plant species may survive in harsh conditions. If this persists, plant diversity may continue to decrease in a few years due to high urbanization without proper town planning or sustainable urban management. We chose only a few plants functional traits and environmental factors to investigate in the research, even though plant species may have other traits and environmental factors that can affect ecosystem functions. Soil factors, for example, play a significant role in plant distribution and diversity. More research on the mechanisms of how environment affects plant traits is required to understand the relationship between plant functional diversity and ecosystem processes while preserving urban development in Malaysia.

Supplementary Materials: The following supporting information can be downloaded at: https://www.mdpi.com/article/10.3390/d14121024/s1, Table S1: Family, species name, nativity status, and native range of plants in Ipoh, Perak, Malaysia.

Author Contributions: Conceptualization, N.S.H., Z.M.S. and F.M.; methodology, N.S.H. and Z.M.S.; validation, F.M. and Z.M.S.; formal analysis, B.A.A., N.S.H. and K.I.; investigation, B.A.A.; resources, N.S.H. and K.I.; data curation, F.M. and Z.M.S.; writing—original draft preparation, B.A.A., N.S.H., F.M., Z.M.S. and K.I.; writing—review and editing, B.A.A. and N.S.H.; visualization, B.A.A.; supervision, N.S.H. and F.M.; project administration, N.S.H.; funding acquisition, N.S.H. All authors have read and agreed to the published version of the manuscript.

Funding: This research was funded by the Ministry of Higher Education Malaysia, grant number FRGS/1/2018/WAB13/UPSI/03/1.

Institutional Review Board Statement: Not applicable.

Data Availability Statement: The data presented in this study are openly available in Figshare at https://doi.org/10.6084/m9.figshare.21609456.v1.

Acknowledgments: We would like to acknowledge PLANMalaysia for providing the land use map of Ipoh City Council. We also thank the Department of Biology, Faculty of Science and Mathematics, and Geography and Environment Department, Faculty of Human Science, Universiti Pendidikan Sultan Idris for providing adequate research facilities.

Conflicts of Interest: The authors declare no conflict of interest.

References

1. Ali, R.; Bakhsh, K.; Yasin, M.A. Impact of urbanization on CO2 emissions in emerging economy: Evidence from Pakistan. *Sustain. Cities Soc.* **2019**, *48*, 101553. [CrossRef]
2. Zhao, S.; Shi, X.M.; Niu, Y.F. Evaluation of the overall level of new urbanization in Hebei province based on factor analysis and GIS. *Asian Agric. Res.* **2014**, *6*, 44–49.
3. Song, G.; Wang, J.; Han, T.; Wang, Q.; Ren, H.; Zhu, H.; Hui, D. Changes in plant functional traits and their relationships with environmental factors along an urban-rural gradient in Guangzhou, China. *Ecol. Indic.* **2019**, *106*, 105558. [CrossRef]
4. Petersen, T.K.; Speed, J.D.M.; Grøtan, V.; Austrheim, G. Competitors and ruderals go to town: Plant community composition and function along an urbanisation gradient. *Nord. J. Bot.* **2021**, *39*, njb.03026. [CrossRef]

5. Trisos, C.H.; Merow, C.; Pigot, A.L. The projected timing of abrupt ecological disruption from climate change. *Nature* **2020**, *580*, 496–501. [CrossRef] [PubMed]
6. Yang, J.; Li, S.; Lu, H. Quantitative Influence of Land-Use Changes and Urban Expansion Intensity on Landscape Pattern in Qingdao, China: Implications for Urban Sustainability. *Sustainability* **2019**, *11*, 6174. [CrossRef]
7. Kadeba, A.; Nacoulma, B.M.I.; Ouedraogo, A.; Bachmann, Y.; Thiombiano, A.; Schmidt, M.; Boussim, J.I. Land cover change and plants diversity in the Sahel: A case study from northern Burkina Faso. *Ann. For. Res.* **2015**, *58*, 109. [CrossRef]
8. Cheng, F.; Liu, S.; Hou, X.; Wu, X.; Dong, S.; Coxixo, A. The effects of urbanization on ecosystem services for biodiversity conservation in southernmost Yunnan Province, Southwest China. *J. Geogr. Sci.* **2019**, *29*, 1159–1178. [CrossRef]
9. Zisadza-Gandiwa, P.; Mango, L.; Gandiwa, E.; Goza, D.; Parakasingwa, C.; Chinoitezvi, E.; Muvengwi, J. Variation in woody vegetation structure and composition in a semi-arid savanna of Southern Zimbabwe. *Int. J. Biodivers. Conserv.* **2013**, *5*, 71–77.
10. Ikyaagba, T.E.; Tee, T.N.; Dagba, B.I.; Ancha, U.P.; Ngibo, K.D.; Tume, C. Tree composition and distribution in Federal University of Agriculturema Kurdi, Nigeria. *J. Res. For. Wildl. Environ.* **2015**, *7*, 147–157.
11. Bowler, D.E.; Buyung-Ali, L.; Knight, T.M.; Pullin, A.S. Urban greening to cool towns and cities: A systematic review of the empirical evidence. *Landsc. Urban Plan.* **2010**, *97*, 147–155. [CrossRef]
12. Abhijith, K.V.; Kumar, P.; Gallagher, J.; McNabola, A.; Baldauf, R.; Pilla, F. Air pollution abatement performances of green infrastructure in open road and built-Up street canyon environments—A review. *Atmos. Environ.* **2017**, *162*, 71–86. [CrossRef]
13. Wolf, K.L.; Robbins, A.S.T. Metro nature, environmental health, and economic value. *Environ. Health Perspect.* **2015**, *123*, 390. [CrossRef] [PubMed]
14. Guitart, D.; Pickering, C.; Bryne, J. Past results and future directions in urban community gardens research. *Urban For. Urban Green.* **2012**, *11*, 364–373. [CrossRef]
15. Botzat, A.; Fisher, L.K.; Kowarik, I. Unexploited opportunities in understanding live able and biodiverse cities. A review on urban biodiversity perception and valuation. *Glob. Environ. Chang.* **2016**, *39*, 220–233. [CrossRef]
16. Aronson, M.F.J.; Handel, S.N.; La Puma, I.P.; Clemants, S.E. Urbanization promotes non-native woody species and diverse plant assemblages in the New York metropolitan region. *Urban Ecosyst.* **2015**, *18*, 31–45. [CrossRef]
17. Wang, M.; Li, J.; Kuang, S.; He, Y.; Chen, G.; Huang, Y.; Łowicki, D. Plant Diversity along the Urban–Rural Gradient and Its Relationship with Urbanization Degree in Shanghai, China. *Forests* **2020**, *11*, 171. [CrossRef]
18. Čeplová, N.; Kalusová, V.; Lososová, Z. Effects of settlement size, urban heat island and habitat type on urban plant biodiversity. *Landsc. Urban Plan.* **2017**, *159*, 15–22. [CrossRef]
19. Jha, R.K.; Nölke, N.; Diwakara, B.N.; Tewari, V.P.; Kleinn, C. Differences in tree species diversity along the rural-urban gradient in Bengaluru, India. *Urban For. Urban Green.* **2019**, *46*, 126464. [CrossRef]
20. Blouin, D.; Pellerin, S.; Poulin, M. Increase in non-native species richness leads to biotic homogenization in vacant lots of a highly urbanized landscape. *Urban Ecosyst.* **2019**, *22*, 879–892. [CrossRef]
21. Godefroid, S.; Koedam, N. Distribution pattern of the flora in a peri-urban forest: An effect of the city–forest ecotone. *Landsc. Urban Plan.* **2003**, *65*, 169–185. [CrossRef]
22. Ranta, P.; Viljanen, V. Vascular plants along an urban-rural gradient in the city of Tampere, Finland. *Urban Ecosyst.* **2011**, *14*, 361–376. [CrossRef]
23. Zerbe, S. Biodiversity in Berlin and its potential for nature conservation. *Landsc. Urban Plan.* **2003**, *62*, 139–148. [CrossRef]
24. Yan, Z.; Teng, M.; He, W.; Liu, A.; Li, Y.; Wang, P. Impervious surface area is a key predictor for urban plant diversity in a city undergone rapid urbanization. *Sci. Total Environ.* **2019**, *650*, 335–342. [CrossRef] [PubMed]
25. Vakhlamova, T.; Rusterholz, H.-P.; Kanibolotskaya, Y.; Baur, B. Changes in plant diversity along an urban–rural gradient in an expanding city in Kazakhstan, Western Siberia. *Landsc. Urban Plan.* **2014**, *132*, 111–120. [CrossRef]
26. Cameron, G.N.; Culley, T.M.; Kolbe, S.E.; Miller, A.I.; Matter, S.F. Effects of urbanization on herbaceous forest vegetation: The relative impacts of soil, geography, forest composition, human access, and an invasive shrub. *Urban Ecosyst.* **2015**, *18*, 1051–1069. [CrossRef]
27. Albrecht, H.; Haider, S. Species diversity and life history traits in calcareous grasslands vary along an urbanization gradient. *Biodivers. Conserv.* **2013**, *22*, 2243–2267. [CrossRef]
28. Melliger, R.L.; Braschler, B.; Rusterholz, H.-P.; Baur, B. Diverse effects of degree of urbanisation and forest size on species richness and functional diversity of plants, and ground surface-active ants and spiders. *PLoS ONE* **2018**, *13*, e0199245. [CrossRef] [PubMed]
29. Tian, Z.; Song, K.; Da, L. Distribution patterns and traits of weed communities along an urban-rural gradient under rapid urbanization in Shanghai, China: Weed communities on urban-rural gradient. *Weed Biol. Manag.* **2015**, *15*, 27–41. [CrossRef]
30. Yang, J.; Li, X.; Li, S.; Liang, H.; Lu, H. The woody plant diversity and landscape pattern of fine-resolution urban forest along a distance gradient from points of interest in Qingdao. *Ecol. Indic.* **2021**, *122*, 107326. [CrossRef]
31. Rahmad, Z.B.; Akomolafe, G.F. Distribution, Diversity and Abundance of Ferns in a Tropical University Campus. *Pertanika J. Trop. Agric. Sci.* **2018**, *41*, 1875–1887.
32. Suratman, M.N. Tree Species Diversity and Forest Stand Structure of Pahang National Park, Malaysia. *Biodivers. Enrich. A Divers. World* **2012**, *19*, 45–56.
33. Onrizal, O.; Ismail, S.N.; Mansor, M. Diversity of plant community at Gunung Ledang, Malaysia. In Proceedings of the International Conference on Agriculture, Environment, and Food Security 2018, Medan, Indonesia, 24–25 October 2018. IOP Conference Series: Earth and Environmental Science, 2019.

34. Ghollasimood, S.; Hanum, I.F.; Nazre, M.; Kamziah, A.K.; Awang Noor, A.G. Vascular Plant Composition and Diversity of a Coastal Hill Forest in Perak, Malaysia. *J. Agric. Sci.* **2011**, *3*, 111. [CrossRef]
35. Nabilla, F.; Devi, K.K.; Siong, H.C. Analysis of fragmented green spaces in kuala lumpur, Malaysia. *Chem. Eng. Trans.* **2019**, *72*, 457–462.
36. Kanniah, K.D. Quantifying green cover change for sustainable urban planning: A case of Kuala Lumpur, Malaysia. *Urban For. Urban Green.* **2017**, *27*, 287–304. [CrossRef]
37. Kanniah, K.D.; Ho, C.S. Urban Forest cover change and sustainability of malaysian cities. *Chem. Eng. Trans.* **2017**, *56*, 673–678.
38. Kleyer, M. Enhancing landscape planning: Vegetation-mediated ecosystem services predicted by plant traits. *Landsc. Urban Plan.* **2021**, *215*, 104220. [CrossRef]
39. Palma, E.; Catford, J.A.; Corlett, R.T.; Duncan, R.P.; Hahs, A.K.; McCarthy, M.A.; McDonnell, M.J.; Thompson, K.; Williams, N.S.G.; Vesk, P.A. Functional trait changes in the floras of 11 cities across the globe in response to urbanization. *Ecography* **2017**, *40*, 875–886. [CrossRef]
40. Funk, J.L.; Larson, J.E.; Ames, G.M.; Butterfield, B.J.; Cavender-Bares, J.; Firn, J.; Wright, J. Revisiting the Holy Grail: Using plant functional traits to understand ecological processes: Plant functional traits. *Biol. Rev.* **2017**, *92*, 1156–1173. [CrossRef]
41. van Grunsven, L.; Benson, M. Urban Development in Malaysia: Towards A New Systems Paradigm. *Urban Policy Ser.* **2020**. Available online: www.thinkcity.com.my/wuf10 (accessed on 20 November 2022).
42. Wan Mohd Jaafar, W.S.; Abdul Maulud, K.N.; Muhmad Kamarulzaman, A.M.; Raihan, A.; Md Sah, S.; Ahmad, A.; Razzaq Khan, W. The Influence of Deforestation on Land Surface Temperature—A Case Study of Perak and Kedah, Malaysia. *Forests* **2020**, *11*, 670. [CrossRef]
43. Wong, K.M. Plant endemism patterns in Borneo and Malay Peninsula and Patterns of Plant Endemism and Rarity in Borneo and the Malay Peninsula. *For. Res. Cent.* **1998**, 139–169.
44. Ashton, P.S. Plant conservation in the Malaysian region. *Malay. Nat. Soc.* **1992**, 86–93.
45. Razak, H.A.; Wahid, N.B.A.; Latif, M.T. Anionic Surfactants and Traffic Related Emission from an Urban Area of Perak, Malaysia. *Arch. Environ. Contam. Toxicol.* **2019**, *77*, 587–593. [CrossRef]
46. Marzluff, J.M.; Bowman, R. *Avian Ecology and Conservation in an Urbanizing World*; Donnelly, R., Ed.; Kluwer Academic: Boston, MA, USA; Springer: Berlin/Heidelberg, Germany, 2001; p. 581.
47. Yang, T.-R.; Lam, T.Y.; Su, S.-H. A simulation study on the effects of plot size and shape on sampling plant species composition for biodiversity management. *J. Sustain. For.* **2019**, *38*, 116–129. [CrossRef]
48. Pérez-Harguindeguy, N.; Díaz, S.; Garnier, E.; Lavorel, S.; Poorter, H.; Jaureguiberry, P.; Cornelissen, J.H.C. Corrigendum to: New handbook for standardised measurement of plant functional traits worldwide. *Aust. J. Bot.* **2016**, *64*, 715. [CrossRef]
49. Cochard, A.; Pithon, J.; Braud, F.; Beaujouan, V.; Bulot, A.; Daniel, H. Intraspecific trait variation in grassland plant communities along urban-rural gradients. *Urban Ecosyst.* **2019**, *22*, 583–591. [CrossRef]
50. Li, H.; Meng, H.; He, R.; Lei, Y.; Guo, Y.; Ernest, A.; Tian, G. Analysis of Cooling and Humidification Effects of Different Coverage Types in Small Green Spaces (SGS) in the Context of Urban Homogenization: A Case of HAU Campus Green Spaces in Summer in Zhengzhou, China. *Atmosphere* **2020**, *11*, 862. [CrossRef]
51. Gatto, E.; Buccolieri, R.; Aarrevaara, E.; Ippolito, F.; Emmanuel, R.; Perronace, L.; Santiago, J.L. Impact of Urban Vegetation on Outdoor Thermal Comfort: Comparison between a Mediterranean City (Lecce, Italy) and a Northern European City (Lahti, Finland). *Forests* **2020**, *11*, 228. [CrossRef]
52. Huang, L.; Chen, H.; Ren, H.; Wang, J.; Guo, Q. Effect of urbanization on the structure and functional traits of remnant subtropical evergreen broad-leaved forests in South China. *Environ. Monit. Assess.* **2013**, *185*, 5003–5018. [CrossRef]
53. Magurran, A.E. *Ecological Diversity and Its Measurement*; Princeton University Press: Princeton, NJ, USA, 1988.
54. Shannon, C.E.; Weiner, V. A Mathematical Theory of Communication University Press. *Ill. Urban* **1949**, *5*, 101–107.
55. Atsbeha, G.; Demissew, S.; Woldu, Z.; Edwards, S. Floristic composition of herbaceous flowering plant species in Lalay and Tahtay Michew districts, central zone of Tigray, Ethiopia. *Afr. J. Ecol. Ecosyst.* **2015**, *2*, 159–169.
56. Turner, I.M. A catalogue of the vascular plants of Malaya. *Gard. Bull. (Singap.)* **1995**, *47*, 346.
57. McKinney, M.L. Urbanization as a major cause of biotic homogenization. *Biol. Conserv.* **2006**, *127*, 247–260. [CrossRef]
58. Fahrig, L.; Baudry, J.; Brotons, L.; Burel, F.G.; Crist, T.O.; Fuller, R.J.; Martin, J.-L. Functional landscape heterogeneity and animal biodiversity in agricultural landscapes: Heterogeneity and biodiversity. *Ecol. Lett.* **2011**, *14*, 101–112. [CrossRef] [PubMed]
59. Jentsch, A.; Steinbauer, M.J.; Alt, M.; Retzer, V.; Buhk, C.; Beierkuhnlein, C. A systematic approach to relate plant-species diversity to land use diversity across landscapes. *Landsc. Urban Plan.* **2012**, *107*, 236–244. [CrossRef]
60. Huang, Y.; Chen, Y.; He, Y.-J.; Wang, M.; Kuang, S.-J.; Chen, G.-J.; Li, J.-X. Effects of socioeconomic factors on urban plant diversity of residential areas in Shanghai, China. *J. Appl. Ecol.* **2019**, *30*, 3403–3410.
61. Aronson, M.F.J.; La Sorte, F.A.; Nilon, C.H.; Katti, M.; Goddard, M.A.; Lepczyk, C.A.; Winter, M. A global analysis of the impacts of urbanization on bird and plant diversity reveals key anthropogenic drivers. *Proc. R. Soc. B Biol. Sci.* **2014**, *281*, 20133330. [CrossRef] [PubMed]
62. Uchiyama, Y.; Kohsaka, R. Spatio-temporal Analysis of Biodiversity, Land-use Mix and Human Population in a Socio-ecological Production Landscape: A Case Study in the Hokuriku Region, Japan. *Procedia Eng.* **2017**, *198*, 219–226. [CrossRef]
63. Lee, C.; Jitaree, W. Structural Change and Urbanization: The Case of Peninsular Malaysia. *Thail. World Econ.* **2019**, *37*, 26–38.

64. Hosni, A.T.; Teriman, S.; Adnan, N.A.; Kamar, M.A.A. Time Series Analysis of Land Cover Using Remote Sensing Technology. *J. Comput. Theor. Nanosci.* **2020**, *17*, 902–910. [CrossRef]
65. McKinney, M.L. Urbanization, Biodiversity, and Conservation. *BioScience* **2002**, *52*, 883. [CrossRef]
66. Peng, Y.; Liu, X. Research progress in effects of urbanization on plant biodiversity. *Biodivers. Sci.* **2007**, *15*, 558.
67. Yuan, Z.; Xu, J.; Wang, Y.; Yan, B. Analyzing the influence of land use/land cover change on landscape pattern and ecosystem services in the Poyang Lake Region, China. *Environ. Sci. Pollut. Res.* **2021**, *28*, 27193–27206. [CrossRef] [PubMed]
68. Li, X.; Jia, B.; Zhang, W.; Ma, J.; Liu, X. Woody plant diversity spatial patterns and the effects of urbanization in Beijing, China. *Urban For. Urban Green.* **2020**, *56*, 126873. [CrossRef]
69. Butchart, S.H.M.; Walpole, M.; Collen, B.; van Strien, A.; Scharlemann, J.P.W.; Almond, R.E.A.; Watson, R. Global Biodiversity: Indicators of Recent Declines. *Science* **2010**, *328*, 1164–1168. [CrossRef] [PubMed]
70. Rosenfield, M.F.; Müller, S.C. Plant Traits Rather than Species Richness Explain Ecological Processes in Subtropical Forests. *Ecosystems* **2020**, *23*, 52–66. [CrossRef]
71. Zirbel, C.R.; Bassett, T.; Grman, E.; Brudvig, L.A. Plant functional traits and environmental conditions shape community assembly and ecosystem functioning during restoration. *J. Appl. Ecol.* **2017**, *54*, 1070–1079. [CrossRef]
72. Morris, K.I.; Chan, A.; Morris, K.J.K.; Ooi, M.C.; Oozeer, M.Y.; Abakr, Y.A.; Nadzir, M.S.M. and Mohammed, I.Y. Urbanisation and urban climate of a tropical conurbation, Klang Valley, Malaysia. *Urban Clim.* **2017**, *19*, 54–71. [CrossRef]
73. Saha, S.; Saha, A.; Das, M.; Saha, A.; Sarkar, R.; Das, A. Analyzing spatial relationship between land use/land cover (LULC) and land surface temperature (LST) of three urban agglomerations (UAs) of Eastern India. *Remote Sens. Appl. Soc. Environ.* **2021**, *22*, 100507. [CrossRef]
74. Lüttge, U.; Buckeridge, M. Trees: Structure and function and the challenges of urbanization. *Trees* **2020**, 1–8. [CrossRef]
75. Hatfield, J.L.; Prueger, J.H. Temperature extremes: Effect on plant growth and development. *Weather Clim. Extrem.* **2015**, *10*, 4–10. [CrossRef]
76. Williams, N.S.G.; Hahs, A.K.; Vesk, P.A. Urbanisation, plant traits and the composition of urban floras. *Perspect. Plant Ecol. Evol. Syst.* **2015**, *17*, 78–86. [CrossRef]
77. Gong, H.; Gao, J. Soil and climatic drivers of plant SLA (specific leaf area). *Glob. Ecol. Conserv.* **2019**, *20*, e00696. [CrossRef]
78. Vallet, J.; Daniel, H.; Beaujouan, V.; Rozé, F.; Pavoine, S. Using biological traits to assess how urbanization filters plant species of small woodlands: Biological traits to assess how urbanization filters plant species. *Appl. Veg. Sci.* **2010**, *13*, 412–424. [CrossRef]
79. Dwyer, J.M.; Hobbs, R.J.; Mayfield, M.M. Specific leaf area responses to environmental gradients through space and time. *Ecology* **2014**, *95*, 399–410. [CrossRef]
80. Gauld, J.H.; Robertson, J.S. Soils and Their Related Plant Communities on the Dalradian Limestone of Some Sites in Central Perthshire, Scotland. *J. Ecol.* **1985**, *73*, 91. [CrossRef]
81. Zakaria, R.; Nizam, M.S.; Faridah, H. Association of Tree Communities with Soil Properties in a Semi Deciduous Forest of Perlis, Peninsular Malaysia. *BIOTROPIA-Southeast Asian J. Trop. Biol.* **2020**, *27*, 69–79.
82. Doan, V.Q.; Kusaka, H. Projections of urban climate in the 2050s in a fast-growing city in Southeast Asia: The greater Ho Chi Minh City metropolitan area, Vietnam. *Int. J. Climatol.* **2018**, *38*, 4155–4171. [CrossRef]
83. Kubota, T.; Lee, H.S.; Trihamdani, A.R.; Phuong, T.T.T.; Tanaka, T.; Matsuo, K. Impacts of land use changes from the Hanoi Master Plan 2030 on urban heat islands: Part 1. Cooling effects of proposed green strategies. *Sustain. Cities Soc.* **2017**, *32*, 295–317. [CrossRef]
84. Grimm, N.B.; Faeth, S.H.; Golubiewski, N.E.; Redman, C.L.; Wu, J.; Bai, X.; Briggs, J.M. Global change and the ecology of cities. *Science* **2008**, *319*, 756–760. [CrossRef]
85. Ciceu, A.; Popa, I.; Leca, S.; Pitar, D.; Chivulescu, S.; Badea, O. Climate change effects on tree growth from Romanian forest monitoring Level II plots. *Sci. Total Environ.* **2020**, *698*, 134129. [CrossRef] [PubMed]
86. Schwoertzig, E.; Poulin, N.; Hardion, L.; Trémolières, M. Plant ecological traits highlight the effects of landscape on riparian plant communities along an urban–rural gradient. *Ecol. Indic.* **2016**, *61*, 568–576. [CrossRef]
87. Lososová, Z.; Tichý, L.; Divíšek, J.; Čeplová, N.; Danihelka, J.; Dřevojan, P.; Fajmon, K.; Kalníková, V.; Kalusová, V.; Novák, P.; et al. Projecting potential future shifts in species composition of European urban plant communities. *Divers. Distrib.* **2018**, *24*, 765–775. [CrossRef]
88. Zhang, L.; Yang, L.; Zohner, C.M.; Crowther, T.W.; Li, M.; Shen, F.; Guo, M.; Qin, J.; Yao, L.; Zhou, C. Direct and indirect impacts of urbanization on vegetation growth across the world's cities. *Sci. Adv.* **2022**, *8*, eabo0095. [CrossRef]
89. Nelson, S.C. Morinda citrifolia *(noni), Ver 4, Species Profiles for Pacifc Island Agroforestry*; Permanent Agricultural Resources: Holualoa, HI, USA, 2006; pp. 1–13.
90. Francis, J.K. *Wildland Shrubs of the United States and Its Territories: Thamnic Descriptions, Volume 1*; General Technical Report IITF-GTR-26; US Department of Agriculture, Forest Service, International Institute of Tropical Forestry, US Department of Agriculture, Forest Service, Rocky Mountain Research Station: Fort Collins, CO, USA, 2004; 830p.
91. Rojas-Sandoval, J. *Morinda citrifolia (Indian mulberry)*; Invasive Species Compendium: Wallingford, UK, 2020.
92. Bally, I.S.E. *Mangifera indica (mango)*; Species Profiles for Pacific Psland Agroforestry: Mareeba, Australia, 2006; pp. 1–25.
93. Derese, S.; Guantai, E.M.; Souaibou, Y.; Kuete, V. Mangifera indica L. (Anacardiaceae). In *Medicinal Spices and Vegetables from Africa*; Academic Press: Cambridge, MA, USA, 2017; pp. 451–483.

94. Hung, T.; Dang, N.; Dat, N. Methanol extract from Vietnamese *Caesalpinia sappan* induces apoptosis in HeLa cells. *Biol. Res.* **2014**, *47*, 20. [CrossRef] [PubMed]
95. Walentowski, H.; Falk, W.; Mette, T.; Kunz, J.; Bräuning, A.; Meinardus, C.; Zang, C.; Sutcliffe, L.M.E.; Leuschner, C. Assessing future suitability of tree species under climate change by multiple methods: A case study in southern Germany. *Ann. For. Res.* **2017**, *60*, 101–126. [CrossRef]
96. Purwaningsih; Kintamani, E. The Diversity of Shorea spp. (Meranti) at Some Habitats in Indonesia. *IOP Conf. Ser. Earth Environ. Sci.* **2018**, *197*, 012034. [CrossRef]

Article

Evaluation of Biological Characteristics of Soil as Indicator for Sustainable Rehabilitation of a Post-Bauxite-Mining Land

Aurelia Oneț [1], Radu Brejea [1], Lucian Dincă [2], Raluca Enescu [2], Cristian Oneț [1,*] and Emanuel Besliu [2,*]

[1] Faculty of Environmental Protection, University of Oradea, 26 General Magheru Street, 410048 Oradea, Romania

[2] National Institute for Research and Development in Forestry "Marin Drăcea", 13 Cloșca Street, 500040 Brașov, Romania

* Correspondence: cristyonet@yahoo.com (C.O.); emanuel.besliu@icas.ro (E.B.)

Abstract: This paper presents a study of the microbial abundance in post-bauxite-mining land soil from Zece Hotare, Bihor county, Romania. The soil samples were collected from 12 soil variants, in the year 2020, after 15 years of long-term restoration. Some chemical parameters and bacterial numbers of six groups of microorganisms were determined in the restored mining land, and these characteristics were compared with those of the soil from a beech forest situated in an adjacent area unaffected by bauxite exploitation. On the basis of the total number of microorganisms belonging to each group studied, the bacterial potential of the soil quality was assessed, calculating the bacterial soil quality index (BSQI), while the Shannon diversity index and the Jaccard distance were applied to show the level of bacterial diversity. The characteristics of the studied chemical and microbiological parameters determined in the beech adjacent area were very similar to those observed in the high-level plateau, low-level plateau, and Black locust areas, indicating similar soil conditions; therefore, the ecological reconstruction 15 years ago, had a very favorable impact on restoration in some affected areas.

Keywords: ecological reconstruction; bacteria diversity; sustainable rehabilitation

Citation: Oneț, A.; Brejea, R.; Dincă, L.; Enescu, R.; Oneț, C.; Besliu, E. Evaluation of Biological Characteristics of Soil as Indicator for Sustainable Rehabilitation of a Post-Bauxite-Mining Land. *Diversity* **2022**, *14*, 1087. https://doi.org/10.3390/d14121087

Academic Editor: Michel Baguette

Received: 4 October 2022
Accepted: 6 December 2022
Published: 9 December 2022

1. Introduction

The main processes leading to soil degradation in former mining areas are as follows:

- Physical: structure destruction, compaction, crust formation, and heavy-metal pollution (Fe, Al, Mg, and Mn);
- Chemical: acidic vulnerability (levels 2–4) due to dissolution and large-scale circulation of contaminated soil and underground waters;
- Biological: reduction in microorganisms, mesofauna, and macrofauna;
- Removal through intense erosion by surface and underground waters, landslides, excavation, and covering with dumped sterile or waste material [1].

Mining activities lead to destructive changes in ground structure and biodiversity, triggering considerable environmental problems [2,3]. Bauxite mining involves serious threats to ecosystems. It causes alteration in species diversity, changes in soil composition, habitat loss, and various social threats. Most of the mining occurs around eco-sensitive areas due to climate change, nutrient imbalance, biodiversity loss, and interrupted ecosystem services [4]. It has been found that mining presents a serious challenge for physical, chemical, and biological restoration. Comprehensive knowledge of the ecology of the landscape structure and configuration, soil type, physical, chemical, and biological properties, dispersal mode, and identification and quantification/inventory of plant communities is critically important for planning restoration programs [5].

The soil degraded through social and economic activities is subjected to a complex system of measures (technical/mining, hydro-amelioration, and soil management) that are meant to rehabilitate the degraded soil's fertility and create a land shaft proper for agriculture, forestry, and other socioeconomic activities [6–8].

The rehabilitation of post-mining land is required to repair damage to local environments. Various methods are employed to achieve this, such as landscape reclamation, planting ground cover crops, utilization of fast-growing plants, and remediation of water and soil contaminants [9,10]. The technology used for the recultivation of degraded soils constitutes two stages: the technical/mining stage to manage the exploitation material, as well as transportation and storage, and the measures needed in order to fight the negative phenomena (soil acidity and soil erosion) of the primitive technogenic soils capable of ensuring durable plant formation with self-tuning capabilities [11–14].

Ecological reconstruction of bauxite residue disposal areas is regarded as an effective approach to eliminating potential environmental risks. Soil microorganisms are very sensitive to environmental perturbation. Bauxite mining is a major threat to soil productivity by inducing perturbations to the soil microbiota that drive nutrient cycles [15]. The study of the abundance, diversity, and dynamics of some groups of bacteria involved in biogeochemical circuits is particularly important in the context of the scientific need to include the biological properties of soils in environmental impact assessment studies and soil quality monitoring programs. Thus, microbial communities and the total number of microorganisms in the soil can serve as a tool for assessing soil quality change. Some studies concluded that unmined soils consisted of higher numbers of microorganisms than rehabilitated sites [15]. The rehabilitation of mines requires an understanding of both soil chemistry and microbiological activity. Soil microorganisms play a key role in soil physicochemical properties, which permits the re-establishment of soil quality after mining [16].

In the present study, we investigate some chemical and microbiological characteristics of soils from a natural forest and post-mining sites, such as pH, humus content, mobile P, mobile K, and numbers of aerobic heterotrophic bacteria, fungi, ammonifiers, nitrifiers represented by nitrate bacteria and nitrite bacteria, and denitrifiers. The microorganism abundance was also used to evaluate the biological potential of the soil quality. The principal aim was to study how chemical and microbiological soil characteristics respond to restoration (tree planting and fertilization) by comparing post-mining sites with a natural forest site.

2. Materials and Methods

2.1. Study Site and Soil Sampling

This study took place in Bihor county, in Apuseni Mountains which belong to the Western Carpathians. The research area covered 10 ha and was located in Pădurea Craiului Mountains, in a restored bauxite mining land (Figure 1). The exploitation of bauxite ended in 1998, while extensive development works were carried out in 2004–2005: leveling, wattle-works on slopes, and planting of Black locust trees in the leveled area and Norway spruce in the higher zones. The Craiului Forest Mountains represent the area between the Crişul Repede River and the Crişul Negru River in the Apuseni Mountains. The Craiului Forest resembles a relatively suspended plateau, mainly formed from Mesozoic limestones with heights over 1000 m in some parts (1027 m in Hondringuşa) in the east; as we proceed westward, the heights decrease to 400–500 m close to Vârciorog and Bucuroaia (525 in Dealul Poiana and 442 m southwest of Vârciorog). These formations consist of Mesozoic limestones associated with the metamorphic schists, Permian conglomerates, sandstone, and rhyolites. Bauxite can be found in the Pădurea Craiului, Remeţi, and Pietrosul regions of the Apuseni Mountains.

The bauxite resources are located in the area enclosed by the Crişul Repede River, the Crişul Negru River, and the Roşia River. The bauxite exploitation was performed on the surface and in underground mines. The annual rainfall average is 615.1 mm: 585.2 mm in 2005, 872.0 mm in 2006, and 585.2 mm in 2007. The multiannual average temperature is 10.2 °C [19].

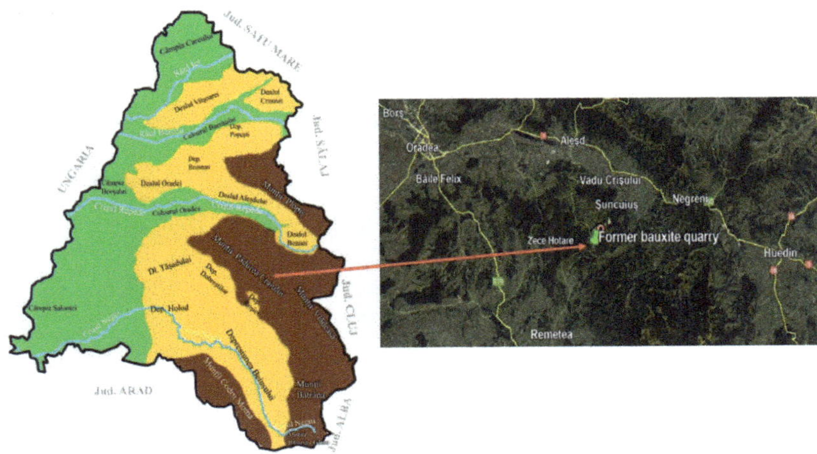

Figure 1. Location of the research area [17,18].

The vegetation of the Craiului Forest includes pasture, a forest of *Picea abies*, and a forest of broad-leaved trees: *Fagus sylvatica*, *Carpinus betulus*, *Acer pseudoplatanus*, *Ulmus montana*, *Fraxinus excelsior*, *Prunus avium*, *Betula pendula*, *Sorbus aucuparia*, *Salix caprea*, and *Juglans regia*. Along the rivers, some specifical species can be encountered: *Salix alba*, *Salix purpurea*, *Salix triandra*, *Populus nigra*, *Alnus glutinosa*, *Carex* sp., *Juncus inflexus*, and *Iris pseudacorus*. There are different vegetative species associated with different areas of the mountain [20–23].

The wattle-work on the hillside determined a better growth rhythm of *Pinus sylvestris*. The following spontaneous species were installed in the post-bauxite-mining land: *Tussilago farfara*, *Cirsium arvense*, *Poa pratensis*, *Rubus caesius*, *Hypericum perforatum*, *Equisetum arvense*, *Polygonum persicaria*, and *Juncus inflexus* [24].

The soil samples were collected in March 2020 from 12 experimental sites. A soil profile was placed in each experimental field. Sampling was conducted using the square method (100 m^2 surface in each area), and the depth of sample collection was 0–20 cm. Chemical and microbiological parameters were determined on the soil samples taken in a natural layout from all 12 profiles using cylinders of 100 cm^3 each in five replicates.

The pH soil values were determined using the potentiometric method, whereas soil organic matter was calculated using the method of wet oxidation and titrimetric dosing (Walkey–Black method). The content of macroelements (mobile phosphorus and potassium) was determined through extraction with ammonium acetate lactate and dosing using a spectrophotometer.

2.2. Microbiological Analysis

The foreign materials (plant debris, calcareous concretions, and other impurities) were removed, and the soil was passed through a sieve with a diameter of <2 mm. The analytical sample (10 g) was extracted from the sifted soil by quartering (this method of tillage was applied for each variant of soil separately, starting with 10 g of soil in each case).

In the studied area, six eco-physiological groups of microorganisms were determined: aerobic heterotrophic bacteria, fungi, ammonifying bacteria, nitrifying bacteria, and denitrifiers.

The microbial abundance of the six groups of microorganisms was determined in 12 experimental fields with different characteristics in terms of fertilization factor, slope influence, and natural vegetation installation:

1. Black locust without fertilization factor (N_0P_0) (BlF0);
2. Black locust with fertilization factor ($N_{60}P_{60}$) (BlF60);
3. Black locust with fertilization factor ($N_{120}P_{120}$) (BlF120);

4. Slope (10%) without wattle-work and fertilization (N_0P_0), planted with Norway spruce (Ns10%F0);
5. Slope (10%) without wattle-work, with fertilization factor ($N_{60}P_{60}$), planted with Norway spruce (Ns10%F60);
6. Slope (10%) without wattle-work, with fertilization factor ($N_{120}P_{120}$), planted with Norway spruce (Ns10%F120);
7. Slope (40%) without fertilization (N_0P_0), planted with Norway spruce (Ns40%F0);
8. Slope (40%) with fertilization factor ($N_{60}P_{60}$), planted with Norway spruce (Ns40%F60);
9. Slope (40%) with fertilization factor ($N_{120}P_{120}$), planted with Norway spruce (Ns40%F120);
10. High-level plateau planted with Black locust (hlpBl);
11. Low-level plateau planted with Black locust (llpBl);
12. Beech tree adjacent area (unaffected by the bauxite exploitation) (btaa).

For the microbiological determinations, six successive decimal dilutions (10^{-1}–10^{-6}) of the soil samples (10 g) were conducted. The first dilution of the soil samples (10^{-1}) was homogenized using a magnetic stirrer at 110 rpm for 5 min.

The total number of aerobic heterotrophic bacteria was determined using the plate culture method [25] (three repetitions) with a solid nutrient medium containing meat extract (incubation: 7 days at 30 °C). Yeasts and molds were grown on *Sabouraud* agar (incubation: 4–5 days at 25 °C). After the plate incubation period, a colony count was performed using a colony counter. Total microflora was expressed as CFU (colony-forming units)/g soil (dry weight).

The "most likely number" (MPN) method was used to determine the total number of ammonifying, nitrifying (nitrite bacteria and nitrate bacteria), and denitrifying microorganisms. A liquid culture medium with peptone water was used. The culture medium used for the determination of nitrite bacteria and nitrate bacteria contained Winogradski's saline solution, diluted at 1:20 [26]. Denitrifying bacteria were determined on "de Barjac" liquid culture medium. For each decimal dilution of the soil samples, five test tubes were used. Each test tube was inoculated with 1 mL of the respective dilution. After an incubation period of 3 weeks at 28 °C, the typical reaction product was analyzed in each test tube as follows:

- Cultures of ammonifying bacteria were analyzed using the Nessler reagent to highlight the ammonia produced by these bacteria. To 1 mL of culture medium was added 1 mL of Nessler reagent. The appearance of a yellow-orange coloration in the inoculated medium indicated the formation of ammonia and, therefore, the presence of ammonifiers;
- In the case of nitrite bacteria, the presence of nitrites was tested with diphenylamine sulfuric reagent, and, in the case of nitrate bacteria, nitrates were detected in the presence of urea in the sulfuric medium (the appearance of a blue coloration was observed);
- The cultures of the denitrifying bacteria were analyzed to highlight the nitrites produced by these bacteria following the reduction of nitrates, using diphenylamine sulfur reagent.

The most probable number of ammonifying bacteria, nitritbacteria, nitrate bacteria, and denitrifiers was determined using McCrady's statistical table. The number of microorganisms was determined by counting the number of tubes giving positive reactions (color change) and comparing the pattern of positive results (the number of tubes showing growth at each dilution) with standard statistical tables [26].

On the basis of the total number of microorganisms belonging to each group studied, the bacterial potential of the soil quality was assessed, calculating the bacterial soil quality index (BSQI) [27] according to the following formula:

$$BSQI = 1/n \times \sum SQI = N, \quad (1)$$

where BSQI is the bacterial index of soil quality, n is the number of bacterial groups, and N is the number of bacteria belonging to each group studied.

2.3. Shannon Diversity Index and Statistical Analysis

For analyses of the diversity of microbial groups identified in the 12 experimental variants, we used the Shannon diversity index and the Jaccard distances, both computed using the vegan package [28] in the R environment.

$$H' = -\sum_{i=1}^{S} pi \ln pi, \quad (2)$$

where H′ is the Shannon diversity index, pi is the relative abundance of species i, S is the total number of species present, and ln is the natural logarithm [29].

The statistical analysis of the collected data, including Pearson correlation, ANOVA, and Kruskal–Wallis test, was performed in the R environment [30].

3. Results

Chemical and Microbiological Analysis

The soil from btaa is a skeletic calcic luvosol, while that in the former bauxite quarry (from the other 11 variants) is a leptosol (according to WRB).

The ANOVA test, which was performed among the 12 sites for the investigated chemical and microbiological characteristics of soils (pH, humus content, mobile P, and mobile K), revealed no significant differences between sites or between repetitions regarding the chemical characteristics of soils.

The microbiological analysis was computed as a function of the microbial abundance. In addition, the results for the microbial abundance were interpreted as a logarithmic expression of the number of microorganisms belonging to each microbial group studied (Table 1).

Table 1. Logarithmic expression of the number of microorganisms belonging to each microbial group studied (log cells/g dry soil).

Experimental Variants	Heterotrophic Bacteria	Fungi	Ammonifying Bacteria	Nitribacteria	Nitratbacteria	Denitrifying Bacteria
BlF0	7.09	6.57	2.56	3.44	2.96	2
BlF60	5.73	5.63	2.96	3.89	1.87	2.07
BlF120	7.25	6.30	2.30	2.38	1.56	2.74
Ns10%F0	5.23	6.47	2.74	-	-	2.83
Ns10%F60	5.91	6.60	3.30	-	-	2.65
Ns10%F120	5.82	6.16	3	-	-	2.96
Ns40%F0	6.54	5.82	2.78	-	-	2.30
Ns40%F60	6.50	6.75	2.86	-	-	2.25
Ns40%F120	7.26	6.92	3.07	-	-	2.91
hlpBl	7.27	6.44	3.30	5.11	1.60	3.04
llpBl	6.90	6.43	2.91	5.36	1.78	2.78
btaa	5.79	5.96	2.96	3.32	2.89	2.55

The Pearson correlation among the six groups of microorganisms showed that the presence of some bacterial species was related to the presence of another species (Figure 2). The strongest positive correlations ($r = 0.77$) were identified between ammonifying bacteria and nitrite bacteria, indicating that an increase in the number of ammonifying bacteria would also increase the number of nitrite bacteria.

The BISQ values shown in Figure 3 indicate a high presence of microorganisms in hlpBl, llpBl, and BlF0, with the exception of soil samples collected from the following experimental areas: Ns10%F120, Ns40%F0, and Ns10%F0, where the BISQ values indicate a moderate abundance of microbial groups. The Kruskal–Wallis test revealed no significant differences among the 12 experimental variants regarding BISQ values ($p = 0.443$), but with a slight superiority of hlpBl and llpBl, which recorded a higher number of bacterial species.

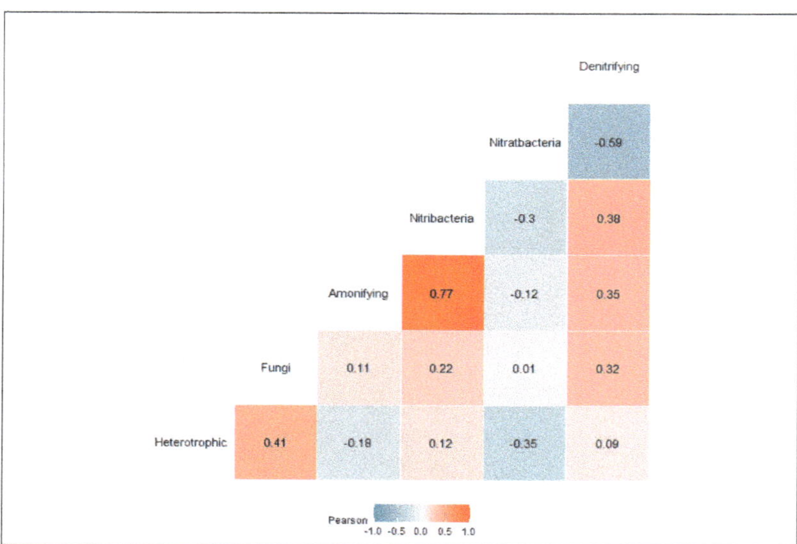

Figure 2. The Pearson correlation computed for the six microbial groups identified in the 12 experimental variants.

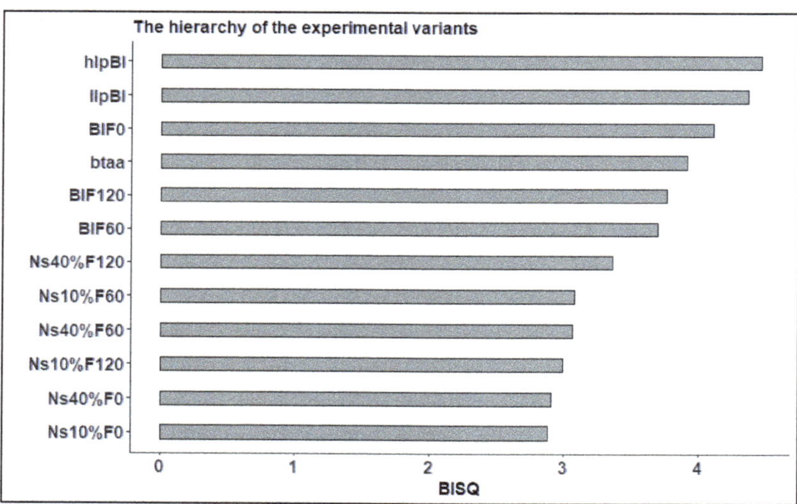

Figure 3. The hierarchy of the experimental variants according to the values of the bacterial indicator of soil quality.

The 12 experimental sites were arranged into two groups as a function of the Shannon diversity index, and the differences among sites were small and insignificant (Figure 4). Higher values for this diversity index were recorded in the btaa, suggesting that this site contained the highest number of bacterial species. The lowest value for this diversity index was identified in Ns40%F60, which seemingly contained the lowest number of bacterial species.

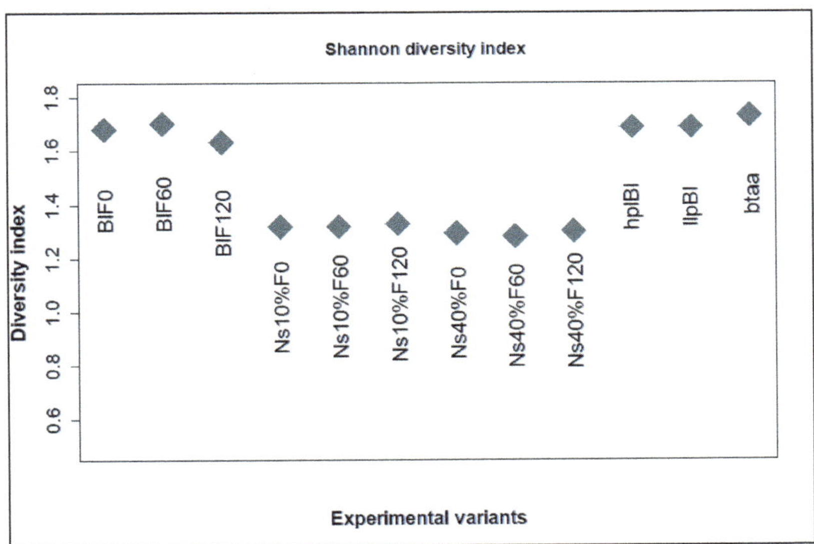

Figure 4. Values of Shannon diversity computed for the 12 experimental variants.

The Jaccard similarity also grouped the 12 experimental sites into two groups (Figure 5). Moreover, in the case of each group, some differences were identified. In the case of group A, the experimental sites were arranged into two groups; the sites Ns10%F0, Ns10%F60, and Ns10%F120 were seemingly homogeneous, as they were characterized by a 10% slope without wattle-work, but differences were observed when considering the fertilization factor.

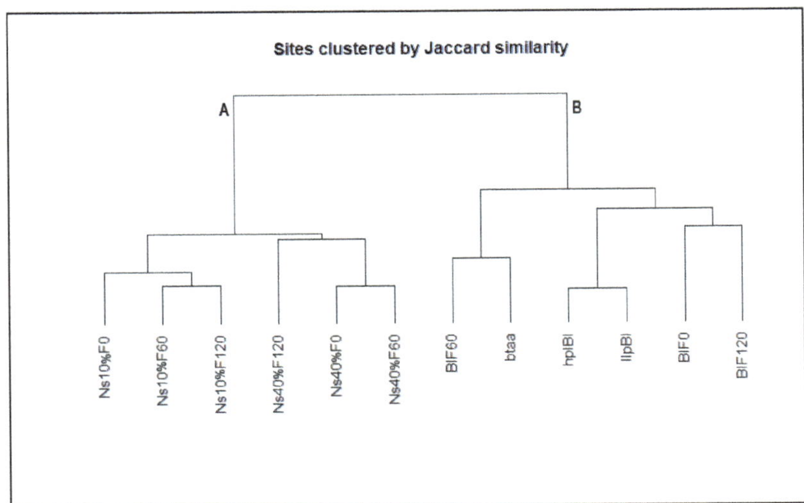

Figure 5. The Jaccard similarity computed for the results obtained by the Shannon diversity index.

Site Ns10%F0 belonged to one group, while sites Ns10%F60 and Ns10%F120 belonged to another. Identical results were recorded in the case of sites Ns40%F0, Ns40%F60, and Ns40%F120, which were similar in terms of slope (40%) but different in terms of fertilization factor. Group B was arranged into three subgroups. In the first subgroup, btaa was similar to BlF60 in terms of bacterial diversity. In the second subgroup, hlpBl and llpBl were also similar in terms of bacteria diversity and slope (0%) but differed in terms of altitude. The third subgroup (BlF0 and BlF120) was similar with respect to the presence of Black locust, which seemingly influenced the diversity of bacterial species.

4. Discussion

The luvisol from the research area is widespread in Romania, accounting for 23% of the total forest soils [31]; it is well supplied with carbon [32–34], water [35,36], and nutritive elements [37–39]. Increasing knowledge about the chemical and bacterial numbers of the rehabilitated sites affected by long-term bauxite exploitation is crucial to assess the soil quality. In this research, we specifically studied how chemical and microbiological soil characteristics respond to restoration by comparing rehabilitated sites with an un-mined site.

Some authors found that restoration was associated with the recovery of the activity and diversity of soil bacterial communities, reaching similar levels to those observed in the conserved forest. Within a few years, restoration allowed recovering crucial physical, chemical, and microbiological soil attributes, reaching levels comparable to those found in conserved forests [40]. Some chemical parameters and bacterial numbers of six groups of microorganisms were determined in the restored mining land. These characteristics were compared with those of the soil from a beech forest situated in an adjacent area unaffected by bauxite exploitation.

Microorganisms from soils affected by degradation processes [41–45] due to human activities represent the form of life that adapts the fastest to new environmental conditions [46]. In other studies on bauxite residue, the bacterial communities were observed to be similar to those from normal soil only after the restoration process [47], as proven in this study. On the other hand, some studies showed that a larger concentration of bauxite residues affected only the activity of the microbial community and not its structure [14]. A smaller number of bacterial species was statistically established in areas with a high slope in this study, as also observed by other authors [48–52].

Vachova et al. [51] investigated the relationship between vegetation and selected soil characteristics as part of landscape restoration. Microorganisms are most important in the phytoremediation process of mining sites because they can contribute to the mobilization or immobilization of metals and metalloids in soil, thus facilitating vegetation development [53]. Our findings are in accordance because we observed that experimental sites with a high abundance of microorganisms (BlF60, hlpBl, and llpBl) were quite similar to sites unaffected by bauxite exploitation (btaa). Moreover, we found no significant differences between the 12 sites or between repetitions regarding the chemical characteristics of soils.

Wu et al. [54] found that the establishment of microbial communities and associated functions may improve the physical and chemical properties of soil, as well as stimulate its formation in bauxite residue. The authors highlighted that the microbiota was significantly developed after long-term natural restoration, as also observed in our research. The long-term restoration created microbial community diversity in bauxite residue.

Bauxite mining activities have the potential to impact the environment, including changes in soil fertility, low soil pH, reduced ability of the soil to hold water, inadequate supply of nutrients for plants, erosion, and exposure to rocks containing sulfides, resulting in the potential for acid mine drainage and disruption of the ecosystem. The authors found significant differences in the chemical properties of soil according to the age of reclamation [55]. The results obtained were not the same as the present study, where the ANOVA test revealed no significant differences in the chemical properties of the soil across the 12 sites, thus indicating their similarity to soil from the natural forest site (btaa).

5. Conclusions

The results of this study emphasize the importance of microorganisms as biological indicators of changes taking place in reclaimed soils. Little is known about the microbial abundance and chemical properties, as well as their role in the characterization process of soil formation in bauxite residue.

After 15 years of long-term restoration, we found that bacterial numbers were similar to those in unexploited soil. In addition, the Pearson correlation among the six groups of microorganisms showed relationships between the presence of some bacterial species.

As bacterial indicators of soil quality, we observed that a high presence of microorganisms was recorded in hlpBl and llpBl, as well as in black locust areas (BlF0, BlF60, and BlF120), indicating their high potential to sustain the life of the microorganisms. In the remaining sites, a moderate presence of microorganisms was recorded, indicating a low potential for microorganisms to adapt and live in these conditions.

The Shannon diversity index showed that experimental sites btaa, hlpBl, and llpBl, as well as black locust areas (BlF0, BlF60, and BlF120), recorded a higher diversity than in the other areas, thus sustaining that these experimental sites were able to support the life of microorganisms and, hence, the growth of the tree species.

Another aspect that needs to be mentioned is that the studied chemical and microbiological parameters determined in btaa were very similar to those observed on the plateau (hlpBl and llpBl) and at experimental sites, BlF0, BlF60, and BlF120, indicating the similarity of the soil conditions in these areas. Therefore, the ecological reconstruction 15 years ago had a very favorable impact on restoring some affected areas.

The Jaccard similarity showed that the factors underlying the differences among the 12 experimental sites regarding the diversity index were the slope, fertilization factor, and Black locust species installed in some areas, which had a positive influence on the land rehabilitation affected by bauxite mining. Additionally, this analysis revealed that BlF60 was the site most similar to the natural experimental site (btaa), thus representing one of the best-reclaimed sites regarding the level of bacterial diversity.

The results obtained in this study offer a biogeochemical perspective describing soil formation in bauxite residue. We found that soil conditions were similar across all 12 experimental sites, enabling us to conclude that the ecological reconstruction 15 years ago had a very favorable impact on the restoration of areas affected by long-term bauxite exploitation.

Author Contributions: Conceptualization, A.O. and L.D.; methodology, A.O., L.D. and R.B.; software, E.B. and R.E.; validation, L.D., A.O., R.B. and C.O.; formal analysis, A.O., E.B. and R.E.; investigation, A.O.; resources, A.O.; data curation, L.D.; writing—original draft preparation, A.O., E.B. and R.E.; writing—review and editing, A.O., E.B., R.E., R.B., C.O. and L.D.; visualization, A.O. and E.B.; supervision, L.D.; project administration, A.O.; funding acquisition, A.O. and L.D. All authors have read and agreed to the published version of the manuscript.

Funding: This research was funded by the BIOSERV program, financed by the Romanian Ministry of Research, Innovation, and Digitization, and project "Increasing the institutional capacity and performance of INCDS" Marin Drăcea under the activity of RDI—CresPerfInst (Contract no. 34PFE./30.12.2021) financed by the Ministry of Research, Innovation, and Digitization through Program 1—Development of the national research, development system, Subprogram 1.2—Institutional performance, projects to finance excellence in RDI.

Institutional Review Board Statement: Not applicable.

Informed Consent Statement: Not applicable.

Data Availability Statement: Not applicable.

Conflicts of Interest: The authors declare no conflict of interest.

References

1. Wuana, R.; Okieimen, F. Heavy Metals in Contaminated Soils: A Review of Sources, Chemistry, Risks and Best Available Strategies for Remediation. *Int. Sch. Res. Not.* **2011**, *2011*, 20. [CrossRef]
2. Crisan, V.; Dinca, L.; Enescu, R.; Onet, A.; Deca, S. Depolluting the slime deposit from Brasov, Romania—A case study. *J. Environ. Prot. Ecol.* **2020**, *21*, 579–587.
3. Xu, Q.; Xia, G.; Wei, Y.; Aili, A.; Yuan, K. Responses of Vegetation and Soil to Artificial Restoration Measures in Abandoned Gold Mining Areas in Altai Mountain, Northwest China. *Diversity* **2022**, *14*, 427. [CrossRef]

4. Yadav, S.K.; Banerjee, A.; Jhariya, M.K.; Meena, R.S.; Khan, N.; Raj, A. Eco-restoration of bauxite mining: An ecological approach. In *Natural Resources Conservation and Advances for Sustainability*; Jhariya, M.K., Meena, R.S., Eds.; Elsevier: Amsterdam, The Netherlands, 2022; pp. 173–193.
5. Lewis, S.; Rosales, J. Restoration of Forested Lands under Bauxite Mining with Emphasis on Guyana during the First Two Decades of the XXI Century: A Review. *J. Geosci. Environ.* **2020**, *8*, 41–67. [CrossRef]
6. Damian, F.; Damian, G.; Lăcătuşu, R.; Macovei, G.; Iepure, G.; Năprădean, I.; Chira, R.; Kollar, L.; Rață, L.; Zaharia, D.C. Soils from the Baia Mare zone and the heavy metals pollution. *Carpathian J. Earth Environ. Sci.* **2008**, *3*, 85–98.
7. Constandache, C.; Peticila, A.; Dinca, L.; Vasile, D. The usage of Sea Buckthorn (*Hippophaerhamnoides* L.) for improving Romania's degraded lands. *AgroLife Sci. J.* **2016**, *5*, 50–58.
8. Silvestru-Grigore, C.V.; Dinulică, F.; Spârchez, G.; Hălălișan, A.F.; Dincă, L.C.; Enescu, R.E.; Crișan, V.E. The radial growth behavior of pines (*Pinus sylvestris* L. and *Pinus nigra* Arn.) on Romanian degraded lands. *Forests* **2018**, *9*, 213. [CrossRef]
9. Whery, A.; Pantu, H. *Hydromelioration Works*; Publishing House Aprilia Print: Timisoara, Romania, 2008.
10. Prematuri, R.; Turjaman, M.; Sato, T.; Tawaraya, K. Post Bauxite Mining Land Soil Characteristics and Its Effects on the Growth of *Falcatariamoluccana* (Miq.) Barneby&JW Grimes and *Albiziasaman* (Jacq.) Merr. *Appl. Environ. Soil Sci.* **2020**, *2020*, 8.
11. Nitu, I.; Dracea, M.; Rauta, C.; Milache, M. *Agricultural Meliorative Works*; Publishing House Agris: Bucharest, Romania, 2000.
12. Mărăcineanu, F.; Căzărescu, S.; Constantion, E.; Bozianu, C. The anthropic degradation of natural environment and their remedy. In *Scientifical Papers*; Faculty of Agriculture, USAMVB Timişoara: Timisoara, Romania, 2008; pp. 117–124.
13. Courtney, R.; Mullen, G.; Harrington, T. An Evaluation of Revegetation Success on Bauxite Residue. *Restor. Ecol.* **2009**, *17*, 350–358. [CrossRef]
14. Fourrier, C.; Luglia, M.; Hennebert, P.; Foulon, J.; Ambrosi, J.P.; Angeletti, B.; Keller, C.; Criquet, S. Effects of increasing concentrations of unamended and gypsum modified bauxite residues on soil microbial community functions and structure—A mesocosm study. *Ecotoxicol. Environ. Saf.* **2020**, *201*, 110847. [CrossRef]
15. Lewis, D.E.; White, J.R.; Wafula, D.; Athar, R.; Dickerson, T.; Williams, H.N.; Chauhan, A. Soil Functional Diversity Analysis of a Bauxite-Mined Restoration Chronosequence. *Microb. Ecol.* **2010**, *59*, 710–723. [CrossRef] [PubMed]
16. Silva, G.O.A.; Southam, G.; Gagen, E.J. Accelerating soil aggregate formation: A review on microbial processes as the critical step in a post-mining rehabilitation context. *Soil Res.* **2022**. [CrossRef]
17. Consiliul Județean Bihor. Available online: www.cjbihor.ro (accessed on 10 September 2022).
18. Google Earth. Available online: https://earth.google.com/web/ (accessed on 10 September 2022).
19. Dragastan, O.; Damian, R.; Csiki, Z.; Lazăr, I. Review of the Bauxite-Bearing Formations in the Northern Apuseni Mts. Area (Romania) and Some Aspects of the Environmental Impact of the Mining Activities. *Carpathian J. Earth Environ. Sci.* **2009**, *4*, 5–24.
20. Barklay, I. Risk Assesement for Tailings Impoundments. In Proceedings of the Mining Environment Congress, Baile Felix, Romania, 25–30 June 2001.
21. Popovici, L.; Moruzi, C.; Toma, I. *Botanical Book*; Pedagogical Publishing House: Bucharest, Romania, 2002.
22. Brejea, R.; Domuta, C.; Bara, V.; Sandor, M.; Bara, C.; Bara, L.; Sabau, N.C.; Domuta, A.; Samuel, A.; Borza, I.; et al. Researches regarding the vegetation reconstruction in a former bauxite quarry in the Padurea Craiului Mountain, Romania. *J. Environ. Prot. Ecol.* **2011**, *12*, 1873–1882.
23. Sabau, N.C.; Domuta, C.; Berchez, O. *Genesis, Degradation and Pollution of the Soil, Part II. Degradation and Pollution of the Soil*; University of Oradea Publishing House: Oradea, Romania, 2002.
24. Domuta, C. *Practicum of Soil Management*; University of Oradea Publishing House: Oradea, Romania, 2007.
25. Atlas, R.M. *Handbook of Microbiological Media*, 3rd ed.; Taylor & Francis Inc.: Boca Raton, FL, USA, 2004; 2056p.
26. Ferguson, M.A.; Ihrie, J. MPN: Most Probable Number and Other Microbial Enumeration Techniques. Available online: https://CRAN.R-project.org/package=MPN (accessed on 30 November 2022).
27. Carpa, R.; Butiuc-Keul, A.; Lupan, I.; Barbu-Tudoran, L.; Muntean, V.; Dobrotă, C. Poly-β-hydroxybutyrate accumulation in bacterial consortia from different environments. *Can. J. Microbiol.* **2012**, *58*, 660–667. [CrossRef] [PubMed]
28. Oksanen, J.; Gavin, L.; Simpson, F.; Guillaume, B.; Roeland, K.; Pierre, L.; Peter, R.; Minchin, R.B.; O'Hara, P.S.; Henry, M.; et al. Vegan: Community Ecology Package. R Package Version 2.6-2. Available online: https://CRAN.R-project.org/package=vegan (accessed on 10 August 2022).
29. Flynn, D.; Lin, B.; Bunker, D. Measuring Biological Diversity in R. Available online: http://traits-dgs.nceas.ucsb.edu/workspace/r/r-tutorial-for-measuring-species-diversity/Measuring%20Diversity%20in%20R.pdf/attachment_download/file (accessed on 15 August 2022).
30. R Core Team. *R: A Language and Environment for Statistical Computing*; R Foundation for Statistical Computing: Vienna, Austria, 2022. Available online: https://www.r-project.org/ (accessed on 10 June 2022).
31. Dincă, L.; Spârchez, G.; Dincă, M. Romanian's forest soil GIS map and database and their ecological implications. *Carpathian J. Earth Environ. Sci.* **2014**, *9*, 133–142.
32. Maia, S.M.F.; Xavier, F.A.S.; Oliveira, T.S.; Mendonça, E.S.; AraújoFilho, J.A. Organic carbon pools in a Luvisol under agroforestry and conventional farming systems in the semi-arid region of Ceará, Brazil. *Agrofor. Syst.* **2007**, *71*, 127–138. [CrossRef]
33. Crișan, V.; Dincă, L.; Oneț, A.; Bragă, C.I.; Enescu, E.; Teușdea, A.C.; Oneț, C. Impact of windthrows disturbance on chemical and biological properties of the forest soils from Romania. *Environ. Eng. Manag. J.* **2021**, *20*, 1163–1172.

34. Dinca, L.; Sparchez, G.; Dincă, M.; Blujdea, V. Organic Carbon Concentrations and Stocks in Romanian Mineral Forest Soils. *Ann. For. Res.* **2012**, *55*, 229–241.
35. Sun, L.; Chang, S.X.; Feng, Y.S.; Dyck, M.F.; Puurveen, D. Nitrogen fertilization and tillage reversal affected water-extractable organic carbon and nitrogen differentially in a Black Chernozem and a Gray Luvisol. *Soil Tillage Res.* **2015**, *146*, 253–260. [CrossRef]
36. Dinca, L.; Badea, O.; Guiman, G.; Braga, C.; Crisan, V.; Greavu, V.; Murariu, G.; Georgescu, L. Monitoring of soil moisture in Long-Term Ecological Research (LTER) sites of Romanian Carpathians. *Ann. For. Res.* **2018**, *61*, 171–188. [CrossRef]
37. Spârchez, G.; Dincă, L.; Marin, G.; Dincă, M.; Enescu, R.E. Variation of eutriccambisols' chemical properties based on altitudinal and geomorphological zoning. *Environ. Eng. Manag. J.* **2017**, *16*, 2911–2918.
38. Enescu, C.M.; Dincă, L.; Bratu, I.A. Chemical characteristics of the forest soils from Prahova County. *Sci. Pap. Ser. Manag. Ec. Eng. Agric. Rural Dev.* **2018**, *18*, 109–112.
39. Michopoulos, P.; Kostakis, M.; Bourletsikas, A.; Kaoukis, K.; Pasias, I.; Grigoratos, T.; Thomaidis, N.; Samara, C. Concentrations of three rare elements in the hydrological cycle and soil of a mountainous fir forest. *Ann. For. Res.* **2022**, *65*, 155–164. [CrossRef]
40. Bizuti, D.T.G.; Robin, A.; Soares, T.M.; Moreno, V.S.; Almeida, D.R.A.; Andreote, F.D.; Casagrande, J.C.; Guillemot, J.; Herrmann, L.; van Melis, J.; et al. Multifunctional soil recovery during the restoration of Brazil's Atlantic Forest after bauxite mining. *J. Appl. Ecol.* **2022**, *59*, 2262–2273. [CrossRef]
41. Oneț, A.; Dincă, L.; Teușdea, A.; Crișan, V.; Bragă, C.; Enescu, R.; Oneț, C. The influence of fires on the biological activity of forest soils in Vrancea, Romania. *Environ. Eng. Manag. J.* **2019**, *18*, 2643–2654.
42. Onet, A.; Dincă, L.C.; Grenni, P.; Laslo, V.; Teusdea, A.C.; Vasile, D.L.; Enescu, R.E.; Crisan, V.E. Biological indicators for evaluating soil quality improvement in a soil degraded by erosion processes. *J. Soils Sediments* **2019**, *19*, 2393–2404. [CrossRef]
43. Enescu, R.E.; Dinca, L.; Zup, M.; Davidescu, S.; Vasile, D. Assessment of Soil Physical and Chemical Properties among Urban and Peri-Urban Forests: A Case Study from Metropolitan Area of Brasov. *Forests* **2022**, *13*, 1070. [CrossRef]
44. Sousa, M.G.; Araujo, J.K.S.; Fracetto, G.G.M.; Ferreira, T.O.; Fracetto, F.J.C.; de AraújoFilho, J.C.; Otero, X.L.; dos Santos, J.C.B.; da Silva, A.H.N.; de Souza Junior, V.S. Changes in organic carbon and microbiology community structure due to long-term irrigated agriculture on Luvisols in the Brazilian semi-arid region. *Catena* **2022**, *212*, 106058. [CrossRef]
45. Siebyła, M.; Hilszczanska, D. Next Generation Sequencing genomic analysis of bacteria from soils of the sites with naturally-occurring summer truffle (*Tuber aestivum* Vittad.). *Ann. For. Res.* **2022**, *65*, 97–110. [CrossRef]
46. Torsvik, V.; Øvreås, L. Microbial diversity, life strategies, and adaptation to life in extreme soils. In *Microbiology of Extreme Soils*; Springer: Berlin/Heidelberg, Germany, 2008; Volume 13, pp. 15–43.
47. Schmalenberger, A.; O'Sullivan, O.; Gahan, J.; Cotter, P.D.; Courtney, R. Bacterial Communities Established in Bauxite Residues with Different Restoration Histories. *Environ. Sci. Technol.* **2013**, *47*, 7110–7119. [CrossRef] [PubMed]
48. Xu, Z.; Yu, G.; Zhang, X.; Ge, J.; He, N.; Wang, Q.; Wang, D. The variations in soil microbial communities, enzyme activities and their relationships with soil organic matter decomposition along the northern slope of Changbai Mountain. *Appl. Soil Ecol.* **2015**, *86*, 19–29. [CrossRef]
49. Bardelli, T.; Gómez-Brandón, M.; Ascher-Jenull, J.; Fornasier, F.; Arfaioli, P.; Francioli, D.; Egli, M.; Sartori, G.; Insam, H.; Pietramellara, G. Effects of slope exposure on soil physico-chemical and microbiological properties along an altitudinal climosequence in the Italian Alps. *Sci. Total Environ.* **2017**, *575*, 1041–1055. [CrossRef] [PubMed]
50. Probst, M.; Gómez-Brandón, M.; Bardelli, T.; Egli, M.; Insam, H.; Ascher-Jenull, J. Bacterial communities of decaying Norway spruce follow distinct slope exposure and time-dependent trajectories. *Environ. Microbiol.* **2018**, *20*, 3657–3670. [CrossRef]
51. Vachova, P.; Vach, M.; Skalicky, M.; Walmsley, A.; Berka, M.; Kraus, K.; Hnilickova, H.; Vinduskova, O.; Mudrak, O. Reclaimed Mine Sites: Forests and Plant Diversity. *Diversity* **2022**, *14*, 13. [CrossRef]
52. Enescu, R.; Vasile, D.; Dincă, L.; Davidescu, S.; Breabăn, I.G. The favorability of orographic and edaphic factors for the main species that comprise urban forests from Brasov city. *Present Environ. Sustain. Dev.* **2022**, *16*, 139–153. [CrossRef]
53. Bruneel, O.; Mghazli, N.; Sbabou, L.; Héry, M.; Casiot, C.; Filali-Maltouf, A. Role of microorganisms in rehabilitation of mining sites, focus on Sub Saharan African countries. *J. Geochem. Explor.* **2019**, *205*, 106327. [CrossRef]
54. Wu, H.; Tang, T.; Zhu, F.; Wei, X.; Hartley, W.; Xue, S. Long term natural restoration creates soil-like microbial communities in bauxite residue: A 50-year filed study. *Land Degrad. Dev.* **2021**, *32*, 1606–1617. [CrossRef]
55. Gunawan, A.B.; Kembarawati, H.R. Changes in Soil Chemical Properties Due to Reclamation of Bauxite Mining. *Nat. Volatiles Essent. Oils* **2021**, *8*, 2700–2711.

MDPI

Article

The Necessity of Maintaining the Resilience of Peri-Urban Forests to Secure Environmental and Ecological Balance: A Case Study of Forest Stands Located on the Romanian Sector of the Pannonian Plain

Serban Chivulescu [1], Nicolae Cadar [2,*], Mihai Hapa [1,3], Florin Capalb [1,3], Raul Gheorghe Radu [4] and Ovidiu Badea [1,3]

1 National Institute for Research and Development in Forestry "Marin Drăcea", 128 Eroilor Boulevard, 077190 Voluntari, Romania
2 National Institute for Research and Development in Forestry "Marin Drăcea", Timisoara, 8 Pădurea Verde Str., 300310 Timișoara, Romania
3 Faculty of Silviculture and Forest Engineering, "Transilvania" University of Brașov, Șirul Beethoven 1, 500123 Brașov, Romania
4 National Institute for Research and Development in Forestry "Marin Drăcea", Brașov St., 13 Cloșca Street, 500040 Brașov, Romania
* Correspondence: nicu_cadar@yahoo.com

Abstract: Climate change's negative effects, such as rising global temperatures and the disruption of global ecological ecosystems as a direct effect of rising carbon emissions in the atmosphere, are a significant concern for human health, communities, and ecosystems. The condition and presence of forest ecosystems, especially those in peri-urban areas, play an essential role in mitigating the negative effects of climate change on society. They provide direct benefits to the residents of large cities and their surrounding areas, and they must be managed sustainably to protect all their component ecosystems. This research was carried out in the forests of Lunca Muresului Natural Park and Bazos Arboretum, located in the Romanian sector of the Pannonian Plain, near urban agglomerations. The results showed high variability in the stands. Using the height-to-diameter ratio indicator concerning dbh and species, a strong Pearson correlation was registered (between 0.45 and 0.82). These values indicate the high stability of these stands, providing positive human–nature interactions such as recreational or outdoor activities (and a complementary yet indirect use value through attractive landscape views). Protecting these ecosystems offers a so-called insurance policy for the next generations from a climate change standpoint.

Keywords: climate change effects; peri-urban forest management; forest stability; urban expansion; greenhouse gas emissions; forest resilience

Citation: Chivulescu, S.; Cadar, N.; Hapa, M.; Capalb, F.; Radu, R.G.; Badea, O. The Necessity of Maintaining the Resilience of Peri-Urban Forests to Secure Environmental and Ecological Balance: A Case Study of Forest Stands Located on the Romanian Sector of the Pannonian Plain. *Diversity* **2023**, *15*, 380. https://doi.org/10.3390/d15030380

Academic Editors: Lucian Dinca and Miglena Zhiyanski

Received: 31 January 2023
Revised: 27 February 2023
Accepted: 27 February 2023
Published: 6 March 2023

1. Introduction

Nowadays, one of the most critical environmental concerns of society is the increasing impact of greenhouse gas emissions, with the most essential being CO_2, which subsequently influences the greenhouse effect in the atmosphere [1]. Many consider it climate change [2–4] or global warming [5]. As Europe has been projected to be the third-largest carbon emitter in the world [6] and forests act as carbon sinks [7], the paradigm is seen to shift slowly but steadily toward a socio-climate focused on national and international policies with increasing pressure on forests in many cases [8–11]. While climate change has been seen to influence drastic changes in current policy frameworks, policy-makers have also considered other factors, such as urban expansion, migration, and the increasing need for sustainable resource provision and agriculture [12,13]. As migration and urban expansion increase so does the need for more food and resources; thus, forests are faced with an old yet controversial problem of conversion to the agricultural land [14,15] more commonly seen in plain areas.

Forests absorb and store CO_2 through photosynthesis, reducing the amount of greenhouse gases in the atmosphere and mitigating the effects of climate change. Forests also store carbon in the form of wood, leaves, and other organic matter, helping to keep carbon locked away for long periods. When forests are cut down or burned, the carbon stored in the trees and other vegetation is released into the atmosphere, contributing to the buildup of greenhouse gases. Thus, forests can also act as a source of greenhouse gas emissions when disturbed through deforestation, forest fires, and other land use changes. Therefore, it is essential to manage forests sustainably to maximize their ability to act as a sink for greenhouse gases and minimize their contribution as a source. In this context, a possible solution for the sustainable management of forest ecosystems and the fulfillment of global policy guidelines resides in more effective management of forest areas [16–18]. Last but not least, forest management must also consider preserving the beneficial effects of forest ecosystems [19,20], such as clean water or recreation.

Forests with special protection functions, in some cases, predominantly social or water, land, and soil protection, have an essential role in protecting forest ecosystems [21,22] while also securing material resources and exercising the provision function of a forest ecosystem [23,24]. According to the forest legislative framework in Romania, only conservation cuttings are allowed in these areas. To a certain extent, without influencing the main protective role of the forest, these types of cuttings are defined by certain interventions, which are applied in forest stands, that fulfill special protection functions and aim to allow the establishment of a new tree group using low-intensity actions with a maximum of 10–15% of standing volume per decade [25]. Since 1954, Romania's area of forests with special protection functions has increased from 12.7% of national forest land to approximately 66% in 2021.

Natural resources suffer the most in the face of urban land expansion into the surroundings mainly due to being often affected or displaced. Moreover, urban expansion is believed to directly influence the transformation or loss of forest areas and their management. While urbanization near forested areas increases, it affects the lives of most people in the transition, as forests act as a home for social and ecological services that improve human health and well-being [26]. At the same time, Henwood et al. [27] suggested that management can evolve into ensuring extended benefits regarding trees, woods, and forests which society might need at a personal level with deep implications toward communities and the cultures in which they live.

Addressing the protection of stable forests, measurable adequate evidence-based support is believed to enhance forest stability, which is similar to having a life insurance policy providing certain security for future generations [28]. Forests can partially mitigate natural hazards, such as erosion, floods, or rock falls [29]; out of these, river and coastal floods are the most critical hazards in many floodplains and coastal stretches of Central and Eastern Europe, with an expected annual damage (EAD) in Romania of USD 289 million up to 2050 [4]. These natural hazards can be predicted by analyzing tree and stand stability, and forest managers can take appropriate measures to mitigate or prevent such events [30–32]. Therefore, stand stability is a critical characteristic [33,34] regarding forest resilience. The research in this area focused on a clear correlation and prediction of height-to-diameter ratio (h/d ratio), in which species mixtures were also considered to provide insights into the overall stability generated by human–forest interaction [35,36]. Recent research [35] has shown that this relationship between diameter and height is an important factor in forest research. Through this relationship, stand volumes and biomass can be estimated [37–39], but it can also give information on tree and stand growth [40], as well as information on stand productivity [41].

The aim of this study was to highlight the importance of peri-urban forests and their role in society. The objectives of this research were (i) to analyze the stand structure of several peri-urban forests located in the eastern sector of the Pannonian Plain in Romania;

(ii) to highlight resilience through the stability of these forest ecosystems; and (iii) to generally assess the urban land expansion in the region and its implications for current and future generations.

2. Materials and Methods

This research was conducted in the Lunca Muresului Natural Park and the Bazos Arboretum (Figure 1) located in the western part of Romania on the Western Plains, which is the Romanian sector of the Pannonian Plain. The study area consists of forests located in the peri-urban region, composed of common oak (*Quercus robur* sp.) with special protection interest and scientific interest in conserving the gene pool and other ecosystems characterized by natural elements with special value.

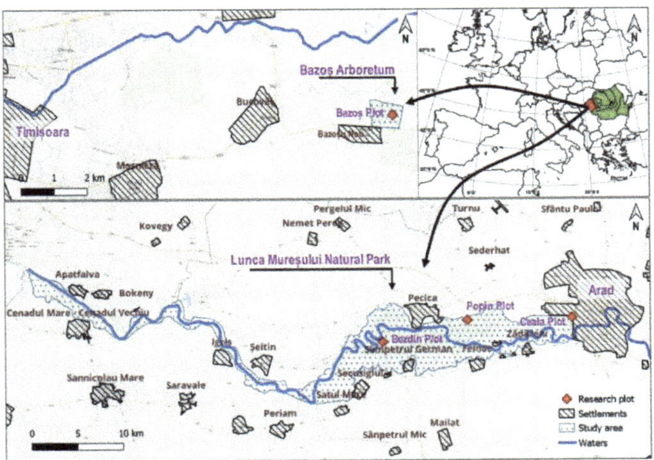

Figure 1. Research area location.

According to their geographical position, these forest stands are found in the area of the continental climate, with sub-Mediterranean influence, where winters are milder, shorter, and relatively cold, while the summers are humid and long. According to the Kőppen Climatic Classification, they are located in the Cfax climate type and described by a mild, humid climate with rainfall all year, where temperatures in the hottest month of the year (June) exceed 22 °C, and maximum rainfall is recorded at the end of spring with the minimum rainfall falling at the end of winter.

The humidity index has an average value of 60–80%, with lower values recorded during fall and winter, while higher values were reported in spring and summer. The average temperature of the year ranged between 9 and 11 °C, with amplitudes rising steadily to 22–23 °C. The average temperature in January exceeds −2 °C, while the average in July reaches 20–21 °C; these temperatures are characteristic for the forest-steppe climate.

The forest stands in the areas studied are included in zones with special protection functions in which only special conservation and sanitary cuttings are allowed to some extent. The management of these types of forests is focused on the conservation of the living and nonliving ecosystem components, which results, in general, to less human influence over time.

In order to conduct this study, four sample plots of one hectare each (100 m × 100 m) were installed and used to measure forest stands. The research plots were installed between 2014 and 2018 and are located in areas with no slope and at an altitude between 107 and 111 m above sea level. The dominate soil type is preluvosols, and over time, it has not been affected by gelling processes following the drop in groundwater levels. We collected information about the tree species, as well as the position and shape of the trees; biometric characteristics, such as diameter at breast height (dbh) of all the trees that exceed 2 cm; and the total height and assessed wood quality class of the tree. In order to assess the wood quality, 4 quality classes were taken into consideration (I, II, III, and IV) in relation to the percentage of industrial wood from total height of the tree (I > 0.5%; II = 0.25–0.50%; III = 0.1–0.25%; IV < 0.1%).

The total above ground tree volume was determined (including branches above five centimeters) using Equation (1) [42–44]. Moreover, the total stand volume was determined by adding the individual tree volumes:

$$\log v = a_0 + a_1 \log d + a_2 \log d^2 + a_3 \log h + a_4 \log h^2 \quad (1)$$

where h represents the tree's total height expressed in meters, d represents the dbh expressed in centimeters, v represents the volume of the tree expressed in cubic meters, and a_0, a_1, a_2, a_3, and a_4 are regression coefficients customized for each species from yield tables [42,45].

To summarize the samples and analyses to be carried out, the main statistical parameters of stand characterization were determined using the R PASTECS package from R software version 1.3.21 [46].

In order to evaluate the forest stand stability and adapt current forest management to the concerns and needs of society, an analysis of the optimal structures of the stands in the areas studied is required to establish the most appropriate theoretical functions to describe these stands. This was undertaken using the theoretical functions Normal [47], the 2-parameter Gamma [47], and Weibull [48] because they are characterized by high flexibility and accuracy in describing the stands [49]. The goodness of fit was tested using the chi-square criterion and the Kolmogorov–Smirnov [50] test.

The graphical analyses were performed using R Studio's plot function [51].

3. Results

3.1. Descriptive Statistics

The statistical indicators reveal that in all four plots, there is a relatively large variation, evidenced by the values of standard deviation, variance, and coefficient of variation (Table 1). In all four research plots, the number of trees (per hectare) varies from 203 (Bazos) to 722 (Bezdin); most of them contain more than 607 trees, except for the Bazos research plot where the stand is older (160 years old according to the management plan), larger in size, and has fewer trees. The minimum diameter varies from 2.35 (Ceala plot) to 11.50 cm (Bazos plot), the maximum diameter varies from 55.00 (Popin plot) to 108.00 cm (Bazos plot), and the average diameter ranges from 18.20 (Ceala plot) to 47.07 cm (Bazos plot). The standard deviation ranged from 11.49 (Popin plot) to 22.69 (Bazos plot), and the minimum coefficient of variation was 55% (Popin plot), and the maximum was 98% (Bazos plot). The variance values ranged from 132.08 (Popin plot) to 514.97 (Bazos plot).

Table 1. Descriptive statistics of diameters.

Research Plot	Coordinates	Number of Trees per Hectare	Minimum dbh (cm)	Maximum dbh (cm)	Average dbh (cm)	Standard Deviation of dbh (s)	Variance of dbh (s^2)	Coefficient of Variance (s %)
Bazos	46°08′07″ N 21°00′32″ E	203	11.50	108.00	47.07	22.69	514.97	48
Bezdin	46°09′39.2″ N 21°07′39.8″ E	722	2.67	87.5	18.65	18.34	336.59	98
Ceala	46°10′04.9″ N 21°16′33.1″ E	664	2.35	68.25	18.20	15.52	240.93	85
Popin	45°45′18.9″ N 21°25′47.3″ E	607	4.00	55.00	20.87	11.49	132.08	55

3.2. Fitting of Experimental Diameter Distribution

The experimental and theoretical diameter distributions of the four research surfaces are graphically represented. From Figure 2, it can be seen that the four stands possess different structures, according to this analysis. Using theoretical functions with a high degree of diversity, it was possible to capture as well as possible the shape of the experimental distribution of the stands, while it could be seen that the structure of none of the stands was described by the theoretical normal function, showing that the stands do not have the characteristics of an even-aged stand.

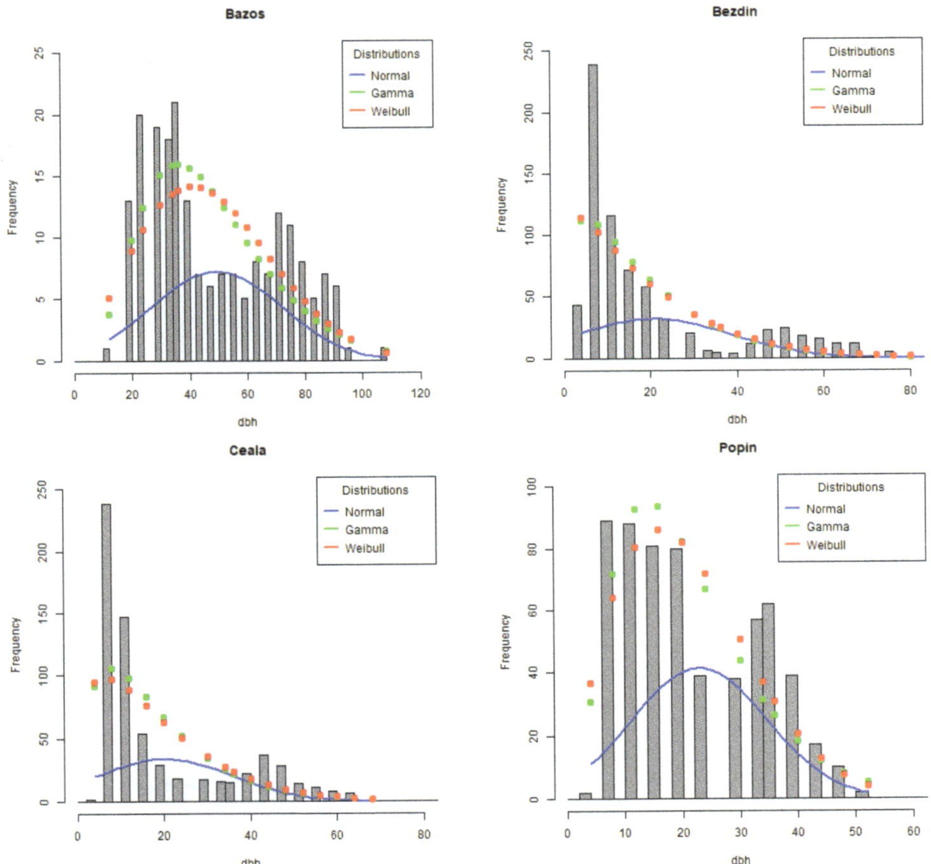

Figure 2. Fitting of experimental dbh distribution.

Using goodness-of-fit tests, the results obtained indicate that following the application of the chi-square criterion within the Bazos and Popin plots, there were no significant differences between the experimental and theoretical values (Table 2). Applying the same significance test for the Bezdin and Ceala plots, significant differences were recorded only for the normal distribution. No significant differences were recorded for the other distributions (Weibull and Gamma).

Table 2. Results of applying statistical tests to the experimental distribution of diameters.

Research Plot	Distribution	Testing the Null Hypothesis with the Test					
		Chi-Square Criterion			Kolmogorov–Smirnov		
		Experimental Value	Theoretical Value (α = 0.05)	Differences	Experimental Value	Theoretical Value (α = 0.05)	Differences
Bazos	Normal	81.87	237.24	insignificant	0.16	0.09	significant
	Weibull	56.52		insignificant	0.13		significant
	Gamma	48.51		insignificant	0.11		significant
Bezdin	Normal	1200.83	785.62	significant	0.22	0.05	significant
	Weibull	357.49		insignificant	0.12		significant
	Gamma	350.50		insignificant	0.13		significant
Ceala	Normal	1606.92	724.01	significant	0.25	0.05	significant
	Weibull	738.48		insignificant	0.18		significant
	Gamma	689.18		insignificant	0.19		significant
Popin	Normal	127.75	662.28	insignificant	0.10	0.05	significant
	Weibull	75.86		insignificant	0.08		significant
	Gamma	91.87		insignificant	0.09		significant

Applying the Kolmogorov–Smirnov significance test to all the plots for standardized data, significant differences between the theoretical and experimental values were recorded. However, it could be observed that the differences between the experimental and theoretical values were lower for the Weibull and Gamma functions, but it should be mentioned that the Kolmogorov–Smirnov test is much more rigorous than the chi-square criterion.

3.3. Stand Stability Height-to-Diameter Ratio (h/d Ratio) in Relation to Diameter and Species

Our analysis revealed that the description of the curve between h/d ratio in relation to diameter and species for the four survey areas shows a decreasing trend (Figure 3). In conjunction with the value of the correlation coefficient with values between 0.45 (Bezdin) and 0.82 (Bazos), it can be affirmed, with some certainty, that there is a high to very high correlation between the mentioned characteristics for almost all stands. An exception is the Bezdin plot, where ash (*Fraxinus excelsior* sp.) and elm (*Ulmus minor* sp.) trees (mixture species) are more distributed in the lower diameter categories, and only oak trees are found in the upper diameter categories, which are older, and as they were the first to settle in the research area, they manage to overlap other species.

3.4. Stand Stability Height-to-Diameter Ratio (h/d Ratio) in Relation to Wood Quality

For all the research areas, the comparative analyses between the h/d ratio and wood quality were performed (Figure 4) and showed a very close relationship between the two stand characteristics, expressed by correlation coefficient values ranging between 0.975 and 0.977. In addition, the results revealed that for all research plots, a higher wood quality was recorded for lower values of the h/d ratio. A small exception was observed in the Ceala research plot, where the fourth (lower) quality class recorded slightly lower h/d ratio values compared to the third (middle to lower) quality class, which is largely due to the species in the stand composition, which has lower longevity (e.g., *Populus* sp.).

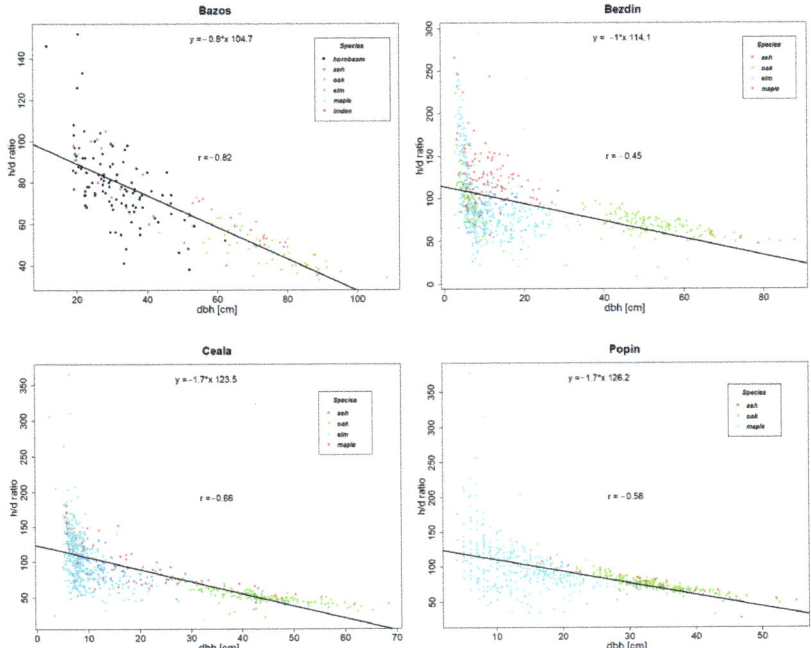

Figure 3. Diameter distribution in relation to species and height-to-diameter ratio.

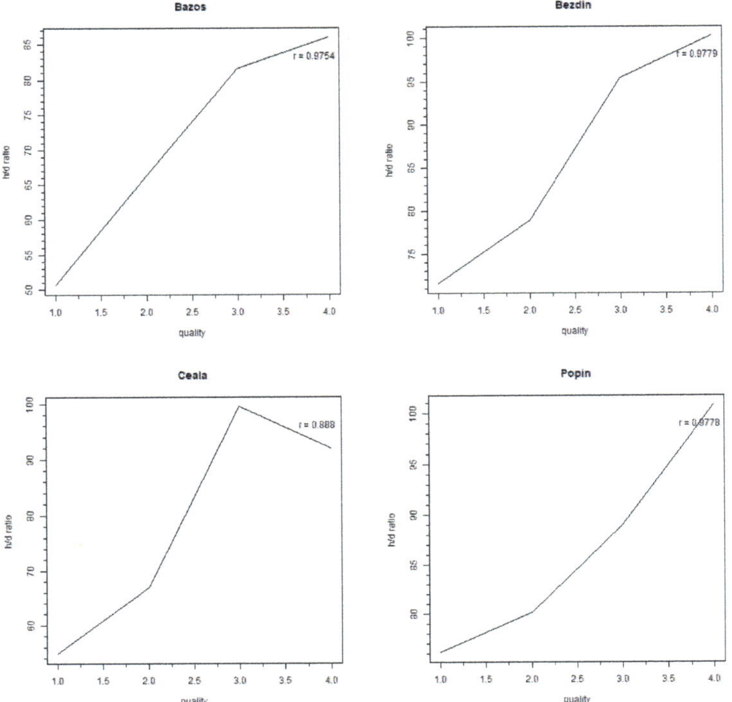

Figure 4. Relationship between h/d ratio and wood quality.

It can also be added that the h/d ratio values recorded for wood quality class 1 (superior) ranged from 50 (Bazos plot) to 76 (Popin plot). For the fourth (lower) quality class, the minimum value recorded for this indicator was 81 (Ceala plot), and the maximum was 101 (Popin plot). The optimal values recorded for this characteristic of the researched stands ranged from 70 (Bazos plot) up to 85 (Bezdin and Popin plots).

4. Discussion

The analysis of the main statistical indicators revealed high variability in diameters in all four research plots. Hence, the values of the coefficient of variation of the diameters were recorded to be 20–30% higher than the specific values for the even-aged stands [52]. At the same time, the amplitude between minimum and maximum dbh indicates an uneven-aged structure of these stands [53–55]. It should be added that uneven-aged stands are generally known to confer a high degree of stability compared to even-aged stands [56], thus providing protection against natural hazards [4] and increasing the resilience of forests overall.

By fitting experimental distributions of diameters, a descending trend of diameters was observed, with most of them found in the small diameter categories. This pattern is characteristic of uneven-aged stands [57]. Further confirmation of the uneven-aged structure is the fact that the experimental distribution of diameters differed significantly from the theoretical Normal distribution, which is considered an indicator of the uneven-aged structure [58].

As Weibull distribution is known to provide a more flexible diameter distribution description [59], it has been possible to capture the characterization of the trees in the stands in more detail. Despite the fact that the results showed differences in the null hypothesis test using the Kolmogorov–Smirnov test, it should be pointed out that this test is more demanding than the chi-square criterion [60]. At the same time, the differences between the theoretical and experimental values recorded were not very large. However, the tree diversity recorded is high, highlighting their importance, mostly because it is known that urban and peri-urban forests with low tree diversity have a high risk of potential ecological disturbances [61].

Even though the term stability has no generally accepted definition, authors such as Grimm and Wissel [62] suggested that constancy, resilience, and persistence are the main stability properties, whereas [63] stressed that the vital stability properties in protective forests are resistance and elasticity and looked to assess the forests' stability properties through such forest responses to disturbances using silvicultural means.

A tree shape, expressed by the h/d ratio, is a good indicator of tree susceptibility to windthrow and snowthrow [64,65] and has been successfully applied in the past for this description [66]. As in other studies [67,68], a trunk shape in the surveyed plots was influenced by silvicultural treatment.

Stand stability plays an important role in the evolution of forest ecosystems, and each tree contributes to the stability of the whole stand. The elements that significantly influence stand stability include component species, growth stage, density, structure, height-to-diameter ratio (h/d ratio), wood quality, phytosanitary conditions, and the presence of wounds caused by biotic factors [34,69]. Recent research has shown a close relationship between two of these elements, especially the height-to-diameter ratio (h/d ratio) and wood quality [67,70,71].

These elements are easy to obtain, especially the height-to-diameter ratio (h/d ratio), and can also provide valuable information on stand stability [67,70] and, for managed stands, indicate the appropriate application of silvicultural treatments [72].

However, perceiving stability as a general ecosystem function might raise some issues regarding forest management due to its diverse implications. In addition to the stability they offer, the peri-urban forests investigated offer direct use value (recreation) and indirect use value (scenic views) [73]. Thus, stability becomes a factor of specific forest characteristics of high importance to society, acting toward protection against various drivers of disturbances, such as changes in the structure and quality of the environment and issues related to biotic factors [4].

The stability of trees and forests near urban areas comes as an informal request of society to access the forest for different recreational reasons safely. In Eastern Europe, they not only exercise their perceived rights but act as a consequence of changes in public access to forests since the dawn of communism and the modern influences of the EU policies. Romania, along with Croatia and Serbia, had no regulations during the 1990s regarding public access to private forests, yet after the communist era and since 2008, private forest owners in Romania gained the legal right to exclusion, prohibiting public access to private forests, even on forest roads [74]. However, among the other rights of owners (access, withdrawal, management, exclusion, and alienation), exclusion rights are the only real freedom that the private owner has; this issue is stressed by Nichiforel et al. [75], suggesting that private owners are becoming creative through rent-seeking activities which imply innovation in the case of Romania and are indirectly producing "rights" for themselves.

Nevertheless, this study is located in a state-owned forested area under protection; the public access is open to a certain extent and is restricted only under extreme conditions (forest fires, floods, wind throws, etc.). Whilst private owners are seeking opportunities to exercise and extend their rights, the state provides ecosystem services, such as recreational, aesthetic, and other similar societal needs, free of cost in most cases, based on the services provided. Whether private owners would take the opportunity and set aside part of their forest due to increased urbanization and societal needs is problematic. Setting aside part of one's property requires incentives under the form of compensations, whereas Romania's compensation schemes are highly regulated, overly bureaucratic, and time-consuming, and forest owners are becoming reticent to such voluntary actions.

Interestingly enough, by forecasting global urban expansion, Seto et al. [76] generally predicted for Europe that the average city would have an annual urban expansion rate of 2.50%, with around 86% attributed to GDP (gross domestic product) per capita growth and 4% attributed to population growth. Globally, the above-mentioned "average" city in the recorded study exhibited an urban population growth rate of 2.18% and an urban land expansion growth rate of 4.84%. This indicates that each city in the study added almost 46,000 urban dwellers per year and approximately 13.5 km^2 of new urban land. According to the Romanian National Institute of Statistics database [77], the overall population growth in urban areas in the studied regions decreased by more than 1% on average, while in rural areas, it increased by 1.5% on average. Regarding the dwellings nearby cities, the overall population growth in rural areas in a 10-year span exceeds 8%. However, in Romania, the 39% growth in GDP per capita in the Western region since 2009 [78] might have influenced urban expansion into rural areas.

Taking into account that forests are constantly mentioned within the ecological development discourse, specifically for rural areas with urban characteristics, the role of forests becomes more diverse rather than just production or afforestation/reforestation. Near urban areas, forestry falls out of its traditionalism and purposes and reinvents itself as urban forestry, where the quality of life aspects shade out the forest impact aspects [79].

The preservation of these ecosystems and the role that their stability plays in providing a so-called life insurance policy for future generations must be considered, mainly due to the low predictability of natural hazards (e.g., windthrow and snowthrow); however, an increase in tree and stand stability can mitigate the impact of natural hazards. By acquiring this knowledge, forest managers can adapt their management to mitigate the impact of such natural hazards.

5. Conclusions

Seen as a significant challenge to countries all over the globe, greenhouse gas (GHG) emissions have gained increased attention over recent decades due to the social problems caused, such as climate change, higher sea levels, and certain fauna extinction. Researchers are trying to find practical solutions to limit their negative effects and impacts on society while, at the same time, addressing the challenges raised by recent greenhouse gas policies on a national and international level.

In the past few years, climate change has gained the notoriety of a "climate crisis"; it became a social issue due to its social impact, with the right to access forests being highly considered in policy formulation. The ongoing increase in large-scale droughts, numerous forest disturbances, and, more importantly, land use changes has led to severe human migration, creating or expanding communities. Therefore, the management of these ecosystems must be adaptive and, at the same time, long-term.

Urbanization and land expansion have led to and will intensify migration from urban to rural areas. As in other European countries, in Romania, urban areas are expected to become larger and overpopulated, and it is believed this will harm people's well-being and health.

Altering the dynamics of human–nature interactions, urbanization influences changes in spatial (overpopulation in cities), temporal (increasing negative interactions over time), and socio-economic inequalities (increased positive interactions among wealthier individuals). The population growth in rural areas comes as a reaction to increased urbanization, and based on the growth in GDP per capita, the environment is faced with much more pressure, especially in terms of ecological and social functions, as households from more socio-economically advantaged backgrounds are expected to make the transition to rural areas and are usually known to be involved in solving environmental and social issues in the local community and ask for their rights by law.

Due to the increase in human–nature interactions in the past decade, beneficial environmental effects are commonly sought in the form of resources, protection from predators, stability and resilience to disturbances, well-being, health, and safety. However, these benefits usually come with a cost, as in the case of recreational activities in protected areas, which can exert negative impacts on local biodiversity. The future management of these peri-urban forests must consider their conservation status and the expansion of these areas where possible concerning the communities. Thus, it implies the establishment of integrated forest management where all ecosystem services are balanced through a participatory approach and fair involvement, satisfying all user needs and mitigating conflicts.

Author Contributions: Conceptualization, S.C. and N.C.; methodology, S.C., M.H. and N.C.; software, S.C.; validation, S.C. and R.G.R.; formal analysis, S.C. and F.C.; investigation, N.C.; resources, N.C. and O.B.; data curation, F.C.; writing—original draft preparation, S.C. and N.C.; writing—review and editing, S.C., M.H. and R.G.R.; visualization, M.H.; supervision, S.C. and O.B; project administration, S.C.; funding acquisition, S.C. and O.B. All authors have read and agreed to the published version of the manuscript.

Funding: This research was funded partially by the BIOSERV Programme, Project ID PN19070103, through the CresPerfInd project—"Increasing the institutional capacity and performance of INCDS "Marin Drăcea" in the activity of RDI"(Contract No. 34PFE./30.12.2021), financed by the Ministry of Research, Innovation, and Digitization through Program 1—Development of the national research—development system, Subprogram 1.2—Institutional performance—Projects to finance excellence in RDI, and partially by Contract 47/N/2019 between the Ministry of Waters and Forests and the National Institute for Research and Development in Forestry "Marin Drăcea" in supporting activities for estimating and reporting national greenhouse gas emissions and retention in forest land use as part of the Land Use and Land Use Change and Forestry sector.

Institutional Review Board Statement: Not applicable.

Informed Consent Statement: Not applicable.

Data Availability Statement: Not applicable.

Conflicts of Interest: The authors declare no conflict of interest.

References

1. Wang, X.; Khurshid, A.; Qayyum, S.; Calin, A.C. The Role of Green Innovations, Environmental Policies and Carbon Taxes in Achieving the Sustainable Development Goals of Carbon Neutrality. *Environ. Sci. Pollut. Res.* **2022**, *29*, 8393–8407. [CrossRef]
2. Whyte, K. Indigenous Climate Change Studies: Indigenizing Futures, Decolonizing the Anthropocene. *Engl. Lang. Notes* **2017**, *55*, 153–162. [CrossRef]
3. Mora, C.; Spirandelli, D.; Franklin, E.C.; Lynham, J.; Kantar, M.B.; Miles, W.; Smith, C.Z.; Freel, K.; Moy, J.; Louis, L.V.; et al. Broad Threat to Humanity from Cumulative Climate Hazards Intensified by Greenhouse Gas Emissions. *Nat. Clim. Chang.* **2018**, *8*, 1062–1071. [CrossRef]
4. Forzieri, G.; Bianchi, A.; Silva, F.B.E.; Marin Herrera, M.A.; Leblois, A.; Lavalle, C.; Aerts, J.C.J.H.; Feyen, L. Escalating Impacts of Climate Extremes on Critical Infrastructures in Europe. *Glob. Environ. Chang.* **2018**, *48*, 97–107. [CrossRef]
5. Ouyang, Z.; Sciusco, P.; Jiao, T.; Feron, S.; Lei, C.; Li, F.; John, R.; Fan, P.; Li, X.; Williams, C.A.; et al. Albedo Changes Caused by Future Urbanization Contribute to Global Warming. *Nat. Commun.* **2022**, *13*, 3800. [CrossRef]
6. Muntean, M.; Guizzardi, D.; Schaaf, E.; Crippa, M.; Solazzo, E.; Olivier, J.; Vignati, E. *Fossil CO2 Emissions of All World Countries*; Publications Office of the European Union: Luxembourg, 2018; Volume 2.
7. Gurmesa, G.A.; Wang, A.; Li, S.; Peng, S.; de Vries, W.; Gundersen, P.; Ciais, P.; Phillips, O.L.; Hobbie, E.A.; Zhu, W.; et al. Retention of Deposited Ammonium and Nitrate and Its Impact on the Global Forest Carbon Sink. *Nat. Commun.* **2022**, *13*, 880. [CrossRef] [PubMed]
8. Baumgartner, R.J. Sustainable Development Goals and the Forest Sector-A Complex Relationship. *Forests* **2019**, *10*, 152. [CrossRef]
9. Börner, J.; Schulz, D.; Wunder, S.; Pfaff, A. The Effectiveness of Forest Conservation Policies and Programs. *Annu. Rev. Resour. Econ.* **2020**, *12*, 45–64. [CrossRef]
10. Mansoor, S.; Farooq, I.; Kachroo, M.M.; Mahmoud, A.E.D.; Fawzy, M.; Popescu, S.M.; Alyemeni, M.N.; Sonne, C.; Rinklebe, J.; Ahmad, P. Elevation in Wildfire Frequencies with Respect to the Climate Change. *J. Environ. Manag.* **2022**, *301*, 113769. [CrossRef]
11. Moosmann, L.; Siemons, A.; Fallasch, F.; Schneider, L.; Urrutia, C.; Wissner, N.; Oppelt, D. The COP26 Climate Change Conference. In Proceedings of the Glasgow Climate Change Conference, Glasgow, Scotland, 31 October–12 November 2021.
12. Krishnan, R.; Agarwal, R.; Bajada, C.; Arshinder, K. Redesigning a Food Supply Chain for Environmental Sustainability – An Analysis of Resource Use and Recovery. *J. Clean. Prod.* **2020**, *242*, 118374. [CrossRef]
13. Wen, C.; Dong, W.; Zhang, Q.; He, N.; Li, T. A System Dynamics Model to Simulate the Water-Energy-Food Nexus of Resource-Based Regions: A Case Study in Daqing City, China. *Sci. Total Environ.* **2022**, *806*, 150497. [CrossRef] [PubMed]
14. Lambin, E.F.; Gibbs, H.K.; Ferreira, L.; Grau, R.; Mayaux, P.; Meyfroidt, P.; Morton, D.C.; Rudel, T.K.; Gasparri, I.; Munger, J. Estimating the World's Potentially Available Cropland Using a Bottom-up Approach. *Glob. Environ. Chang.* **2013**, *23*, 892–901. [CrossRef]
15. Kurowska, K.; Kryszk, H.; Marks-Bielska, R.; Mika, M.; Leń, P. Conversion of Agricultural and Forest Land to Other Purposes in the Context of Land Protection: Evidence from Polish Experience. *Land Use Policy* **2020**, *95*, 104614. [CrossRef]
16. Bastin, J.F.; Finegold, Y.; Garcia, C.; Gellie, N.; Lowe, A.; Mollicone, D.; Rezende, M.; Routh, D.; Sacande, M.; Sparrow, B.; et al. Response to Comments on "The Global Tree Restoration Potential". *Science* **2019**, *366*, eaay8108. [CrossRef] [PubMed]
17. Holl, K.D.; Brancalion, P.H.S. Tree Planting Is Not a Simple Solution. *Science* **2020**, *368*, 580–581. [CrossRef] [PubMed]
18. Hernández-Morcillo, M.; Torralba, M.; Baiges, T.; Bernasconi, A.; Bottaro, G.; Brogaard, S.; Bussola, F.; Díaz-Varela, E.; Geneletti, D.; Grossmann, C.M.; et al. Scanning the Solutions for the Sustainable Supply of Forest Ecosystem Services in Europe. *Sustain. Sci.* **2022**, *17*, 2013–2029. [CrossRef] [PubMed]
19. Watson, J.E.M.; Evans, T.; Venter, O.; Williams, B.; Tulloch, A.; Stewart, C.; Thompson, I.; Ray, J.C.; Murray, K.; Salazar, A.; et al. The Exceptional Value of Intact Forest Ecosystems. *Nat. Ecol. Evol.* **2018**, *2*, 599–610. [CrossRef] [PubMed]
20. Makovníková, J.; Kološta, S.; Flaška, F.; Pálka, B. Potential of Regulating Ecosystem Services in Relation to Natural Capital in Model Regions of Slovakia. *Sustainability* **2023**, *15*, 1076. [CrossRef]
21. Frank, G.; Müller, F. Voluntary Approaches in Protection of Forests in Austria. *Environ. Sci. Policy* **2003**, *6*, 261–269. [CrossRef]
22. Bončina, A.; Simončič, T.; Rosset, C. Assessment of the Concept of Forest Functions in Central European Forestry. *Environ. Sci. Policy* **2019**, *99*, 123–135. [CrossRef]
23. Pilli, R.; Pase, A. Forest Functions and Space: A Geohistorical Perspective of European Forests. *IForest* **2018**, *11*, 79. [CrossRef]
24. Tiemann, A.; Ring, I. Towards Ecosystem Service Assessment: Developing Biophysical Indicators for Forest Ecosystem Services. *Ecol. Indic.* **2022**, *137*, 108704. [CrossRef]
25. Nicolescu, V.-N. 1.1 Romanian Forests and Forestry: An Overview. In *Plan B for Romania's Forests and Society*; Universitatea "Transilvania": Braşov, Romania, 2022; pp. 39–48.
26. Nowak, D.J.; Walton, J.T.; Dwyer, J.F.; Kaya, L.G.; Myeong, S. The Increasing Influence of Urban Environments on US Forest Management. *J. For.* **2005**, *103*, 377–382.
27. Henwood, K.; Pidgeon, N. Talk about Woods and Trees: Threat of Urbanization, Stability, and Biodiversity. *J. Environ. Psychol.* **2001**, *21*, 125–147. [CrossRef]
28. Funk, J.M.; Aguilar-Amuchastegui, N.; Baldwin-Cantello, W.; Busch, J.; Chuvasov, E.; Evans, T.; Griffin, B.; Harris, N.; Ferreira, M.N.; Petersen, K.; et al. Securing the Climate Benefits of Stable Forests. *Clim. Policy* **2019**, *19*, 845–860. [CrossRef]
29. Berger, F.; Rey, F. Mountain Protection Forests against Natural Hazards and Risks: New French Developments by Integrating Forests in Risk Zoning. *Nat. Hazards* **2004**, *33*, 395–404. [CrossRef]

30. Wonn, H.T.; O'Hara, K.L. Height:Diameter Ratios and Stability Relationships for Four Northern Rocky Mountain Tree Species. *West. J. Appl. For.* **2001**, *16*, 87–94. [CrossRef]
31. O'Hara Kevin, L. What Is Close-to-Nature Silviculture in a Changing World? *Forestry* **2016**, *89*, 1–6. [CrossRef]
32. Qin, Y.; He, X.; Lei, X.; Feng, L.; Zhou, Z.; Lu, J. Tree Size Inequality and Competition Effects on Nonlinear Mixed Effects Crown Width Model for Natural Spruce-Fir-Broadleaf Mixed Forest in Northeast China. *For. Ecol. Manag.* **2022**, *518*, 120291. [CrossRef]
33. Bosela, M.; Lukac, M.; Castagneri, D.; Sedmák, R.; Biber, P.; Carrer, M.; Konôpka, B.; Nola, P.; Nagel, T.A.; Popa, I.; et al. Contrasting Effects of Environmental Change on the Radial Growth of Co-Occurring Beech and Fir Trees across Europe. *Sci. Total Environ.* **2018**, *615*, 1460–1469. [CrossRef]
34. Cukor, J.; Vacek, Z.; Vacek, S.; Linda, R.; Podrázský, V. Biomass Productivity, Forest Stability, Carbon Balance, and Soil Transformation of Agricultural Land Afforestation: A Case Study of Suitability of Native Tree Species in the Submontane Zone in Czechia. *Catena* **2022**, *210*, 105893. [CrossRef]
35. Zhang, L.; Bi, H.; Cheng, P.; Davis, C.J. Modeling Spatial Variation in Tree Diameter-Height Relationships. *For. Ecol. Manag.* **2004**, *189*, 317–329. [CrossRef]
36. Raptis, D.I.; Kazana, V.; Kazaklis, A.; Stamatiou, C. Mixed-Effects Height–Diameter Models for Black Pine (Pinus Nigra Arn.) Forest Management. *Trees -Struct. Funct.* **2021**, *35*, 1167–1183. [CrossRef]
37. Crecente-Campo, F.; Tomé, M.; Soares, P.; Diéguez-Aranda, U. A Generalized Nonlinear Mixed-Effects Height–Diameter Model for Eucalyptus Globulus L. in Northwestern Spain. *For. Ecol. Manag.* **2010**, *259*, 943–952. [CrossRef]
38. Ozcelik, R. Tree Species Diversity of Natural Mixed Stands in Eastern Black Sea and Western Mediterranean Region of Turkey. *J. Environ. Biol.* **2009**, *30*, 6.
39. Kang, H.; Seely, B.; Wang, G.; Cai, Y.; Innes, J.; Zheng, D.; Chen, P.; Wang, T. Simulating the Impact of Climate Change on the Growth of Chinese Fir Plantations in Fujian Province, China. *N. Z. J. For. Sci.* **2017**, *47*, 1–14. [CrossRef]
40. Huang, S.; Wiens, D.P.; Yang, Y.; Meng, S.X.; Vanderschaaf, C.L. Assessing the Impacts of Species Composition, Top Height and Density on Individual Tree Height Prediction of Quaking Aspen in Boreal Mixedwoods. *For. Ecol. Manag.* **2009**, *258*, 1235–1247. [CrossRef]
41. Vanclay, J.K. Tree Diameter, Height and Stocking in Even-Aged Forests. *Ann. For. Sci.* **2009**, *66*, 1–7. [CrossRef]
42. Giurgiu, V.; Decei, I.; Drăghiciu, D. *Metode şi Tabele Metode şi Tabele Dendrometrice [Methods and Yield Tables]*; Editura Ceres: Bucharest, Romania, 2004; pp. 27–575.
43. Budeanu, M.; Apostol, E.N.; Popescu, F.; Postolache, D.; Ioniţă, L. Testing of the Narrow Crowned Norway Spruce Ideotype (Picea Abies f. Pendula) and the Hybrids with Normal Crown Form (Pyramidalis) in Multisite Comparative Trials. *Sci. Total Environ.* **2019**, *689*, 980–990. [CrossRef]
44. Pascu, I.-S.; Dobre, A.-C.; Badea, O.; Andrei Tanase, M. Estimating Forest Stand Structure Attributes from Terrestrial Laser Scans. *Sci. Total Environ.* **2019**, *691*, 205–215. [CrossRef]
45. Chivulescu, S.; Ciceu, A.; Leca, S.; Apostol, B.; Popescu, O.; Badea, O. Development Phases and Structural Characteristics of the Penteleu-Viforata Virgin Forest in the Curvature Carpathians. *Iforest* **2020**, *13*, 389–395. [CrossRef]
46. Grosjean, P.; Ibanez, F. Package for Analysis of Space-Time Ecological Series. PASTECS, R Package, Version 1.2-0 for R v. 2.0. 0 & Version 1.0-1 for S+ 2000 rel. 2004. Available online: https://cran.r-project.org/web/packages/pastecs/pastecs.pdf (accessed on 17 January 2023).
47. Hogg, R.V.; Craig, A.T. Some Special Distributions. *Introd. Math. Stat.* **1978**, *1*, 156–168.
48. Sharif, M.N.; Islam, M.N. The Weibull Distribution as a General Model for Forecasting Technological Change. *Technol. Forecast. Soc. Chang.* **1980**, *18*, 247–256. [CrossRef]
49. Chivulescu, Ş.; Pitar, D.; Apostol, B.; Leca, Ş.; Badea, O. Importance of Dead Wood in Virgin Forest Ecosystem Functioning in Southern Carpathians. *Forests* **2022**, *13*, 409. [CrossRef]
50. Stephens, M.A. Tests of Fit for the Logistic Distribution Based on the Empirical Distribution Function. *Biometrika* **1979**, *66*, 591–595. [CrossRef]
51. Team, R.C. *R: A Language and Environment for Statistical Computing*; R Foundation for Statistical Computing: Vienna, Austria, 2009; Available online: https://cir.nii.ac.jp/crid/1570854175843385600 (accessed on 17 January 2023).
52. Giurgiu, V. *Metode ale Statisticii Matematice Aplicate în Silvicultură [Mathematical Statistical Methods Applied in Forestry]*; Ceres: Bucharest, Romania, 1972.
53. Monserud, R.A.; Sterba, H. A Basal Area Increment Model for Individual Trees Growing in Even- and Uneven-Aged Forest Stands in Austria. *For. Ecol. Manag.* **1996**, *80*, 57–80. [CrossRef]
54. Roessiger, J.; Ficko, A.; Clasen, C.; Griess, V.C.; Knoke, T. Variability in Growth of Trees in Uneven-Aged Stands Displays the Need for Optimizing Diversified Harvest Diameters. *Eur. J. For. Res.* **2016**, *135*, 283–295. [CrossRef]
55. Bayat, M.; Bettinger, P.; Heidari, S.; Khalyani, A.H.; Jourgholami, M.; Hamidi, S.K. Estimation of Tree Heights in an Uneven-Aged, Mixed Forest in Northern Iran Using Artificial Intelligence and Empirical Models. *Forests* **2020**, *11*, 324. [CrossRef]
56. Pretzsch, H. The Course of Tree Growth. Theory and Reality. *For. Ecol. Manag.* **2020**, *478*, 118508. [CrossRef]
57. Zhang, L.; Gove, J.H.; Liu, C.; Leak, W.B. A Finite Mixture of Two Weibull Distributions for Modeling the Diameter Distributions of Rotated-Sigmoid, Uneven-Aged Stands. *Can. J. For. Res.* **2001**, *31*, 1654–1659. [CrossRef]
58. Bergeron, Y.; Leduc, A.; Harvey, B.D.; Gauthier, S. Natural Fire Regime: A Guide for Sustainable Management of the Canadian Boreal Forest. *Silva Fenn.* **2002**, *36*, 81–95. [CrossRef]

59. Bebbington, M.; Lai, C.D.; Zitikis, R. A Flexible Weibull Extension. *Reliab. Eng. Syst. Saf.* **2007**, *92*, 719–726. [CrossRef]
60. Mitchell, B. A Comparison of Chi-Square and Kolmogorov-Smirnov Tests. *Area* **1971**, *3*, 237–241.
61. Blood, A.; Starr, G.; Escobedo, F.; Chappelka, A.; Staudhammer, C. How Do Urban Forests Compare? Tree Diversity in Urban and Periurban Forests of the Southeastern US. *Forests* **2016**, *7*, 120. [CrossRef]
62. Grimm, V.; Wissel, C. Babel, or the Ecological Stability Discussions: An Inventory and Analysis of Terminology and a Guide for Avoiding Confusion. *Oecologia* **1997**, *109*, 323–334. [CrossRef]
63. Motta, R.; Haudemand, J.-C. Protective Forests and Silvicultural Stability. *Mt. Res. Dev.* **2000**, *20*, 180–187. [CrossRef]
64. Wilson, J.S.; Oliver, C.D. Stability and Density Management in Douglas-Fir Plantations. *Can. J. For. Res.* **2000**, *30*, 910–920. [CrossRef]
65. Harrington, T.B.; Harrington, C.A.; DeBell, D.S. Effects of Planting Spacing and Site Quality on 25-Year Growth and Mortality Relationships of Douglas-Fir (Pseudotsuga Menziesii Var. Menziesii). *For. Ecol. Manag.* **2009**, *258*, 18–25. [CrossRef]
66. Valenzuela, C.E.; Ballesta, P.; Maldonado, C.; Baettig, R.; Arriagada, O.; Mafra, G.S.; Mora, F. Bayesian Mapping Reveals Large-Effect Pleiotropic QTLs for Wood Density and Slenderness Index in 17-Year-Old Trees of Eucalyptus Cladocalyx. *Forests* **2019**, *10*, 241. [CrossRef]
67. 67. Chivulescu, S.; Leca, S.; Ciceu, A.; Pitar, D.; Apostol, B. Predictors of wood quality of trees in primary forests in the Southern Carpathians. *Poljoprivreda i Sumarstvo* **2019**, *65*, 13. [CrossRef]
68. Brazier, J.D. The Effect of Forest Practices on Quality of the Harvested Crop. *Forestry* **1977**, *50*, 49–66. [CrossRef]
69. Bodin, P.; Wiman, B.L.B. The Usefulness of Stability Concepts in Forest Management When Coping with Increasing Climate Uncertainties. *For. Ecol. Manag.* **2007**, *242*, 541–552. [CrossRef]
70. Pretzsch, H.; Rais, A. Wood Quality in Complex Forests versus Even-Aged Monocultures: Review and Perspectives. *Wood Sci Technol* **2016**, *50*, 845–880. [CrossRef]
71. Drew, D.M.; Downes, G.M.; Seifert, T.; Eckes-Shepard, A.; Achim, A. A Review of Progress and Applications in Wood Quality Modelling. *Curr. For. Rep.* **2022**, *8*, 317–332. [CrossRef]
72. Boncina, A.; Cavlovic, J.; Curovic, M.; Govedar, Z.; Klopcic, M.; Medarevic, M. A Comparative Analysis of Recent Changes in Dinaric Uneven-Aged Forests of the NW Balkans. *Forestry* **2014**, *87*, 71–84. [CrossRef]
73. Tu, G.; Abildtrup, J.; Garcia, S. Preferences for Urban Green Spaces and Peri-Urban Forests: An Analysis of Stated Residential Choices. *Landsc. Urban Plan.* **2016**, *148*, 120–131. [CrossRef]
74. Lawrence, A.; Deuffic, P.; Hujala, T.; Nichiforel, L.; Feliciano, D.; Jodlowski, K.; Lind, T.; Marchal, D.; Talkkari, A.; Teder, M.; et al. Extension, Advice and Knowledge Systems for Private Forestry: Understanding Diversity and Change across Europe. *Land Use Policy* **2020**, *94*, 104522. [CrossRef]
75. Nichiforel, L.; Schanz, H. Property Rights Distribution and Entrepreneurial Rent-Seeking in Romanian Forestry: A Perspective of Private Forest Owners. *Eur. J. For. Res.* **2011**, *130*, 369–381. [CrossRef]
76. Seto, K.C.; Fragkias, M.; Güneralp, B.; Reilly, M.K. A Meta-Analysis of Global Urban Land Expansion. *PLoS ONE* **2011**, *6*, e23777. [CrossRef]
77. INSSE Oficial Website of Romanian National Institute of Statistics. Available online: https://insse.ro/cms/en (accessed on 15 January 2023).
78. Timis County Council, Romania Timis County Council Information Report. Available online: https://www.cjtimis.ro/infocjt/admin/upload/9wv26dmxrpb3.pdf (accessed on 7 December 2022).
79. Elands, B.H.M.; O'Leary, T.N.; Boerwinkel, H.W.J.; Wiersum, K.F. Forests as a Mirror of Rural Conditions; Local Views on the Role of Forests across Europe. *Proc. For. Policy Econ.* **2004**, *6*, 469–482. [CrossRef]

Article

Functional Traits and Local Environmental Conditions Determine Tropical Rain Forest Types at Microscale Level in Southern Ecuador

Omar Cabrera [1,2,*], Pablo Ramón [2], Bernd Stimm [2], Sven Gunter [3] and Reinhard Mosandl [2]

1 Department of Biological and Natural Sciences, Universidad Técnica Particular de Loja, San Cayetano Alto s/n, Loja 1101608, Ecuador
2 Department of Life Science Systems, Technical University of Munich, Hans-Carl-von- Carlowitz-Platz 2, 85354 Freising, Germany
3 Institut für Internationale Waldwirtschaft und Forstökonomie, Leuschnerstraße 91, 21031 Hamburg-Bergedorf, Germany
* Correspondence: hocabrera@utpl.edu.ec; Tel.: +593-7-3701444

Abstract: The main objective of this study was to determine the heterogeneity of tropical mountain rain forests along a micro-altitudinal gradient scale, integrating species functional traits in the separation of communities. To achieve this, a forest area of 13 ha in the Biological Reserve of San Francisco was monitored. First, we performed non-metric multidimensional analyses, and afterwards, we looked for correlations between plot altitude and characteristics of the forest (basal area, the number of species, the number of trees ≥20 cm diameter at breast height, per hectare, the forest canopy opening) were associated. To determine which characteristics significantly influence the separation of forest "communities", we used a multivariate canonical correspondence analysis (CCA). Finally, we carried out the "Four Corners" analysis, combining abundance matrices, traits and environmental variables. We confirmed that the altitude and some associated characteristics are the key factors for the formation of two forest types. In addition, we determined that the inclusion of species functional traits confirms the separation of forest communities, and that elevation and its associated environmental variables function over relatively small areas and scales.

Keywords: diversity; four corner; microscale elevation; correlation; elevation effects

Citation: Cabrera, O.; Ramón, P.; Stimm, B.; Gunter, S.; Mosandl, R. Functional Traits and Local Environmental Conditions Determine Tropical Rain Forest Types at Microscale Level in Southern Ecuador. *Diversity* **2023**, *15*, 420. https://doi.org/10.3390/d15030420

Academic Editors: Lucian Dinca, Miglena Zhiyanski and Mario A. Pagnotta

Received: 24 January 2023
Revised: 18 February 2023
Accepted: 8 March 2023
Published: 13 March 2023

1. Introduction

Tropical forests (TFs) cover 7% of the earth's surface. Worldwide, 28% of dense forests are tropical mountain forests (TMFs). TFs contain 50% of the world's forest biomass and are considered the most important natural carbon sinks, with a paramount importance in managing the global climate change [1]. TMF are one of the most diverse and threatened ecosystems on earth; this is especially true for the eastern Andean forests [2]. The last map of global diversity of vascular plant species [3] emphasized the areas of TMFs as the most important hotspots of the world.

Furthermore, TFs generate 36% of the net primary terrestrial production, contributing to the regulation of carbon dioxide (CO_2) concentration in the atmosphere [4,5]. TMFs harbor hydrographic river basins and therefore, they are an essential component in the water regime regulation [6]. Some other functions of these forests are producing wood and non-wood products, catching and storing precipitations and humidity, maintaining the quality of the water, and also reducing erosion and protecting against landslides and floods.

The so-called "Tropical Andes" hotspot includes various types of forests known worldwide for their high diversity [2]. In this type of ecosystem, altitude plays a fundamental role in the distribution of diversity [7]. Several authors recognized that altitude and its associated variables determine richness, floristic composition and structure of the forests in these montane environments [8,9]. In many cases, these variables have a synergic and direct effect on species richness, which generally reaches a peak at approximately 1000 m a.s.l.

In Ecuador, it is estimated that the total number of vascular plant species is between 18,000 and 22,000, one of the highest in the world. Additionally, along the Andes, there is a large number of endemic species restricted to just the middle elevations (900–3000 m) [10]. Despite their global importance, the TMFs in Ecuador are the most threatened type of ecosystems, mainly due to change in land use [11]. The latest reports indicate that already in 2005, 51% of the forest area was lost, with a deforestation rate of 1.7%, which is equivalent to 198,000 ha.

The TMFs distributed along the Andes Real Cordillera in southeastern Ecuador, also termed "montane cloud forest" [12] or "evergreen montane forest" [13], can be partitioned into "mountain rainforest" located between 1800–2800 m a.s.l. and "high mountain evergreen forest" located between 2800 and 3100 m a.s.l. [14]. In both cases, the classification was based on widely used physiognomic patterns of the vegetation in response to macroscale geographic regions. Generally, hierarchical vegetation classification models only work at large scales, and elevation data and species distribution models are too broad to be useful. A classification based on structural parameters, diversity and functional traits of the species is still incipient; this is due to the fact that, at a geographic microscale, the response of the community to the high environmental heterogeneity is much more complicated, and the attempts to classify vegetation into physiognomic distinct categories are not straightforward. In this context, our study is based on the hypothesis that by using functional traits of the species, along with environmental variables, we would be able to delimit the different types of forests present in our study area.

2. Materials and Methods

2.1. General Description of the Study Site

The study was carried out in the Reserva Biologica San Francisco (RBSF) 03°58′ S, 79°04′ W; 1850 m a.s.l. [15], located to the North of Podocarpus National Park (PNP) in Southern Ecuador. RBSF is situated within the eastern cordillera of the Andes [16]. This area is geographically located between Loja and Zamora-Chinchipe provinces (Figure 1).

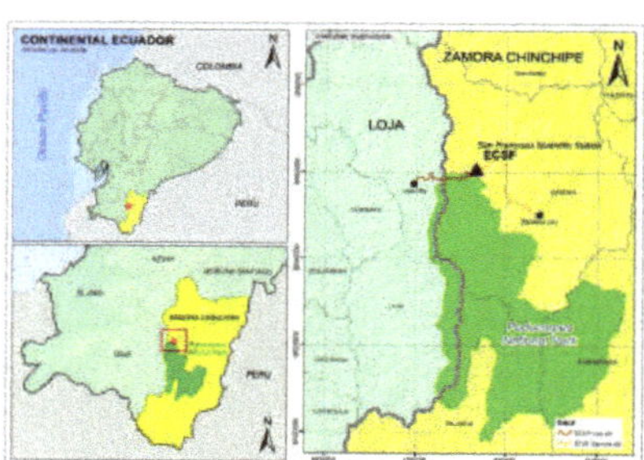

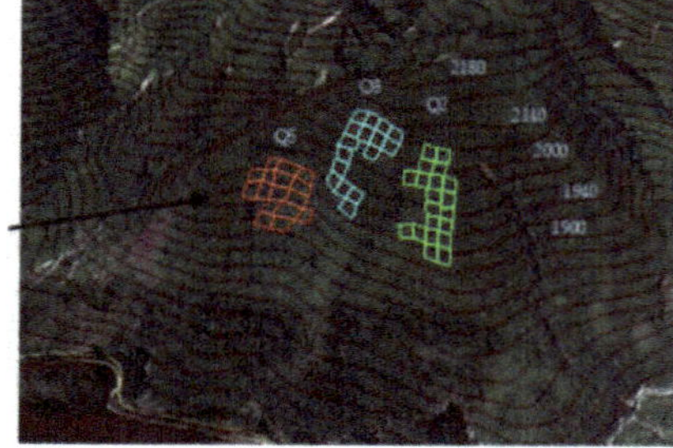

Figure 1. Location of Reserva Biologica San Francisco and Podocarpus National Park. Blocks and plots ubication.

The soils in the RBSF mainly belong to the order of Inceptisols. At the lower parts of the slopes, Dystrudepts are more frequent, whereas in the upper parts, Humaquepts and Petraquepts dominate [17]. The soils in tropical mountain forests are characterized by thick organic layers, which store large contents of biomass and nutrients [18].

The climate is perhumid, with marked altitudinal gradients in air temperature, humidity and rainfall [19]. The annual mean air temperature ranges from 19.4 °C in the valley to 9.4 °C on the highest mountain tops. The distribution of rainfall is linked to altitude, due to orographic precipitation formation. The average annual rainfall amounts vary between 2300 mm in the valley and 6700 mm on the mountain tops [19].

2.2. Installation of Plots and Forest Inventory

We established 50 × 50 m permanent plots in three different watersheds of the RBSF, named Q2, Q3, and Q5. Twenty plots were installed in Q2 (5 ha), sixteen plots in Q3 (4 ha) and sixteen plots in Q5 (4 ha). In all plots, trees with a diameter at breast height (DBH) ≥ 20 cm were labelled, identified and measured; in addition, 52 central internal plots were installed in each of the large plots, where all the individuals with a DBH of 5–19.9 cm were labelled, identified and measured. Samples from the trees that could not be identified in the field were collected; identification was made by comparison with other botanical specimens collected in the LOJA and HUTPL herbaria of the universities closest to the site.

2.3. Data Acquisition

The data for the floristic and structural analyses were taken directly from the inventories carried out in the field; this allowed, afterwards, the calculation of the average DBH per ha, basal area/ha, average total height of the stand. The canopy openness (defined as the unweighted fraction of unobscured sky) was measured on hemispherical pictures using Gap Light Analyzer software. All terrain data, such as altitude of the midpoint of the plot, inclination and exposure, were acquired in the field and used directly in the environmental matrix. The functional traits of each species used to determine their influence on the division of forest types were: (1) wood density (WD) [20–23]; (2) average diameter (DBH); (3) growth (annual diametric increase); (4) leaf type; (5) ecological guild [21,23]; (6) dispersal syndrome. Each of these traits is related to different resource use strategies, through acquisitive mechanisms and conservative mechanisms. For example, wood density is related to growth rate, carbon allocation, disease or pest resistance, and primary production [8,20,24]. Leaf type predicts the growth of tropical trees [25] and they present adaptations that allow plants to live in various environmental conditions [26]. As conservative traits, we used the ecological guild or succession type and the "dispersal syndrome", which is closely related to the relative abundance of species [27] (Table 1).

Table 1. Functional (conservative and acquisitive) traits, including categories, units and codes used to separate forest types.

Type of Trait	Functional Trait	Categories	Unit
Acquisitive	Wood density	-	g cm^3
	Stem density	-	number
	Diametric growth	-	cm $year^{-1}$
	Leaf type	Simple Compound	
Conservative	Ecological guild	Shade tolerant Partial shade tolerant Partial light tolerant Light tolerant	
	Dispersal syndrome	Anemochory Zoochory Barochory	

2.4. Data Analyses

To characterize the forests state in RBSF, structural and floristic parameters were calculated using the following equations (Equations (1)–(5)) [28–35].

$$\text{Relative Density (RD)} = \left(\frac{\text{Individuals per species}}{\text{Total individuals in the plot}}\right) * 100 \quad (1)$$

$$\text{Relative Dominance(RDm)} = \left(\frac{\text{Basal area per species}}{\text{Total Basal Area in the plot}}\right) * 100 \quad (2)$$

$$\text{Relative Diversity(RDi)} = \left(\frac{\text{\#species per Family}}{\text{Total species}}\right) * 100 \quad (3)$$

$$\text{Importance Value Index(IVI)} = (\text{Relative density} + \text{Relative Dominance} \quad (4)$$

$$\text{Shannon} - \text{Wiener Index } (H') = -\sum_{i=1}^{S} p_i \ln p_i \quad (5)$$

S = number of species (richness);

P_i = proportion of individuals of species with respect to total individuals (i.e., relative abundance of the species) n1/N;

ni = number of individuals of the species.

n1 = number of individuals of the species.

N = number of all the individuals of all the species.

2.5. Statistic Analysis

We generated three matrices. The first consisted of the abundance of individuals in each of the plots, obtained from the post-inventory count. The second matrix consisted of the traits of each of the species involved in the study (Table 2). Finally, for each plot, both environmental and terrain data were included in the third matrix. These three matrices were used for all statistical analyses of floristic grouping and to measure the influence of functional traits and environmental factors on vegetation grouping.

Table 2. CCA model p-values for each variable used.

Variable	Var.N	LambdaA	*p*	*F*-Ratio
Altitude	4	0.61	0.001 **	7.09
DBH	7	0.24	0.001 **	3.01
Species	3	0.15	0.001 **	1.76
Trees/ha	2	0.14	0.030 *	1.75
Canopy Openness	8	0.11	0.044 *	1.38
Basal Area/ha	1	0.08	0.279	1.09

Signif. codes: ** 0–0.001; * _0.01.

2.5.1. Non-Metric Multidimensional Scaling (NMDS)

To calculate the resemblance matrix between plots, the Bray–Curtis dissimilarity distance was used. In order to build a dissimilarity matrix between plots, a non-metric multidimensional scaling (NMDS) was performed, to visualize the main environmental factors that influence the grouping of these forest communities. The results were plotted in an NMDS ordination diagram. Values of species abundance, basal area, trees density and environmental variables were fitted onto the first two axes of the NMDS ordination.

The averages at plot level and group level of Shannon diversity index SDI and other characters were compared using a Kruskall–Wallis test between each group assimilable to a vegetation unit. Ordination was performed with the package 'vegan' [36] in R software.

2.5.2. Canonical Correspondence Analyses

Canonical correspondence analyses (CCA) were used to test whether the same parameters employed in the correlation analysis influenced the grouping (based on species abundance) of floristic sample plots distributed on the three blocks. This analysis is a multivariate technique that allows representing low-dimensional geometric space proximity between a set of objects influenced by a series of predictor variables. The lambda value corresponds to the eigenvalue of each extracted variable in each axis of the array. F–ratio is calculated using the trace or the sum of all the eigenvalues, while the p value indicates the significance of variables ($p \leq 0.05$).

Normally, the CCA involves two matrices: the matrix of dependent variables (e.g., a matrix of sites × species) and the matrix of independent variables (e.g., a matrix of environmental variables). The connection between the two matrices is carried out by means of multivariate regression techniques [37]. Parameters were elevation (m a.s.l.), number of trees ha^{-1}, basal area ha^{-1}, canopy openness (%), average DBH and species richness (total number ha^{-1}).

2.5.3. Correlation

To assess whether there was a significant correlation between the number of trees ha^{-1}, basal area ha^{-1}, DBH ha^{-1}, canopy openness and diversity in terms of Shannon Index and altitudinal gradient, Spearman correlation coefficient was used. If the correlation was significant ($r \geq 0.4$; $p \leq 0.05$), we considered that the altitude had an impact on the structure and composition of RBSF, determining two or more floristic groups.

2.5.4. Fourth Corner

To complete the analyses that related the characteristics of the forests (growth, DBH, basal area, canopy openness, richness, composition, etc.) and the attitudinal gradient to determine communities or, in this case, determine the types of forests existing in areas and gradients at microscale, we included the functional traits of the species that theoretically were those that respond to altitudinal changes. For this, we used a model-based approach of the fourth-corner analysis [38,39]. This method relates an R matrix of environmental variables to a Q matrix of species traits, by means of an L matrix of species abundance [39]. Depending on the type of variable to relate, a statistic for each pair of traits and environmental variables was computed (Pearson chi-square for two qualitative variables, pseudo-F and correlation ratio for one quantitative and another qualitative variable). Furthermore, a global multivariate statistic linking R and Q arrays was computed as the sum of all statistics mentioned above, in the fourth-corner matrix [40]. The significance of all fourth-corner statistics was tested using permutation model [40]. Here, we used a model where cell values in the L matrix were permuted within each column, testing the null hypothesis that the species were randomly distributed with respect to environmental conditions [41]. All calculus was developed using the "ade4" package [42] in the R software (R Core Team 2021).

3. Results

3.1. Grouping Plots

Based on plots of the three sampling sites with floristic similarity and a strong correlation with various attributes of the forest (basal area ha^{-1}, canopy openness, trees ha^{-1} and alpha diversity), two types of forest could be determined, clearly different in structure and species composition. The spatial distribution of sample plots over the altitudinal gradient implies a change in the structure and diversity of each of the plots. Structural groups were defined based on a correlation between altitude and different attributes of each of the study plots.

The results of NMDS analysis showed that the sampled plots were divided into two clearly defined groups. The floristic and abundance data used in the matrix formed two groups that we called "forest types". The first group, named "Valley Forest" (VF), consisted of 15 plots from the Q2 site, and all the plots from the Q5 site. The other group, named "Ridge Forest" (RF), was made up of all the plots of the Q3 site and five plots of the Q2 site (Figure 2).

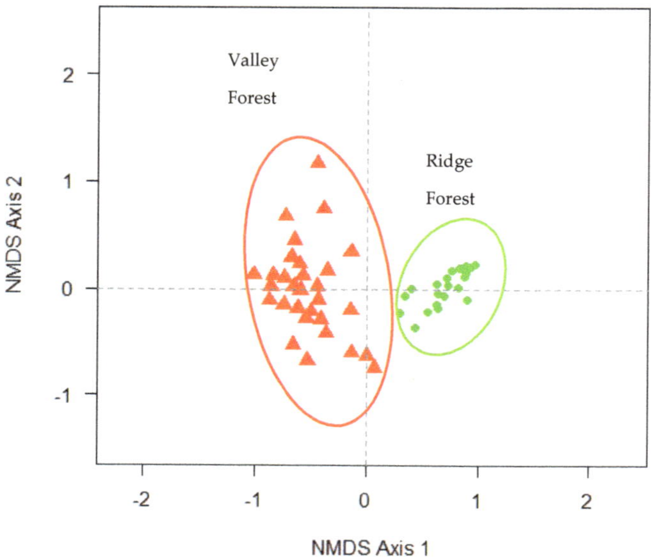

Figure 2. Non-metric multidimensional scaling analysis of species composition for the samples (plots).

3.2. *Influence of Altitudinal and Structural Parameters on Grouping Communities*

After canonical correspondence Analyses (CCA), altitude was a significative variable in the grouping of the plots, along with basal area and trees per hectare (Figure 3). To a smaller degree, species diversity (expressed as Shannon index) and canopy openness were also factors that determined the grouping. The analyses of the values of the canonical axes explained 18.1% of variance in species data and 84% of its relationship with environmental variables (Table 2). This indicates that the structural and floristic characteristics were the result of the influence of the altitudinal gradient and all environmental variables correlated to this (temperature environmental, precipitation, wind, soil, etc.).

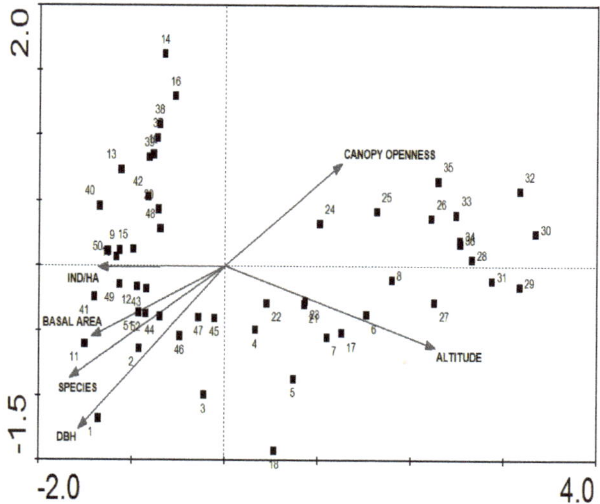

Figure 3. Biplot showing the ordination of 52 sample plots, points show the grouping and the vectors indicate the variables that mostly influenced the grouping.

3.3. Correlation between Elevation and Characteristic of Forest

The basal area decreases ad altitude increased; at 1900 m a.s.l., the average basal area can reach 44 m^2 per hectare (includes only the trees >20 cm DBH). In the plots at 2100 m a.s.l., the average basal area had values of 5 to 14 m^2 per hectare. The same applied to the number of individuals and to the diversity of the sampled sites. This confirmed the trends found in tropical mountain forests, which means that as altitude increases, the diversity of tree species decreases. Canopy openness was higher with the increasing altitude of the plots. The graphs showed a strong correlation between altitude and variable characteristics of each plot (Figure 4).

The VF is characterized by the presence of *Tabebuia chrysantha* (Jacq.) G. Nicholson, *Cedrela montana* Moritz ex Turcz., *Inga acreana* Harms. and *Ficus citrifolia* Mill. There are also other species such as *Cecropia montana* Warb. ex Snethl., *Guarea pterorhachis* Harms and *Heliocarpus americanus* L. that are unique to this group.

The RF is characterized by the presence of *Podocarpus oleifolius* D. Don ex Lamb., *Hyeronima moritziana* Mull. Arg, *Clusia ducuoides* Engl., which were species selected as PCT. Other species that characterize the group are *Purdiaea nutans* Planch., *Graffenrieda emarginata* (Ruiz and Pav.) and *Alchornea grandifolia* Triana and Mull Arg.

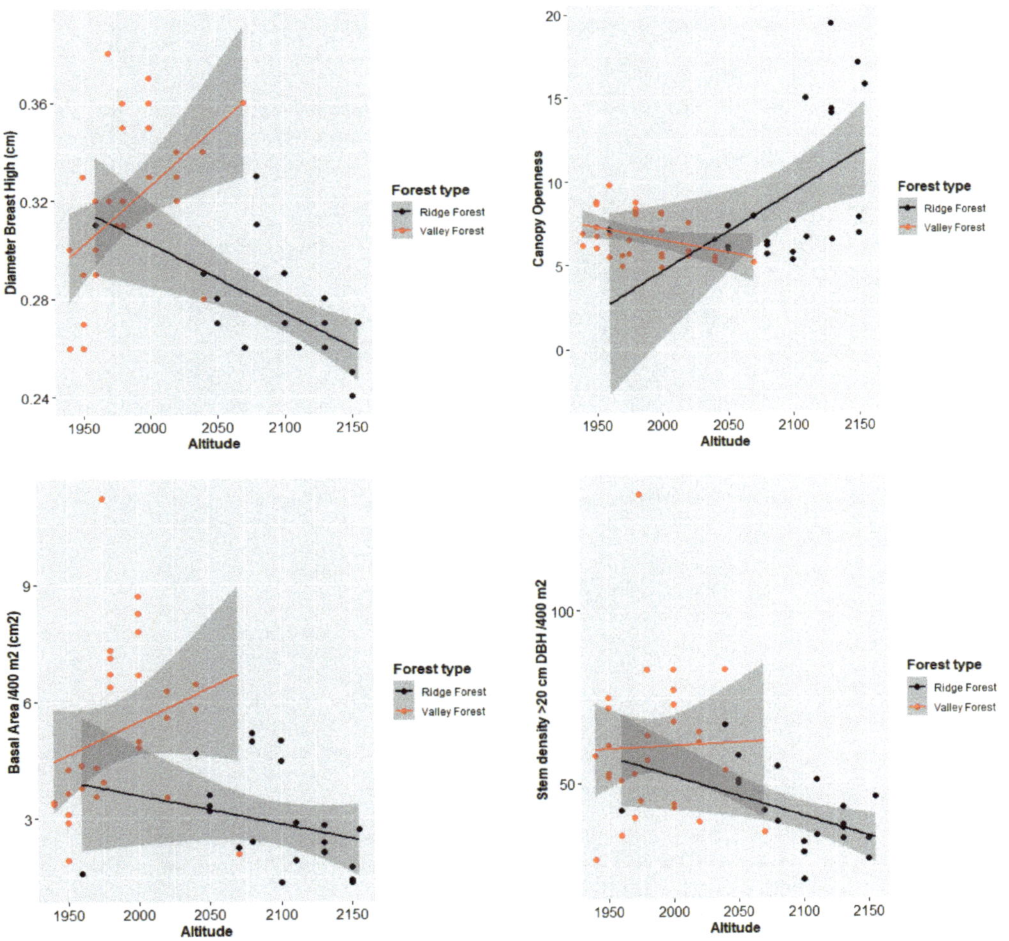

Figure 4. *Cont.*

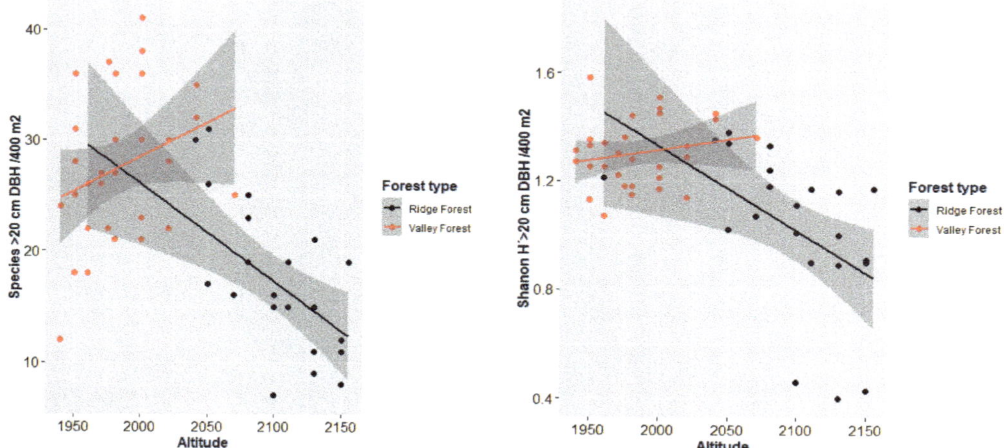

Figure 4. Changes of diversity and structure in altitudinal gradient. Significant correlations ($p < 0.05$) are shown in regression line. For all tests, n = 52 plots.

3.4. *Structural Parameters of Floristic Groups*

In VF, we encountered 141 species, which belonged to 51 families, while in RF, the diversity was represented by 86 species belonging to 51 families. Table 3 shows the relative diversity values of the most important families in each of the determined forest types.

Table 3. Values of relative diversity in both forest types.

Valley Forest			Ridge Forest		
FAMILIES	**Species**	**Relative Diversity (%)**	**FAMILIES**	**# Species**	**Relative Diversity (%)**
Lauraceae	17	12.06	Lauraceae	23	27.7
Moraceae	13	9.22	Euphorbiaceae	7	8.4
Euphorbiaceae	10	7.09	Rubiaceae	5	6
Melastomataceae	9	6.38	Melastomataceae	4	4.8
Meliaceae	9	6.38	Myrtaceae	4	4.8
Cecropiaceae	4	2.84	Clusiaceae	3	3.6
Mimosaceae	4	2.84	Cunnoniaceae	3	3.6
Myrtaceae	4	2.84	Aquifoliaceae	2	2.4
Aquifoliaceae	2	2.13	Arecaceae	2	2.4
Other families (41)	1–2	0.71–2.13	Asteraceae	2	2.4
			Meliaceae	2	2.4
			Mimosaceae	2	2.4
			Moraceae	2	2.4
			Myrsinaceae	2	2.4
			Sapindaceae	2	2.4
			Sapotaceae	2	2.4
			Other families (16)	1	1.2

Table 4 shows the values of density and relative dominance of the most important VF species. Table 5 shows the values of density and relative dominance of the most important RF species.

Table 4. Values of relative density and relative dominance in the VF.

Valley Forest		
Species	**Relative Density (%)**	**Relative Dominance (%)**
Cecropia montana	10.52	15.04
Tabebuia chrysantha	4.82	13.53
Guarea pterorhachis	4.46	10.54
Cecropia gabrielis	4.15	5.58
Heliocarpus americanus	3.63	6.13
Hyeronima asperifolia	3.42	7.54
Piptocoma discolor		6.63
Tapirira obtuse		5.57
Sapium glandulosum	3.27	4.58
Miconia quadripora	3.11	4.61
Inga acreana	2.64	
Nectandra membranacea	2.54	
Other species	0.5–2.64	0.05–3.57

Table 5. Values of relative diversity in both forest types.

Ridge Forest		
Species	**Relative Density (%)**	**Relative Dominance (%)**
Alchornea grandiflora	11.95	13.37
Alzatea verticillata	10.67	15.03
Clusia ducoides	10.44	10.06
Graffenrieda emarginata	7.77	5.37
Purdiaea nutans	5.57	5.45
Hyeronima moritziana	3.83	3.52
Podocarpus oleifolius	3.71	4.23
Tapirira obtusa	3.36	3.86
Myrcia sp.	3.25	2.35
Dictyocaryum lamarckianum	2.67	
Naucleopsis glabra	2.2	
Persea ferruginea	1.86	
Alchornea pearcei	1.74	2.65
Clusia elliptica	1.74	
Nectandra sp.	1.74	
Vismia tomentosa	1.62	
Persea sp.	1.51	
Myrsine coriácea	1.39	
Hyeronima asperifolia	1.28	
Matayba inelegans	1.16	
Other species (58)	0.93–0.12	
Other species (76)		0.06–2.27

The RF had a total basal area of 54.3 m^2, with an average of 10.3 $\pm$ 3.1 m^{2}ha^{-1}, while the VF had a total basal area of 168.7 m^2, with an average of 21.8 $\pm$ 7.9 m^{2}ha^{-1}, considering only the trees >20 cm DBH.

In the RF, there was a total of 862 individuals > 20 cm DBH and an average of 164.2 $\pm$ 35.3 trees ha^{-1}. In the Valley Forest, there was a total of 1933 individuals >20 cm DBH and an average of 248.8 $\pm$ 81.4 trees ha^{-1}.

The total basal area of trees in the class 5.1–20 cm DBH in the RF was 3.47 m^2, which averaged 11.5 $\pm$ 3.9 m^2 ha^{-1}. In VF, the total basal area was 3.7 m^2 in the class 5.1–20 cm DBH, which means an average of 8.8 $\pm$ 3.8 m^2 ha^{-1}. In RF, there was a total of 443 individuals in the class 5.1–20 cm DBH, with an average of 1464 $\pm$ 461.8 trees ha^{-1}. In VF, there was a total of 392 individuals in the class 5.1–20 cm DBH with an average of 906.8 $\pm$ 383 ind./ha.

The figures below show the distributions of individuals in all diametric classes in the two forest types (Figure 5).

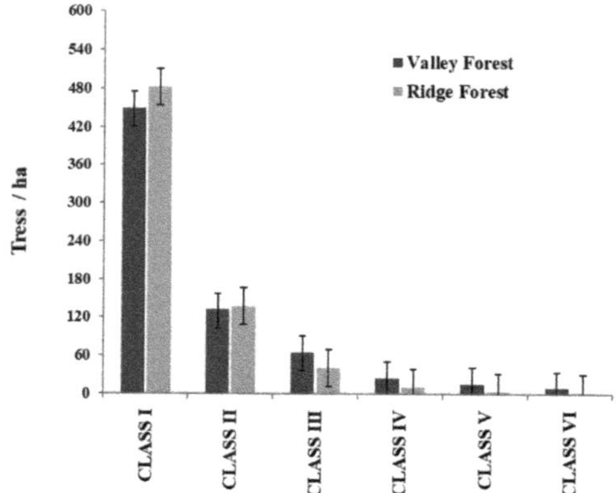

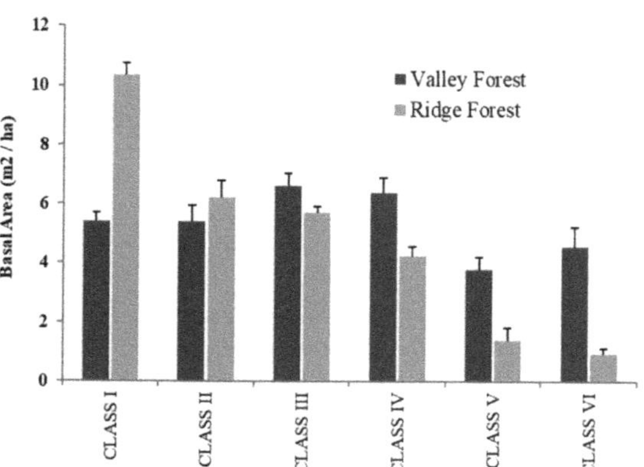

Figure 5. Average basal area ha^{-1} and trees ha^{-1} by diameter classes on each forest type.

The structural difference between the two floristic groups was low, but most evident in the first two diametric classes. In RF, there were more individuals per hectare in the first two classes, while in the Valley Forest, there were more individuals per hectare in the higher diameter classes.

In VF, there were more trees in the larger diameter classes, allowing for harvest at a higher intensity. In RF, the small number of large trees suggests lower harvest intensity, the large number of small individuals and the structure of natural regeneration being a typical feature of these forests at this altitude. Regarding the higher number of individuals in lower diameter classes, the RF contained higher basal area values in these classes than the other group (Figure 5).

The diameter distributions of trees in RBSF indicate that the trees were distributed in large quantities in smaller diameter classes and the numbers decreased in a negative exponential way for higher diameter classes. This inverse J-shape distribution type represents a good structure which is typical for natural forests (Figure 5). Each group has its unique species. In the VF, there were 87 exclusive species, representing 47.5% of the total species identified in the study.

3.5. Functional Traits Influence Forest Separation

The previous results confirmed that the elevation and its associated variables were the main factors that drive the composition and structure of the determined forest types. Likewise, there was a significant correlation at the community level between environmental factors and the plant functional traits as a response to these environmental changes. When incorporating the functional species traits in the determination of the forest types, we found that the conservative traits, such as the dispersal syndrome, were significantly related to the species ($p = 0.01$) and the alpha diversity of the forest types ($p = 0.04$). The ecological guild to which the species belongs was strongly related to the elevation ($p = 0.05$) and the forest type ($p = 0.01$).

Species acquisitive traits also play a significant role in characterizing forest types. Diametric growth, as the main acquisitive trait, was significantly related to altitude ($p = 0.004$) and its related variables that, as we observed previously, play an important role in the separation of forest communities.

Both the acquisitive and the conservative traits are important in the conformation of the forest communities; the Four Corner analysis allowed us to include the characteristics of the species that in other analyses went unnoticed, but were significant when classifying the vegetation. Figure 6 shows the correlations between the traits and the environmental variables and Table 6 shows the correlation values between the functional traits of the species and the environmental variables.

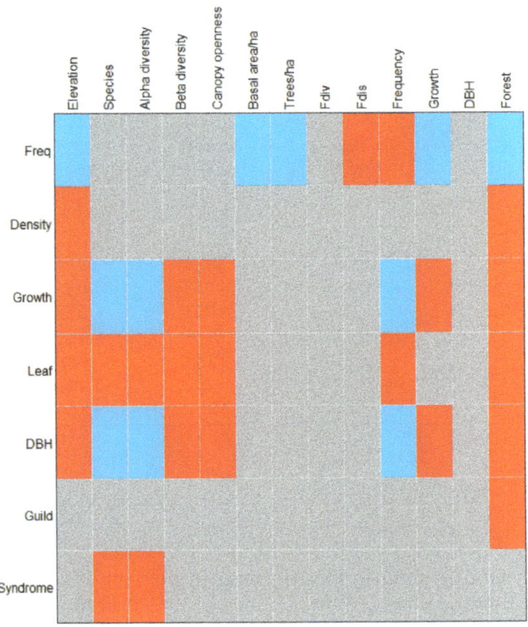

Figure 6. The relationship between trees functional traits and habitat variables. Colors represent the strength of interactions (shading) and their direction. Red indicates a positive association and blue indicates a negative association. The scale bar indicates the values of fourth-corner coefficients. Elev elevation, TD tree density. See Table 2 for definition of traits.

Table 6. *p* values and type of correlation used in the Four Corner analysis between the functional traits of the species and the environmental variables, based on the species abundance matrix in each of the determined forest plots.

	Traits	Test	Stat. Obs	Std. Obs	Alter	*p*-Value	*p*-Value .adj
Altitude	FREC	R	0.103	3.021	two-sided	0.007	0.019 *
AB. HA	FREC	R	0.127	3.716	two-sided	0.001	0.004 **
ARB.HA	FREC	R	−0.126	−3.713	two-sided	0.001	0.004 **
Fdis	FREC	r	0.093	2.609	two-sided	0.019	0.049 *
FREC	FREC	r	0.156	4.515	two-sided	0.001	0.004 **
Growth	FREC	r	−0.129	3.589	two-sided	0.003	0.011 *
Forest	FREC	r	0.102	3.032	two-sided	0.005	0.016 *
Altitude	Density	r	0.140	4.090	two-sided	0.002	0.008 **
Forest	density	r	0.115	3.327	two-sided	0.004	0.014 *
Altitude	Growth	r	0.254	7.673	two-sided	0.001	0.004 **
Species	Growth	r	0.159	4.652	two-sided	0.001	0.004 **
diver.A	Growth	r	0.188	5.568	two-sided	0.001	0.004 **
diver.B	growth	r	0.185	5.449	two-sided	0.001	0.004**
CO	growth	r	0.136	4.073	two-sided	0.001	0.004 **
FREC	Growth	r	0.142	4.132	two-sided	0.002	0.008 **
Forest	Growth	r	0.238	7.201	wo-sided	0.001	0.004 **
Altitude	LEAF	F	5.027	9.009	Greater	0.001	0.004 **
Species	LEAF	F	6.133	1.133	Greater	0.001	0.004 **
diver.A	LEAF	F	4.652	8.274	Greater	0.002	0.008 **
diver.B	LEAF	F	3.064	5.131	greater	0.007	0.019 *
CO	LEAF	F	2.027	3.387	Greater	0.013	0.035 *
FREC	LEAF	F	2.887	4.742	Greater	0.006	0.018 *

Table 6. *Cont.*

	Traits	Test	Stat. Obs	Std. Obs	Alter	*p*-Value	*p*-Value .adj
DBH	LEAF	F	1.821	3.053	Greater	0.022	0.05.
Forest	LEAF	F	8.230	1.433	Greater	0.001	0.004 **
altitud	DBH	r	0.298	8.638	two-sided	0.001	0.004 **
Species	DBH	r	0.215	6.213	two-sided	0.001	0.004 **
diver. A	DBH	r	0.272	7.984	two-sided	0.001	0.004 **
diver. B	DBH	r	0.262	7.776	two-sided	0.001	0.004 **
CO	DBH	r	0.183	5.216	two-sided	0.001	0.004 **
FREC	DBH	r	0.117	3.438	two-sided	0.004	0.014 *
growth	DBH	r	0.105	3.035	two-sided	0.005	0.016 *
Forest Type	DBH	r	0.275	8.205	two-sided	0.001	0.004 **
Altitude	GUILD	F	8.883	2.810	Greater	0.023	0.05.
Forest	GUILD	F	1.161	4.166	greater	0.007	0.019 *
Species	SYND	F	21.493	5.452	Greater	0.005	0.016 *
diver. A	SYND	F	14.803	3.061	Greater	0.016	0.042 *

Signif. codes: 0 -' 0.001 '**' 0.01 '*' 0.05 '.' 0.1 ' ' 1.

4. Discussion

In the proposed model for classification of continental vegetation of Ecuador, Sierra et al. [14] used the regional division of the territory and included the concept of subregions. The hierarchical model divided the regions into sectors, i.e., each region had subregions, and these in turn were divided into sectors, with nesting vegetation physiognomy (Mangrove Forest, Shrubland, Espinar, Savannah, Paramo and Gelidofitia) and other hierarchical criteria such as environmental criteria (dry, wet, fog), phenological criteria (evergreen, deciduous, semi deciduous) and floristic altitudinal levels (lowland, premontane, lower montane, montane, upper montane), determining 62 cover types, 27 on the coast, 24 in the Andes and 11 in the Amazonia.

However, the classification at the plant formation level included large areas that were evidently heterogeneous with respect to key factors, mainly in the composition and structure of the forests. Other authors, such as Valencia et al. [12], called this formation "montane cloud forest" in southeastern Ecuador. This is also known as "Evergreen Montane Forest" [13], using climatic ("cloud") and functional ("evergreen") elements in the naming of the formations.

Other classifications proposed altitudinal limits as the main factor in the separation of forests, along with a geomorphological factor [43], asserting that the forest in the southern region of Ecuador can be divided into evergreen lower montane forests (up to 2100–2200 m a.s.l.) and upper montane forests, up to the tree line. Above ~2700 m, a shrub-dominated sub-paramo prevails, where small patches of Elfin Forest, the so-called Ceja Andina, dominate the landscape. Both montane forest types can be subdivided into a lower slope (ravine) forest and an upper slope (ridge) forest [44,45].

A new approach of vegetation classification at smaller scales is based on using variables of structure, taxonomic and functional diversity (traits of the species) and environmental factors (including elevation); as a result, four types of forest can be distinguished by combining different types of classification [8]. For example, [46] studied structural parameters; [47] the floristic trees composition; [48] the bryophytes species composition and [49] making a synopsis of previous studies.

All these studies that classified vegetation using environmental parameters associated with elevation were consistent with the categories of vegetation in the RBSF. Even the topographic variables at smaller scales play an important role in the classification of the vegetation, as demonstrated by the studies carried out by [50] in the same study area and [51] who combine topographic variables and functional features to determine small-scale species association in tropical forests of China.

In the Andean vegetation classifications, the elevation gradient and its associated variables play an essential role in separating forests. This is also true in our case, where elevation works at the microscale, that is, the elevational influence occurs in relatively small areas and over relatively short distances, influencing not only the structure and composition of forests, but also the variation in the composition of traits in vascular plants [52], such as leaf size [53], seed dispersal [54] and wood density [55], which are basically the same traits that we used for our analysis.

Although we did not include topographic variables in our analysis, there were reports that these correlated with vegetation classification, especially at a small scale, and were significant in identifying ecological units that included vegetation structure and composition [56], delimiting forest lowland and montane formations [57]. The topographic variables also affect the functional traits of the species, especially the mass of seeds and the density of the wood [51]; in our case, species wood density was also correlated with altitude and forest type.

Regarding the scale of influence, it should be noted that, although we determined in our study that the influence of elevation and associated variables is also significant at microscales, there are cases in which elevation works at larger scales and its evaluation was also significant [58]. This emphasizes the fact that, in addition to the environmental factors associated with elevation, there are other factors intrinsic to the species that help the differentiation of plant communities.

In a specific approach, altitude and its associated environmental factors are crucial when determining and differentiating forest types. In the VF up to 2050 m a.s.l., the number of trees tends to remain relatively stable, while in the RF, the decline in the number of trees at higher altitudes was evident. This pattern is not strict along the Cordillera de los Andes, but it may be subject to changes in temperature and humidity over relatively short intervals [59].

Finally, when referring to the groups determined by our study, we can indicate that the VF were characterized by lower stem density, but with greater basal areas (tree diameters) and higher canopies compared to the RF, where less tree species were also present. The differences in forest structure are mainly due to the climatic conditions and prevailing soil types [60,61]. At the phytosociological level, the floristic structure and composition of the VF and the RF are coupled to more widely distributed floristic formations, such as the order *Alzateetalia verticillatae* and *Purdiaeaetalia nutantis*, respectively [62], which validates the floristic analysis carried out in our study area, by the coincidence of indicator species of each floristic association.

For both floristic groups, the microclimatic and topographic conditions cause the species to find suitable sites and share habitat and topographic preferences of occurrence [50], an argument that also reinforces the grouping of the species present in the forest of the RBSF.

Author Contributions: Conceptualization, O.C., S.G., B.S. and R.M.; methodology, O.C., R.M. and P.R.; formal analysis, O.C. and P.R.; data curation, O.C. and P.R.; writing—original draft preparation, O.C., P.R., S.G., B.S. and R.M.; writing—review and editing, O.C., P.R. and R.M. All authors have read and agreed to the published version of the manuscript.

Funding: This research received no external funding.

Institutional Review Board Statement: Not applicable.

Informed Consent Statement: Not applicable.

Data Availability Statement: Data sharing is not applicable to this article.

Acknowledgments: The first two authors thank the UTPL for allowing the publication of the manuscript to be completed. All the authors thank the DFG (Deutsche Forschungsgemeinschaft) for the support to carry out the studies in southern Ecuador. We thank Diana Szekely for the English revision of the final version of the manuscript.

Conflicts of Interest: The authors declare no conflict of interest.

References

1. Saatchi, S.S.; Harris, N.L.; Brown, S.; Lefsky, M.; Mitchard, E.T.; Salas, W.; Zutta, B.R.; Buermann, W.; Lewis, S.L.; Hagen, S. Benchmark. Map of forest carbon stocks in tropical regions across three continents. *Proc. Natl. Acad. Sci. USA* **2011**, *108*, 9899–9904. [CrossRef] [PubMed]
2. Myers, N.; Mittermeier, R.A.; Mittermeier, C.G.; Da Fonseca, G.A.; Kent, J. Biodiversity hotspots for conservation priorities. *Nature* **2000**, *403*, 853–858. [CrossRef] [PubMed]
3. Kier, G.; Mutke, J.; Dinerstein, E.; Ricketts, T.H.; Küper, W.; Kreft, H.; Barthlott, W. Global patterns of plant diversity and floristic knowledge. *J. Biogeogr.* **2005**, *32*, 1107–1116. [CrossRef]
4. Pan, Y.; Birdsey, R.A.; Fang, J.; Houghton, R.; Kauppi, P.E.; Kurz, W.A.; Phillips, O.; Shvidenko, A.; Lewis, S.L.; Canadell, J.G.; et al. A large and persistent carbon sink in the world's forests. *Science* **2011**, *333*, 988–993. [CrossRef]
5. Malhi, Y.; Phillips, O.L. Tropical forests and global atmospheric change: A synthesis. *Philos. Trans. R. Soc. B* **2004**, *359*, 549–555. [CrossRef]
6. FAO. *Global Forest Resources Assessment 2005. Main Repor. Progress Towards Sustainable Forest Management*; FAO Forestry Papers, FAO: Rome, Italy, 2006; p. 147.
7. Cirimwami, L.; Doumenge, C.; Kahindo, J.M.; Amani, C. The effect of elevation on species richness in tropical forests depends on the considered lifeform: Results from an East African mountain forest. *Trop. Ecol.* **2019**, *60*, 473–484. [CrossRef]
8. Homeier, J.; Breckle, S.W.; Günter, S.; Rollenbeck, R.T.; Leuschner, C. Tree Diversity, Forest Structure and Productivity along Altitudinal and Topographical Gradients in a Species-Rich Ecuadorian Montane Rain Forest. *Biotropica* **2010**, *42*, 140–148. [CrossRef]
9. Apaza-Quevedo, A.; Lippok, D.; Hensen, I.; Schleuning, M.; Both, S. Elevation, topography, and edge effects drive functional composition of woody plant species in tropical montane forests. *Biotropica* **2015**, *47*, 449–458. [CrossRef]
10. Valencia, R.; Churchill, S.P.; Balslev, H.; Forero, E.; Luteyn, J.L. Composition and structure of an Andean Forest fragment in eastern Ecuador. In *Biodiversity and Conservation of Neotropical Montane Forests. Proceedings of a Symposium, New York Botanical Garden*; New York Botanical Garden: New York, NY, USA, 1995.
11. Mosandl, R.; Günter, S.; Stimm, B.; Weber, M. Ecuador suffers the highest deforestation rate in South America. In *Gradients in a Tropical Mountain Ecosystem of Ecuador*; Springer: Berlin/Heidelberg, Germany, 2008.
12. Valencia, R.; Cerón, C.; Palacios, W. Las formaciones naturales de la Sierra del Ecuador. In *Propuesta Preliminar de un Sistema de Clasificación de Vegetación Para el Ecuador Continental*; Sierra, R., Ed.; Proyecto INEFAN/GEF-BIRF and EcoCiencia: Quito, Ecuador, 1999; pp. 81–108.
13. Balslev, H.; Øllgaard, B. Mapa de vegetación del sur de Ecuador. In *Botánica Austroecuatoriana*; Ediciones Abya-Yala: Quito, Ecuador, 2002; pp. 51–64.
14. Sierra, R. *Propuesta Preliminar de un Sistema de Clasificación de Vegetación Para el Ecuador Continental*; Proyecto Inefan/Gef-Birf y Ecociencia: Quito, Ecuador, 1999.
15. Ohl, C.; Bussmann, R. Recolonisation of natural landslides in tropical mountain forests of Southern Ecuador. *Feddes Repert.* **2004**, *115*, 248–264. [CrossRef]
16. Brehm, G.; Homeier, J.; Fiedler, K. Beta diversity of geometrid moths (Lepidoptera: Geometridae) in an Andean montane rainforest. *Divers. Distrib.* **2003**, *9*, 351–366. [CrossRef]
17. Schrumpf, M.; Guggenberger, G.; Valarezo, C.; Zech, W. Tropical montane rain forest soils. Development and nutrient status along an altitudinal gradient in the south Ecuadorian Andes. *Erde* **2001**, *132*, 43–59.
18. Wilcke, W.; Yasin, S.; Schmitt, A.; Valarezo, C.; Zech, W. Soils along the altitudinal transect and in catchments. In *Gradients in a Tropical Mountain Ecosystem of Ecuador*; Springer: Berlin/Heidelberg, Germany, 2008; pp. 75–85.
19. Fries, A.; Rollenbeck, R.; Bayer, F.; Gonzalez, V.; Oñate-Valivieso, F.; Peters, T.; Bendix, J. Catchment precipitation processes in the San Francisco valley in southern Ecuador: Combined approach using high-resolution radar images and in situ observations. *Meteorol. Atmos. Phys* **2014**, *126*, 13–29. [CrossRef]
20. Zanne, A.E.; Lopez-Gonzalez, G.; Coomes, D.A.; Ilic, J.; Jansen, S.; Lewis, S.L.; Chave, J. Global Wood Density Database. Dryad. Identifier. 2009. Available online: http://datadryad.org/handle/10255/dryad.235 (accessed on 23 January 2023).
21. Chave, J.; Coomes, D.A.; Jansen, S.; Lewis, S.L.; Swenson, N.G.; Zanne, A.E. Towards a worldwide wood economics spectrum. *Ecol. Lett.* **2009**, *12*, 351–366. [CrossRef]
22. MAE. *Resultados de la Evaluación Nacional Forestal*; Ministerio del Ambiente: Quito, Ecuador, 2014.
23. Palacios, W.A. Forest species communities in tropical rain forest of Ecuador. *Lyonia* **2004**, *7*, 33–40.
24. Osazuwa-Peters, O.L.; Jiménez, I.; Oberle, B.; Chapman, C.A.; Zanne, A.E. Selective logging: Do rates of forest turnover in stems, species composition and functional traits decrease with time since disturbance? A 45 year perspective. *For. Ecol. Manag* **2015**, *357*, 10–21. [CrossRef]
25. Dos Santos, V.A.H.F.; Ferreira, M.J. Are photosynthetic leaf traits related to the first-year growth of tropical tree seedlings? A light-induced plasticity test in a secondary forest enrichment planting. *For. Ecol. Manag* **2020**, *460*, 117900. [CrossRef]
26. Wang, C.; Zhou, J.; Xiao, H.; Liu, J.; Wang, L. Variations in leaf functional traits among plant species grouped by growth and leaf types in Zhenjiang, China. *J. For. Res.* **2017**, *28*, 241–248. [CrossRef]
27. Jara-Guerrero, A.; De la Cruz, M.; Méndez, M. Seed dispersal spectrum of woody species in south Ecuadorian dry forests: Environmental correlates and the effect of considering species abundance. *Biotropica* **2011**, *43*, 722–730. [CrossRef]

28. Aguirre Calderón, Ó.A.; Jiménez Pérez, J.; Kramer, H.; Akça, A. Análisis estructural de ecosistemas forestales en el Cerro del Potosí, Nuevo León, México. *Ciencia* **2003**, *6*.
29. Cerón, C. *Manual de Botánica Ecuatoriana, Sistemática y Métodos de Estudio, Escuela de Biología*; Universidad Central del Ecuador: Quito, Ecuador, 1993.
30. Madsen, J.E.; Øllgaard, B. Floristic composition, structure, and dynamics of an upper montane rain forest in Southern Ecuador. *Nord. J. Bot.* **1994**, *14*, 403–423. [CrossRef]
31. Madsen, J.E. Floristic Composition, Structure and Dynamics of an Upper Montane Rain Forest in Southern Ecuador. Ph.D. Thesis, Botanical Institute of Aarhus University, Aarhus, Denmark, 1991.
32. Finol, U.H.; Finol, U.H. Estudio fitosociológico de las unidades 2 y 3 de la Reserva Forestal de Caparo, Estado Barinas. *Acta Botánica Venez.* **1976**, *11*, 15–103.
33. Galindo, R.; Betancur, J.; Cadena, J.J. Estructura y composición florística de cuatro bosques andinos del santuario de flora y fauna Guanentá-Alto río Fonce, cordillera oriental colombiana. *Caldasia* **2003**, *25*, 313–335.
34. Mori, S.A.; Boom, B.M.; de Carvalino, A.M. Ecological importance of Myrtaceae in an eastern Brazilian wet forest. *Biotropica* **1983**, *15*, 68–70. [CrossRef]
35. Magurran, A.E. Why diversity? In *Ecological Diversity and Its Measurement*; Springer: Dordrecht, The Netherlands, 1988; pp. 1–5.
36. Petchey, O.L.; Gaston, K.J. Functional diversity: Back to basics and looking forward. *Ecol. Lett.* **2006**, *9*, 741–758. [CrossRef]
37. Cayuela, L. *Una Introduccion a R. EcoLab, Centro Andaluz de Medio Ambiente*; Universidad de Granada: Granada, Spain, 2010; 109p.
38. Legendre, P.; Galzin, R.; Harmelin-Vivien, M.L. Relating behavior to habitat: Solutions to the fourth-corner problem. *Ecology* **1997**, *78*, 547–562. [CrossRef]
39. Dray, S.; Choler, P.; Dolédec, S.; Peres-Neto, P.R.; Thuiller, W.; Pavoine, S.; ter Braak, C.J. Combining the fourth-corner and the RLQ methods for assessing trait responses to environmental variation. *Ecology* **2014**, *95*, 14–21. [CrossRef]
40. Dray, S.; Legendre, P. Testing the species traits–environment relationships: The fourth-corner problem revisited. *Ecology* **2008**, *89*, 3400–3412. [CrossRef]
41. Aubin, I.; Ouellette, M.H.; Legendre, P.; Messier, C.; Bouchard, A. Comparison of two plant functional approaches to evaluate natural restoration along an old-field–deciduous forest chronosequence. *J. Veget. Sci.* **2009**, *20*, 185–198. [CrossRef]
42. Chessel, D.; Dufour, A.B.; Dray, S. Analysis of Ecological Data: Exploratory and Euclidean Methods in Environmental Sciences. Version 1.4-14. 2 October 2010; Lyon, France1: 4, 11. 2009. Available online: http://pbil.univ-lyon1.fr/ADE-4/ade4-html/00 Index.html (accessed on 23 January 2023).
43. Homeier, J.; Leuschner, C.; Bräuning, A.; Cumbicus, N.L.; Hertel, D.; Martinson, G.O.; Spannl, S.; Veldkamp, E. Effects of nutrient addition on the productivity of montane forests and implications for the carbon cycle. In *Ecosystem Services, Biodiversity and Environmental Change in a Tropical Mountain Ecosystem of South Ecuador*; Springer: Berlin/Heidelberg, Germany, 2013; pp. 315–329.
44. Werner, F.A.; Homeier, J. Is tropical montane forest heterogeneity promoted by a resource-driven feedback cycle? Evidence from nutrient relations, herbivory and litter decomposition along a topographical gradient. *Funct. Ecol.* **2015**, *29*, 430–440. [CrossRef]
45. Paulick, S.; Dislich, C.; Homeier, J.; Fischer, R.; Huth, A. The carbon fluxes in different successional stages: Modelling the dynamics of tropical montane forests in South Ecuador. *For. Ecosyst.* **2017**, *4*, 5. [CrossRef]
46. Paulsch, A. Development and Application of a Classification System for Undisturbed and Disturbed Tropical Montane Forests Based on Vegetation Structure. Ph.D. Dissertation, Fakultat fur Biologie, Chemie und Geowissenschaften der Universitat Bayreuth, Bayreuth, Germany, 2002.
47. Homeier, J. Baumdiversität, Waldstruktur und Wachstumsdynamik Zweier-Tropischer Bergregenwälder in Ecuador und Costa Rica. Ph.D. Thesis, University of Bielefeld, Bielefeld, Germany, 2004.
48. Parolly, G.; Kürschner, H. Ecosociological studies in Ecuadorian bryophyte communities. V. Syntaxonomy, life forms and life strategies of the bryophyte vegetation on decaying wood and tree bases in S Ecuador. *Nova Hedwigia* **2005**, *81*, 1–36. [CrossRef]
49. Homeier, J.; Werner, F.A.; Gradstein, S.R.; Breckle, S.; Richter, M. Potential vegetation and floristic composition of Andean forests in South Ecuador, with a focus on the RBSF. *Ecol. Stud.* **2008**, *198*, 87.
50. Kuebler, D.; Hildebrandt, P.; Günter, S.; Stimm, B.; Weber, M.; Mosandl, R.; Muñoz, J.; Cabrera, O.; Aguirre, N.; Zeilinger, J.; et al. Assessing the importance of topographic variables for the spatial distribution of tree species in a tropical mountain forest. *Erdkunde* **2016**, *70*, 19–47. [CrossRef]
51. Liu, J.; Yunhong, T.; Slik, J.F. Topography related habitat associations of tree species traits, composition and diversity in a Chinese tropical forest. *For. Ecol. Manag.* **2014**, *330*, 75–81. [CrossRef]
52. Duivenvoorden, J.F.; Cuello, A.N. Functional trait state diversity of A ndean forests in Venezuela changes with altitude. *J. Veg. Sci.* **2012**, *23*, 1105–1113. [CrossRef]
53. Malhado, A.C.M.; Malhi, Y.; Whittaker, R.J.; Ladle, R.J.; ter Steege, H.; Phillips, O.L.; Butt, N.; Aragão, L.E.O.C.; Quesada, C.A.; Araujo-Murakami, A.; et al. Spatial trends in leaf size of Amazonian rainforest trees. *Biogeosciences* **2009**, *6*, 1563–1576. [CrossRef]
54. Antonio Vazquez, G.J.; Givnish, T.J. Altitudinal gradients in tropical forest composition, structure, and diversity in the Sierra de Manantlán. *J. Ecol.* **1998**, *86*, 999–1020.
55. Chave, J.; Muller-Landau, H.C.; Baker, T.R.; Easdale, T.A.; Steege, H.T.; Webb, C.O. Regional and phylogenetic variation of wood density across 2456 neotropical tree species. *Ecol App.* **2006**, *16*, 2356–2367. [CrossRef]
56. Körner, C.; Paulsen, J. A world-wide study of high altitude treeline temperatures. *J. Biogeogr.* **2004**, *31*, 713–732.

57. Jadán, O.; Donoso, D.A.; Cedillo, H.; Bermúdez, F.; Cabrera, O. Floristic groups, and changes in diversity and structure of trees, in tropical montane forests in the Southern Andes of Ecuador. *Diversity* **2021**, *13*, 400.
58. Treiber, J.; von Wehrden, H.; Zimmermann, H.; Welk, E.; Jäger, E.J.; Ronnenberg, K.; Wesche, K. Linking large-scale and small-scale distribution patterns of steppe plant species—An example using fourth-corner analysis. *Flora* **2020**, *263*, 151553. [CrossRef]
59. Malizia, A.; Blundo, C.; Carilla, J.; Acosta, O.O.; Cuesta, F.; Duque, A.; Aguirre, N.; Aguirre, Z.; Ataroff, M.; Baez, S.; et al. Elevation and latitude drive structure and tree species composition in Andean forests: Results from a large-scale plot network. *PLoS ONE* **2020**, *15*, e0231553. [CrossRef] [PubMed]
60. Moser, G.; Leuschner, C.; Hertel, D.; Graefe, S.; Soethe, N.; Iost, S. Elevation effects on the carbon budget of tropical mountain forests (S Ecuador): The role of the belowground compartment. *Glob. Chang. Biol.* **2011**, *17*, 2211–2226. [CrossRef]
61. Dislich, C.; Huth, A. Modelling the impact of shallow landslides on forest structure in tropical montane forests. *Ecol. Model* **2012**, *239*, 40–53. [CrossRef]
62. Bussmann, R.W. Los bosques montanos de la Reserva Biológica San Francisco (Zamora-Chinchipe, Ecuador)—Zonación de la vegetación y regeneración natural. *Lyonia* **2003**, *3*, 57–72.

Article

Valuation of the Diversity of Native Plants and the Cultural-Archaeological Richness as an Integrative Approach for a Potential Use in Ecotourism in the Inter-Andean Valley of Cusco, Southern Peru

Isau Huamantupa-Chuquimaco [1,*], Yohny Luz Martinez Trujillo [2] and Edilberto Orosco Ucamayta [3]

1 Herbario Alwyn Gentry (HAG), Universidad Nacional Amazónica de Madre de Dios (UNAMAD), AV. Jorge Chávez N°1160, Puerto Maldonado 17001, Madre de Dios, Peru
2 Instituto Científico UAC, Programa ICDS, Universidad Andina del Cusco (UAC), Urb. Ingeniería Larapa Grande A-7, San Jerónimo, Cusco 17001, Cusco Region, Peru
3 Instituto Científico UAC, Universidad Andina del Cusco (UAC), Urb. Ingeniería Larapa Grande A-7, San Jerónimo, Cusco 17001, Cusco Region, Peru
* Correspondence: andeanwayna@gmail.com

Abstract: In recent years, ecological tourism has become very important as it contributes significantly to sustainable development. In order to assess the potential for ecotourism and cultural-archaeological attributes, we studied the plant diversity of 10 traditionally visited natural routes of the valley of Cusco, Peru. Plant gamma diversity was represented by 384 species of vascular plants, with 220 genera, and 69 families; the most diverse were: Asteraceae with 93 species, Poaceae (36), and Fabaceae (15). The species with the highest frequency in the 10 routes are: *Amaranthus caudatus*, *Escallonia resinosa*, *Stenomesson pearcei*, and *Baccharis buxifolia*. Route 2 (Picol-Huaqoto) was the one with the greatest alpha diversity with 120 species. The CHAO-1 richness estimator estimates a gamma diversity of 570 species for all of the Cusco Valley. The Bray–Curtis beta diversity shows a high similarity (55%) and three floristic groups as determined by a non-metric multidimensional analysis (NMDS) and cluster analysis. The highest concentration of flowering plant species is grouped mainly during the rainy season ($R^2 = 0.19$), and this relationship is significantly different from the dry season ($p < 0.005$). The integrative biological–cultural analysis identified routes R8, R10, R6, R7, and R2 as those with the greatest potential for ecotourism use in the Cusco Valley. The plant diversity and cultural–archaeological offerings along the tourist routes documented in this study present significant opportunities for the city of Cusco to attract both national and foreign tourists. Additionally, this study highlights the importance of prioritizing conservation and preservation efforts for these areas.

Keywords: alpha diversity; Andes; tourist circuit; beta diversity; native plants; phenology; species richness

Citation: Huamantupa-Chuquimaco, I.; Martinez Trujillo, Y.L.; Orosco Ucamayta, E. Valuation of the Diversity of Native Plants and the Cultural-Archaeological Richness as an Integrative Approach for a Potential Use in Ecotourism in the Inter-Andean Valley of Cusco, Southern Peru. *Diversity* **2023**, *15*, 760. https://doi.org/10.3390/d15060760

Academic Editors: Lucian Dinca, Miglena Zhiyanski and Mario A. Pagnotta

Received: 29 January 2023
Revised: 8 April 2023
Accepted: 11 April 2023
Published: 9 June 2023

1. Introduction

Peru has world-renowned centers of high diversity such as the tropical Andes [1]. A high species richness of vascular and non-vascular plants was recorded in this region with southern Peru being a sub-region of the tropical Andes [2]. Pre-Columbian civilizations settled in this area, and over time Andean–Amazonian populations developed the use and management of many wild plants, placing the area among the most diverse in the world, similar to China, for example [3].

In the southern Peruvian region and in the Cusco Valley, located in the Huatanay River Basin, various pre-Inca cultures developed, mainly with the Inca culture best represented. The Inca culture, in the process of developing its geopolitical and cultural hegemony, developed extraction and deforestation activities, especially in lands destined for agriculture, as the basis of its expansion. This began with the establishment of the Marcavalle culture, approximately 800 years BC [4]. Socio-cultural development in the Cusco Valley to date has involved various human activities that have displaced natural systems that have been transformed into crop fields, land for livestock, and urban and industrial areas [5]. This resulted in changes in land use, water pollution, soil degradation, habitat fragmentation, and the overexploitation of species, among other developments [6]. Evidently, the historic growth of cities in southern Peru has been predominantly informal and has therefore occurred in a disorderly fashion [7]. An example of this is the current population size of the city of Cusco, which currently has more than 450,000 inhabitants distributed mainly in the districts of Wanchaq, San Sebastián, and Santiago, few of which have adequate urban planning [8].

Regarding plant resources, the loss and massive degradation dates back mainly to colonial times with the overexploitation of tree and shrub species for use as fuel (firewood), construction, the expansion of agricultural and livestock areas, and the introduction of other exotic species, for example eucalyptus. Concerning this situation, some chroniclers, such as Garcilaso 1971 [9] and Valverde 1539 [10], suggest that when the Europeans arrived, the Cusco Valley corresponded to a great stony city with a warm climate surrounded by forests with abundant trees crossed by steamy ravines and rivers, where you could also recognize various forest tree species such as "mulli", "quishuar", "lambran", "pisonay", and "siwis", useful for housing structures and as fuel [11–13].

In recent years, tourism in the Cusco region has been one of the most important activities, which is directly linked to the extraordinary cultural and natural legacy of the Inca and pre-Inca cultures [14]. Furthermore, as an economic activity, ecotourism is one of the activities with less anthropic impact on the Cusco landscape, and has been prioritized worldwide [15]. Tourism has also generated benefits for local populations, as has been recognized in other areas [16].

Recently, in the province of Cusco and its surroundings, the use of biological resources in tourism has had a great growth; proof of this is that the different circuits or tourist routes mainly include cultural resources represented by structures and archaeological complexes such as temples, ruins, cobblestone roads, and housing constructions among others and nature attractions that include flora, fauna, ecosystems, and landscapes. In the valley of Cusco, five routes are recognized among the most frequently used routes for all visitors who travel in the city and valley of Cusco: the Sacsayhuaman Archaeological Park, Qenqo, Huanacaure, Tambomachay Hill, and Inkilltambo [17].

Despite the fact that some routes are traditionally recognized as the most frequently used, there are others that are also being implemented as new alternatives to the growing tourist offering to both locals and foreign visitors. For this reason, it is necessary to know what these routes are and the attributes that each one offers, with an integral criterion, which includes the biological and cultural components, since these are the most decisive when considering ecotourism activities.

The main objective of this study was to identify the potential of ten traditional natural routes for use in ecotourism activities in the inter-Andean valley of Cusco, for which the values of the biodiversity level (diversity of native plants and phenology patterns), anthropic impact, level of accessibility, and cultural components (number of archaeological remains) were used for their assessment.

2. Materials and Methods

2.1. Study Area

The Cusco Valley is located in the Huatanay river basin, within the Choco micro-basin and running for an approximate extension of 42 km to the confluence with the Vilcanota river (Figure 1). Throughout this area, the high Andean zone with woody vegetation is found in a range from 3200 to 4800 m above sea level. According to Marin 1961 [18], this area belongs to the Andean domain and the phytogeographic provinces of "Puneña" and "Mesoandina". Aragon and Chuspe 2018 [19] describe it as the ecoregion of Peruvian inter-Andean valleys, within the vegetation types of lowland forests and highland shrublands of the humid Puna and highland grasslands and scrub of the humid Puna (Figure 2).

Phytosociologically, we can describe the evaluated zones as being represented by the grassland in the upper zone, in the middle zone as a transition between shrubland and grassland, and the lower zones as dominated by trees, mainly by *Escallonia resinosa*. The climate is represented by two well-marked periods known as the dry season that goes from June to October and the rainy season from November to May; the average annual temperature is 10–13 °C and the area has an average total annual rainfall of 574–690 mm (average 12 years 1998–2020, Kayra weather station) [18,20,21].

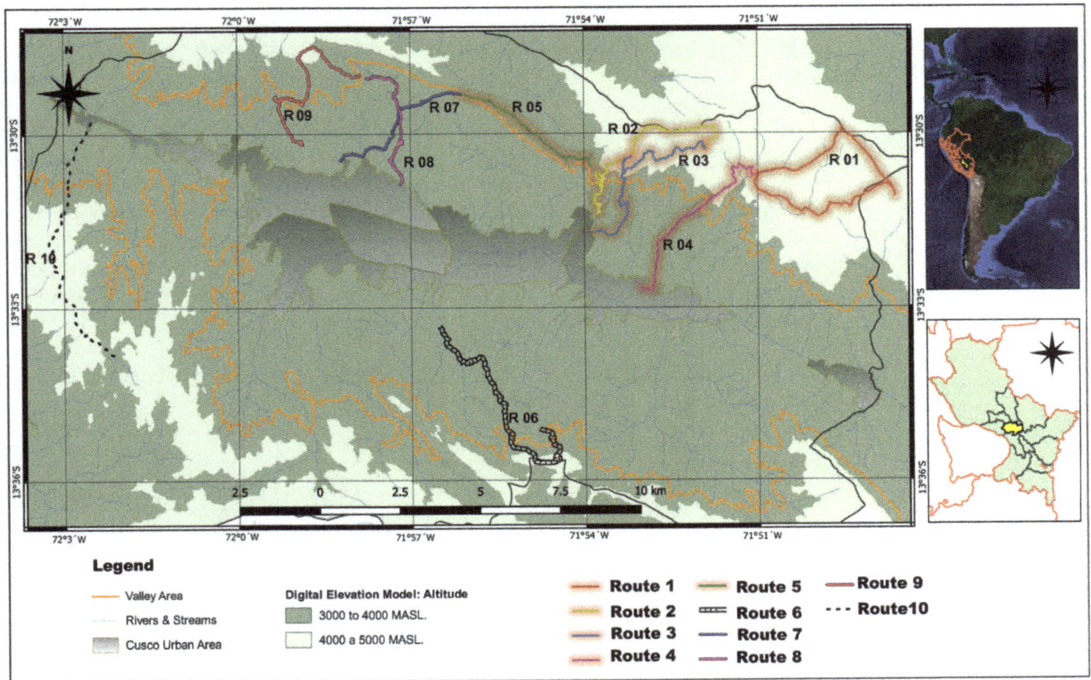

Figure 1. Distribution map of the touristic routes evaluated in the valley of Cusco.

Figure 2. Characteristic high Andean landscape in the valley of Cusco, Peasant Community of Quishuarcancha—Cerro Mámá Simona—Poroy (Route 10).

2.2. Methodology

2.2.1. Selection and Evaluation of Ecotourism Routes

For our study, 10 routes (eco-trails), primarily those that have traditionally been the most frequently traveled by local and foreign visitors, were selected. This traditional selection took into account the following associated criteria over time: frequency of visits by local tourism (residents of the Cusco region), Peruvian nationals, and foreign tourists; observable biological diversity (plants and animals); number of cultural remains (pre-Inca and Inca archaeological complexes), and accessibility.

This study was carried out from October 2019 to July 2021. The 10 studied Routes were: Route 1 (R1) Huaqoto-Canteras de Huaqoto—Relicts of *Polylepis*—Path to Huanca—Huaqoto, Route 2 (R2): Tambillo—Cruz cerro Picol—Bosque de Santa María, Route 3 (R3): Tambillo—Cerro Picol—Picol Fault—Campos de San Jerónimo, Route 4 (R4): Huaqoto—San Jerónimo (along the pilgrimage road to Huanca, Route 5 (R5): Lomo de Iguana—Abra de Ccorao, Route 6 (R6): Rural Community of Kircas—Cerro Huanacaure—Inka Trail, Route 7 (R7): Yuncaypata—Temple of the Moon—Qenqo, Route 8 (R8): Pukapukara—Inkilltambo—Huayraqpunku, Route 9 (R9): Tambomachay—Balcón del Diablo—Sacsayhuaman, and Route 10 (R10): Quishuarcancha Peasant Community—Mama Simona Hill—Poroy. Each route had an extension of between 5370 and 12,700 m (Table 1). The altitudinal ranges of the routes varied from the valley floor between 3300–3400 to the maximum limits of 4500–4600 masl (Table 1, Figure 1).

Table 1. Quantitative and qualitative diversity data of native plants and the Cultural-Archaeological richness, evaluated in the 10 routes of the valley of Cusco.

Routes	Trail Length (km)	Average Altitude	N° Archeological Units	Level of Anthropic Impact	Accessibility Level	Qualitative Richness (Observed)	Quantitative Richness (Sampled)	Individuals	Fisher's Alpha Diversity	Diversity q = 0 (Hill)
R1	12.70	4186	0	2	2	94	51	368	16.07	86
R2	8.70	4081	0	2	2	120	82	326	35.23	110
R3	8.94	3938	0	2	2	79	41	344	12.13	78
R4	7.560	3702	1	2	1	97	50	322	16.57	80
R5	5.420	3869	0	3	3	94	42	322	12.9	77
R6	9.470	3719	2	2	1	108	64	200	32.55	86
R7	5.390	3656	2	2	2	103	66	179	37.77	95
R8	5.370	3594	2	1	1	111	70	130	61.82	124
R9	7.690	3833	1	2	2	89	40	310	12.23	77
R10	10.430	3947	2	1	1	109	71	264	31.86	108

Note: level of anthropic impact (High = 3, Medium = 2, and Low = 1); level of accessibility (Good = 1, Regular = 2, and Bad = 3).

2.2.2. Plant Diversity, Composition, and Phenology

In order to know the total richness of plants in each Route and the total existing in the 10 routes, we evaluated each one (on average 50 m from both sides of the route) of them by means of annotated lists, taking photographs, and, in the cases in which there were doubts about the identifications, we proceeded to collect them for later identification. Live form data (tree, shrub, and grass) and phenology (flowering and fruiting) as well as its vernacular name were recorded.

To estimate the richness, composition, and alpha and beta diversity, quantitative abundance data were collected; for this purpose, three modified Whittaker plots Campbell, 2002 [22] were installed on each route, each one located in the lower, middle, and upper zone of each route. We evaluated the trees in plots of 20 × 50 m (0.1 hectare), the shrubs in 5 × 10 m (50 m^2), and the herbs in 5 quadrats of 0.5 × 2 m^2 (5 m^2 total). The variables evaluated were the number of individuals (adults and developed juveniles), habit, crown area (coverage), phenology, and vernacular name.

For the taxonomic treatment of plant families, we used the reference established by the Angiosperm Phylogeny Group [23], as well as the herbaria CUZ, herbariums, and online digital repositories such as Tropicos (http://www.tropicos.org, accessed on 12 march 2020). For the taxa identifications, taxonomic keys and specialized bibliography were used, such as the Catalog of Angiosperms and Gymnosperms plants of Peru [24].

Alpha Diversity

The alpha diversity was calculated by two indicators: (a) the richness of each route was obtained from the qualitative data, representing the total number of species inventoried in each one, (b) the quantitative data of the total values of the three plots were used to calculate the Fisher's alpha index (α). Fisher's alpha index is an abundance model derived from a logarithmic series that uses only the number of species (S) and the total number of individuals (N) [25]. This index is expressed as $S = \alpha \ln [1 + (N/\alpha)]$, where: S = number of species in the sample and N = number of individuals in the sample.

Furthermore, to estimate the total number of species that can be found on each route, we developed abundance curves using the rarefaction method, with the extrapolation function at the level of each route and subsequently for all routes. For this part of the analysis, we used Hill numbers $q = 0$ to quantify the effective number of all species, including rare species [26]. All these analyses were executed using the iNEXT package developed for the R software [27].

Beta Diversity and Similarity

The percentage of floristic similarity between plots was calculated using the quantitative Bray–Curtis index (obtained from abundance data) [28]. To visually represent these similarities, we used ordination methods and cluster analysis. The cluster analysis used the similarities and the group averages as a linkage method between groups and pairs of groups to build dendrograms. To complement this analysis, we further used a non-metric multidimensional scaling ordination (NMDS) [29]. The statistical analysis was performed in the program PAST v4.11 [30].

Gamma Diversity

To know the richness in the entire valley of Cusco, we summed up the total number of species inventoried in the 10 routes, and to estimate a projected total from the inventoried routes, the CHAO-1 non-parametric estimator was used (considering each of the 10 routes as units). This estimator prioritizes the presence of rare species within a sample (in this case each subplot), with the formula $Chao1 = Sobs + ((n - 1/n)\ F1(F1 - 1)/2(F2 + 1))$, where: Sobs is the number of species observed in the whole valley, n the number of samples, F1 is the number of species observed in a single route (singleton species), and F2 is the number of species observed in two routes (doubleton species) [31,32]. The statistical analyses and graphs were made using the R software [27].

Phenology Patterns

To better understand the periods of greatest and least diversity, we have analyzed the phenological flowering patterns of all the plants on the 10 routes, which are also taken into account by visitors. For this, the differences between flowering individuals of each family, in the rainy seasons and those in the dry seasons, were evaluated with the non-parametric statistical comparison test of two independent Mann–Whitney samples. This analysis was expressed graphically in a box plot. To investigate if there was any relationship with environmental variables obtained from Kayra's meteorological station, located in the valley of Cusco, we carried out a linear regression analysis using the Spearman's non-parametric correlation coefficient [33] using the following dataset: the total species flowering each month, the data of total monthly precipitation, and the average monthly temperatures. The statistical analysis was performed in the program PAST v4.11 [30].

2.2.3. Comprehensive Consideration of the Potential of the Routes for Ecotourism

To obtain a weighting that helps us to identify the potential of each route for ecotourism activity in the Cusco Valley, for the present study we have considered the biodiversity component, mainly from the flora component, accompanied by ecosystem and cultural components. For this study, we have adapted the methodology known as multicriteria evaluation (EMC), frequently used in works related to the identification of biological, cultural, and other potentialities for tourist resources. These criteria are described as "those that at the time of evaluation provide a logical and coherent way the contribution of the suitability criteria" for a potential proposal in ecotourism [34,35].

In order to know the routes with the greatest potential for use in the ecotourism activity in the Cusco Valley, we have considered four important criteria in each route: (a) the total richness of registered vascular plant species, which corresponds to the value of total richness; (b) the number of units of cultural remains; these refer to the presence of archaeological remains corresponding mainly to archaeological complexes including temples, ruins, rooms, and others belonging to the pre-Inca or Inca cultures and all these complexes are evaluated from the each routes; (c) the level of anthropic impact that the route has, which is qualified as low with the value of 1, regular with 2, and high with 3, which translates into whether the route mainly maintains the proportion in each route of the native vegetation and flora and the presence or absence of anthropic activities such as grazing, crops, or deforested areas without vegetation cover; and (d) the accessibility of the routes based on the fact that these routes have conditions for walking, such as appropriate distance and travel time; good accessibility is measured as having a value of 1, regular with 2, and not very adequate with 3 (Table 1).

3. Results

3.1. Diversity and Phenology

3.1.1. Richness, Composition, and Diversity

In the 10 routes, 384 species (qualitative richness) of vascular plants were found, distributed in 220 genera and 69 families, and 95% were identified at the species level and 5% at the morphospecies level. Regarding the habits, 17 species correspond to trees, 54 to shrubs, and 313 to herbs. The five Routes with the greatest floristic richness that exceed 100 species were R2 with 120 species, R8 (111), R10 (109), R6 (108), and R7 with 103 (Figure 1, Table S1). The families with the highest species richness were Asteraceae with 93 species, Poaceae (36), Fabaceae (15), and Rosaceae with 14 (Figure 3). The most diverse genera in terms of species were *Baccharis* with ten species, *Senecio* (nine), *Deyeuxia*, and *Plantago* with eight each. The species with the highest frequency of presence have been *Amaranthus caudatus* L. and *Escallonia resinosa* (Ruiz and Pav.) Pers., present in ten routes; present in nine routes were *Stenomesson pearcei* Baker, *Baccharis buxifolia* (Lam.) Pers., *Baccharis latifolia* (Ruiz and Pav.) Pers., *Baccharis odorata* Kunth, *Mutisia acuminata* Ruiz and Pav. (Figure 3D), *Puya ferruginea* (Ruiz and Pav.) L.B. Sm., and *Polylepis incana* (Kunth); and present in eight routes were *Amaranthus hybridus* L., *Chenopodium ambrosioides* L., *Chenopodium murale* L., *Grindelia boliviana* Rusby, and *Calceolaria myriophylla* (Kraenzl.) (Table S1).

In the quantitative sampling, a total of 2007 individuals were evaluated, comprising 252 species (quantitative richness) belonging to 54 families, the most diverse being Asteraceae with 67, followed by Poaceae (21), and Fabaceae with 12. The most abundant species were *Anatherostipa obtusa* with 173 individuals, *Lachemilla pinnata* (84), *Aciachne pulvinata* (71), *Lachemilla orbiculate* (58), and *Anatherostipa aff. obtusa* and *Muhlenbergia peruviana*, both with 57 individuals.

Figure 3. Species of native plants evaluated in the ecotourism routes in the Cusco Valley. (**A**) *Bomarea dulcis* (Alstroemeriaceae), (**B**) *Barnadesia horrida* (Asteraceae), (**C**) *Gynoxys longifolia*, (**D**) *Mutisia acuminata*, (**E**) *Mutisia cochabambensis*, (**F**) *Lupinus cuzcensis* (Fabaceae), (**G**) *Gentianella ernestii* (Gentianaceae), (**H**) *Hesperoxiphion herrerae* (Iridaceae), (**I**) *Viola aff. pygmaea* (Violaceae).

Alpha Diversity

The route with the greatest diversity of vascular plants was R8, with Fisher's alpha = 61.82, followed by R7 (Fisher's alpha = 37.77), R2 (Fisher's alpha = 35.23), R6 (Fisher's alpha = 32.55), and R10 (Fisher's alpha = 31.86). The other routes obtained values less than 30 Fisher's alpha (Table 1). The estimated richness for each route, through Hill's diversity ($q = 0$), indicates that the Route with the greatest diversity was R8 with 124 species, followed by R2 with 110 and R10 with 108 species; for the other routes the richness was estimated to be less than 100 species (Table 1, Figure 4).

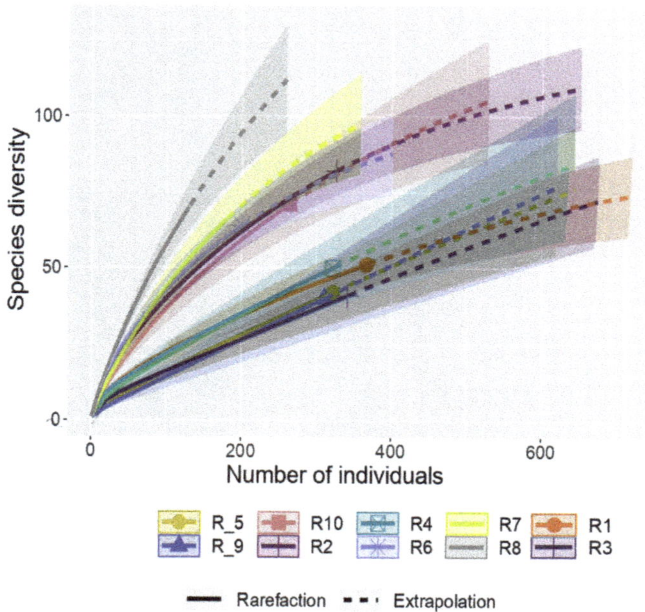

Figure 4. Diversity using the rarefaction method based on the Hill index.

Beta Diversity and Floristic Affinities

The cluster and the NMDS plots show the formation of three main groups of routes. In the cluster with 55% similarity (Figure 5a,b), group 1 is made up of routes R2, R4, R6, R7, R8, R9, and R10; group 2 by R1 and R3; and group 3 is represented only by R5. This same pattern is evident in the NMDS with the three groups with low stress (0.13), which indicates a good fit of the model to the data (Figure 5b). In the first dimension, group 1 is concentrated in the central part, with the most distant group being 2 with R1 and R3, and finally, at the end of the axes, there is only R5 as group 3.

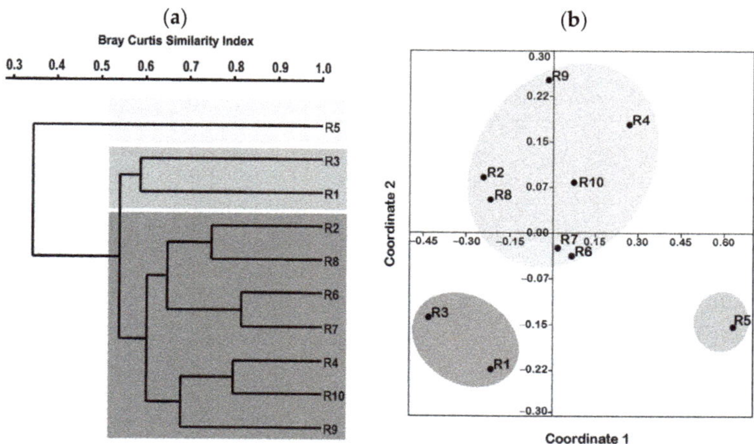

Figure 5. (**a**) Similarity between the 10 sampled routes in the Cusco Valley. (**b**) NMDS showing the grouping of the floristic groups between 10 sampled routes in the Cusco Valley.

Gamma Diversity

The total diversity observed for the Cusco Valley, from the 10 routes, is represented by 384 species. The quantitative data, through the rarefaction curve by means of the CHAO-1 index, estimates that for the entire Cusco Valley there is a richness of 570 species of vascular plants (Figure 6).

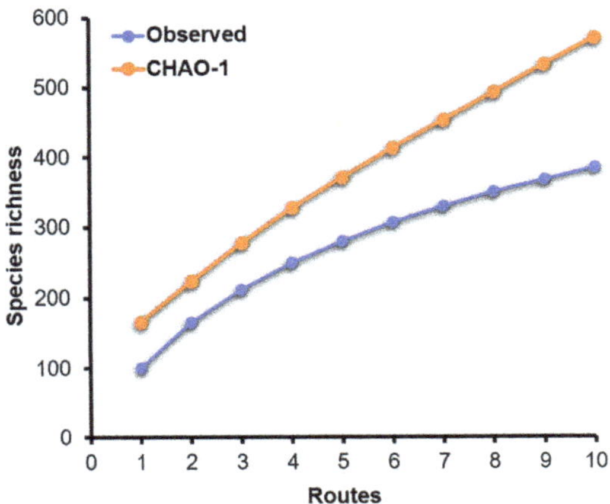

Figure 6. A. CHAO-1 diversity, observed and estimated.

3.1.2. Phenology Patterns

Phenology patterns show the largest number of species are flowering during the months of the rainy season, from February to July. Then, there is a gradual decrease in the dry season, where only some species flower. Few biennials have more than one flowering season within the year (Table 2). Herbs mainly flower from December to May and most of the shrubs also concentrate their flowering during these months. In the case of trees, it is observed that there is a continuous flowering with the exception of June and August, where no species was recorded to flower (Table 2, Figure 7).

Table 2. Number of flowering species per habit and per month during the year 2020. ND = no data available.

Habit	January	February	March	April	May	June	July	August	September	October	November	December
Tree	8	12	8	10	12	ND	2	ND	2	2	2	5
Shrub	19	41	37	36	40	2	4	3	ND	7	7	15
Herb	103	203	206	203	199	26	37	15	13	21	37	107
Total	130	256	251	249	251	28	43	18	15	30	46	127

The months with the greatest diversity of flowering plants are February (256 species), March and May (251), and April (249). The months in which there are fewer species are August and September (Table 2, Figure 6).

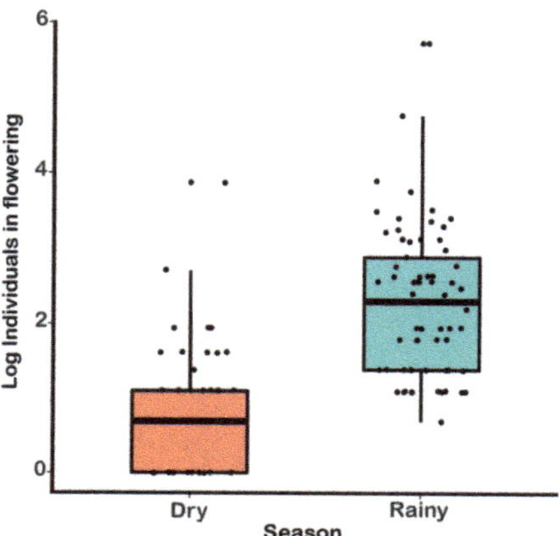

Figure 7. Phenological variation between the two seasons (Mann–Whitney test $p = 0.00001$).

Throughout the year, the botanical families that contributed the most flowering sightings during the year were Asteraceae with 307 observations, Poaceae with 117, Fabaceae 49, Rosaceae 43, and Plantaginaceae 34 (Table 3).

Table 3. List of the 10 most representative families with individuals in flowering, in the dry and rainy seasons (the values correspond to the sums of observations in the dry and rainy months).

Family	Rainy Season	Dry Season
Asteraceae	307	48
Poaceae	117	15
Fabaceae	49	5
Rosaceae	43	7
Plantaginaceae	34	7
Orobanchaceae	33	7
Amaranthaceae	30	5
Amaryllidaceae	30	3
Gentianaceae	29	5
Orchidaceae	27	1

For the total 69 families, in the rainy season 1264 flowering individuals (mean 18.31; standard deviation 38.84) were observed, and for the dry season 180 individuals were observed (mean 2.60; standard deviation 6.05). The Mann–Whitney test shows that there is a significant difference between the number of observations of individuals in flower in the rainy and dry seasons ($p = 0.00001$). Therefore, the rainy season is more favorable to appreciate a high richness of species and individuals in the evaluated routes (Figure 7).

Regression analysis shows us that there is an association between richness and precipitation in the evaluated 12 months, with a $R^2 = 0.19$ Spearman's correlation coefficient of $p = 0.15$, which is considered a low to regular association (Figure 7a).

The relationship between species richness and temperature shows a $R^2 = 0.02$ Spearman's correlation coefficient of $p = 0.48$, which is considered a low association (Figure 8a,b). Between temperature and precipitation during the studied 12 months, we observed that precipitation is the best variable that explains the increase in plant richness in the rainy months of the year.

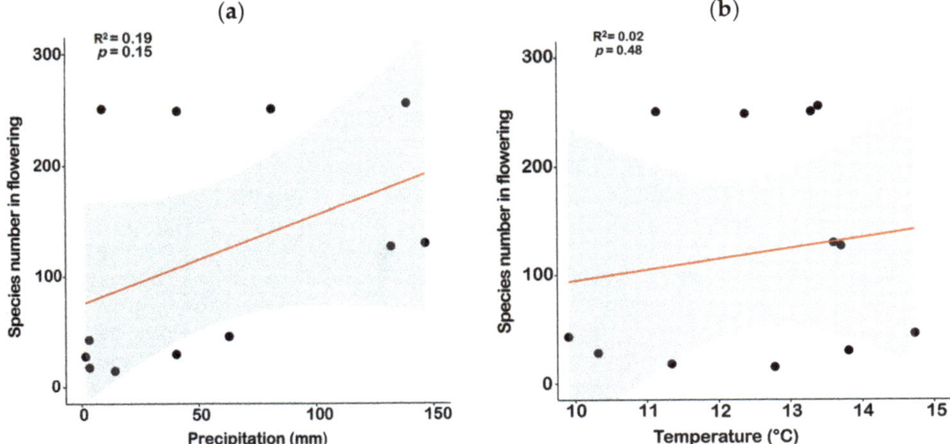

Figure 8. Representation of the annual relationship between environmental variables and the richness. (**a**) Precipitation vs. species richness, (**b**) Temperature vs. species richness.

3.2. *Potential of Routes for Ecotourism*

In the 10 routes, 10 archaeological units were registered. In R4, the Raqayraqayniyuq complex is present (San Jerónimo district); R6 has two units: the Huanacaure hill complex and the Inca trail; R7 has the archaeological complexes of the Temple of the Moon (Amarumarkawasi) and Qenqo; R8 includes the Puka Pukara and Inkiltambo complexes; R10 has the Inca trail to Apu Mama Simona and the Quiswarcancha complex; R4 has a single complex Inca trail to Huacoto; and R9 has part of the Sacsaywaman archeological park.

The five routes with the highest weighting considering the four criteria are Route 8 (total richness = 111 species, archaeological units = 2, low anthropic impact = 1, and with good accessibility = 1); Route 10 (total richness = 109, archaeological units = 2, anthropic impact = 1, and with good accessibility = 1); Route 6 (total richness = 108, archaeological units = 2, anthropic impact = 2, and with good accessibility = 1); Route 7 (total richness = 103, archaeological units = 2, anthropic impact = 2, and with regular accessibility = 2); and Route 2 which, despite not having any archaeological unit, presents the greatest diversity of all the routes with 120 species, regular anthropic impact (2), and regular accessibility (2) (Table 1).

The other routes, with the exception of R4 and R9, do not present any archaeological units. However, in general, they have access conditions, and are regularly impacted by human activities, mainly agricultural activity.

4. Discussion

4.1. *Diversity and Phenology*

4.1.1. Richness, Composition, and Diversity

Gamma diversity is represented by 384 species of vascular plants found in the 10 routes of the valley of Cusco, and shows in general that these types of vegetation, that include the inter-Andean dry forest in the lower zone and the "pajonales" in the highlands, encompass quite representative habitats that include transition areas. This is evident in the lower areas with the presence of forests dominated by populations of native trees: *Escallonia resinosa* (chachacomo) and *Polylepis racemosa* (queuña), and in the grasslands dominated by the grass *Anatherosthipa obtusa* (ichu). These routes can each house up to more than 100 species. The total richness or gamma diversity of the plants registered in the present study is notably higher than that documented by Herrera 1941 [12], who points out that, at that time, in the surroundings of the city of Cusco there were up to 254 species of higher plants. This approximation somewhat calls attention to the fact that, in those years, the city of Cusco still had many remnants of forest and more intact areas near the banks of the micro-watersheds throughout the entire valley. Compared with the study by Galiano et al., 2005 [36], who registered little more than 480 species for the entire Cusco Valley, in our study more than 80% of them are cataloged. It should be noted that the study of these authors, in addition to the Cusco Valley, included the towns of Urcos and Huacarpay, which are not directly part of the Cusco Valley.

Family diversity patterns, both in qualitative and quantitative evaluations, show similar results as found in other regions of the Cusco Valley, as described by the works of Galiano et al., 2005 [36]; Tupayachi 2019 [13]; Herrera 1941 [12]; and Marín 1950 [18]. In these studies, the families with the greatest diversity were Asteraceae, Poaceae, Fabaceae, and Lamiaceae, the exception being the Bromeliaceae family. In the present study, we only registered two species in the different routes. The same authors mention that, in general, for the forests and grasslands of the Cusco Valley and its surroundings, these genera are more representative: *Deyeuxia* (ex *Calamagrostis*) of the Poaceae, *Puya* (Bromeliaceae), *Solanum* (Solanaceae), *Senecio* (Asteraceae), and *Lupinus* (Fabaceae).

In the 10 studied routes, the species considered to be most frequent present interesting distributions, for example *Amaranthus caudatus, A. hybridus, Chenopodium ambrosioides.*, and *C. murale* have a wide range of distribution and can be found even in degraded environments, which shows that along the routes these species were also adapting alongside wild species from habitats that were little disturbed. The other group of frequent species are trees, such as *Escallonia resinosa, Baccharis buxifolia, Barnadesia horrida* (Figure 3B), *B. latifolia., B. odorata, Mutisia acuminata* (Figure 3D), *Polylepis incana*, and the herbaceous species *Stenomesson pearcei, Grindelia boliviana*, and *Calceolaria myriophylla*, which are from little impacted habitats and which are also quite frequent in other inter-Andean valleys of southern Peru. At a quantitative level, in the 30 plots the pattern of abundance is different in terms of the individuals evaluated. The species *Anatherostipa obtusa, Lachemilla pinnata*, and *Aciachne pulvinata* stand out as these are species of greater density and small size. These species are present mainly in the highlands such as grasslands and transition zones within the inter-Andean dry forest.

The gamma diversity values estimated by the CHAO-1 index show that the richness of vascular plants can reach around 570 species for the Cusco Valley; this value is comparable to that recorded in Galiano et al., 2005 [36] with little more than 480 species in the vicinity of the Cusco Valley. At the level of each route, if we compare the values of Fisher Alpha with other Andean zones such as in the departments of Cajamarca and Pasco above 2800 masl, they present indices between 12 and 60 [25] which indicates that the Cusco routes in general contain a similar moderate to low diversity, which are also relatively similar to other Andean regions. The values found are also corroborated with the diversity values of Hill where three routes, R2, R8, and R10, estimate the presence of more than 100 species, which are complemented by the qualitative results, where routes R2, R10, R8, R6, and R7 exceed more than 100 species. These values and estimates show that although the ecotourism routes described are often not well recognized with high species richness, with more detailed sampling, areas with high richness and alpha diversity can be recorded, especially if we make a complete evaluation during a year.

The beta diversity evaluated with the UPGMA and NMDS methods, under the Bray–Curtis index, generally exhibits a high floristic similarity among the 10 routes, that is, more than 50% of species are shared among all of them. However, three consistent groups are distinguished, of which group 1, with 7 routes (R2, R4, R6, R7, R8, R9, and R10), represents those that include valley bottom areas, inter-Andean dry forest, and grassland areas in their routes, while group 2 (R1, R3) corresponds to drier areas with few tree populations in the lowlands, and group 3 is represented only by route R5 and contains less floristic richness since it only covers habitats such as grassland and shrubland with a greater anthropic impact.

4.1.2. Phenology

Apparently, the monthly flowering periods are conditioned by the presence of rain within a year, and it is within this period where more flowering species could be seen. This same pattern is mentioned by Marín 1950 [18], who made some notes on the floral phenology in different months of the year in the surroundings of Cusco of the arboreal, shrubby, and herbaceous flora, reaching the conclusion that the month of March has the largest floral anthesis in Cusco, which is also considered to be a widespread pattern in the Andean region of Peru.

Groups such as the Asteraceae, Poaceae, Fabaceae, and Rosaceae are the ones that show a greater flowering pattern in the rainy months. Species such as *Viguiera procumbens*, *Ageratina cuzcoensis*, *Begonia veitchii*, *Gentianella ernestii* (Figure 3F), the mostly perennial *Barnadesia horrida* (Figure 3C) and *Mutisia acuminata* (Figure 3D), and the endemic *Lupinus cuzcensis* (Figure 3G) are the ones that can be recorded in an abundant way during the flowering season in the areas mainly of the inter-Andean dry forest at the bottom of the valley. These are even used as festive decorations in carnivals, as a symbol of prosperity in the year.

4.2. *Integral Assessment of the Potential and Sustained Management of Routes for Ecotourism*

Sustainable tourism is one of the activities that has been providing opportunities to local populations [16]. Ecotourism in recent years has caused the least negative impacts on the environment, compared with other activities [15]. In the area of the Cusco Valley, tourist activity is represented by routes recognized as "Cusco City tour", "Cusco Valley tourism", and "Sacred Valley tourism". For Cusco, the most traveled routes are Inkiltambo, Qenqo, Sacsayhuamán Archaeological Park, Apu Huanacaure, and Apu Pikol [17].

In the present study, we considered the integral biological, ecosystemic, and cultural criteria, which support our conclusion that the routes R8, R10, R6, R7, and R2 are the ones with the greatest potential for use in ecotourism within the Cusco Valley. These routes also include the most important archeological complexes in the peri-urban area of the Cusco Valley, as indicated in [17]. These findings agree with the hypothesis that the Cusco Valley, with its micro-climatic variability, geological formations, ecosystems, and fertile lands, has been the reason the area was chosen for the capital city by the Incas and other pre-Inca cultures, despite being above 3000 masl, as stated in Galiano 2005 [11] and Marín 1950 [18].

The studied ecotourism routes, apart from the impressive diversity of plants (384 species and an estimated 570), were traditionally areas where resources such as construction material, medicinal plants, and foodstuffs, among others, were obtained in an important way in the daily life of the "cusqueño" inhabitant in practices which are currently still being practiced (field observation).

Despite the fact that these routes have a high potential for ecotourism use, it is important to highlight that some of them are subject to risks and threats, mainly due to the expansion of housing construction, the authorization of cultivation areas, and fires that occur mostly during dry months. To contribute to its conservation, we propose the following measures for the recovery and restoration of the populations of native forests and associated flora on the evaluated routes: (1) an area on the margins at each route that includes forests and remnants in which strict protections are respected; (2) potential restoration areas are identified to recover deforested areas that currently correspond to deteriorated areas such as in the lower areas of R2, R4, and R10; (3) the preparation and implementation of a plan to replace *Eucalyptus* plantations (*Eucalyptus globulus*) by native forest species, taking into account historical records, such as those species present in the R10, which includes highly threatened species such as *Gynoxys longifolia* (Figure 3C) and *Escallonia myrtilloides*, both native forest species that were decimated in the past in the vicinity of the city of Cusco; (4) the implementation of education campaigns, studies, and programs for the reevaluation of the native flora in the valley of Cusco; (5) the consideration of intangible conservation areas to avoid inadequate access by tourist visitors, for example, areas where endemic plants with extinction threats grow, nesting areas for some birds such as partridges and hummingbirds with a degree of threat, and mammal feeding areas, for example, the taruka. Within these areas, we have identified the upper parts of the grassland, where you can find representative animal species with high degrees of threat, such as the taruka "*Hippocamelus antisensis*", the Andean fox "*Lycalopex culpaeus*", and the Andean partridge "*Nothoprocta ornata*".

5. Conclusions

In the present work, we show that in the Cusco Valley there is a high gamma diversity of vascular plants, with a projected 570 species, in 10 ecotourism routes traditionally traveled by local and foreign populations in the peri-urban areas of the city of Cusco. The beta diversity also shows a high plant similarity between the 10 routes, where a group of routes show a consistent grouping corresponding to those with the highest alpha diversity such as R2, R6, R7, R8, and R10. Due to the climatic characteristics of this part of the Andean region, the greatest diversity of plants in flowering season can be observed during the rainy season, which corresponds to the first months of the year.

The potential use of each evaluated route, by applying the integrative method, which in our case was based on biological (represented by the diversity of plants), ecosystem (with the level of anthropic impact, and its accessibility), and cultural values (with the number of cultural units present in each route, which correspond to legacies of the rich Inca and pre-Inca cultures), has allowed us to assign an appropriate weighting to each route. The routes with the highest values were R2, R6, R7, R8, and R10. These routes have a high diversity of plants, the presence of up to two cultural units, a low level of anthropic impact, and adequate accessibility. In general, for the 10 routes, it is recommended that some mitigation measures be implemented, such as the restoration of degraded environments of natural habitats, reforestation, ecological connectivity areas, and strict protection areas. For the sustainable use and management of these areas, local authorities must also consider the establishment of rest areas, signage, the preparation of local guides, and zoning. These areas undoubtedly represent a good recreation alternative and the opportunity for programs for environment education and an appreciation of nature for the population of the city of Cusco as well as national and foreign tourists.

Supplementary Materials: The following supporting information can be downloaded at: https://www.mdpi.com/article/10.3390/15060760/s1, Table S1: General table of species with collections by routes.

Author Contributions: Research conception and design: I.H.-C. and Y.L.M.T.; field study and sampling: I.H.-C., Y.L.M.T. and E.O.U.; analysis and interpretation: I.H.-C. and E.O.U.; drafting of the manuscript: I.H.-C. All authors have read and agreed to the published version of the manuscript.

Funding: The present investigation was developed thanks to the financing obtained by the "Project for the Improvement and Expansion of the Services of the National System of Science, Technology and Technological Innovation" 8682-PE, from the "World Bank" to "CONCYTEC" and "PROCIENCIA" as financing entities of the project "From trekking routes to ecotourism routes/paths: adaptation of trails in Andean and high Andean areas for the recreation of the local population, Case: Cusco—Peru", which was financed by CONCYTEC–PROCIENCIA within the framework of the call E041-01 (N°148-2018 FONDECYT-BM-IADT-AV).

Institutional Review Board Statement: Not applicable.

Data Availability Statement: Not applicable.

Acknowledgments: To Di Yanira Bravo, for her support in managing the viability of the Ecotrails project through CONCYTEC and the research institute of the Andean University of Cusco. To Edwin Bellota, for his collaboration in the identification and selection of the 10 ecotourism routes in the Cusco Valley, and for the suggestions to the manuscript. To the members of the UNSAAC ECOTAXON Research Center: Anne Arias, for her collaboration and assistance in collecting data in the field, and Armando Cuti, Jean Pier Quispe, Rolando Chaparrea, Rodrigo Calvo, Hugo Copa, Fany Cutire, Jhon A, Yuca, Esau Huaman, Elias Quispe, Valentina Palomino, Flora Sucsa, Mario Sanchez, Nidia

Sánchez, and Williams Navarro, for their assistance in field sampling and logistical support. Thanks to Roosevelt García-Villacorta for providing suggestions for an early version of the manuscript. To the inhabitants of the communities immersed in the ecotourism routes of the districts of San Sebastián, San Jerónimo, and Ccorao of the province of Cusco.

Conflicts of Interest: The authors declare no conflict of interest.

References

1. Ulloa-Ulloa, C.; Acevedo-Rodriguez, P.; Beck, S.; Belgrano, M.; Bernal, R.; Berry, P.; Brako, L.; Celis, M.; Davidse, G.; Forzza, R.; et al. An integrated assessment of the vascular plant species of the Americas. *Science* **2017**, *358*, 1614–1617. [CrossRef] [PubMed]
2. León, B.; Roque, J.; Ulloa-Ulloa, C.; Pitman, N.; Jorgensen, P.; Cano, A. Libro Rojo de las Plantas endémicas del Perú. *Rev. Peru Biol. Número Espec.* **2006**, *13*, 971.
3. Brack, A. *Diccionario Enciclopédico de Plantas Útiles del Perú*; Centro de Estudios Regionales Andinos Bartolomé de las Casas: Cusco, Peru, 1999; p. 550.
4. Barreda, M.L. Los medios de Subsistencia Andinos en el Pre ceramico. In *Historia Natural del Valle del Cusco*; SPN-CUSCO, Ed.; SOPRONAC: Cusco, Peru, 2005; pp. 115–120. Available online: https://agris.fao.org/agris-search/search.do?recordID=US201300039906 (accessed on 15 February 2020).
5. Saloma, M.G. La ciudad del Cusco a través del tiempo. In *Historia Natural del Valle del Cusco*; SPN-CUSCO, Ed.; SOPRONAC: Cusco, Peru, 2005; pp. 131–146.
6. Flores, G.J. El valle del Cozco. paisaje domesticado. In *Historia Natural del Valle del Cusco*; SPN-CUSCO, Ed.; SOPRONAC: Cusco, Peru, 2005; pp. 125–129.
7. Espinoza, A.; Fort, R. Mapeo y Tipología de la Expansión Urbana en el Perú. 2020. Available online: http://www.grade.org.pe/publicaciones/mapeo-y-tipologia-de-la-expansion-urbana-en-el-peru/ (accessed on 10 July 2021).
8. INEI-Cusco, Censo 2017. Tú Cuentas para el Perú. (I. N. Informatica. Editor) 2018. Available online: https://censo2017.inei.gob.pe/el-departamento-de-cusco-cuenta-con-1-205-527-habitantes/ (accessed on 10 November 2021).
9. Garcilaso de La Vega. *Comentarios Reales de los Incas. Primera Parte que Tratan del Origen de los Yncas. Reyes que Fueron del Perv de su Idolatria Leyes y Gobierno en paz y en Guerra: De sus Vidas y Conquistas y de Todo lo que fue Aquel Imperio y su República antes que los Españoles Pasaran a el*; Imprenta de Pedro Crasbeeck: Lisboa, Portugal, 1609.
10. Valverde, V. Cusco el valle de los cedros. Cusco. 1539.
11. Galiano, G.S. Cusco: El valle de los Cedros y la biodiversidad en la época inca (el retorno de los siwis). *Rev. SOPRONAC* **2019**, *10*, 111–118.
12. Herrera, L.F. *Sinopsis de la Flora del Cuzco—Tomo I-Parte Sistemática*; Publicado Bajo los Auspicios del Supremo Gobierno: Lima, Peru, 1941.
13. Tupayachi, H.A. Usos y aplicaciones de algunas especies forestales nativas en los bosques andinos del valle del Cusco. *Rev. SOPRONAC* **2019**, 88–111.
14. PROMPERU. Perfil del Turista Extranjero. 2019. Available online: https://www.promperu.gob.pe/TurismoIN/sitio/PerfTuristaExt (accessed on 21 June 2021).
15. OMT. Turismo Internacional 2019 y Perspectivas 2020 Barometro del Turismo Mundial. 2020. Available online: https://webunwto.s3.eu-west-1.amazonaws.com/s3fs-public/2020-01/Presentacion-barometro-jan%202020.pdf (accessed on 17 March 2021).
16. Leal, D.M. *Turismo Ecológico y Sostenible: Perfiles y Tendencias*; Grupo de Investigación Interdisciplinar GRIT-OSTELEA: Barcelona, Spain, 2017.
17. MINCETUR. gob. pe Plataforma Digital del Estado Peruano. 2022. Available online: https://www.gob.pe/institucion/mincetur/noticias/604000-peru-registro-la-llegada-de-242-mil-turistas-internacionales-durante-el-primer-trimestre-2022 (accessed on 12 March 2022).
18. Marín, M.F. *Notas Fenológicas Sobre la Vegetación de los Alrededores del Cusco*; Revista Universitaria, II Semestre; Uiversidad Nacional San Antonio Abad del Cuso: Cusco, Peru, 1950; Volume 99, pp. 322–327.
19. Aragón, J.I.; Chuspe, M.E. *Ecología Geográfica del Cusco*, 1st ed.; Chuspe Zans, M.E., Ed.; Gobierno Regional del Cusco: Cusco, Peru, 2018; Available online: https://www.researchgate.net/publication/327172888_Ecologia_Geografica_del_Cusco (accessed on 8 April 2022).
20. Herrera, L.F. La flora de la provincia del Cercado. In *Historia Natural del Valle del Cusco*; SPN-CUSCO, Ed.; SOPRONAC: Cusco, Peru, 2005; pp. 153–166.
21. Molleapaza, A. Apuntes Para el Estudio de la Flora y Vegetación del Valle del Cusco-Sacsayhuaman. Ph.D. Thesis, UNSAAC, Cusco, Peru, 1963.
22. Campbell, P.; Comiskey, E.J.; Alonso, A.; Dallmeier, F.; Nunez, P.; Beltran, H.; Baldeon, S.; Nauray, W.; De la Colina, R.; Acurio, L.; et al. Modified Whittaker plots as an assessment and monitoring tool for vegetation in a lowland tropical rainforest. *Environ. Monitor. Assess.* **2002**, *76*, 19–41. [CrossRef] [PubMed]
23. Angiosperm Phylogeny Group. An update of the Angiosperm Phylogeny Group classification for the orders and families of flowering plants: APG IV. *Bot. J. Linn. Soc.* **2016**, *181*, 1–20, retrieved 11 June 2016. [CrossRef]

24. Brako, L.; Zarucchi, J. *Catálogo de las Angiospermas y Gymnospermas del Perú*; Missouri Botanical Garden: St. Louis, MO, USA, 1993; EE. UU.
25. Phillips, O.; Miller, P. *Global Patterns of Plant Diversity: Alwyn H. Gentry's Forest Transect Dataset*; Monographs in Systematic Botany; Missouri Botanical Garden: St. Louis, MO, USA, 2002; p. 319.
26. Hsieh, T.C.; Ma, K.H.; Chao, A. iNEXT: An R package for rarefaction and extrapolation of species diversity (Hill numbers). *Methods Ecol. Evol.* **2016**, *7*, 1451–1456. [CrossRef]
27. R Development Core Team. *A Language and Environment for Statistical Computing*; R Foundation for Statistical Computing: Vienna, Austria, 2021; ISBN 3-900051-07-0.
28. Magurran, A.E. *Ecological Diversity and Its Measurement*; Princeton University Press: Princeton, NJ, USA, 1988.
29. McCune, B.; Grace, J.B. *Analysis of Ecological Communities*; MjM Software Design: Cleveland, OH, USA, 2002.
30. Hammer, Ø.; Harper, D.A.T.; Ryan, P.D. Past: Paleontological Statistics Software Package for Education and Data Analysis. *Palaeontol. Electron.* **2001**, *4*, 1–4.
31. Chao, A. Nonparametric estimation of the number of classes in a population. *Scand. J. Stat.* **1984**, *11*, 265–270.
32. Chao, A. Estimating the population size for capture-recapture data with unequal catchability. *Biometrics* **1987**, *43*, 783–791. [CrossRef] [PubMed]
33. Press, W.H.; Teukolsky, S.A.; Vetterling, W.T.; Flannery, B.P. *Numerical Recipes in C*; Cambridge University Press: Cambridge, UK, 1992.
34. Ghahroudi Tali, M.; Sadough, S.H.; Nezammahalleh, M.A.; Nezammahalleh, S.K. Multi-criteria evaluation toselect sites for ecotourism facilities: A case study Miankaleh Peninsula. *Anatolia* **2012**, *23*, 373–394. [CrossRef]
35. Sarook, S.; Wright, J.; Edwards, M. Evaluating consistency of stakeholder input into participatory GIS-based multiple criteria evaluation: A case study of ecotourism development in Kurdistan. *J. Environ. Plan. Manag.* **2016**, *60*, 1529–1553. [CrossRef]
36. Galiano, W.; Tupayachi, A.; Nuñez, P. Flora del Valle del Cusco. In *Historia Natural del Valle del Cusco*; SPN-CUSCO, Ed.; SOPRONAC: Cusco, Peru, 2005; pp. 197–232.

MDPI
St. Alban-Anlage 66
4052 Basel
Switzerland
www.mdpi.com

Diversity Editorial Office
E-mail: diversity@mdpi.com
www.mdpi.com/journal/diversity

www.ingramcontent.com/pod-product-compliance
Lightning Source LLC
LaVergne TN
LVHW071600170726
843515LV00008B/2311

* 9 7 8 3 0 3 6 5 9 4 1 4 9 *